SEX AND COHABITATION AMONG EARLY HUMANS

SEX AND COHABITATION AMONG EARLY HUMANS

Anthropological and Genetic Evidence for Interbreeding Among Early Humans

RENE J. HERRERA
Colorado College, Colorado Springs, CO, United States

RALPH GARCIA-BERTRAND
Colorado College, Colorado Springs, CO, United States

ELSEVIER

ACADEMIC PRESS
An imprint of Elsevier

Academic Press is an imprint of Elsevier
125 London Wall, London EC2Y 5AS, United Kingdom
525 B Street, Suite 1650, San Diego, CA 92101, United States
50 Hampshire Street, 5th Floor, Cambridge, MA 02139, United States
The Boulevard, Langford Lane, Kidlington, Oxford OX5 1GB, United Kingdom

ISBN : 978-0-12-819086-9

For information on all Academic Press publications
visit our website at https://www.elsevier.com/books-and-journals

Publisher: Stacy Masucci
Acquisitions Editor: Peter B. Linsley
Editorial Project Manager: Sara Pianavilla
Production Project Manager: Sreejith Viswanathan
Cover Designer: Miles Hitchen

Typeset by STRAIVE, India

Contents

About the authors

Rene J. Herrera holds degrees from the City College of the City University of New York (BS), New York University (MS), and Fordham University (PhD). He continued his studies in molecular biology/genetic engineering as a postdoctoral associate at the Rockefeller University and New York Medical College. His areas of research include mechanisms underlying control of gene expression at the pre-mRNA splicing level, human evolution, and human forensic DNA fingerprinting. He is a frequent contributor to journals in the field of human genetics. Dr. Herrera has established collaborative efforts with several leading scientists worldwide, attesting to his philosophy of expanding the reaches of his research. His commitment to creating a professional foundation for future scientific generations has led him to contribute to the education of numerous students as a mentor. A total of 15 PhD and 41 MS students obtained their degrees under his supervision, and hundreds of undergraduate students have benefited from his guidance in conducting research. He has been a member of numerous National Institute of Health (NIH), National Science Foundation (NSF), and international research evaluation panels and a grant evaluator for educational and underrepresented student programs for decades. He teaches general courses in biology and genetics as well as advanced courses in human genetics, molecular biology, human evolution, microbiology, forensic DNA fingerprinting, and anatomy/physiology. He routinely acts as a consultant to the legal profession in issues involving forensic DNA fingerprinting.

He is the author of more than 200 full-length articles and 2 previously published books in the fields of human evolution and hominin speciation:

1. *Genomes, Evolution, and Culture; Past, Present, and Future of Humankind* by Rene J. Herrera, Ralph Garcia-Bertrand, and Francisco M. Salzano.
2. *Ancestral DNA, Human Origins, and Migrations* by Rene J. Herrera and Ralph Garcia-Bertrand.

Affiliations and expertise: Visiting Scholar, Colorado College, Colorado Springs, CO, United States.

Ralph Garcia-Bertrand holds degrees from the University of Nevada (BS, MS), and the University of California, Riverside (PhD). He continued his studies as a University of California, Presidents Post-Doctoral Fellow in the Department of Genetics, at the University of California, Berkeley. His research focuses on human population genetics and evolution. Dr. Garcia-Bertrand was a former Chair of the Department of Molecular Biology, and the Verner Reed Professor of Genetics, at Colorado College. He has won several teaching awards and has been the Director of undergraduate education programs funded by IBM, the Howard Hughes Medical Institute, and the Freeman Foundation. He was a subcommittee member for various NIH research, educational, and underrepresented student programs for more than 22 years. He is a lifetime member of the Society for the Advancement of Chicanos and Native Americans in Science, and the National Association of Biology Teachers. He teaches general and advanced genetics, human evolution, and a course on the ethics of genetics.

Affiliations and expertise: Emeritus Professor of Genetics, Molecular Biology Department, Colorado College, Colorado Springs, CO, United States.

CHAPTER 1

Origin and diversity of hominins

"If you look at skulls, you'll see different features of modern humans arising in different locations at different times" "we're a species with multiple African origins."[1]

Eleanor Scerri

Origin of the apes

The earliest apes

The common ancestors of present-day apes and humans were ancient apes. They looked more like current-day apes than anatomically modern humans (AMHs). Just like hominins they originated in sub-Saharan Africa and then migrated out to populate Europe, Asia, and Near Oceania. These early ancestors had their origins in Africa, diverging from monkey-like primates when the continent was an extensive forest, and migrating to Eurasia where they evolved into the organisms with traits we recognize today: long arms, stiff lower backs, long, curved fingers and toes, broad, flat chests, no tails, locomotion by brachiation, and close family ties. The evolution of apes is a tale of back and forth migrations, rapid speciation, and massive extinctions. Some apes remained in Asia and some returned to Africa, where one group became the common ancestor to present-day apes and humans.[2]

High diversity characterized ape species during the Miocene [23–5.3 million years ago (mya)]. Among primates, the Miocene was the age of the apes. More than 100 species of apes are known from the fossil record across Africa, Europe, and Asia. These ancient apes ranged in size from about 10 kg to possibly as much as 80 kg. They inhabited various ecological niches, which is reflected in their diverse anatomies.[3] The earliest apes have been found in Africa and looked very different from extant apes. One early genus of ape was *Proconsul*, also known as *Dryopithecus*. This curious name means "before Consul" and although it was the title given to provincial governors of ancient Rome, the taxonomic designation was derived from the name of a captive, performing chimpanzee named "Consul," who resided in London Zoo. At the time, the name Consul was in vogue for captive chimpanzees, and trained chimpanzees were so famous in Europe that when the resident chimp at Belle Vue Zoo in Manchester, England died in 1894, a honorary poem entitled "The Missing Link" was dedicated to Consul.[4]

Sex and Cohabitation Among Early Humans
https://doi.org/10.1016/B978-0-12-819086-9.00013-0

The name *Proconsul africanus* was coined by Arthur Hopwood in 1933 when he discovered a set of three fossil specimens during an expedition in 1931 with Louis Leakey near Lake Victoria, Kenya.[5] Considering the genesis of the name *Proconsul*, it was supposed to imply the ancestral relation to the chimpanzee, the rest of the apes, and the hominin lineage.

The period of time during which *Proconsul* lived in East Africa varies considerably depending on the literature source. Although most encompass the Early Miocene, times range from 25 mya to as recent as 17 mya. A resilient group, *Proconsul* survived repeated population contractions and expansions resulting from dramatic climatic fluctuations derived from Africa's collision with Eurasia and repeated volcanic activity. To date, a total of four species have been described within the genus and about 1000 sets of remains have been recovered, representing much of the anatomy.[6] These species ranged in size from 15 to 50 kg.

Proconsul is considered an evolutionarily transition stage in primate evolution. *Proconsul* belongs to a time when ancient monkeys and apes began to diverge from a common ancestor. They possessed a mixture of Old World monkey and ape characteristics, thus their classification in the ape superfamily Hominoidea is somewhat controversial, with some primatologists including *Proconsul* before the split of the apes from Old World monkeys, and some after within the ape lineage. If included within the apes, *Proconsul* represents one of the most ancient genera within the Hominoidea from the Early Miocene.

Proconsul was not a genus of suspensory organisms like modern apes.

This early ape did not possess the characteristic traits of present-day apes, such as thick dental enamel and the anatomical adaptation for knuckle-walking. *Proconsul*'s anatomy reflects an arboreal existence. They were lightly built and had short chests and curved fingers. Their forelimbs and hind limbs were about the same size, and the lower spinal column was long and flexible, indicating quadrupedal locomotion. Hands were employed for palm-down walking as seen in monkeys. On the other hand, they possessed a number of ape-like characteristics, including larger brain relative to body mass, no tail, large lateral supracondylar (rounded bulge at the end of the humerus), as well as more flexible hips, shoulders, wrists, ankles, hands, and feet compared to monkeys.

The extinction of *Proconsul* is shrouded in mystery. The population began to dwindle gradually, and it is not clear whether the final disappearance of the genus came about as a result of this progressive decline or whether the final *coup de grâce* to *Proconsul* came about as a result of dramatic events, such as a series of violent and consistent volcanic eruptions that lasted hundreds to thousands of years.[7] At the time of *Proconsul*'s extinction, the Kisingiri volcano in Rusinga Island, Lake Victoria, Kenya was experiencing frequent hydromagmatic eruption episodes involving extensive pyroclastic surges and powerful hydromagmatic explosions. Thus, *Proconsul* at the later stages existed in a frequently disturbed habitat. It has been proposed that the entire residual population of *Proconsul* remaining in East Africa may have been eliminated instantly during a single massive volcanic explosion.[8]

Regardless of its mode of extinction, it is theorized that *Proconsul* overlapped with *Kenyapithecus*, both living in a rather restricted area of East Africa. Although *Kenyapithecus* has been given a number of names such as *Proconsul nyanzae*, *Dryopithecus africanus*, and *Equatorius africanus*, the genus is nevertheless considered a true ape with a number of derived traits, such as thicker enamel, large mandibles, large premolars and upper incisors, as well as extremities equipped for knuckle-walking and semiterrestrial locomotion. The dental adaptations first observed in this group were pivotal for the subsequent exploitation of various niches that facilitated the extensive speciation of early apes after their colonization of Europe and Asia. This new, stronger masticatory machinery allowed for the consumption of harder foods, such as seeds and nuts. These adaptations impacted subsequent hominin evolution. In the words of Louis Leakey about the discovery of the initial remains in Fort Ternan, Kenya in 1961, *Kenyapithecus* was "a very early ancestor of man himself".[9]

Apes out of Africa

Although apes originated in Africa, it was traditionally thought that they also evolved many of their traits in Africa. However, the total absence of fossil apes from Africa between 17 and 15 mya and their presence in Eurasia suggest that most hominoids lived outside Africa and likely evolved their signature characteristics, including their cranial and dental features, in Europe. Currently, many scholars are of the opinion that the traits that characterize African apes today originated in Europe about 2 my before they are seen back in the African fossil record.

Around 18 mya, global sea levels dropped, exposing dry land between Africa and Eurasia. This geological event established land bridges across the Tethys Sea, an ancient body of water that has since evolved into the smaller Mediterranean sea of today. A massive exodus of animals followed northward. Mammals, including the ancestors of European apes, were attracted to the European subtropical forests of the time. It seems that this migration was driven by the luscious, vacant habitats in Europe compared to the hardship involved in existing in Africa. Europe, in this time, was the land of opportunity. As expected, about the same time, many of the Early Miocene apes were becoming extinct in Africa. In about half a million years after the land bridge had formed, it was swallowed up by the sea again, but not before the early apes became established in the new land. *Kenyapithecus* is believed to have been one of the early ape groups that dispersed north from Africa to Eurasia through the land bridge. Other early African apes, such as *Afropithecus* and *Heliopithecus*, were ancestors of species that also made their way into Europe. This rather sudden migration from Africa to Eurasia prompted rapid speciation of mammals in the new land, including primates. Between 17 and 13.5 mya, the first native European ape appear on the scene, *Griphopithecus*. Both already exhibited one of the trademarks typical of apes: thick dental enamel. Approximately 12.5 mya, a number of species that looked like present-day apes emerged in Europe. Their anatomy indicates that they likely were able to swing from branches like modern apes. The ones in Asia are thought to have evolved into the orangutan, and those in Europe into extant African species, such as the gorilla and chimpanzee. For example, the European genera *Dryopithecus* and *Rudapithecus* are linked to extant African apes due to the presence of frontal sinuses (a pair of small, air-filled cavities located behind the eyes in the brow region). The presence of the frontal sinus in contemporary African great apes and hominins clearly indicates that we are related to these ancient European apes.[10] Orangutans lack frontal sinuses.

Repatriated apes

Some of these European lineages of ancient great apes returned to Africa about 9 mya as Europe experienced dramatic climatic changes. Europe became drier at the end of the Miocene. The rise of the Himalayas made Eurasia much cooler, and deciduous woodland replaced the subtropical forests throughout the continent. Many of the great ape species went extinct in Eurasia as a result of these dramatic environmental shifts. In addition, it has been postulated that dietary overspecialization was partially responsible for their demise. It is thought that these ape lineages became totally dependent on very specific diets and subsequently were unable to swift and adapt to diets of leaves when subtropical forest were replaced by seasonal deciduous trees.[11] As further climatic deterioration ensued, the species became cornered into evolutionary dead-ends late in the Miocene.

Of the lineages that managed to migrate south, one flourished and eventually evolved into the common ancestor of hominins and apes. However, prior to 2014 paleoanthropologists were clueless regarding the identity of this migrant ape lineage. All of that changed with the discovery of a great ape fossil from Kenya.[2] The discovery consisted of a 13-million-year-old skull the size of a baseball. It was the fossil of a 16-month-old infant, which was subsequently named *Nyanzapithecus alesi*. The anatomy of the skull revealed that the specimen belonged to a group of fruit-eating, slow-climbing primates that resembled baby gibbons. The presence of teeth much larger than those of other early apes and bony ear tubes, an important derived ape feature, suggested that *alesi* was close to the ancestral line of apes from which hominins and living apes derive. High-resolution X-ray imaging indicated that *alesi* possessed semicircular canals (balance organs in the inner ear) typical of deliberate and slow-moving apes as opposed to fast acrobatic arm-swinging arboreal monkeys. Yet *alesi* does not represent the *immediate common ancestor*, the actual species that gave rise to the great apes and hominin lineages. It is thought that *alesi* lived about 6 mya prior to the actual separation of hominins from apes. From 13 mya when *Nyanzapithecus alesi* existed to 7 mya when the great ape-hominin lineage separation occurred, the number of ape fossils is limited. This dearth in the fossil record is known as the "African ape gap".[12] The paucity in the ape fossil record is often attributed to the detrimental environmental conditions in Africa from around 11 to 5 mya. During this period, Africa experienced considerable rainfall, and since ancient apes lived in wooded habitats, their remains were less likely to fossilize, and their bodies decomposed in the wet and acidic soil. In addition, the global climate during the Miocene is known to have been much warmer than the present, which further contributed to poor fossilization conditions.[13]

The common ancestors

In spite of the limited number of ape fossil remains found from the period when the ape-hominin common ancestor lived, some informative specimens have been recovered. Three important discoveries have contributed to our knowledge of ape evolution during the "African ape gap": *Samburupithecus kiptalami* and *Nakalipithecus nakayamai* from Kenya, and *Chororapithecus abyssinicus* from Ethiopia. These findings indicate that apes were not totally eradicated from Africa as a result of the Early Miocene environmental conditions and exodus to Eurasia, or had re-established themselves back in Africa. These species indicate that a number of ape species coexisted in the same general geographic area, indicating some degree of diversity.[14]

Chororapithecus abyssinicus lived in the southern margin of the Afar rift in Ethiopia approximately 10–10.5 mya. This ape existed during this critical time period, just prior to the common ancestor, and its remains provide important information on early hominin evolution. This African, large-bodied ape possessed a dentition very similar to present-day gorillas, and the teeth of *Chororapithecus* are considered by many experts to be indistinguishable from the ones of contemporary gorillas in size, shape, and proportions. In fact, the morphology of these teeth encompass the range exhibited by gorillas today. The discovery consisted of a canine and eight partial molars from three to six specimens. The dentition possesses thick enamel and clear-cutting edges, suggesting a fibrous-plant diet consisting of hard or abrasive food. Although most experts consider *Chororapithecus* a direct ancestor of contemporary gorillas, some scholars do not rule out the possibility that it represents the remains of an early hominin.[15]

Although a more recent dating of *Chororapithecus* based on geochemical, magnetostratigraphic, and radioisotope technology (see "Absolute dating" section in Chapter 4 for definitions) indicates a more modern age of about 8 mya,[16] which is more in line with the chronology of the traditional timing of the gorilla-hominin split,[15] the older dates of 10–10.5 mya initially reported for the fossils add support for a more ancient African evolution of the gorilla/chimpanzee/human clade and suggest an older separation of the great apes lineage from the hominins. If the older chronology holds, it means that the crucial genesis of hominins also needs to be pushed back in time and the time of certain evolutionary events re-evaluated. Yet a number of authorities caution that the dental likeness between *Chororapithecus* and contemporary apes may be the result of convergent evolution and only reflect independent adaption to similar diets. Thus, *Chororapithecus* may not be immediately related to the common ancestor of great apes and hominins, and may represent a more distant ancient group within the ape lineage.

Another important finding is that of *Samburupithecus kiptalami*, who lived about 9.6 mya in the Samburu Hills of northern Kenya. This species is known only from a portion of a maxilla (upper fixed bone of the jaw). It belonged to a 60 kg terrestrial quadruped ape with a diet of fruits that lived in a wooded habitat surrounded by savannah.[17] Morphologically, the dentition of this ape is equivalent to the one of an extant female gorilla. Unlike those of other Miocene apes, the teeth of *Samburupithecus* are rather elongated, like in gorillas, chimpanzees, and hominins. Using high-resolution micro-CT (see "Technical advances" section in Chapter 5 for definition), the teeth reveal a number of ancestral characters traceable to *Proconsul* from the early Miocene, such as relative small premolars, a smaller molar 1 relative to molar 2, and thick molar enamel, among others. The presence of these ancestral *Proconsul*-like characteristics in *Samburupithecus* indicates the persistence of this lineage for at least 6 my. On the other hand, its large premolars represent derived features that link this species to present-day great apes. The unique combination of ancestral and derived traits in *Samburupithecus* and its highly specialized dental morphology (e.g., molars with highly inflated and centrally located cusps) throws into doubt its consideration as a species close to the African ape-hominin split. And although the exact relationship of this species to present-day apes and humans is uncertain, most experts are of the opinion that it represents only a distant common ancestor to the great ape-hominin clade.

Nakalipithecus nakayamai is the third ape species discovered that dates to the critical and relative unknown "African ape gap" period. Discovered in 2005, this species was dated to approximately 10 mya by the ^{40}Ar–^{39}Ar ratio and reversed polarity methods.[14] This fossil was unearthed from the locality of Nakali, Kenya, about 30 km from the site where *Samburupithecus kiptalami* was found. The proximity in time and space of these two ancient ape species suggest that they coinhabited East Africa, advocating for some degree of diversity in the region.

The remains consist of a partial lower jaw and 11 teeth. The upper premolars of *Nakalipithecus* are similar to the ones of *Samburupithecus* in that both are elongated mesiodistally (distal surface of a tooth). In the latter, the peaks are more enlarged and positioned more centrally, and in the former they are more peripherally positioned. From these features, it seems that *Samburupithecus* was more specialized in its diet compared with other Miocene and extant apes. Also in *Nakalipithecus* the dentine-enamel junction is flatter than in *Samburupithecus*.

Many scholars think that *Nakalipithecus* was closely related to the immediate common ancestor, unlike *Samburupithecus*, who is not considered near the immediate common ancestor to the African ape-hominin bifurcation.

In relation to the issue of ape diversity in Africa, it is noteworthy that in the site where *Nakalipithecus* was isolated, three premolars dissimilar to those of either *Nakalipithecus* or *Samburupithecus* were found, suggesting the presence of another, more primitive large ape in the area.[18] Taken together, the discoveries of *Samburupithecus kiptalami*, *Nakalipithecus nakayamai*, and *Chororapithecus abyssinicus* are indicative of some degree of ape diversity in East Africa during the Late Miocene. The different East African ape species combined with a limited number of fossils recovered during the "African ape gap" period are interpreted by some investigators as poor preservation of fossils and not a lack of ancient apes in Africa from 13 to 7 mya. Thus, the limited fossil record in Africa may just be the result of moist, acidic soil. According to proponents of this hypothesis, there is no need to invoke an Eurasian Miocene ape return to Africa 13 mya to re-establish an ape population that would eventually give rise to the last common ancestor of the African great apes and hominins. In light of these findings, it is possible that some Middle and Late Miocene apes survived and evolved in sub-Saharan Africa, and then gave rise to the last common ancestor of African great apes and humans. Furthermore, it is also possible that although some lineages returned from Eurasia to Africa around 13 mya, this event did not repopulate Africa, but provided additional diversity to persistent indigenous African lineages.

The genesis of hominins

The hominin lineage had its origins about 6–7 mya when our ancestors separated from the line that led to the chimpanzees and bonobos.[19] Unfortunately, the fossil record of potential common ancestors is rather silent after *Samburupithecus*, *Nakalipithecus*, and *Chororapithecus* about 10 mya. This chronology indicates a gap of about 3–4 my in which limited information exists on the origins of hominins. These crucial final stages prior to the ape-hominin split are missing from the fossil record. Yet the end result of this evolutionary process is known: the great ape lineage is today represented by a number of species and subspecies that reside in Africa and Asia. On the other side of this evolutionary black box, hominins are represented by a single extant species: *Homo sapiens*.

In spite of the dearth of fossils from this critical time during the ape-hominin divergence, the field of molecular biology has provided important information regarding the event. For example, today it is known approximately when the separation between the chimpanzee and human lineages took place. This information is derived by comparing the DNA sequence differences between chimpanzees and humans with an estimate of the mutation rate of DNA. Based on DNA data, it is known that chimpanzees share a more recent common ancestor with humans than with gorillas[20] and that chimpanzees are more closely related to contemporary *sapiens* than to gorillas.[21] Since the similarities among living African apes are thought to have been inherited from a common ancestor, the common ancestor that early hominins shared with chimpanzees possessed anatomical and behavioral characteristics in common with contemporary apes.[22]

Independent from the lack of fossils from the time of the ape-hominin separation, it is not totally clear what evolutionary force was responsible for the split of the two lineages. However, it is known that between 20 and 10 mya, Africa was a much wetter region; a single super-rainforest covered the continent from shore to shore, providing an environment well suited for tree-dwelling primates. It is also known that a number of major geological processes led to climatic and habitat changes between 10 and 7 mya in East Africa. During this period of time, the fossil record is virtually unavailable. It has been proposed that these geo-climatic alterations transformed part of the forest habitat to open savannas providing selection pressure in favor of bipedalism and upright posture. These environmental changes were initiated by uplifting of the Earth's crust along the East African Rift that created the East African Mountain Range. To the east of the range, forest ruled and the ape lineages persisted, while to the west, hominins evolved as the habitat became increasingly dryer, transforming the forests to savannas (Fig. 1). Although data based on *Australopithecus africanus* and *Australopithecus afarensis* as well as their associated fauna are indicative that they lived in open savannas,[23] subsequent findings related to ecosystems of earlier hominins suggest that dry open grassland and bipedalism do not always correlate. Thus, at this point the cause and effect link between upright posture among hominins and the geological events triggered by the East African Mountain Range are questionable, at least for certain hominin species.

Motivated in part by interest generated from the ape-hominin divergence times estimated by molecular biologists,[24] archeologists intensified their efforts to find fossil

Fig. 1 Open savannah (https://www.nsf.gov/news/news_summ.jsp?cntn_id=121029).

remains from the critical time period of the split. The archeological efforts were fruitful as fossils from three informative genera/species were discovered in subsequent years. Their taxonomic names are *Sahelanthropus tchadensis*, *Orrorin tugenensis*, and *Ardipithecus kadabba/ramidus*, and many experts consider them the earliest hominin species known to date. Yet, to the surprise of scholars, these groups lacked the features that are associated with later hominins and their status as hominins is based entirely on dentition and certain anatomical traits related to upright posture and bipedal locomotion. Specifically, these groups exhibit marked diminution of canines and lack the C/P3 honing complex (an ancestral condition in which the back side of the upper canine sharpens itself against the lower first premolar). The C/P3 honing complex is not seen in hominins and it is thought that to have been lost after the ape-hominin split. Bipedalism is typically diagnostic of the hominin lineage. However, the hominin status of these three genera is still considered tenuous, with some scholars indicating they are apes.[25] Furthermore, with only three fossil findings each containing only small fragment of the skeleton, this incipient early stage of hominin evolution is poorly known and it is risky to draw strong conclusions.

Was *Sahelanthropus* a hominin?

Sahelanthropus is represented only by a cranium and teeth (Fig. 2), and due to the collage of ancestral and derived traits that it exhibits, many questions remain regarding where this species fits in the hominin evolutionary tree. The specimen was found in 2001 in an arid region just south of the Saharan desert in what is today the west-central nation of Chad.[26] The specimen, nicknamed Toumaï or "hope of life" in the local language, has been dated to 7.2–6.8 mya[27] by the radioisotope ^{10}Be/^{9}Be method (see "Absolute dating" section in Chapter 4 for definitions). Based on the species of other animal fossils found near the specimen, it was determined that Sahelanthropus lived in a wooded habitat in the vicinity of a lake. The vertebrate species found in close proximity to *Sahelanthropus* included elephants, giraffes, antelopes, hippopotamus, crocodiles, lizards, monkeys, fish, and wild boar. At the time of discovery, the combination of ape- and hominin-like characters was a source of confusion among experts, and the general consensus was that *Sahelanthropus* represented an organism in a stage of evolutionary transition due to diverse and changing habitats.

The braincase of *Sahelanthropus* is about 380 cm^3 in volume, comparable to the average size of a chimpanzee brain. Among the ancestral characters, it possesses massive brow bridges, a sloping face, a u-shaped dental arcade, diastema (space between incisors and canines) (Fig. 3), protruding face, marked postorbital constriction, widely shaped eye sockets, ape-like neck muscle attachments, and elongated skull. The derived features reflecting hominin affinities are the small canines and the location of the *foramen magnum* (opening at the base of the skull through which the spinal cord passes). The position of the

Fig. 2 Cast of the *Sahelanthropus tchadensis*'s skull and facial reconstruction (https://en.wikipedia.org/wiki/Sahelanthropus; http://humanorigins.si.edu/evidence/human-fossils/species/sahelanthropus-tchadensis).

foramen magnum underneath the skull is a diagnostic sign of upright posture and bipedalism (Fig. 4). In quadrupeds, such as apes, the *foramen magnum* is located near the back of the skull. The progressive shift of the *foramen magnum* toward the base of the skull allows the spinal column to support the cranium more effectively in an erect position. The significance of a vertical posture is the potential for bipedal locomotion. The basal position of the *foramen magnum* underneath the skull in *Sahelanthropus* dating back to the time of origin of the hominin lineage was surprising since at the time bipedalism was thought to have evolved more recently, with the australopithecines. In addition, considering the many ancestral characteristics seen in the skull and teeth, the expectations were that *Sahelanthropus* was an arboreal quadruped. Taking all of this into account, it is likely that Toumaï was a transitional facultative bipedal as opposed to a habitual bipedal. A further consideration is that since upright posture and bipedal locomotion likely evolved in stages, and possibly independently in various hominin lineages, different groups of hominins may have practiced unique forms of bipedalism.

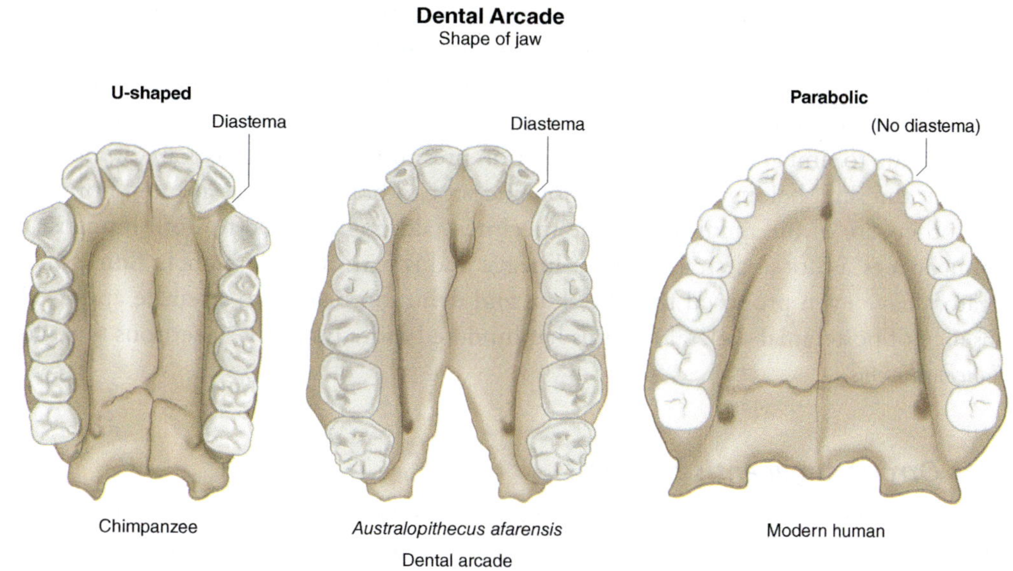

Fig. 3 U-shaped dental arcade and diastema (http://images.slideplayer.com/39/10972464/slides/slide_15.jpg).

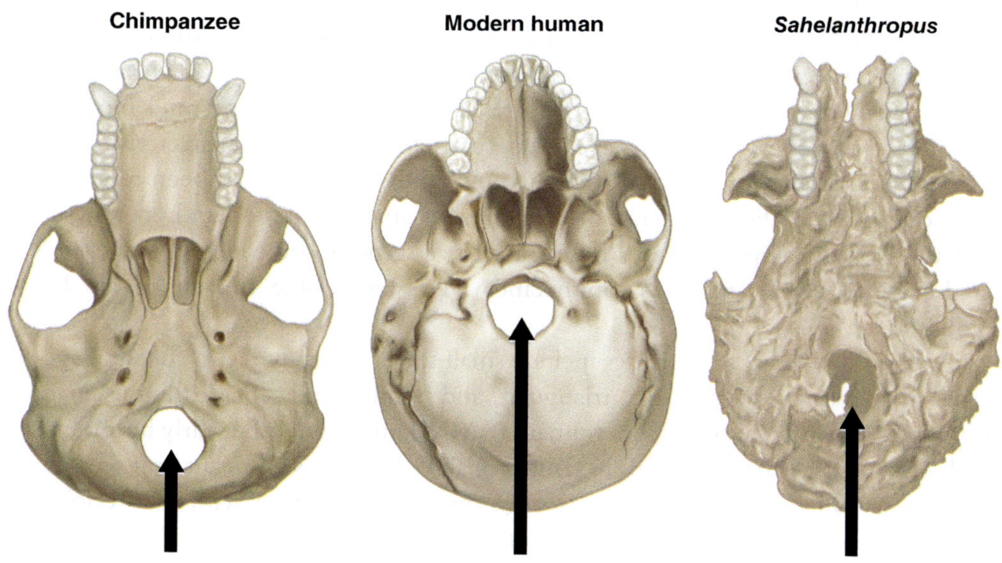

Fig. 4 Position of the *foramen magnum* (https://prezi.com/uxc8hpya41l1/sahelanthropus-tchadensis-and).

In spite of the various advanced characters found in Toumaï, not all experts are convinced that *Sahelanthropus* was a basal species within the hominin lineage.[25] Some authorities claim that the derived characters that *Sahelanthropus* shares with hominins are the result of parallel evolution, as unrelated species at times independently evolved similar features due to similar environmental selection pressures. Other scholars are of the opinion that the skull belonged to a common ancestor to both the great ape and hominins. Still others are of the opinion that the specimen, although related to both humans and chimpanzees, is not ancestral to either lineage and is part of a side branch. Some investigators have even proposed that the skull and dentition belong to a female ape.[28] The absence of postcranial skeletal remains has hindered more definitive assertions regarding *Sahelanthropus*.

Was *Orrorin tugenensis* bipedal?

As of 2007, a total of 20 fossil samples have been discovered and assigned to this species, including parts of a mandible, a number of teeth, a symphysis (fibrocartilaginous fusion between 2 bones), 3 pieces of femora, a partial humerus, a proximal phalanx (digital bones of the hands and feet), and a distal thumb phalanx. The specimens were unearthed from sites in Tugen Hills, Central Kenya and dated to the Lukeino Formation from approximately 6.0 to 5.7 mya.[29] Most scholars are convinced that *Orrorin* is a hominin, yet its position within the human lineage is contentious. Over the years, the combination of ancestral and derived traits in these samples has generated heated controversy regarding whether this species or australopithecines are the ancestors of the genus *Homo*.[30] *Orrorin* shares several apomorphic (unique) features with australopithecines and more recent hominins, but none with extant apes. One particular diagnostic characteristic that has attracted considerable attention is the anatomy of the femur. It has been reported that the morphology of the femura is more like those of anatomical modern humans (AMHs) than those of australopithecines.[31] Based on these assertions, some believe that *Orrorin* led to the genus *Homo* and australopithecines were just a side branch of the human evolutionary tree.

Orrorin's derived characteristics include small teeth, but they retain ancestral ape-like shape such as canines with large, triangular, and pointed ends. Their thinner enamel is reminiscent of anatomical modern humans (AMHs) but not of other early hominins such as *Ardipithecus* (see the next section). But probably the most significant trait seen for the first time in *Orrorin* is the morphology of its thighbone or femur with its interiorly oriented, ball-shaped head and the presence of a prominent, centrally located lesser trochanter (pyramidal protrusion from the proximal part of the shaft of the femur) that projects medially. These characteristics of *Orrorin* are important for locomotion since the femur's head articulates with the pelvis and in this frontal orientation, as seen in *Orrorin*, it allows for upright posture and bipedal locomotion.[32] Considering these features of the *Orrorin*'s thighbone, it is likely that this species practiced some version of facultative bipedalism, yet it was not committed to full habitual erect locomotion. In general, the postcranial skeleton indicates a mosaic of arboreal and terrestrial traits. For example, the hands have long curved fingers, a condition observed repeatedly in early hominins and associated with grasping branches, whereas the thumbs are proportionally smaller, a feature linked to manipulation of objects, dexterity, and tool-making.

Soon after the discovery of *Orrorin*, it became obvious that some reconsideration of australopithecines' position in the hominin lineage was required. It is noteworthy that *Orrorin* lived about 6 mya, about 2 my prior to australopithecines, yet *Orrorin* had more traits in common with AMHs than with *Australopithecus*. A possible explanation for this apparent conundrum is that the long-standing contention that australopithecines were direct ancestors of the genus *Homo* is not correct and that australopithecines were only a side branch in human evolution that eventually died out. To address these issues, a combination of high-resolution geometric morphometric and multivariate analyses (see "Statistical analyses" section in Chapter 4) were conducted employing a number of femora from extinct and extant ape and hominin species.[33] The results demonstrated the following:

(1) The taxonomic groups could be distinguished by means of the major patterns of shape variation of their proximal femora (Fig. 5).

(2) *Orrorin*, as a group, possesses a unique combination of ancestral Miocene ape and derived early hominin traits, which are intermediate between the former and the australopiths.

(3) These features are consistent with its age and the view that it was an incipient biped.

Furthermore, due to the lack of a fully developed, AMH-like abductor muscle apparatus (group of hip muscles involved in moving the thighs apart), it is likely that *Orrorin*'s bipedalism was very different than that of AMHs.

Were ardipithecines arboreal, terrestrial, or both?

Two species have been discovered as part of the genus: *Ardipithecus kadabba* and *Ardipithecus ramidus*, dated at 4.4 and 5.6 mya, respectively. Currently, most experts agree that *A. kadabba* was a direct ancestor of *A. ramidus*. *A. kadabba* is known only from teeth and a few pieces of skeletal bones[34] while *A. ramidus* is represented by about 110 fossilized bones including a skull (Fig. 6), mandible, teeth, and arm bones from a minimum of 36 individuals.[35] Members of both species were discovered from the Middle Awash region of the Ethiopia's Afar Depression at the northern ridge of the East African Rift. The large sample size of the *ramidus* remains is highly unique among early hominins. Part of this precious collection, discovered during a 9-year period from 1994 to 2003, included an exceptionally well-preserved skeleton of an individual that was 45% complete. This particular specimen was a female who weighed about 50 kg (110 lb) and could

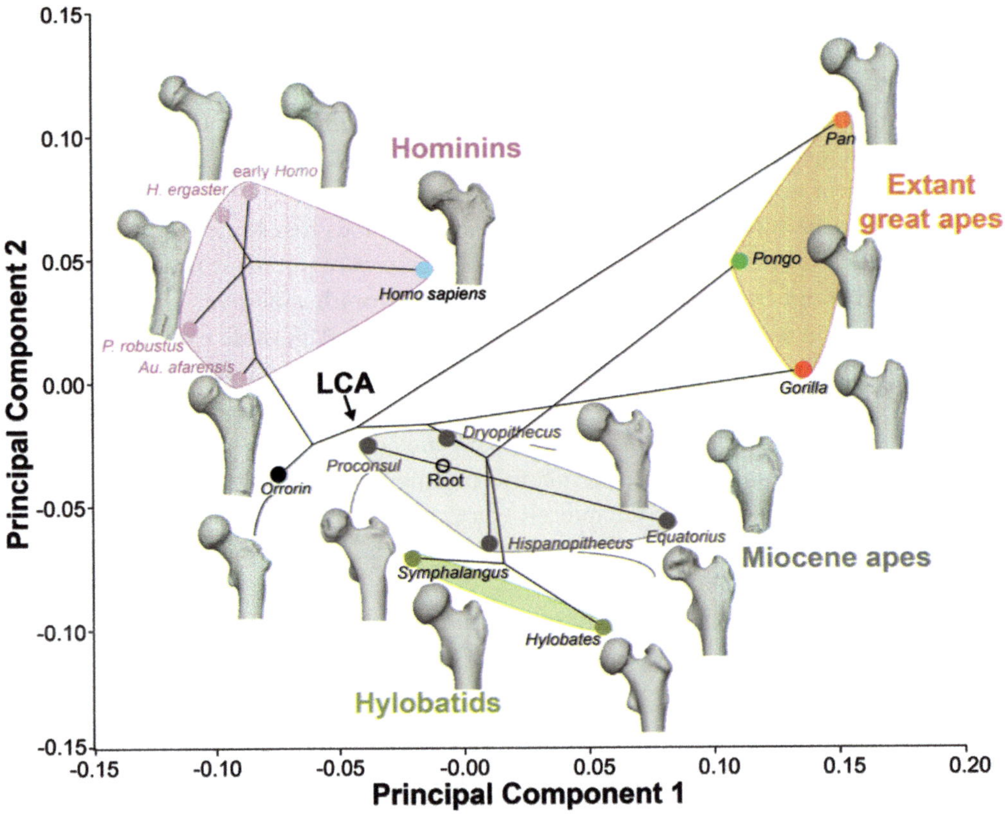

Fig. 5 Principal component analysis based on Femura anatomy (https://www.nature.com/articles/ncomms3888). The phylogeny presented in this figure is projected into a plot defined by the two first principal components (PCs) of the covariance matrix among extant genera and fossil species means (PC1 = 32.9% and PC2 = 18.5% of the total variance, respectively). Internal node morphologies were reconstructed using squared-change parsimony, and branch lengths were derived from estimated divergence times.

have been up to 120 cm (3 ft 11 in.) tall. She was subsequently named Ardi. She was radiometrically dated to 4.4 mya using the two volcanic ash layers that were encapsulating the fossil. Although some experts contest that ardipithecines may be within the great ape lineage, most scholars are convinced that they were hominins.

Ardi represents an exceptional finding. Much of the feet and virtually all of the hands were recovered and reconstructed. The completeness of Ardi's remains allowed scientists, for the first time, to perform detailed comparative biological/anatomical studies of an early hominin. However, it took an international team of researchers 15 years to analyze fully and publish a description of the skeleton. Due to the poor preservation of the remains, extensive reconstruction was required. It seems that when Ardi died,

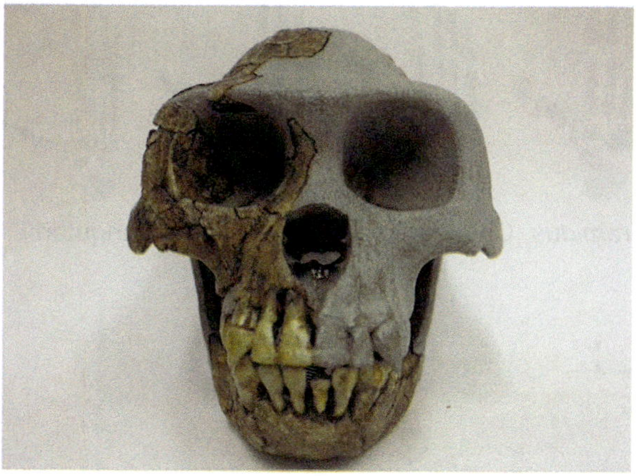

Fig. 6 Ardi (https://australianmuseum.net.au/learn/science/human-evolution/ardipithecus-ramidus).

her body was crushed down into the mud by a heavy animal, causing dislocation and damage to bones. It is likely that subsequent erosion brought the badly crushed and distorted bones back to the surface. The pieces of fossils were fragile and required extremely painstaking work to clean. After years of separating the pieces of fossil from the debris, CT-scanning technology and digital analysis assisted in finding matches and putting the skeletal fragments together.

When the reconstruction was completed, it was evident that Ardi's overall skeletal structure was that of a hominin with unexpected combination of derived and ancestral features. Among the derived traits, *ramidus* possessed small blunt canines, reduced canine size and sexual dimorphism, lack of a functional C/P3 honing complex, anteriorly positioned *foramen magnum*, and a number of traits connected with upright posture and bipedalism.[36] Ancestral characters include a fully opposable big toe (Fig. 7), absence of longitudinal arch in the foot, relatively equal forelimb to hind limb lengths, and ape-like lower pelvis.[37]

In connection with locomotion, it seems that Ardi was able to move among the branches as well as on the ground. This flexibility surely allowed *ramidus* to exploit a number of habitats for food and protection. Ardi's feet, pelvis, legs, and hands indicate that she enjoyed bipedalism while on the ground, but acted as a quadruped when moving about in the trees. The unique collage of derived and ancestral traits suggests that Ardi was a facultative biped and at the same time was able to move among the branches by placing her whole palms down, a mode of locomotion known as arboreal palmigrade quadrupedalism. This mode of arboreal quadruped locomotion was facilitated by the rigid wrist bones and the very flexible wrists and finger joints of *Ardipithecus*, ancestral traits inherited from ancient apes. In addition, Ardi's big toe extended out from the main axis of the foot as in apes. This anatomy facilitated holding to branches. In addition, *Ardipithecus* possessed a small bone inside a foot tendon that keeps the big toe rigid and thus facilitated bipedal

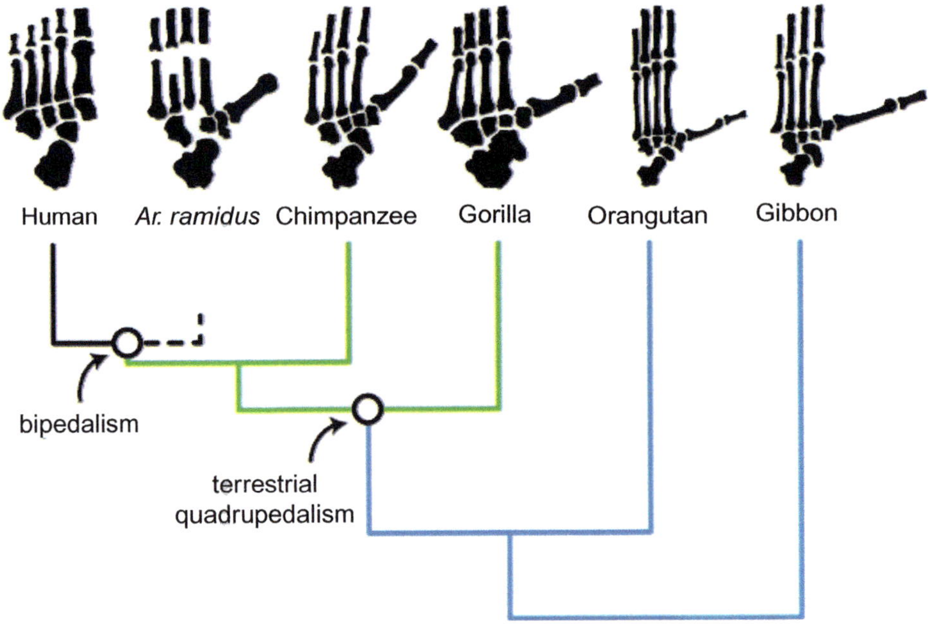

Fig. 7 Fully opposable big toe and variations (https://english.netmassimo.com/2019/05/01/the-role-of-ardipithecus-ramidus-in-the-evolution-of-bipedalism-in-humans).

locomotion. Clearly, the type of bipedal locomotion afforded by these features was very different and less efficient than the ones seen in more recent hominins and AMHs. The pelvis also exhibits a collage of ancestral and derived characters with flaring upper wings that allow bipedal locomotion without rolling from side to side like a chimpanzee, while the lower pelvis retained the ancestral condition that allows attachment of large hind limb muscles necessary for holding to branches.

Scholars also have some ideas regarding Ardi's habitat from the thousands of species found in the vicinity. The findings included antelope, primates, forest-dwelling birds, and seeds from figs and palms, indicative of a wet woodland unlike the current region's dry environment. In addition, wear patterns and the isotope composition of Ardi's teeth indicate a diet of fruits, nuts, and other eatables found in forests.

Ardi's canines and postcanines exhibit a marked reduction in size as well as a dramatic change in morphology to a more feminine type. Ardi's canines are not as pointed as in more ancestral hominins. They are short, broad, and diamond-shaped. These dental alterations in size and shape are part of a trend in sexual dimorphism that includes reduction in body size, which has continued into contemporary humans. It has been suggested that these changes in anatomy are reflected in modifications in social behavior in which males' roles are more along the lines of a provider to the family as opposed to a fighter to gain access to females.[38]

Reassessment of the Paleoenvironment of early hominins

It has been postulated that the creation of the East African Rift (EAR) System and the resulting East Africa Ridge as well as the East African Great Lakes played a pivotal role in early hominin evolution. The EAR began developing around the onset of the Miocene, 22–25 mya, triggering dramatic geological events including progressive ecological changes in the regional habitat, transforming wet forest to arid grasslands and savannas. It has been proposed that the mountains of the East Africa Ridge acted (as they still do) as barriers, keeping rain precipitation on the west side of the cliff. The limited amount of rain to the east of the range transformed the land into the dry savanna seen today. It has been theorized that early arboreal hominin populations gradually experienced strong selection pressure for traits that would allow them to survive in the new grassland environment. The organisms on the dry west side of the ridge needed to respond to the novel challenges to survive as species. They needed to evolve physiological and anatomical adaptations that would allow them to survive in the new semidry plains of Northeast Africa. Many scholars see upright posture and bipedalism as adaptations to this new environment. These ideas are known as the "savanna theory".[39] Scholars cite the medley of traits reflecting both arboreal and bipedal locomotion in early hominins as evidence for the selection pressure generated by the relatively flat, structureless savanna. The correlation between dry savanna and adaptations to upright posture and bipedalism is seen in members of the different species of the genus *Australopithecus*, which lived in the dry grassland of Northeast Africa.

Advocates for the savanna theory suggest that upright posture was advantageous to early hominins since it provided a mechanism to detect and frighten potential predators. This hypothesis is based on the idea that by standing upright, hominins were able to scare attackers away by appearing as large and menacing as possible. An upright posture may also have been advantageous in the savanna as a method of surveillance and for attracting mates. Furthermore, upright posture and bipedalism allow the use of hands for various tasks other than locomotion. For example, free hands facilitate behaviors that characterize hominins such as tool-making, hunting, and gathering/processing food, as well as caring for the young, disabled, or old.

Although the savanna theory is supported by the open habitat data derived from the *Australopithecus africanus* and *Australopithecus afarensis* sites, some of the paleoenvironmental information from earlier hominins indicates that they lived in different landscapes. The fauna recovered in association with these two australopithecine species support the theory that bipedalism and the expansion of the savanna and grassland habitats during the Pliocene (5.3–2.6 mya) are linked. Yet some studies have reported that *Sahelanthropus*, *Orrorin*, and *Ardipithecus* may have lived in a number of other different habitats not characterized by open semidry savannas. The fauna associated with *Sahelanthropus*, for example, reflect a lakeshore environment surrounded by extensive forest. Other investigators indicate that *Sahelanthropus* lived in a medley of habitats, including forest at the edge of lakes, savanna-woodlands, and grasslands, as well as shrub-bushlands and grassy woodlands.[40] On the other hand, the fossil plants and mammals associated with *Orrorin* suggest that this early hominin lived in a dry evergreen forest habitat and not in a savanna as proposed by the savanna theory.[29] A different report suggests that *Orrorin* inhabited open woodlands, dense woodlands, or forests, possibly along lake margins.[41] One investigation indicates that one *Ardipithecus kadabba*'s habitat was woodland on the banks of a river or lake with floodplain grassland along the water margins[42] while a different study on *Ardipithecus ramidus* suggest a closed woodland habitat with possible patches of forest.[43] All of the putative habitats described above for *Sahelanthropus*, *Orrorin*, and *Ardipithecus* have certain elements in common: woodlands and the vicinity to water. Yet not all experts negate the savanna theory as an explanation for the beginnings of our lineage. In one study based on the plant species at the Aramis site in Ethiopia, it was concluded that *ramidus* lived not in a woodland or forest as previously reported, but in a tree- or bush-savanna with about 25% or less woody canopy cover.[44] On the basis of these findings, it is noteworthy that during the first 2 my of hominin evolution (5–3 mya), during the time periods that the early hominins lived, characteristics related with upright posture and bipedalism evolved at a high rate (Fig. 8). Then, amid the transition between early australopithecines and early *Homo habilis*, the rate of change involving characters related to bipedalism decelerated and eventually reached a plateau about 2.5 mya with *H. habilis*. It is possible that early hominins such as *Sahelanthropus*, *Orrorin*, and *Ardipithecus*, being species in anatomical and behavioral transition, exploited diversified habitats including arboreal as well as terrestrial locomotion that provided the flexibility to utilize different resources that would allow them to survive.

How diversified were hominins—Is the tree linear or bushy?

Although the idea of the tree of life dates back to biblical times and is described in Chapter 2 of the Book of Genesis (Old Testament) as a religious metaphor for good and evil, it was not until 1809 that the French naturalist Jean-Baptiste Lamarck (1744–1829) conceived the first tree of life in the context of biology in his *Philosophie Zoologique*.[45] However, Lamarck was not of the opinion that living things derived from a common ancestor. He thought simple forms of life were created continuously by spontaneous generation and that organisms changed from simple to complex within independent lineages represented in the tree of life made up of parallel lines. Charles Darwin (1809–82) was the first one to apply the scheme of the tree of life to explain how all life is related by common descent. The idea behind Darwin's representation is that all living things are at the ends of branches and that all extinct organisms are located within the branches at various distances from a hypothetical genesis indicating the origin of life. Early forms of the tree portrayed organisms arranged only along the *Y*-axis, which

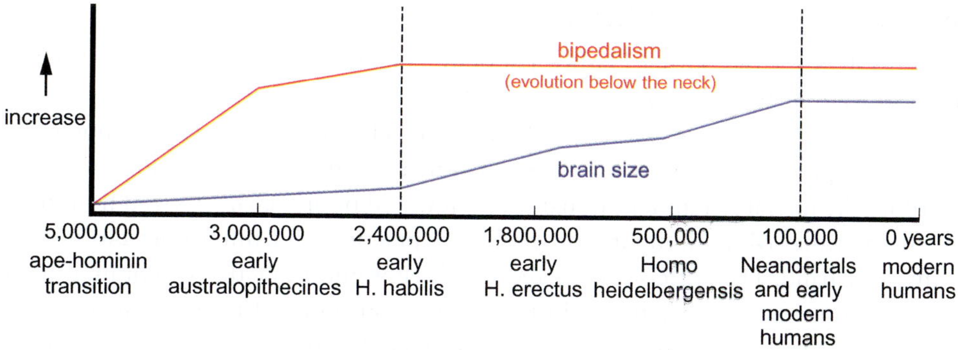

Fig. 8 Bipedalims and brain size (http://anthro.palomar.edu/homo2/mod_homo_2.htm).

indicated the passage of evolutionary time. Since Darwin's time, the tree of life has been employed to illustrate (in a rather simplistic way) how species are related and how they change as a function of time. The perception in most people's mind has been a type of one-to-one relationship between species, with one group of organisms somehow leading to another. Today most trees of life are more complicated, showing horizontal gene transfer, which indicates instances of interbreeding or introgression (Fig. 9).

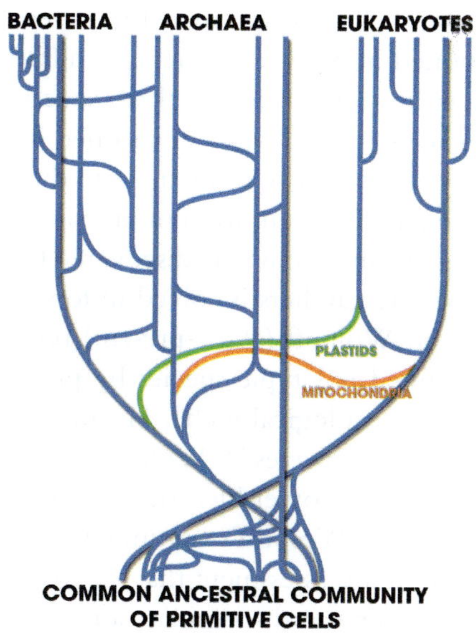

Fig. 9 Tree of life showing horizontal transfer (https://upload.wikimedia.org/wikipedia/commons/thumb/1/1b/Tree_Of_Life_%28with_horizontal_gene_transfer%29.svg/1782px-Tree_Of_Life_%28with_horizontal_gene_transfer%29.svg.png).

The single species hypothesis

In the 1950 XV Cold Spring Harbor Laboratory Conference on the topic of the Evolution of Man in Huntington, New York, the German-American leading evolutionary biologist Ernst Mayr (1904–2005) (see Fig. 3 and "The new physical anthropology" section of Chapter 5) expressed his consternation to a group of evolutionary biologists regarding the liberal assignment of names to new and poorly characterized hominin fossils. At the time, Mayr thought that the liberal practice of naming hominin fossils was getting out hand. His criticism centered on the common practice of coining different genus and species names to hominin fossils based on minimal fossil numbers and little justification. Mayr said in an interview, "Let's cut all this mass of names. Let's cut it down to 2 or 3 genera, and maybe 5 or 6 species." He added, "Everybody was totally outraged [at me]. And part of it was, of course, nonsense." He went on, "*Pithecanthropus* is nonsense! That's just *Homo*. That's not a different genus!" (http://library.cshl.edu/oralhistory/interview/cshl/symposia/1950-CSHL-symposium-origin-and-evolution/). The CSHL meeting, and many of the ideas expressed in it, initiated a reductionist trend among many anthropologists that advocated for a conservative approach to naming new hominin taxa, particularly when the fossil evidence available was minimal. Mayr further proposed the principle of biologically defined polytypic species, which he used to explain the diversity of fossil hominins as being the result of geographic variation within a single species of *Homo*.[46] He put forward the notion that there has been no speciation within hominin evolution involving reproductive isolation. Mayr went on to state that among AMHs, Java man, and the South African ape-man, there were no anatomical features that require them to be partitioned into several genera. In addition, within what is known today as the hominin clade, he recognized just three successive species: *Homo transvaalensis*, *Homo erectus*, and *Homo sapiens*.[46] Mayr envisioned hominin evolution as a linear array of populations, with one giving rise to the next in direct succession. In retrospect, his conservative and draconian ideas seem to stem, in part, from the adverse reaction to the excessive liberal coining of names to newly discovered fossils, a tendency of overconfidence experienced by many researchers involved in fossil discovery.

Later in 1968, Milford H. Wolpoff (1942–present) reinstated and extended Mayr's principle of biologically defined polytypic species by proposing that only one species at a time can occupy a specific ecological niche. His ideas were based on the principle of competitive exclusion among species.[47] Wolpoff subsequently explained in 1971 how competitive exclusion among Lower Pleistocene hominins could explain the single species hypothesis.[48] In general, Wolpoff's arguments in support of his competitive exclusion principle are based on his assessment that the fossil record did not provide distinct and unique diagnostic criteria to discriminate among populations, and he was not convinced that the fossils were synchronic (occurring or existing in the same time period). This second point represents a contentious argument, since strict synchrony is often difficult to assess from the paleontological record since individual findings are usually far between in time and location.

The notion of a simple, unilinear depiction of human evolution, as envisioned by Mayr in the 1950s, gradually became unsustainable as discoveries of contemporary hominins started accumulating during the decades that followed. Specifically, in the early 1970s a number of findings began to report on fossils that were clearly contemporaneous. In one such article, in 1976 Richard Leakey (1944–present) and Alan C. Walker (1938–2017) wrote that "an enormous wealth of early hominid [hominin] remains has been discovered over the past few years by expeditions within eastern Africa. Evidence has been presented for the existence over a considerable period of time of at least two contemporaneous hominid species." At the time, scholars were becoming aware that an early form of the genus *Homo* was sharing the same territory with at least one species of *Australopithecus*.[49] Leakey and Walker, for example, reported on the finding of a *Homo erectus* cranium and an *Australopithecus boisei* cranium from the Koobi Fora Formation, east of Lake Turkana.[50] These two fossils were so different that they could not be members of the same genus. The body of evidence in favor of different hominins coexisting was mounting and new ideas were reflected in more complicated phylogenies in which hominins were depicted as in Stephen Jay Gould's (1941–2002) "bushy" hominin tree.[51] It was clear that scholars needed to face hominin evolutionary complexity.[52]

Since then, the view of the hominin evolutionary tree has changed from simple and linear to bushy and complex as the numbers of distinct fossils were discovered and the methods of analysis increased and became more sophisticated. The concept of nonoverlapping succession of species became impossible to contemplate. It was becoming clear that in many instances, among the hominin clades, the notion of one species evolving entirely into another species with the disappearance of the ancestral group, a process known as anagenesis, was not the main mechanism by which hominin evolution occurred.

Today, discoveries from a number of disciplines are providing a radical new image of hominin evolution. As the data are emanating from recent archeological sites worldwide, they are also motivating the reassessment of remains previously collected. This fresh outlook on hominin evolution is largely fueled by the discovery of an unprecedented number of identified remains, underscoring the diversity of groups and species that have lived since the genesis of hominins. It is becoming clearer now that the hominin evolutionary landscape is not only bushy, but also complex, with a multitude of species living in close proximity and interacting. The fossil discoveries coupled with the novel advanced technologies to analyze them have clearly demonstrated that the simplistic linear view of hominin evolution in which one species was replaced by another was the result of limited and fragmentary data.

It is likely that the habitat heterogeneity afforded by the open savannas and woodlands in various parts of Africa provided for extensive biodiversity of flora and fauna species, including hominins. In other words, the environmental diversity and climatic fluctuation experienced in Africa likely provided the selection pressure that drove at least part of the observed hominin heterogeneity.

Compounding the impact of new fossil discoveries and revolutionary technologies and methods for the analyses of fossils and archeological remains is the realization that we may have been underestimating hominin diversity, as fragmentary skeletal remains may have been misclassified as part of the same taxa due to mosaicism. Mosaicism at times may be misleading, as different anatomical features from different body parts reflect various ancestral and derived conditions from putative source populations. Scholars have uncovered that many extinct populations and species are indistinguishable from each other in many parts of the skeleton, despite being very different in others. Furthermore, technology transfer among distinct groups and species as seen in tools, artifacts, and art may obscure the true diverse source of their creators. Thus, many of the recovered remains may have been initially identified as part of the same group or species when, in fact, examination of a more complete skeleton and/or additional artifacts would have revealed unique taxa. New discoveries of more complete specimens, the prevalence of mosaicism among hominin species, and the realization that technology as well as art transfer among populations may have occurred justify caution in interpreting fragmentary specimens.

The realization of potential widespread hybridization among hominin populations and species underscores another source of genetic and anatomical diversity. It also reinforces the notion that overall hominin diversity has been underestimated. Furthermore, the archeological and anthropological records together indicate that the ancestors of AMHs evolved in Africa amid a diversity of hominin populations and species. Some of these groups exhibited a number of traits that we see in AMHs today.[53] These characteristics from our evolutionary past have been modified as part of adaptive or random changes. Still other traits have been lost in lineages that became evolutionary dead-ends.

Genetically, AMHs are a collage of numerous ancestries. It is likely that during our evolutionary past, the lineage that gave raise to *sapiens* was the recipient of DNA from various hominin groups. As new fossils are discovered and DNA technology advances, analysis of ancient DNA reveals the genetic signature of species and populations that no longer exist, yet they still "survive" within us as part of our genome. Even the DNA imprint of "ghost" species is revealing that ancient populations, of which there is no fossil or artifact evidence, persist within the DNA of AMHs. Thus, *sapiens*, as the sole surviving hominin, represents a reservoir of DNA fragments from extinct groups. Recent genetic analysis of extinct and extant groups argues for complex scenarios involving separation, migration, speciation, reunions, and interspecies gene flow that resulted in diversity.

All of these issues are motivating scholars to reassess previously discovered fossils and artifacts, and to reconsider prevailing notions relating to cohabitation and interactions among populations at specific times. Furthermore, many investigators are also reexamining with an open mind the artistic and technological potential of ancient species.

Hominin diversity in time and space

A multiplicity of hominin taxa has been demonstrated in time and space. For example, the current fossil record demonstrates that different species of hominins existed at similar time periods, and in some instances the data clearly indicates they were contemporaneous. However, fossils unearthed from different sites some distance apart do not address the question of whether the species in question actually interacted (e.g., technology/artistic transfer, competition, interbreeding). Additional information is required—for example, how close these ancient groups needed to be in order to interact? How eager were hominin populations to interact, and were these interactions amicable or violent? These are issues impacted by a number of biological and behavioral characteristics specific to the species in question. The presence of different species at distant sites is just an indication of diversity within a time period. Examples of instances that meet both the location and temporal requirements for coexistence are fewer than cases of just temporal overlap.

Temporal overlap

Prior to approximately 2 mya, hominins only existed in sub-Saharan Africa. Most authorities are convinced that *Homo sapiens* originated in Africa and later—likely in different waves—colonized Eurasia, America, and Oceania. Thus, when the subject of early hominin evolution (first 5 my after the ape-hominin split) is discussed, it is entirely an African phenomenon.

Fig. 10 from a 2016 publication[54] illustrates realistic estimates of the diversity and temporal ranges of known hominin species. The degree of overlap as a function of time among species is considerable. It is likely that this diagram represents only a fraction of the actual hominin diversity that existed, and as additional archeological discoveries and advances in dating technology are made, it is probable that additional diversity will be realized.

For the purpose of reviewing the available information about hominin diversity, the time line of specific species is represented in Table 1. The time periods correspond to certain phases in hominin evolution characterized by specific traits and trends. By subdividing the 7-mya time span of hominin evolution into phases, it is not implied that hominin evolutionary change occurred in irregular intervals or in "staccato" mode. Much of the discontinuity perceived today in the evolutionary continuum is just a reflection of the fragmentary nature of the fossil record.

Fig. 10 and Table 1 indicate that the hominin evolutionary tree has been quite populated starting 4 mya.[54] According to Table 1, as of 2016, the number of hominin species identified for specific periods are 3 for 7.0–5.0 mya, 2 for 5.0–4.0 mya, 3 for 4.0–3.0 mya, 3 for 3.0–2.5 mya, 5 for 2.5–2.0 mya, 7 for 2.0–1.5 mya, 5 for 1.5–1.0 mya, 5 for 1.0–0.25 mya and 6 for 0.25 mya to the present. The degree of taxonomic diversity (number of species) in Table 1 is based on high to moderate stringency criteria using anatomical traits. Notice

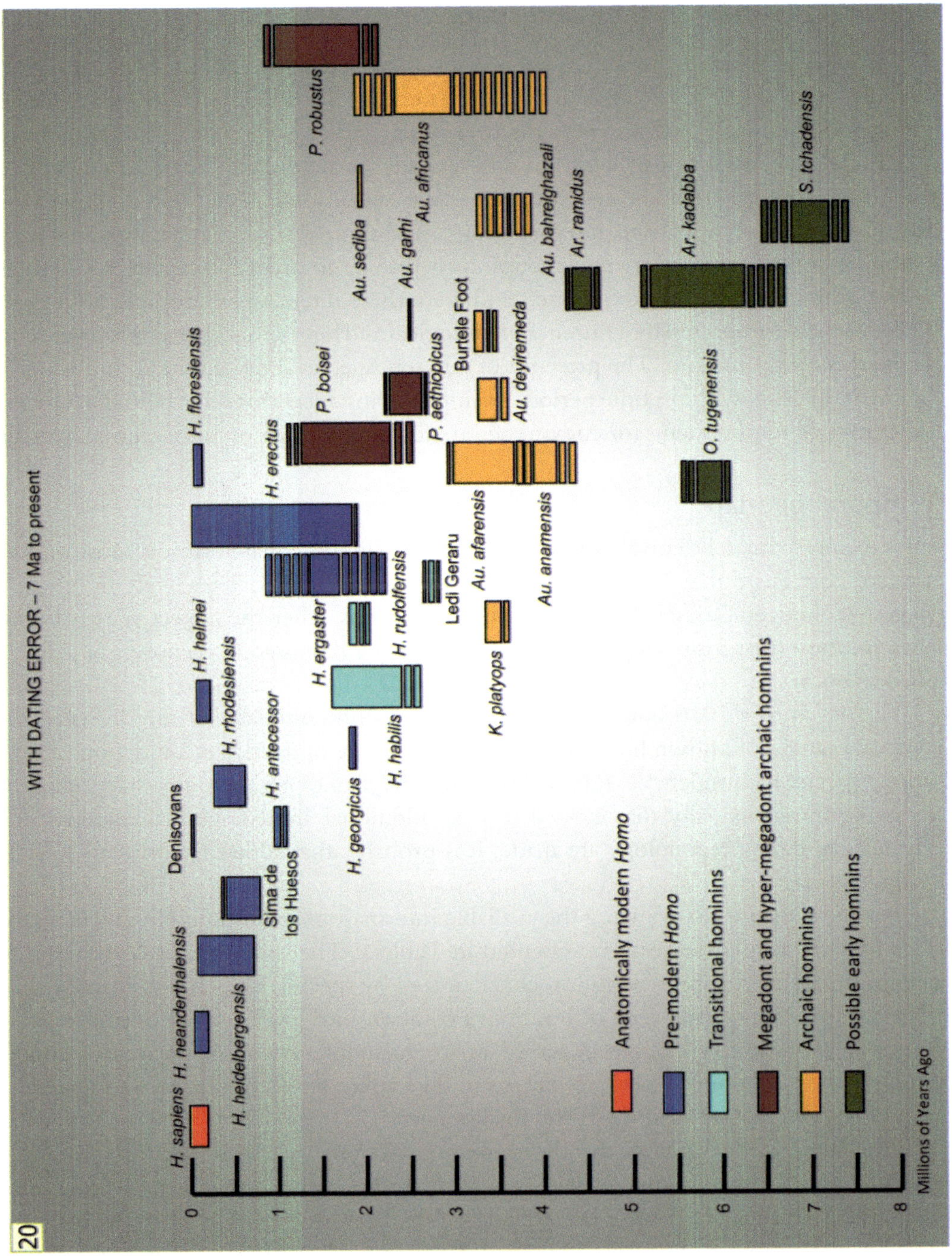

Fig. 10 Hominin diversity (https://onlinelibrary.wiley.com/doi/full/10.1002/ajpa.22902). A more realistic estimate of the temporal ranges of the hominin species, site collections, and individual fossils referred to in this review. The bottoms and tops of the continuous columns represent, respectively, the published first and last appearances dates. The tiles below and above many of the columns reflect the various sources of error that should be added to the mean ages that were used to locate the bottom and the top of each continuous column.

Table 1 Time line of specific hominin species (https://onlinelibrary.wiley.com/doi/full/10.1002/ajpa.22902).

- 7–5 Ma
  - *High confidence*
    - N/A
  - *Moderate confidence*
    - *Orrorin tugenensis* vs *Sahelanthropus tchadensis*
    - *Orrorin tugenensis* vs *Ardipithecus kadabba*
    - *Sahelanthropus tchadensis* vs *Ardipithecus kadabba*
  - *Low confidence*
    - N/A
- 5–4 Ma
  - *High confidence*
    - *Ardipithecus ramidus* vs *Australopithecus anamensis*
  - *Moderate confidence*
    - N/A
  - *Low confidence*
    - N/A
- 4–3 Ma
  - *High confidence*
    - *Australopithecus afarensis* vs Burtele foot
  - *Moderate confidence*
    - *Australopithecus afarensis* vs *Kenyanthropus platyops*
  - *Low confidence*
    - *Australopithecus afarensis* vs *Australopithecus bahrelghazali*
    - *Australopithecus afarensis* vs *Australopithecus deyiremeda*
    - *Australopithecus bahrelghazali* vs *Australopithecus deyiremeda*
    - *Australopithecus bahrelghazali* vs *Kenyanthropus platyops*
    - *Australopithecus bahrelghazali* vs Burtele foot
    - *Kenyanthropus platyops* vs *Australopithecus deyiremeda*
    - *Kenyanthropus platyops* vs Burtele foot
    - Burtele foot vs *Australopithecus deyiremeda*
- 3.0–2.5 Ma
  - *High confidence*
    - *Australopithecus africanus* vs *Paranthropus aethiopicus*
    - *Australopithecus africanus* vs LD 350-1
    - *Paranthropus aethiopicus* vs LD 350-1
  - *Moderate confidence*
    - N/A
  - *Low confidence*
    - N/A
- 2.5–2.0 Ma
  - *High confidence*
    - *Australopithecus africanus* vs *Paranthropus boisei*
    - *Paranthropus boisei* vs *Homo habilis* sensu *lato*
    - *Australopithecus garhi* vs *Homo habilis* sensu *lato*

Continued

Table 1 Time line of specific hominin species (https://onlinelibrary.wiley.com/doi/full/10.1002/ajpa.22902)—cont'd

- *Moderate confidence*
  - *Australopithecus africanus* vs *Homo habilis* sensu *lato*
  - *Australopithecus africanus* vs *Australopithecus garhi*
  - *Paranthropus boisei* vs *Australopithecus garhi*
  - *Homo habilis stricto stricto* vs *Homo rudolfensis*
- *Low confidence*
  - N/A

2.0–1.5 Ma

- *High confidence*
  - *Homo erectus* vs *Paranthropus robustus*
  - *Homo erectus* vs *Paranthropus boisei*
  - *Homo erectus* vs *Australopithecus sediba*
  - *Paranthropus robustus* vs *Homo habilis* sensu *lato*
  - *Paranthropus robustus* vs *Australopithecus sediba*
  - *Paranthropus boisei* vs *Homo habilis* sensu *lato*
  - *Paranthropus boisei* vs *Australopithecus sediba*
  - *Homo habilis* sensu *lato* vs *Australopithecus sediba*
- *Moderate confidence*
  - *Homo erectus* vs *Homo habilis* sensu *lato*
  - *Paranthropus robustus* vs *Paranthropus boisei*
  - *Homo habilis* sensu stricto vs *Homo rudolfensis*
- *Low confidence*
  - *Homo erectus* vs *Homo ergaster*
  - *Homo erectus* vs *Homo georgicus*

1.5–1.0 Ma

- *High confidence*
  - *Homo erectus* vs *Paranthropus robustus*
  - *Homo erectus* vs *Paranthropus boisei*
  - *Paranthropus robustus* vs *Homo habilis* sensu *lato*
  - *Paranthropus robustus* vs *Homo antecessor*
  - *Paranthropus boisei* vs *Homo antecessor*
  - *Homo habilis* sensu *lato* vs *Homo antecessor*
- *Moderate confidence*
  - *Homo erectus* vs *Homo habilis* sensu *lato*
  - *Paranthropus robustus* vs *Paranthropus boisei*
- *Low confidence*
  - *Homo erectus* vs *Homo ergaster*
  - *Homo erectus* vs *Homo antecessor*

1.0–0.25 Ma

- *High confidence*
  - *Homo erectus* vs *Homo heidelbergensis*
  - *Homo erectus* vs *Homo rhodesiensis*
  - *Homo erectus* vs *Homo helmei*

Table 1 Time line of specific hominin species (https://onlinelibrary.wiley.com/doi/full/10.1002/ajpa.22902)—cont'd

- *Homo erectus* vs Sima de los Huesos
- *Homo heidelbergensis* vs Sima de los Huesos
- *Homo rhodesiensis* vs Sima de los Huesos
- *Homo helmei* vs Sima de los Huesos

Moderate confidence
- N/A

Low confidence
- *Homo heidelbergensis* vs *Homo rhodesiensis*
- *Homo heidelbergensis* vs *Homo helmei*

0.25 Ma to present

High confidence
- *Homo sapiens* vs *Homo neanderthalensis*
- *Homo sapiens* vs *Homo erectus*
- *Homo sapiens* vs *Homo heidelbergensis*
- *Homo sapiens* vs *Homo floresiensis*
- *Homo sapiens* vs Denisovans
- *Homo erectus* vs *Homo heidelbergensis*
- *Homo erectus* vs *Homo floresiensis*
- *Homo neanderthalensis* vs *Homo erectus*
- *Homo neanderthalensis* vs *Homo floresiensis*
- *Homo heidelbergensis* vs *Homo floresiensis*

Moderate confidence
- *Homo neanderthalensis* vs Denisovans
- *Homo erectus* vs Denisovans

Low confidence
- *Homo neanderthalensis* vs *Homo heidelbergensis*
- *Homo heidelbergensis* vs Denisovans

For each of the nine time intervals relevant to the hominin fossil record, we assess the strength of the evidence for taxic diversity among hominin species by allocating each pairwise comparison to one of three categories of confidence, high, moderate or low.

that although the three previously discussed early hominins—*Sahelanthropus tchadensis*, *Orrorin tugenensis*, and *Ardipithecus kadabba/ramidus*—exhibited temporal overlap, they lived in different locations: Northern Chad, Central Western Kenya, and Eastern Ethiopia, respectively.

Cases of coexistence among hominins

Instances indicating potential coexistence among hominins are numerous. The cases described below represent only specific cases chosen from various geographical areas in which the evidence is particularly compelling. Care was taken to select cases from various time periods. These examples are not intended to be exhaustive or comprehensive.

Ardipithecus ramidus/*Australopithecus anamensis*/*Australopithecus afarensis*

Only about 200,000 years separate the published time ranges of *Ardipithecus ramidus* and *Australopithecus anamensis* in the Afar region of Eastern Ethiopia. If various sources of error are added to the mean ages that were used to locate the bottom and the top of each continuous column,[54] these two genera overlapped by approximately 200,000 years (Fig. 10). Fossils of these two genera have been unearthed from sites only 9.7 km apart.[55] Thus the likelihood that these two hominins knew of each other is high. This temporal and space overlap is significant since it involves two very different genera, one considered one of the earliest hominins (*Ardipithecus*) and the other an archaic hominin (*Australopithecus*) with very different anatomical characteristics. *Ardipithecus* still retained pedal and pelvic characteristics of an arboreal hominin while *Australopithecus* was clearly a bipedal species.[56] Although these two genera coexisted in Ethiopia, it is not clear whether *Ardipithecus ramidus* was ancestral to *Australopithecus anamensis*. The fact that they coexisted does not rule out the possibility that a population of the former evolved into the latter or that *Ardipithecus ramidus* was a relict species that became extinct, an evolutionary dead-end.

An issue of interest considering that these two genera coexisted is the degree to which they were in direct competition. *Ardipithecus ramidus* was a group in mid-transition from arboreal quadrupedal locomotion using their palms down (Fig. 11) to some form of bipedalism. On the other hand, it is known from the humerus and a partial tibia that *Australopithecus anamensis* practiced bipedalism as primary locomotion. Furthermore, in terms of diet, they targeted different foods. It is likely that *Ardipithecus ramidus* was an omnivore

Fig. 11 Arboreal locomotion of *Ardipithecus ramidus* (https://pixels.com/featured/early-hominid-ardipithecus-ramidus-mauricio-anton.html).

that consumed a generalized diet of plants, meat, and fruits while avoiding tough, abrasive foods like nuts and tubers. *Australopithecus anamensis*, on the other hand, was a very specialized eater surviving on open field food such as seeds, grasses, fruits, and tubers. Thus it is possible that they managed to avoid competition by exploiting different habitats and resources.

At the other end of the *Australopithecus anamensis* time line, about 3.8 mya, signs of another time-geographic space overlap are evident. *Australopithecus anamensis* is the oldest species known of the very successful Australopithecine group, and traditionally it was thought to be the ancestral species to *Australopithecus afarensis* who also lived in the Afar region of Eastern Ethiopia. For many years the paradigm was that *A. anamensis* and *A. afarensis* were one evolving species or chronospecies that resulted from anagenesis, in which one species totally evolves into another. In other words, scholars favor the notion that *afarensis* evolved by phyletic transformation and gradualism within an unbranched lineage. The two Australopithecines, *anamensis* and *afarensis*, are very different from each other. The notion of phyletic transformation involving these two species suffered a setback when, in 2019, a nearly complete *anamensis* skull from Woranso-Mille (Ethiopia) dated to 3.8 mya was discovered. Since *afarensis* remains have been dated to 3.9 mya, this means that *anamensis* could not have totally evolved into *afarensis* and disappeared completely. Thus, these two species could not have been chronospecies and they overlapped.[57] Remarkably, this temporal and spatial overlap lasted for at least 100,000 years.

Lucy and company

It is also becoming apparent that slightly later during the Middle Pliocene, 3.8–3.3 mya in the Afar region of Ethiopia, a group of Australopithecines shared the landscape. This was the time that Lucy (*Australopithecus afarensis*) lived. Fossil evidence indicates that at least two genera and five hominin species coexisted during this time period in the central and northeast region of sub-Saharan Africa. These include *Australopithecus bahrelghazali* from Chad in Central Africa, *Kenyanthropus platyops* from West Turkana, Kenya, and *Australopithecus afarensis*, *Australopithecus deyiremeda*, and the Burtele foot (a taxonomic name has not been assigned but it has similar locomotion to *Ardipithecus ramidus*) from the Woranso-Mille area in the small Afar region of northeast Ethiopia.

Given that coexistence of multiple, closely related hominin species occurred in Central/Northeast Africa between 3.8 and 3.0 mya, what prevented direct competition for resources and allowed species survival? There is precedence for coexistence among closely related nonhominoid primates as well as a number of other mammal species from around 5 mya to 12,000 ya in Africa.[58] Yet the question remains as to what degree, if at all, hominins such as *Australopithecus afarensis*, *Australopithecus deyiremeda*, and the Burtele foot species interacted.

In the case of *Australopithecus afarensis* and the Burtele foot specimen, they exhibited two different modes of locomotion. The anatomy of the Burtele foot, with an opposable great toe, indicates that it practiced a more ancestral type of bipedalism, which was more similar to that of the earlier *Ardipithecus ramidus* and may have included an arboreal component. *Australopithecus afarensis* and *Australopithecus deyiremeda* also were very different species living in direct sympatry. Three jawbones and teeth likely to be from three *A. deyiremeda* individuals exhibit traits distinctly different from *afarensis*.[59] These australopithecine species were eurytopic, being capable of tolerating a wide range of environmental conditions, habitats (from savanna-like grassland to woodland), and resources.[60] Thus, by being generalists, *deyiremeda* and *afarensis* may have been flexible in their habitat and food consumption, thereby preventing direct competition. It is also possible that the wide breadth of their shared niche allowed one or both species to target an alternative diet during food scarcity while sharing their food of choice when this was abundant. Another strategy often used among sympatric species to avoid direct competition and allow coexistence is niche partitioning, in which different foraging territories and/or dietary resources are targeted by multiple species. Even predation pressure has been demonstrated to drive diversity and allow coexistence by reducing competition among species sharing the same habitat.[61] Currently, all of these possibilities are viable and only additional fossil discoveries and/or technologies would help us to discern the strategies used by australopithecines to prevent direct intense competition and allow coexistence.

Two's company, four's a crowd

Back in the 1970s, when R. E. F. Leakey was reporting on newly discovered hominin fossils from Koobi Fora in the Turkana basin of northern Kenya, it was becoming clear that the region was a center of hominin diversity. One of those initial findings was discovered by Bernard Ngeneo and was designated KNM-ER 1470. Curiously, Ngeneo is not listed as an author in the publication reporting on the fossil finding in 1973.[62] This was the most complete early *Homo* skull unearthed up to that point. Upon initial evaluation after reconstruction, the Leakeys thought that the specimen was about 3 my old and since *Homo habilis* was the hominin from that period and the idea of coexistence was not in vogue, they were convinced that the fossil belonged to the same species, *habilis*. Yet the anatomical characteristics of the fossil clearly did not fit *habilis*. The skull was very large, with a large brain, and its face was long, wide, and flat while the teeth were uncharacteristically large. This morphology was not in accordance with the expectations of a *habilis* fossil. However, many scholars were still fixed on *habilis* and attributed the differences to sexual differences and/or geographical variation within the single species. Confusion about the KNM-ER 1470 fossil persisted for about 40 years due to the fact that the remains did not include teeth (only roots were recovered) and a lower jaw was lacking. In 2012, three additional fossils were discovered that finally allowed scholars to address these issues and provided information on the identity of KNM-ER 1470.[63] One of the newly discovered fossils, designated KNM-ER 62000, was found by Elgite Lokorimudang in 2008 and is very similar to KNM-ER 1470. This parallelism was important because it demonstrated that the original specimen was not an oddity. This new fossil included the upper jaw and almost all of its cheek teeth, allowing the investigators to visualize the lower jaw of KNM-ER 1470 and get a better idea of how the original skull looked. KNM-ER 62000 was identified as a juvenile and named *Homo rudolfensis*. These new findings not only provided information on how KNM-ER 1470 may have looked, but also clearly demonstrated that a different species of early *Homo*—not *habilis*—lived alongside *Homo erectus* in the small area encompassing Koobi Fora. The other two specimens were also good matches and provided confirmation for KNM-ER 1470. KNM-ER 60000, for example, provided a lower jaw with a full set of teeth with the characteristic small incisors and flat profile of *rudolfensis*. This lower jaw matched nicely the upper jaws of KNM-ER 1470. The third *Homo rudolfensis* fossil found by the Leakey team included a piece of mandible with some roots and molar crowns belonging to an adult.

Today, almost 50 years after Ngeneo's discovery of fossil KNM-ER 1470, the Koobi Fora region is credited as the home of four distinct hominin species. As of January 2020, the Smithsonian Institution acknowledges that *Homo rudolfensis*, *Homo habilis*, *Homo erectus* (*Homo ergaster*), and *Paranthropus boisei* coexisted in Koobi Fora sometime between 2.0 and 1.5 mya (https://humanorigins.si.edu/evidence/human-fossils/fossils/knm-er-1470). Koobi Fora is a small area and the four species lived within a radius of just over 10 km in diameter. Within such a limited area, it is likely that the four groups were aware of each other.

As with the australopithecines of the Woranso-Mille region in the Afar of northeast Ethiopia, the questions remain of how these four species avoided direct intense competition to allow coexistence, and how they could have partitioned the small region of Koobi Fora into distinct niches. Comparison of the dentition of *Homo rudolfensis* and *Homo habilis*, for example, indicates that the former had larger molars compared to the latter. In addition, the wear patterns between the two are different: in *rudolfensis* the patterns are horizontal and extensive while in *habilis* the molar surfaces are less flat and more structurally complex, exhibiting considerable relief. The *rudolfensis* dental profile suggests a herbivorous organism consuming a diet of seeds, nuts, and rough fruits and plants, while *habilis* probably ate a more diverse omnivorous diet. Thus these two early *Homo* species had different dietary strategies. It is possible that by consuming different foods, these hominins were able to avoid direct competition and survive as separate species. *Homo erectus*, a more derived hominin, was a very different organism compared to *Homo habilis*, *Homo rudolfensis*, and *Paranthropus boisei* (Fig. 12). The available mechanisms to avoid direct competition with *Homo rudolfensis*, *Homo habilis*, and *Paranthropus boisei* would have been likely to involve extensive behavioral differences that stemmed from being more intellectually capable and physically different. For example, *erectus* was an active runner,

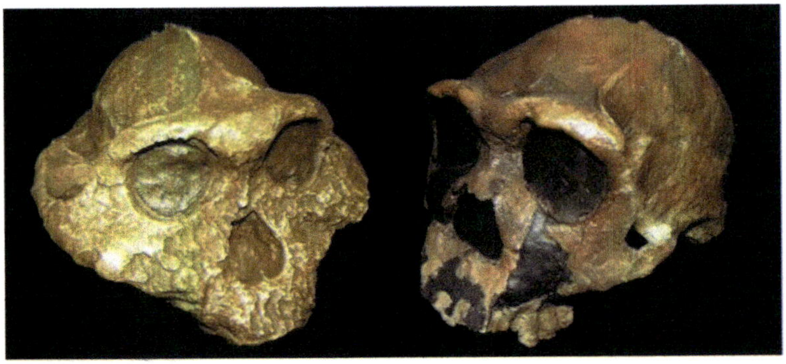

Fig. 12 Two very different specimens that coexisted: *Paranthropus boisei* ER 406 and *Homo erectus* ER 3733 (https://nutcrackerman.com/2016/03/10/two-very-different-specimens-that-coexisted-p-boisei-er-406-and-h-ergaster-er-3733).

produced more sophisticated tools, used fire, cared for living members, and probably, as reflected in their small teeth, included more meat in their diet. *Paranthropus boisei* was a more ancestral hominin and also a very different organism compared to *Homo rudolfensis* and *Homo habilis*. Nicknamed the "Nutcracker Man" for its robust characteristics, *boisei* was a generalist in terms of its diet and thus it is thought that *boisei* did not depend on a single, specific food source. However, its massive dentition and strong jaw muscles indicate that its diet consisted of hard items, such as nuts, seeds, and hard fruits.[64] Wear analyses also point to a diet of hard foodstuffs and constant chewing. In addition, carbon isotope measurements to access the type of food consumed by *boisei* indicate that its diet was based primarily on grasses or sedges with tubers to help establish caloric intake requirements. It has been reported that *boisei* ate the most amount of plant material compared to any other hominin species examined.[65] Although it is not clear what the exact nature was of the mechanism(s) that allowed for the partition of niches among these four hominin species at Koobi Fora, their distinct dietary profiles would have provided for some degree of separation to avoid intense competition.

Homo naledi and other recent hominins in South Africa

In more recent times, another instance of hominin coexistence likely occurred in South Africa. It involved a species named *Homo naledi*, which was discovered in October 2013 near the city of Johannesburg. The initial discovery yielded a total of 1550 hominin specimens belonging to about 15–18 individuals. The fossils were found in exceptionally good condition and included every body part from individuals ranging in age from infants to the elderly. Of particular interest were the precious findings of an almost-complete hand and foot, which provided important insight into the behavior of the species. Together with the Sima de los Huesos site in Atapuerca, Spain, the *naledi* assemblage represents the most significant archeological discovery of the past 50 years. Initially, the fossils were thought to be much older than they actually are. With no direct dating to guide them, the investigators estimated the remains at about 2 my old.[66,67] This dating was based solely on the anatomical characteristic of the fossils. To the surprise of everyone, 2 years later, the age of the fossils was directly determined to be between 236,000 and 335,000 years old.[68] This time period was unexpectedly recent—so recent that AMHs were already in existence in the area. A second finding of additional *naledi* individuals occurred in 2017 in an adjacent chamber.[69] This second assemblage of fossils most probably belonged to the same population as the original one.

The *naledi* discoveries are remarkable in a number of ways. They represent the most extensive and complete assemblage of hominin fossils from Africa. The location, position, and distribution of the remains indicate that they were not laid there all at once but over a period of time. The bodies were deposited in the minuscule atria by some unknown force and were found randomly arranged in 6 in. of sediment. In addition, the fossils were found in very inaccessible locations of the cave and except for one owl skeleton, only *naledi* remains were recovered.

It is likely that bodies were deposited or dropped in the chambers by clan members, possibly as part of a burial ceremony. Yet it is challenging to get to the chambers and thus difficult to explain how the bodies got to their resting place. In one area, bodies had to be squeezed through a 12 m tunnel that was restricted in certain places to only 8 in. in diameter. It is possible that the cave looked different when the bodies were deposited in the chambers, and subsequent geological activity restricted the passage to the present condition. Alternatively, a chute at the top of the cave may have allowed tribal members simply to drop the bodies into the cavities, and the opening now no longer exists. It is also plausible that since *naledi* bodies were small (adult males about 5 ft tall and 100 pounds) by modern human standards, their passage through the maze of narrow tunnels may have been easier. The likelihood of predators dragging the bodies into the enclosures is doubtful, since other than the dead owl, only *naledi* remains were found. Other species' remains would be expected to result from predatory activity. In connection with this issue, it is interesting that most of the bones were disarticulated (separated at the joints), possibly the result of forcibly pulling the inanimate bodies through the narrow openings and/or intentionally done postmortem to facilitate dragging the dead into the chambers. Based on these facts, it is likely that clan members somehow placed or threw the bodies into the chamber from an opening above that no longer exists. Was this done as part of ceremonial interments? It is noteworthy that although the relative distribution of the bones pointed to previous visitors—possibly clan members and/or cavers—disturbing the bones in their resting area, there is no indication that the bodies were deposited in any particular order or orientation. Furthermore, no artifacts, such as stone tools and personal effects that are usually part of more modern hominins' ceremonial assemblies, were found. No tools have been found at the site and if it is a burial site, it may have not have been customary to include tools.

As is the case with other hominins, *Homo naledi* possessed a mosaic of ancestral and derived features. Some of the traits are reminiscent of more ancient groups, such as australopithecines, while other morphologies are characteristic of more recent groups, such as *Homo erectus*, and others are unique to *naledi*. *Homo naledi*'s advanced traits include temporal and occipital bulges (characteristic of *erectus*) and reduced postorbital constriction (narrowing of the skull behind the eye sockets). In early AMHs, postorbital constriction is barely visible and in contemporary humans it has disappeared altogether. The teeth are small with crowns and five cusps, like in modern humans. The mandibular muscles are small, indicating a soft diet requiring less mastication compared with the ancestral condition. The rib cage, on the other hand, is ancestral and conical in shape as in australopithecines and great apes. The hands exhibit a combination of ape-like and human-like traits. Specifically, the thumb, wrist, and palm bones are human-like, whereas the fingers are still curved, an ancestral morphology designed to hold on to branches during arboreal locomotion. The pelvis also represents a collage of ancestral and derived characteristics. The base of the pelvis looks like that in AMHs, whereas the blades spread outward as in australopithecines. Remarkably, the feet look like those of modern humans, whereas the shoulders are ape-like.

The brain of *naledi* was small, strikingly primitive in size ($\sim$560 cm^3 for males and $\sim$465 cm^3 for females), within the range of australopithecines. Yet in spite of its small brain, the skull was not robust as in more ancestral hominins. Furthermore, recent investigations on the internal anatomy of *naledi*'s brain indicate that despite its overall small size, it shares some advanced aspects of human brain organization as seen in the inferior frontal lobe and lateral orbital gyri (convolutions situated on the ventral frontal lobe).[70] These areas of the brain are known to be responsible for dexterity, language, and social behavior, and may indicate that *Homo naledi* possessed certain attributes in common with AMHs, behaviors present in modern humans, in spite of a small brain size. This type of data also suggests that the typical morphological information provided by cranial size may be insufficient in accessing phylogenetic relationships.[70]

Regarding the issue of coexistence among hominins, it is noteworthy that the corrected young age of *Homo naledi* means that this species was living in what is now South Africa with a number of other members of the genus *Homo*, including *Homo sapiens* and *Homo heidelbergensis*. Recent estimates based on ancient DNA from South African specimens indicate that *Homo sapiens* originated as early as 350,000 ya.[71] Specifically, the Florisbad *Homo sapiens* skull discovered about 300 km from the *naledi* site in Florisbad, South Africa has been dated to about 259,000 ya.[72] Furthermore, using computerized assessments based on CT scans of fossilized skulls, the virtual image of the last common human ancestor to all modern humans (representative of the earliest AMH) suggests that *sapiens* originated between 260,000 and 350,000 ya through a fusion of ancient populations in East and South Africa.[73] These dates coincide remarkably with the ages reported for *Homo naledi*.

The time span of *Homo heidelbergensis* is roughly from 700,000 to 200,000 ya. The species lived in South Africa, as attested by the discovery of the Saldanha skull from

Hopefield. Although it has not been dated directly, its age has been estimated to be roughly 500,000 years old.[74] Furthermore, the discoverers of the *naledi* fossils argue that the lineage is remote in time and originated about 2 mya, long enough to have coexisted with *Homo erectus* as well. African *Homo erectus* became extinct approximately 500,000 ya. So how a small-brained species, such as *Homo naledi*, lingered on in southern Africa, seemingly alongside more advanced, big-brained *sapiens* and *heidelbergensis* and competed with them is a mystery. This is especially intriguing when considering that this happened at a time when the other, more "advanced" *Homo* species living alongside them were manufacturing sophisticated Acheulean stone tools during the Middle Stone Age. How did *naledi* avoid such competition? Although the dentition and lower body skeleton suggest that *naledi* may have had a lifestyle and diet similar to those of other hunter-gatherers present elsewhere in Africa at the same time, such as late *Homo heidelbergensis* and early *Homo sapiens*, the anatomy of its lower body skeleton and curved fingers suggest that its niche included spending time and feeding among the trees. It is possible that a partial arboreal existence may have provided a mechanism to avoid direct competition with the other hominins in the area.

Dmanisi: Extreme anatomical differences within a species or coexistence among different populations?

In the 2010s, a number of fossils corresponding to five individuals were discovered in the village of Dmanisi, a small town in the Republic of Georgia in Southwest Asia. The five skulls recovered exhibit extreme anatomical differences that are difficult to explain (Fig. 13). Presented with this degree of morphological unlikeness among the five specimens, the natural reaction would be to think they are different hominin species. Yet the Dmanisi skulls, as a group, represent an unprecedented discovery since the remains were found in the same location and from the same time period.[75] Confronted with this scenario, it is unclear how the existence of such a diverse group of individuals could be explained. One possibility is that all five highly contrasting skulls belonged to a single species with an elevated level of physical variance (highly polymorphic). However, no precedent for such a

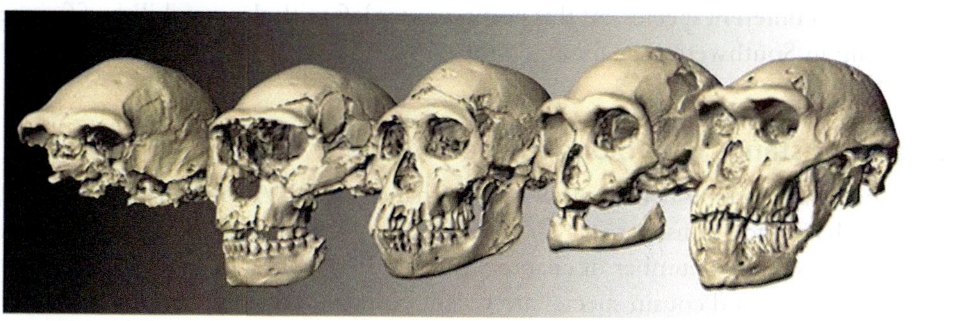

Fig. 13 The five Dmanisi skulls (https://www.livescience.com/40503-dmanisi-ancient-human-skull-photos.html).

degree of intraspecies heterogeneity has ever been reported. In addition, this type of conclusion would prompt scholars to consider the classification of remains that, in the past, have been considered distinct species, such as *Homo erectus*, *Homo habilis*, *Homo rudolfensis*, and *Homo heidelbergensis*, as a single species. All of these distinct fossils have been previously characterized as separate species. This interpretation may sound unconvincing and unlikely, but it was exactly the conclusion reached by the discoverers of the five skulls. Lordkipanidze and colleagues in their 2013 paper, after the discovery of the fifth skull, argued that the variation within the Dmanisi skulls indicate that *all* Early Pleistocene (2.6–0.8 mya) *Homo* specimens should be collapsed into a single species, *Homo erectus*.[76] An alternative potential explanation is that the skulls belong to distinct populations of hominins living in close proximity to each other.

A key factor in this dilemma is whether the degree of observed differences among the five Dmanisi skulls is equivalent or less than the differences seen among the various previously characterized Early Pleistocene *Homo* species classified as unique species. Lordkipanidze and coauthors indicate that the Dmanisi skulls exhibit a cranial shape variation consistent with that seen in chimpanzees or AMHs individuals, thus the differences can be accommodated within a single species.[76] However, Lordkipanidze et al.'s proposition to amalgamate *Homo habilis*, *Homo rudolfensis*, and *Homo erectus* into a single species, *H. erectus*, has been criticized by a number of experts who have collectively indicated that the evidence is weak. For example, it has been pointed out that the examination of overall cranial morphology carried out by Lordkipanidze and collaborators was biased and does not have the diagnostic power to discriminate between closely related species.[77] Taxa are distinguished by a variety of explicit traits and not *general* skull morphology. Critics also highlight that it would have been helpful if Lordkipanidze and colleagues had indicated specific traits delineating which are derived (absent in the last common ancestor) and which are ancestral (already present in the last common ancestor). This latter point is a distinction that is considered essential to assess phylogenetic relationships.[77] When the exact mandibular, dental, and cranial traits were evaluated by Schwartz and collaborators, they concluded that the variation represented differences among species and not within-species variation.[78] They established that the five Dmanisi skulls represent as many as four different species. At this point, we are left with the possibility of four species coexisting in Southwest Asia about 1.8 mya.

One-night stands or prolonged coexistence?

It has been well established by now that exchange of DNA took place among *sapiens*, Neanderthals, and Denisovans.[71,79–82] The topic of interbreeding among these species will be discussed in a number of chapters later in this book. The events of introgression involving *sapiens* and cousin species are recent compared to the previous examples provided in this chapter involving earlier hominins. These recent gene flow events occurred during a period of time between 120,000 and 25,000 ya in the Levant, Europe, and Asia when the various species were in the same general area. But transfer of DNA by itself does not provide the whole story. Other questions remain to be answered. For instance, which circumstances led to DNA transfer? DNA data alone does not provide information on the events that lead to introgression. For example, was intercourse consensual? Were the interactions that led to gene flow among these species the result of transient encounters (e.g., involving rape) or more prolonged coexistence? Were these species involved in technical and artistic exchange and collaboration, or was it just physical attraction, sex? The answers to these questions probably cannot be answered by molecular genetics approaches; they require additional information within the realm of archeology.

Anatomically modern humans migrated out of Africa on at least two separate occasions, but it is highly likely that numerous dispersals occurred. It is known that an early dispersal took place approximately 100,000–177,000 ya. This incursion vanished from the Levant about 80,000 ya.[83] It is suspected that these AMHs became extinct or retreated from the Levant back to Africa as a result of increasing cold weather or competition with Neanderthals. However, AMH fossils from the Zhiren Cave in China dating to about 125,000 ya suggest that at least some members of this early *sapiens* Levant enclave survived and traveled to continental Southeast Asia.[84]

A subsequent dispersal took place approximately between 50,000 and 60,000 ya. The AMH fossil remains at Misliya Cave (177–194 kya), Qafzeh and Skhul Caves (90–120 kya), and Manot Cave (50–60 kya) in current-day Israel signal distinct expansion events from Africa.[85] It is theorized that the migration of 50–60 kya out of Africa went beyond the Near East and, after some delay, successfully populated Europe. Neanderthals, on the other hand, resided in the Levant during two time intervals, as glaciers and cold climatic conditions to the north forced them to migrate south. The first Neanderthal tenure in the Levant occurred from 120,000 to 100,000 ya and the second one between 61,000 and 48,000 ya. Neanderthals were absent from the Levant between 39,000 and 41,000 ya.[86] Thus, in the Levant, AMHs and Neanderthals shared the region twice. In Europe, Neanderthals occupied an extensive range (Fig. 14A) that eventually was reduced to a number of refugia (Fig. 14B), and ultimately their last stance was in Southern Iberia 28,000–24,000 ya.[87]

These time lines suggest that *sapiens* and Neanderthals had the opportunity to coexist in the Levant—and evidence does indeed exist. Archeological data indicate that AMHs and Neanderthals shared living quarters in caves of the Near East. Specifically in the Tabun, Jamal, el-Wad, and Skhul caves in the Mount Carmel range near the Sea of Galilee in northern Israel, some of the earliest modern human remains outside Africa have been found. These AMH fossils date to the Early Middle Paleolithic with ages ranging between 100,000 and 177,000 years.[83] Therefore, these AMHs must have migrated from Africa to the Levant as part of the first incursion that vanished from the Near East about 80,000 ya.

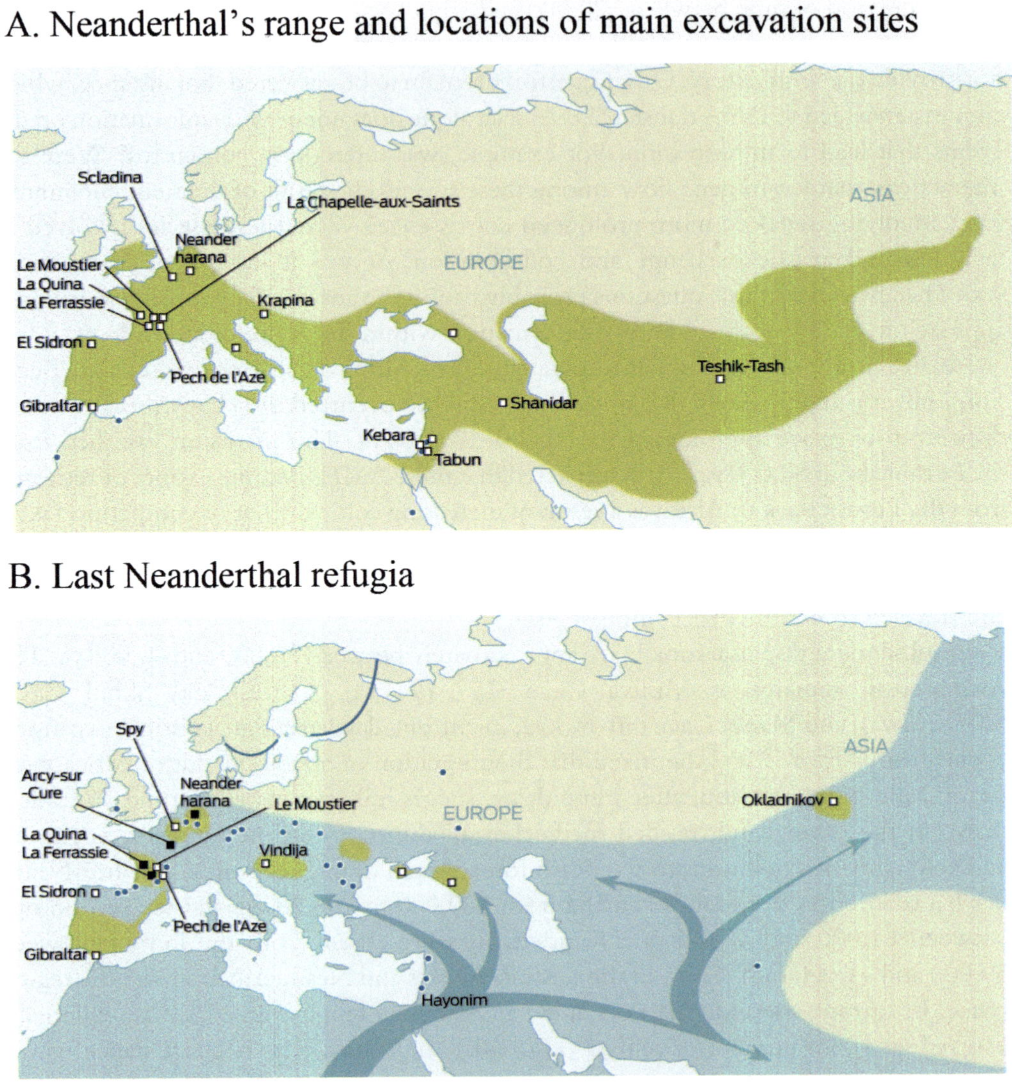

Fig. 14 Expansion of *Homo sapiens* and coexistence with Neanderthal (https://en.wikipedia.org/wiki/Neanderthal_extinction#/media/File:Expansion_of_early_modern_humans_from_Africa.jpg). (A) Neanderthal's range and locations of main excavation sites. (B) Last Neanderthal refugia.

Most scholars think that this was an unsuccessful dispersal out of Africa that barely managed to reach beyond the borders of the African continent. Notably, Neanderthal bones have been discovered from the Tabun Cave and the same age strata.[88] In addition, a second location at Manot Cave, a site located 40 km northeast of the Mount Carmel cave dig (Fig. 15), has been identified as a site of cohabitation and *sapiens*-Neanderthal interbreeding due to the combination of ancestral and derived anatomical features in a

Fig. 15 Manot Cave and nearby caves in the Mount Carmel range (https://www.researchgate.net/publication/271598068).

fossil. The fossil has been named Manot 1 and is a partial calvarium (skull cap or dome of the skull) that dates to 54,700 ya. Unlike the previous remains from the Mount Carmel caves, this age indicates that the calvarium belonged to an individual whose ancestors were part of the 50–60 kya migration out of Africa that repopulated the Levant. In addition, its descendants colonized Europe about 45,000 ya. Manot 1 provides evidence that during the Middle to Upper Paleolithic interface, both modern humans and Neanderthals contemporaneously inhabited the southern Levant.[85] According to the senior author of the study, it is likely that the Manot people dispersed into the Near East and encountered an established population of Neanderthals, with whom they must have interacted, shared technology, and interbred. Together with the Neanderthal specimens recovered from Kebara and Amud at the Mount Carmel Range,[89] Manot 1 and various others *sapiens* specimens strongly suggest coexistence.

Europe was another region in which AMHs and Neanderthals coexisted. It is known that after some delay subsequent to leaving Africa, Paleolithic AMHs eventually reached Europe. The lag of time traditionally quoted is about 30,000 years; this is based on an exit from Africa around 70,000 ya and an incursion into Europe approximately 40,000 ya. However, this delay in penetrating Europe could be extended if the presence of AMHs in East Asia dating back to 125,000 ya is confirmed. This new date of AMHs in Asia is based on a collection of teeth discovered in Fuyan Province in Southeast China that suggests that *sapiens* exited Africa and successfully colonized Asia 85,000 years prior to entering Europe.[90] As previously stated, if corroborated, this finding suggests that the first migration out of Africa (100,000–177,000 ya) did not perish in the Levant and was fertile in colonizing East Asia. Independent of the extent of the time delay in repopulating the Levant, Europe was colonized late in comparison to Asia. It is likely that the cold and inhospitable climatic conditions at the time of the early exodus from Africa hindered migration into Europe. Those early AMHs were members of a tropical species ill-adapted to the subfreezing winters and harsh habitats of Europe. Furthermore, Neanderthals probably made it difficult for the newcomers. Competition for food was a factor, and resources were likely to have been limited. Neanderthals were better adapted to the weather conditions prevalent in Europe at the time of the first migrations out of Africa.

Eventually, AMHs reached Europe and encountered the Neanderthals. According to an extensive and comprehensive study based on improved radio-isotope ^{14}C techniques, these species coexisted for 2600–5400 years.[91] In this study, robust chronologies from 40 geographically (from Spain to Russia) targeted Neanderthal archeological sites (Mousterian tool tradition) were assessed. The Neanderthal dates were then compared to the ages of the earliest dated AMH sites in Europe associated with the Uluzzian and Aurignacian traditions. This allowed the researchers to estimate the temporal overlap between the two species in various regions. Overall, the data portray a region-specific complex pattern of species replacement. Since the incursion of *sapiens* into Europe, the continent was a mosaic of the two species, with pockets of Neanderthal populations shrinking gradually and ultimately disappearing one by one. The results indicate that Neanderthals vanished at different times in different regions. Their demise was thus not abrupt; they survived in dwindling populations throughout their range. In other words, Neanderthals survived in isolated refuges well after the arrival of modern humans in Europe.[92] In this regard, some scholars today believe that the extinction of Neanderthals was not related to active competition with AMHs, but was the direct result of deteriorating climatic conditions. Yet while all of this was happening, there was ample time for the transmission of technology, culture, art, and DNA.

However, the question remains: how close together did Neanderthals and AMHs live? And to what degree did the two species interact? The Iberian Peninsula provides a good case scenario to answer some of these questions at least partially. The Iberian Peninsula has been traditionally regarded as the last bastion of Neanderthal resiliency. For example, the long-held notion of an Ebro Frontier (Fig. 16) posits that most of Iberia, south of the Ebro River on the Spanish side of the Pyrenees, was a refugium where Neanderthals survived for several thousands of years longer than in the rest of Europe.[93] Although this hypothesis has been questioned on the grounds of bias in the form of unrepresentative site selection, many scholars consider that there is an element of truth in the idea that the Ebro River delineated a region to the south where Neanderthals persisted longer than throughout the rest of Europe. It is estimated that the chronology indicates a lag of at least 3000 years relative to the rest of Europe.

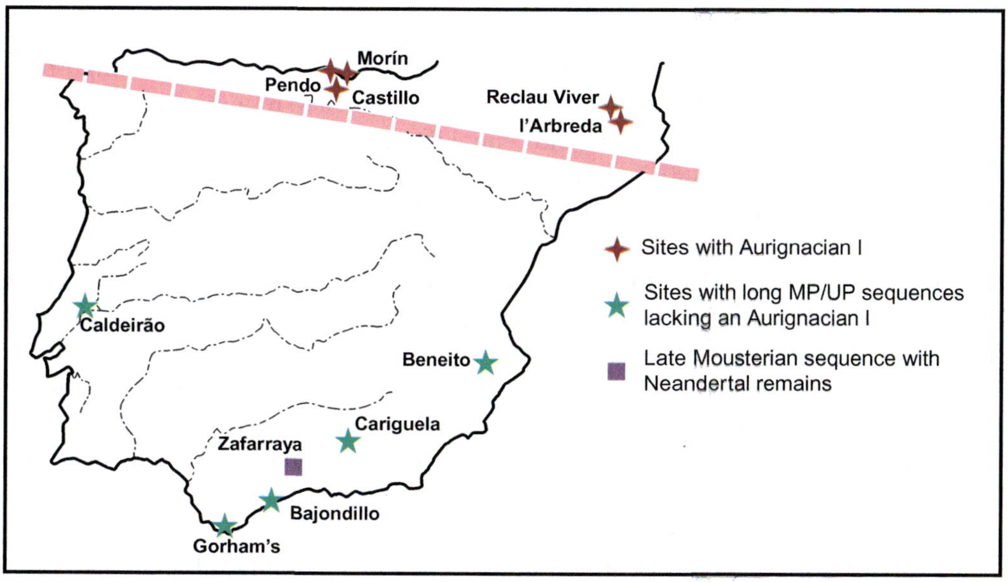

Fig. 16 The Ebro Frontier (file:///Users/ReneJHerrera/Downloads/EbroFrontierCalpe982000%20(1).pdf).

In addition, the occupation distribution pattern of Neanderthals seems not to have been uniform within the Iberian Peninsula, with the southern and southern coastal regions being the most resilient. An often-cited example is Gorham's Cave in Gibraltar, in the southern tip of Iberia. This location has provided dates as recent as 24,000 ya for the last Neanderthal enclave.[92] Unfortunately, the age estimates were not direct, but based on ash remains. At the same time, archeological data from AMH sites from nearby sites, such as Bajondillo in Malaga, approximately 100 km east of Gibraltar, have been dated to about 32,000 ya.[94] Other locations, such as Cueva Antón in the province of Murcia in southeastern Spain (about 400 km from Gibraltar), provide dates of Neanderthal occupation as recent as 37,100 ya, and less than 2 km away at Abrigo de La Boja, AMH lived 36,500 ya. Other locations in southern Iberia also demonstrate that both species were in close proximity for several thousand years.[95]

This coexistence has been described as a mosaic of AMHs and Neanderthal groups sparsely distributed in the region that persisted for an extended period.[92] Thus, it seems that in southern Iberia there was not a dramatic replacement resulting in the extinction of Neanderthals, but a protracted coexistence of the two species. In this type of scenario, with the limited archeological context available, it is difficult to ascertain to what degree and in what types of interaction the two species could have been involved. Under those conditions, it is likely that much would depend on whether there was a willingness to reach out or avoid each other—if they had something to gain from each other, for example, involving collaboration, possibly through trade.

In terms of the rest of Iberia, which includes the central and northern parts of the Peninsula, the time overlap between the two species has been contentious, especially in light of numerous reports of contradictory date estimates, disputed dating methods, and contamination with recent carbon sources.[96] Subsequent to the re-dating of sites to conform with improved experimental protocols designed to minimize contamination, it has been estimated that in northern and central Iberia, Neanderthals went extinct about 39,000–38,000 ya.[97] Yet some exceptions to this more recent survival of Neanderthals in central and northern Iberia have been reported, including the 12 de Octubre and Cañaveral sites in the Madrid basin (center of the Peninsula) with luminescence dates between 40,000 and 33,000 ya at 12 de Octubre, and 33,000 ya at Cañaveral.[96] Thus, considering an arrival of AMHs around 37,000 ya in this part of the Peninsula, a period of at least 1000 years of coexistence is estimated.

Indication of coexistence involving Neanderthals and AMHs is not limited to the Iberian Peninsula. In other parts of Europe, instances of coexistence have been documented. Although in certain sites such as Roc-de-Combe and Le Piage in southern France and El Pendo in northern Spain the documentation of coexistence is controversial due to potential mixing of strata/artifacts resulting from soil disturbances,[98] in other locations the evidence in the form of interstratifications (cooccurrence) of Neanderthal and AMH material is unquestionable. One of these clear-cut instances of coexistence is seen in the Grotte des Fées de Châtelperron site in central France, where Châtelperronian (Neanderthal-specific) and Aurignacian (AMH-specific) artifacts are contained within the same sediment layer.[99] At this site the Neanderthal-specific and AMH-specific material occur together because they were deposited about the same time by both species. It is widely accepted that Neanderthals are responsible for the tool-making Châtelperronian tradition in Western Europe while the classic Aurignacian style is a trademark of AMHs. Thus, interstratification of these traditions is indicative of coexistence and possible cohabitation. Noteworthy in connection with the lithic artifacts found at the Grotte des Fées de Châtelperron site is that the Aurignacian blades and carinated scrapers made by AMHs were manufactured from high-quality blue-gray or pale brown "Chalcedony" flint imported into the site from distant flint sources at least 100 km to the north, while the Châtelperronian pieces were produced by the Neanderthals from poor-quality flint that broke easily and had to be replaced.

Further support for this coexistence was provided by high-precision dating using radiocarbon accelerator mass spectrometer measurements. The dates for the five layers (B5–B1) in question are as follows: the basal Châtelperronian (Neanderthal) level (layer B5) gave dates of 40,650 and 39,150 ya; later Châtelperronian levels (layers B1–B3) were dated at 36,340 and 34,550 ya; and the intervening level B4 (which contained the Aurignacian or AMH-specific material) produced dates of 39,780 and 35,540 ya.[100] These dates indicate that AMH deposits are sandwiched in between two Neanderthal layers, suggesting coexistence.

Further east, in north central Russia at the western foothills of the Polar Urals and near the Arctic Circle in a location known as Byzovaya, Late Mousterian lithic artifacts and animal bones were excavated and reported in 2011.[101] Radiocarbon and optical-stimulated luminescence dates of fossils and sand grains indicate that the site was inhabited during a short span of time approximately 31,000–34,000 ya. Since the Mousterian style of manufacturing stone tools is strictly associated with Neanderthals, this finding suggests that the residents of this arctic enclave were Neanderthals. If confirmed, this discovery at Byzovaya is remarkable in a number of ways. It underscores the fundamental resiliency of ancient humans to survive at 65° latitude about 100 miles below the Arctic Circle with limited technology and protection from very harsh weather. It also represents one of the latest Neanderthal sites known, demonstrating that they survived for longer than average not only in Iberia in the extreme south of their range in relative mild weather, but also in the far north in the harsh weather of the Arctic Circle. It indicates that Neanderthals persisted outside Iberia longer than previously thought. Thus Byzovaya represents a late northern refuge for Neanderthals, about 1000 km north of any other known Mousterian site. If indeed these Arctic residents were Neanderthals, the premise that cold weather was the cause of their demise throughout their range needs to be reconsidered.

Recent archeological evidence indicates that AMHs sporadically inhabited regions up to 70° N in the Russian arctic during the period of the Byzovaya settlement. These occupations included the region of the Polar Urals where Byzovaya is located.[102] Based on this information, it is reasonable to wonder if these populations of AMHs were using Mousterian technology transmitted to them by more southern Neanderthal groups. If the answer is no, this means that Neanderthals persisted in those high latitudes and coexisted with AMHs for a longer time than was previously thought.

Transitional industries and technology transfer

AMHs brought with them from Africa a technological tradition for making tools, known as Aurignacian, that lasted approximately from 43,000 to 26,000 ya. The name derives from the Cave of Aurignac in Haute-Garonne, southwestern France. It consisted mainly of stone flake tools rather than blades. As part of this style, flakes were retouched to make scrapers. The tradition also included worked bone or antler points. The flint tools included fine blades and bladelettes struck from prepared cores rather than using crude flakes. At the same time as the AMHs arrived in Europe, the Neanderthals were fabricating tools using a different tradition, the Mousterian (160,000–40,000 ya), named after the Le Moustier site in the Dordogne region of southwestern France. Mousterian is a technocomplex defined primarily by flake-based assemblages made using the Levallois technique. With this technique, Neanderthals produced large flakes with the shape of an inverted tortoise shell. So distinctive were these traditions that some investigators believe the tools they generated serve as markers for the presence of these hominins.

A subsequent tool tradition known as Châtelperronian (44,500–36,000 ya) was discussed earlier in this chapter in connection to Neanderthal sites. Although most scholars consider Châtelperronian a product of Neanderthal ingenuity, it is also regarded as a transition technology because it appears in an intermediate position between the Middle and Upper Paleolithic, and because only Neanderthal fossils have been found associated with it at the later stages of their existence. Another technological/cultural tradition considered to be transitional is the Uluzzian (45,000–40,000 ya) of central-southern Italy and Greece, which scholars argue to be the product of AMHs or of Neanderthals with cultural influence from *H. sapiens* who lived nearby.[103]

As is the case with the highly contested archeological dates described earlier as they relate to interstratification and coexistence, the authorship of specific artifacts and lithic tools are currently intensely debated. In other words, scholars have not agreed that Neanderthals and/or AMHs made specific tools. The arguments surrounding this debate stem from whether Neanderthals were able to create specific tools and ornaments independent of *sapiens*' influence or whether it is necessary to invoke their know-how and/or inspiration to account for the artifacts. The second possibility would necessitate direct (person to person) or indirect (e.g., copying artifacts left behind by AMHs or stolen from them) transfer of knowledge. Just as with the question of coexistence, arguments involving disturbances of sediments causing erroneous assessments of time and admixture of sediments of different ages play a role in many of the debated archeological sites. In addition, there is the often-confronted parallelism in styles observed between Neanderthal- and *sapiens*-associated sophisticated tools and ornaments in dig sites. Other than contamination of the Neanderthal strata with more recent AMH material, or vice versa, different scenarios have been proposed to explain their association with sophisticated Châtelperronian tools and decorative personal items of no utilitarian value. Some scholars posit that Neanderthals developed the Châtelperronian tradition independently of AMHs. Others think that Neanderthals were influenced by AMHs. Still others have theorized that trading and/or procurement of tools explains the discovery of *sapiens*-like lithics in Neanderthal camps. What is clear is that Châtelperronian items appeared at about the same time that AMHs entered Europe. This coincidence has led many archeologists to believe that Neanderthals were copying the types of utensils that they observed AMHs use.

Three archeological sites have been at the center of these debates: Saint Césaire (also known as La Roche a Pierrot), Grotte du Renne (or Arcy-sur-Cure), and Quinçay. In all three locations, Neanderthal fossils have been found in direct association with Châtelperronian artifacts. Of the three sites, Grotte du Renne represents the most convincing argument that Neanderthals produced Châtelperronian items including sophisticated bone tools and body ornaments prior to the arrival of AMHs, suggesting that no technological transmission occurred at least at that site. Yet doubts were cast in 2010 after radiological dates suggested possible contamination of AMH artifacts with Neanderthal remains.[104] The assertions of contamination based on radiological age were invalidated a year later on the grounds of mixing of different-aged layers *after collection* and statistical analyses.[105] These results suggested that the association of Châtelperronian material and the Neanderthal fossils is legitimate. Similarly, Saint Césaire has yielded well-identified Neanderthal remains. However, unfortunately, their association with Châtelperronian artifacts is unclear. Recent studies using confocal microscopy (an imaging technique that employs laser point scans) suggest that alterations in the location of artifacts have taken place, substantiating previous arguments concerning the unreliability of the Neanderthal-Châtelperronian association.[106] The degree of uncertainly at Saint Césaire is such that at this point in time it is not possible to link the Châtelperronian artifacts to Neanderthals. The Quinçay site in west central France is the only Châtelperronian site where personal ornaments have been discovered in the absence of a younger Upper Paleolithic layer on top.[107] Thus, the possibility of postdepositional mixing of strata with AMH material into the Châtelperronian layer is diminished as an explanation for the Neanderthal-ornament association.

Neanderthals: Acculturated or the acculturators?

In addition to the fossil and lithic evidence suggesting coexistence and technological exchange between Neanderthals and AMHs, a number of recent reports indicate that

Fig. 17 Red scalariform sign, panel 78 in hall XI of La Pasiega gallery C. Square delineate calcite deposit used for dating (https://science.sciencemag.org/content/359/6378/912).

Neanderthals were the artists of cave painting at least 65,000 ya (Fig. 17).[108] The dates were generated using the uranium-thorium (U-Th) method directed at the calcite deposits covering the paintings in three Spanish caves: a rectangular sign at La Pasiega (Cantabria, north central Spain), hand figures at Maltravieso (Caceres, western Spain's Extremadura region), and red traces on stalagmites at Ardales (Malaga, south Spain). Soon after the original article was published, Randall White and numerous collaborators published a strong critique directed at the dating method and lack of scientific rigor used in the study, indicating that it was unreliable and an overestimation of the real age of the cave paintings.[109] They pointed out, for example, that the calcite deposits originally tested by Hoffmann et al.[108] were not closed systems and that carbonate can crystallize, causing a preferential loss of uranium, which would inflate the values and make the samples register as older.

Although the U-Th dating method is scientifically sound and the work performed by Hoffmann and collaborators is of good quality, the absence of cave drawings for more than 25,000 years subsequent to the dates reported has caused experts to wonder why cave art is not seen again until 40,000 ya. What happened to the artists during this large interim? That is a difficult question to answer. In addition, many scholars think that since Hoffmann et al.'s claims are remarkable and unprecedented, they would require very strong supporting data in the form of other independent dating systems for confirmation and acceptance. Subsequent to the original report, an independent research group reanalyzed the data and concluded that there was robust support for an age of 47,000 years for the painting at Ardales, but less support for the 65,000 age as stated by Hoffmann et al..[110] If the Ardales paintings, dated conservatively at 47,000 years old, are confirmed to be anthropologic representations as opposed to just red deposits, this would indicate that some late Neanderthal population created them, since at the time there were no AMHs in Europe. This type of information prompts the question: were the Neanderthals the acculturated or the acculturators?

References

1. Yong E. *The New Story of Humanity's Origins in Africa*. Boston, USA: The Atlantic; 2018.
2. Nengo I, Tafforeau P, Gilbert C, et al. New infant cranium from the African Miocene sheds light on ape evolution. *Nature*. 2017;548:169–174.
3. Böhme M, Spassov N, Fuss J, et al. A new Miocene ape and locomotion in the ancestor of great apes and humans. *Nature*. 2019;575:489–493.
4. Walker A, Shipman P. *The Ape in the Tree: An Intellectual & Natural History of Proconsul*. Cambridge, MA, London, UK: The Belknap Press of Harvard University Press; 2005.
5. Morell V. *Ancestral Passions: The Leakey Family and the Quest for Humankind's Beginnings*. New York, NY, USA: Simon and Schuster; 1996.
6. Tuttle RH. Seven decades of east African Miocene anthropoid studies. In: Ishida H, Tuttle R, Pickford M, Ogihara N, Nakatsukasa M, eds. *Human Origins and Environmental Backgrounds*. Germany: Springer, Berlin; 2006.
7. Bestland EA, Krull ES. Palaeoenvironments of Early Miocene Kisingiri volcano *Proconsul* sites: evidence from carbon isotopes, palaeosols and hydromagmatic deposits. *J Geol Soc Lond*. 1999;156: 965–976.
8. The Free Library. *Volcano May Have Wiped out our African Ancestors*. London, UK: Geographical; 1999.
9. Zimmer C. Kenyan skeleton shakes ape family tree. *Science*. 1999;285:1335–1337.
10. Begun DR, Nargolwalla MC, Kordos L. European Miocene hominids and the origin of the African ape and human clade. *Evol Anthropol*. 2012;21:10–23.
11. DeMiguel D, Alba DM, Moya-Sola S. Dietary specialization during the evolution of Western Eurasian hominoids and the extinction of European great apes. *PLoS ONE*. 2014;9, e97442.
12. Bernor RL. New apes fill the gap. *PNAS*. 2007;104:19661–19662.
13. Herbert TD, et al. Late Miocene global cooling and the rise of modern ecosystems. *Nat Geosci*. 2016;9:843–847.
14. Kunimatsu Y, et al. A new Late Miocene great ape from Kenya and its implications for the origins of African great apes and humans. *Proc Natl Acad Sci*. 2007;104:19220–19225.
15. Dalton R. Oldest gorilla ages our joint ancestor. *Nature*. 2007;448:844–845.
16. Katoh S, et al. New geological and palaeontological age constraint for the gorilla-human lineage split. *Nature*. 2016;530:215–218.
17. Tsujikawa H. The Palaeoenvironment of *Samburupitheus kiptalami* on its associated fauna. *Afr Study Monogr*. 2005;32(Suppl):51–62.
18. Morita W, Morimoto N, Kunimatsu Y, et al. A morphometric mapping analysis of lower fourth deciduous premolar in hominoids: implications for phylogenetic relationship between *Nakalipithecus* and *Ouranopithecus*. *C R Palevol*. 2017;116:655–669.

19. Steiper ME, Young NM. Primate molecular divergence dates. *Mol Phylogenet Evol*. 2006;41:384–394.
20. Bailey WJ. Hominoid trichotomy: a molecular overview. *Evol Anthropol*. 1993;2:100–108.
21. Wildman DE, et al. Implications of natural selection in shaping 99.4% nonsynonymous DNA identity between humans and chimpanzees: enlarging genus Homo. *Proc Natl Acad Sci U S A*. 2003;100: 7181–7188.
22. Lovejoy CO. Reexamining human origins in light of *Ardipithecus ramidus*. *Science*. 2009;326: 74e1–74e8.
23. Dart RA. The osteodontokeratic culture of *Australopithecus prometheus*. Transvaal Museum (Pretoria). *Memoir*. 1957;10:1–105.
24. Easteal S, Herbert G. Molecular evidence from the nuclear genome for the time frame of human evolution. *J Mol Evol*. 1997;44:S121–S132.
25. Wood B, Harrison T. The evolutionary context of the first hominins. *Nature*. 2011;470:347–352.
26. Brunet M, et al. A new hominid from the Upper Miocene of Chad, Central Africa. *Nature*. 2002;418:145–151.
27. Lebatard A-E, et al. Cosmogenic nuclide dating of *Sahelanthropus tchadensis* and *Australopithecus bahrelghazali*: Mio-Pliocene hominins from Chad. *Proc Natl Acad Sci U S A*. 2008;105:3226–3231.
28. Wolpoff MH, et al. An ape or the ape: is the Toumaï cranium TM 266 a hominin? *PaleoAnthropology*. 2006;2006:36–50.
29. Senut B, et al. First hominid from the Miocene (Lukeino Formation, Kenya). *C R Acad Sci*. 2001;332:137–144.
30. Reynolds SC, Gallagher A. African genesis: perspectives on hominin evolution. In: *African Genesis: Perspectives on Hominin Evolution*. Cambridge, UK: Cambridge University Press; 2012.
31. Pickford M, Gommery D, Treil J. Bipedalism in *Orrorin tugenensis* revealed by its femora. *C R Palevol*. 2002;1:191–203.
32. Richmond BG, Jungers WL. *Orrorin tugenensis* femoral morphology and the evolution of hominin bipedalism. *Science*. 2008;319:1662–1665.
33. Almécija S, et al. The femur of *Orrorin tugenensis* exhibits morphometric affinities with both Miocene apes and later hominins. *Nat Commun*. 2013;4:2888.
34. Ann G. A new kind of ancestor: *Ardipithecus* unveiled. *Science*. 2009;326:36–40.
35. White TD, Suwa G, Asfaw B. *Australopithecus ramidus*, a new species of early hominid from Aramis, Ethiopia. *Nature*. 1994;371:306–312.
36. Suwa G, et al. The *Ardipithecus* skull and its implications for hominin origins. *Science*. 2009;326:68e1–68e7.
37. Lovejoy CO, et al. The great divides: *Ardipithecus ramidus* reveals the postcrania of our last common ancestors with African apes. *Science*. 2009;326:100–106.
38. Suwa G, Kono RT, Scott W, Simpson SW, Asfaw B. Paleobiological implications of the *Ardipithecus ramidus* dentition. *Science*. 2009;326:69–99.
39. Lovejoy CO. The origin of man. *Science*. 1981;211:341–350.
40. Vignaud P, et al. Geology and palaeontology of the upper Miocene Toros-Menalla hominin locality, Chad. *Nature*. 2002;418:152–155.
41. Pickford M, Senut B. The geological and faunal context of late Miocene hominin remains from Lukeino, Kenya. *C R Acad Sci*. 2001;332:145–152.
42. Su DF, et al. Paleoenvironment. In: Haile-Selassie Y, Wolde G, eds. *Ardipithecus kadabba: Late Miocene Evidence From the Middle Awash, Ethiopia*. Berkeley, CA, USA: University of California Press; 2009.
43. White TD, et al. Macrovertebrate paleontology and the Pliocene habitat of *Ardipithecus ramidus*. *Science*. 2009;326:87–93.
44. Cerling TE, et al. Comment on the paleoenvironment of *Ardipithecus ramidus*. *Science*. 2010;328:1105.
45. Burkhardt RW. Lamarck, evolution, and the politics of science. *J Hist Biol*. 1970;3:275–298.
46. Mayr E. Taxonomic categories in fossil hominids. *Cold Spring Harb Symp Quant Biol*. 1950;15:109–118.
47. Wolpoff MH. Telanthropus and the single species hypothesis. *Am Anthropol*. 1968;70:477–493.
48. Wolpoff MH. Competitive exclusion among Lower Pleistocene hominids: the single species hypothesis. *Man*. 1971;6:601–614.

49. Leakey REF. Further evidence of Lower Pleistocene hominids from East Rudolf, North Kenya. *Nature*. 1971;231:241–245.
50. Leakey REF, Walker A. *Australopithecus, Homo erectus* and the single species hypothesis. *Nature*. 1976;261:572–574.
51. Gould SJ. Ladders, bushes, and human evolution. *Nat Hist Mag*. 1976;85:24–31.
52. Wood BA. Facing up to complexity. *Nature*. 2012;488:162–163.
53. Bobe R, Carvalho S. Hominin diversity and high environmental variability in the Okote Member, Koobi Fora Formation, Kenya. *J Hum Evol*. 2019;126:91–105.
54. Wood B, Boyle EK. Hominin taxic diversity: fact or fantasy? *AJPA*. 2016;159:37–78.
55. Ward C, Frederick M. New fossils of *Australopithecus anamensis* from Kanapoi, Kenya and evolution within the *A. anamensis*-Afarensis Lineage. *J Vertebr Paleontol*. 2008;28:157A.
56. Kimbel WH, et al. *Ardipithecus ramidus* and the evolution of the human cranial base. *Proc Natl Acad Sci U S A*. 2014;111:948–953.
57. Haile-Selassie M, et al. A 3.8-million-year-old hominin cranium from Woranso-Mille, Ethiopia. *Nature*. 2019;573:214–219.
58. Frost SR. African Pliocene and Pleistocene cercopithecid evolution and global climatic change. In: Bobe R, Alemseged Z, Behrensmeyer AK, eds. *Hominin Environments in the East African Pliocene: An Assessment of the Faunal Evidence*. The Netherlands: Springer; 2007.
59. Haile-Selassie Y, et al. New species from Ethiopia further expands Middle Pliocene hominin diversity. *Nature*. 2015;52:483–488.
60. Levin NE, Haile-Selassie Y, Frost SR, Saylor BZ. Dietary change among hominins and cercopithecids in Ethiopia during the early Pliocene. *Proc Natl Acad Sci U S A*. 2015;112:12304–12309.
61. Terborgh JW. Toward a trophic theory of species diversity. *Proc Natl Acad Sci U S A*. 2015; 112:11415–11422.
62. Leakey REF. Evidence for an advanced plio-pleistocene hominid from East Rudolf, Kenya. *Nature*. 1973;242:447–450.
63. Meave G, et al. New fossils from Koobi Fora in northern Kenya confirm taxonomic diversity in early Homo. *Nature*. 2012;488:201.
64. Cerling TE, et al. Diet of *Paranthropus boisei* in the early Pleistocene of East Africa. *Proc Natl Acad Sci*. 2011;108:9337–9341.
65. Ungar PS, et al. Dental microwear and diet of the Plio-Pleistocene hominin *Paranthropus boisei*. *PLoS ONE*. 2008;3, e2044. Petraglia M (ed.).
66. Berger LR, et al. *Homo naledi*, a new species of the genus *Homo* from the *Dinaledi* Chamber, South Africa. *elife*. 2015;1–35(4), e09560.
67. Dembo M, et al. The evolutionary relationships and age of *Homo naledi*: an assessment using dated Bayesian phylogenetic methods. *J Hum Evol*. 2016;97:17–26.
68. Dirks PHG, et al. The age of *Homo naledi* and associated sediments in the rising star Cave, South Africa. *elife*. 2017;6, e24231.
69. Hawks J, et al. New fossil remains of *Homo naledi* from the Lesedi Chamber, South Africa. *elife*. 2017;6, e24232.
70. Holloway RL, et al. Endocast morphology of *Homo naledi* from the Dinaledi Chamber, South Africa. *PNAS*. 2018;115:5738–5743.
71. Schlebusch CM, et al. Southern African ancient genomes estimate modern human divergence to 350,000 to 260,000 years ago. *Science*. 2017;358:652–655.
72. Scerri ML, et al. Did our species evolve in subdivided populations across Africa, and why does it matter? *Trends Ecol Evol*. 2018;33:582–594.
73. Mounier A, Lahr M. Deciphering African late middle Pleistocene hominin diversity and the origin of our species. *Nat Commun*. 2019;10:3406.
74. Singer R. The saldanha skull from hopefield, South Africa. *Am J Phys Anthropol*. 1954;12:345–362.
75. Ferring R, et al. Earliest human occupations at Dmanisi (Georgian Caucasus) dated to 1.85-1.78 Ma. *Proc Natl Acad Sci*. 2011;108:10432–10436.
76. Lordkipanidze D, et al. A complete skull from Dmanisi, Georgia, and the evolutionary biology of early *Homo*. *Science*. 2013;342:326–331.

77. F. Spoor F. Palaeoanthropology small-brained and big-mouthed. *Nature*. 2013;502:452–453.
78. Schwartz JH, et al. Comment on a complete skull from Dmanisi, Georgia, and the evolutionary biology of early *Homo*. *Science*. 2014;344:360.
79. Prufer K, et al. The complete genome sequence of a Neanderthal from the Altai Mountains. *Nature*. 2014;505:43–49.
80. Green RE, et al. A draft sequence of the Neandertal genome. *Science*. 2010;328:710–722.
81. Sankararaman S, et al. The combined landscape of Denisovan and Neanderthal ancestry in present-day humans. *Curr Biol*. 2016;26:1241–1247.
82. Lohse K, Frantz LA. Neandertal admixture in Eurasia confirmed by maximum-likelihood analysis of three genomes. *Genetics*. 2014;196:1241–1251.
83. Hershkovitz I, et al. The earliest modern humans outside Africa. *Science*. 2018;359:456–459.
84. Liu W, et al. Human remains from Zhirendong, South China, and modern human emergence in East Asia. *Proc Natl Acad Sci U S A*. 2010;107:19201–19206.
85. Hershkovitz I, et al. Levantine cranium from Manot Cave (Israel) foreshadows the first European modern humans. *Nature*. 2015;520:216–219.
86. Fu Q, et al. An early modern human from Romania with a recent Neanderthal ancestor. *Nature*. 2015;524:216–219.
87. Finlayson C. Gorham's Cave, Gibraltar—the persistence of a Neanderthal population. *Quat Int*. 2008;181:64–71.
88. Coppa A, et al. Newly recognized Pleistocene human teeth from Tabun cave Israel. *J Hum Evol*. 2005;49:301–315.
89. Rak Y, et al. Découverte d'un squelette humain dans les niveaux moustériens de la grotte de Kebara (Israël). *Paléorient*. 1983;9:53–54.
90. Liu W, et al. The earliest unequivocally modern humans in southern China. *Nature*. 2015; 526:696–699.
91. Higham T, Douka K, Wood R, et al. The timing and spatiotemporal patterning of Neanderthal disappearance. *Nature*. 2014;512:306–309.
92. Finlayson C, et al. Late survival of Neanderthals at the southernmost extreme of Europe. *Nature*. 2006;443:850–853.
93. Zilhão J, et al. Precise dating of the Middle-to-Upper Paleolithic transition in Murcia (Spain) supports late Neanderthal persistence in Iberia. *Heliyon*. 2017;3(1–51), e00435.
94. d'Errico F, Sanchez Goni MF. Neanderthal extinction and the millennial scale climatic variability of OIS 3. *Quat Sci Rev*. 2003;22:769–788.
95. Finlayson C. *Neanderthals and Modern Humans*. Cambridge, UK: Cambridge Univ. Press; 2004.
96. Alcaraz-Castaño M, et al. A context for the last Neandertals of interior Iberia: Los Casares cave revisited. *PLoS ONE*. 2017;12, e0180823.
97. Maroto J. Current issues in late Middle Palaeolithic chronology: new assessments from Northern Iberia. *Quat Int*. 2012;247:15–25.
98. Zilhão J, et al. Analysis of Aurignacian interstratification at the Châtelperronian-type site and implications for the behavioral modernity of Neandertals. *Proc Natl Acad Sci U S A*. 2006;103:12643–12648.
99. Gravina B, Mellars P, Ramsey CB. Radiocarbon dating of interstratified Neanderthal and early modern human occupations at the Châtelperronian type-site. *Nature*. 2005;438:51–56.
100. Mellars P, Gravina B, Bronk RC. Confirmation of Neanderthal/modern human interstratification at the Châtelperronian type-site. *Proc Natl Acad Sci U S A*. 2007;104:3657–3662.
101. Slimak L, et al. Late Mousterian persistence near the arctic circle. *Science*. 2011;332:841–845.
102. Kuzmin YV, Keates SG. Siberia and neighboring regions in the Last Glacial Maximum: did people occupy northern Eurasia at that time? *Archaeol Anthropol Sci*. 2018;10:111–124.
103. Villa P, et al. From Neandertals to modern humans: new data on the Uluzzian. *PLoS ONE*. 2018;13, e0196786.
104. Mellars P. Neanderthal symbolism and ornament manufacture: the bursting of a bubble? *Proc Natl Acad Sci U S A*. 2010;107:20147–20148.
105. Caron F, et al. The reality of Neanderthal symbolic behavior at the Grotte du Renne, Arcy-sur-Cure, France. *PLoS ONE*. 2011;6, e21545.

106. Galland A. Quantifying lithic surface alterations using confocal microscopy and its relevance for exploring the Châtelperronian at La Roche-à-Pierrot (Saint-Césaire, France). *J Archaeol Sci*. 2019;104:45–55.
107. White R. Personal ornaments from the Grotte du Renne at Arcy-sur-Cure. *Athena Rev*. 2001;2:41–46.
108. Hoffmann DL, et al. U-Th dating of carbonate crusts reveal Neanderthal origin of Iberian cave art. *Science*. 2018;359:912–915.
109. White R, et al. Still no archaeological evidence that Neanderthals created Iberian cave art. *J Hum Evol*. 2019. https://doi.org/10.1016/j.jhevol.2019.102640.
110. Ludovic S, et al. Comment on "U-Th dating of carbonate crusts reveals Neandertal origin of Iberian cave art". *Science*. 2018;361:eaau1371.

CHAPTER 2

The species concept and the reality of introgression

Every paleontologist knows that most new species, genera, and families, and that nearly all categories above the level of family appear in the record suddenly and are not led up to by known, gradual, completely continuous transitional sequences.

George Gaylord Simpson

Evolution and the species concept

Resistance to evolutionary theory continues even in these days of tremendous scientific advancements. One barrier to evolutionary acceptance is the amount of time needed for evolutionary events to take place versus humans' perception of events that occur in a limited period of days, months, and our relatively short lifetime. The perception of millions or billions of years is beyond human comprehension, and is therefore meaningless to most people when explaining the concept of gradual evolution and speciation. This difference between evolutionary time and human time perception affects how we define species, because species are generally longer-lived than their human observers. In addition, organism phenotypes and genotypes are transient, making it almost impossible for humans to perceive the development of species by merely observing them.[1]

Because of the time span and the complexities of the genetic processes involved in evolution, trying to understand evolutionary events requires a leap of faith, imagination, and an understanding of research data that have been collected. The inability to understand the complexities of evolutionary data, however, should not be a barrier to the acceptance of evolutionary theory, just as a lack of understanding of complex physics does not preclude our acceptance of some of its results. For example, most people do not understand the complexities of wireless technologies or the mechanics that gave rise to them, but they know that the science exists and they use the results of those scientific investigations regularly. Similarly, numerous tests of evolutionary theory and scientific data continue to support the idea that biology and life only make sense when examined through the theory of evolution, although not everyone understands all the scientific details behind evolution.

Sex and Cohabitation Among Early Humans
https://doi.org/10.1016/B978-0-12-819086-9.00008-7

Arguments over evolution often occur at the level of religious and philosophical beliefs, implying that if evolution exists, then humans are not unique, and are nothing more than animals. In addition, evolution has been used to influence biological and social divisions by associating evolution with certain undesirable traits among individual groups. Instead of dividing us, evolutionary theory and the data we have accumulated to support it should unite us in knowing that at the molecular level, all humans are genetically similar. The concern that evolution implies gender, ethnic, and social inequality and encourages stereotypes is unfounded. It should be understood perceived inequalities are all products of our own making. In addition, evolutionary data have shown not only that we are related to other humans, but also that from a genetic standpoint, we are interrelated with other organisms who share many of our genes.

Although we understand a great deal about evolution, there are still many unanswered questions about the specific details: what constitutes a species? What is the importance of classifying organisms into species? From a historical perspective, how did we reach the current biological methods of organism classification? How could life have gone from a single cell organism that divided asexually to evolve into multicellular humans that reproduce sexually? How could humans have evolved from different primates into different genera and eventually into *Homo sapiens*? Do Neanderthals and humans belong to the same species?

What constitutes a species

One of the major questions that occurs when describing evolutionary transitions is: what constitutes a species? The biological species concept describes a species as individuals who are part of a reproductive population, a gene pool, and a genetic system. However, even with this classic definition, there are still gaps in how to delineate different species. Do these interbreeding organisms have to share a recent common ancestor? Must they exhibit specific characters? As a result, the definition of a species has perplexed scientists and philosophers for more than a century.

Although the Latin word *species* translates to "kind," definitions encompass a wide range: a group of similar or closely related individuals; a specific taxonomic unit; a group of individuals capable of breeding and reproducing viable offspring; or a unit of biodiversity. The difficulty in defining a species comes from the fact that a species is just a snapshot of a group of individuals over evolutionary time, so that even members of the same species do not always look alike, act alike, or interbreed. Although biologists disagree on a specific definition of species, there is agreement that the key events in evolution are those that create new species. Thus, change within a species, "speciation," is a fundamental aspect of evolution. Speciation is an opportunity to form a new species, an end point of variation at a particular point in time, with an unknown long-term outcome.

Biological change has resulted in a complex and diverse group of living organisms, and a biological classification system where species or subspecies are the lowest rank in the hierarchical classification scheme. In this biological classification, species are described as belonging to a genus (a group of organisms with common characteristics), that can contain more than one species. A genus is then part of a family, a family part of an order, an order part of a class, a class part of a phylum, and a phylum part of a kingdom. For example, humans are part of the animal kingdom (*Animalia*), the phylum *Chordata*, the class *Mammalia*, the order *Primates*, the family *Hominidae*, and the genus and species *Homo sapiens*.

Thousands of new species are named every year, and the scientists that identify the species get to choose the name. All species are given a two-part name (with the exception of viruses) that is italicized or underlined in the literature. The first name is the genus to which the species belongs; this is always capitalized. The second name (the species) is not capitalized and is usually Latinized. The species name varies and can indicate: (1) something specific about the features of the organism or its environment; (2) the person describing it; or (3) a famous person admired by the collector. For example, *Thermophilus aquaticus* is a bacterial organism that lives in the water of hot springs. *Materpiscis attenboroughi* is the first fossil of an organism known to have given birth to live young and is named after the person who discovered it, Sir David Attenborough. *Gnathia marleyi* is a Caribbean crustacean that parasitizes fish, named after Bob Marley, a famous Jamaican musician. In many cases the genus is abbreviated as just a capital letter. For example, the genus *Homo* is abbreviated *H.* when describing *H. sapiens*.

The use of genus and species names avoids confusion over the use of common names. Many organisms can have a common name, like "frog." Common names are used to describe types of organisms; however, they can vary greatly from place to place—for example, a frog in South America can be vastly different from a frog from North America. The dyeing dart frog of South America, *Dendrobates tinctorius*, is a poisonous diurnal and brightly colored frog. In contrast, the northern leopard frog *Rana pipiens*, native to North American, varies in color from green to brown and produces a chemical called ranpirnase (a ribonuclease enzyme), which is a potential treatment for cancer. Unlike the common names, which can vary regionally, the scientific names avoid confusion by being unique and universal.

Historical definitions of species

The species definition problem has become a philosophical debate, and as molecular and genetic technology increases, it remains unclear what data are most important in distinguishing one species from another.

The philosophical arguments of species date back to the time of Aristotle and his idea of essentialism. Aristotle's view was that species have inherent characteristics (*genos*) and have innately different natures or outlooks (forms) shared by all their members, with offspring inheriting all characteristics of the parents and that all forms were unchanging. This concept of essentialism has since been disproven by modern genetics, epigenetics, and studies in the social sciences. However, Aristotle did come up with a hierarchy of organisms called the *Scala Naturae*, which formed a model of classification for other taxonomists.[2]

Aristotle's ideas were accepted until the 1600s, when John Ray attempted a more biological approach to defining species. He recognized that variations existed within a group, but also that a species was a group of individuals that would "breed true" and not change. He recognized that parents gave rise to some variation within offspring within a species, but one species never arose from another group or vice versa. His idea of variation was different from that proposed by Aristotle.

Carolus Linnaeus was the first person to have a significant impact on scientific classification and the first to come up with a species concept based on morphology. He was born in 1707 and was a Swedish scientist who laid the foundations for modern taxonomy. His interest in botany drove his ideas about the classification of plants, based on morphology and natural relationships. His interest in categorizing numerous organisms that he found in nature led to his work on taxonomy entitled the *Systema Naturae*. He believed that there was an organismal hierarchy of categories based on observable characteristics, but that there was no connection between species, no matter how similar they might be. Because of his strong religious beliefs, he felt species were divinely fixed and could vary through hybridization and acclimation, but that one species could not emerge from another. Many of his commonsense classifications are still used today—for example, the class *Mammalia* (Latin for breast), or the genus and species *Homo sapiens* (wise man), or *Homo troglodytes* (caveman). Linnaeus's ideas are still used today by some taxonomists to define typological or morphological species where a group of organisms conforms to certain fixed properties and phenotypes.

Many taxonomists still agree with Linnaeus and his morphological analysis (if an organism looks different and its measured characteristics are different, then it is probably a new species; if not, then it is the same species). This species concept categorizes a species as an individual that represents morphological characteristics that differentiate it from all other individuals. One problem with this type of classification is that these taxonomists usually measure only a few specific characters, and these are not necessarily inclusive of the whole organism, such as only measuring leaves, sepals, petals, and reproductive structures in a plant. In addition, this type of classification ignores population-level variation (see below).

In the 1800s, Jean-Baptiste Lamarck and Charles Darwin proposed new ideas about species. Lamarck (a French naturalist) proposed that species could change over time. He called this phenomenon "transmutation," and it was a radical idea that preceded the theory of evolution, helping to set the stage for Darwin's theory. Darwin believed that species and evolution arose within populations, not individuals. He argued that the force behind the evolution of species was natural selection, which arose from naturally occurring variation among individuals. His idea was that the term "species" was only useful for naming groups of individuals that resembled each other and interacted.[3]

In 1937, Theodosius Dobzhansky, a prominent Russian-American geneticist, developed the biological species concept and defined speciation as the evolution of isolating mechanisms induced by mechanisms that reduced gene flow between populations. Later, Ernst Mayr proposed reproductive isolation as a method for distinguishing different species. His definition went beyond hybrid sterility and defined isolation as any mechanical, biochemical, ecological, genetic, or behavioral mechanism that reduced gene flow and reduced hybrid viability. He emphasized that the first step to becoming a species was the separation of a small group into a unique environment, away from the parental group, until through adaptation, mutation, and natural selection, the group became reproductively isolated: no longer able to mate and produce viable offspring with the parent group.[4]

Ernst Mayr's emphasis on reproductive isolation is still used today as a measure of species identification between populations. However, mating potential and reproductive isolation are sometimes difficult to test. Behavioral differences, anatomical variations, and geographical isolation alone could prevent organisms from mating, and therefore it is not always possible to determine if these organisms could no longer produce fertile offspring.[5]

For many years the definition of a species was a group of individuals capable of reproducing viable offspring. However, with the use of DNA analysis, morphological and physiological analysis, and the concept of species inhabiting a specific ecological niche, Mayr's definition of a species became outdated. In 1995, Mallet proposed the genotypic cluster concept for a species.[5] He proposed a single genotypic cluster (or species) were composed of groups of individuals from genetic or morphological individuals with few or no intermediates to other clusters. He suggested that these clusters could be recognized if there was a single group of individuals that had multiple, unlinked inherited characters or genetic markers. A pair of such genotypic cluster species would be such if the frequency distribution of genotypes showed a bimodal distribution. The presence of more than one species or genotypic cluster can then be recognized when genotypes are multimodal, and low heterozygosity and linkage disequilibria (the nonrandom association of alleles at two or more loci in a population resulting in haplotypes with unexpected frequencies) are evident between the clusters. Within each genotypic cluster in a local region, allele frequencies will conform to Hardy-Weinberg equilibrium and different unlinked loci will be in approximate linkage equilibrium.

DNA barcoding is one method of distinguishing species by genotypic clusters using mitochondria DNA sequences. Once the sequences are established, statistical methods like principle component analysis, which summarizes the variability in the data sets in two or three dimensions, can be used to plot the samples either close together when they are very similar or far apart when they are very different.

Morphological variation can also be used to develop species clusters, but one must be careful to choose the proper characters. Sexual dimorphism (like the male peacock's colorful tail feathers versus the dull ones of the females), characters influenced by environmental conditions (coral colors and skeletal structure influenced by depth and light), or characters that change during ontogeny (juvenile versus adult) are not useful for distinguishing species. Morphological distinctions and analysis should be carefully evaluated under similar environmental conditions and developmental stages. In these cases, morphological variability established using image analysis or careful examination can result in data being used to plot similar samples close together and those that are different far apart. Major differences can result in different clusters and the designation of different species.

Years of exploration and identification of species have led to the conclusion that there is an unequal distribution of species worldwide. More species occur in the tropics than in temperate zones regardless of whether they have a sexual or asexual method of reproduction. The increase in species diversity from the poles to the tropics is referred to as the latitudinal diversity gradient. The pattern appears to be ubiquitous and overall speciation rates are higher, and extinction rates lower, in the tropics than in temperate regions. Because evolution creates and sustains biodiversity by way of adaptive changes in ecologically relevant traits, understanding how rates of speciation and extinction vary with latitude is central to explaining this pattern. However, it is difficult to estimate diversity rates associated with specific geographic locations, and research productivity in tropical ecosystems is especially low. Yet the genetics of ecological speciation is becoming increasingly popular and studied across a variety of taxa and environments, especially in the tropics. Several investigators have come up with ideas of how to measure diversity accurately in the tropics by simultaneously addressing biogeographical and taxonomic questions, while effectively assessing relationships between reproductive isolation and ecological divergence. For example, Beheregaray et al.[6] have proposed a new framework for ecological speciation in the tropics that integrates information from phylogenetics, phylogeography, population genomics, and simulations in evolutionary landscape genetics (combining population genetics and landscape ecology).

Phenotypic variation, geographical distribution, and biodiversity all contribute to speciation patterns. These patterns can provide clues to how speciation occurs, the origins of variation among different species, and the worldwide distribution of biodiversity.[7]

Species classification and the variety of species concepts

Close observations of an organism's cells, DNA, behavior, nonreproductive morphology, ecology, and ancestry have led to reclassification of organisms that were once categorized as a single species. From here, the problem then becomes where does one start and stop with the methods and tools used in species classification? The answer is that the mechanisms for classification do not stop. As we acquire new technology, that technology will be used to study variation and, in some cases, this will result in new species identification or a reduction in the number of species. Thus, classification mechanisms are transient and the species concept continues to be redefined.

Various levels of species distinction have resulted in taxonomists being classified as "lumpers" or "splitters," depending on their approach to recognizing differences or similarities in organisms and classifying them as different or the same species. Lumpers tend to classify large numbers of individuals within the same species, sometimes ignoring slight intermediate variations, and only rarely find genuinely unique characteristics that compel them to separate a group into a different species. As a result, there would be an overall smaller number of species. Splitters, on the other hand, tend to break up species based on small differences that may be the result of unusual environmental conditions, or small sample size, resulting in an overall larger number of species.

The naming of species follows the nomenclature guide described above but does not define specific rules for how species are identified (morphology, ecological niche, genetic makeup, behavioral characteristics, etc.). Therefore, as new data are used, or as new evidence is discovered, designations could change the boundaries used in the classification and even lead to the description of a subspecies.

A subspecies is a category that ranks below a species, and is often a group of permanently geographically isolated individuals that are genetically distinct but still capable of interbreeding with the designated species. Subspecies are named by a Latin trinomial (three names) in animals. An example would be the Siberian tiger, *Panthera tigris altaica*. Plants and bacteria use the trinomial form but with a subspecies designation. An example would be the bacterial plant pathogen *Xylella fastidiosa* subsp. *Multiplex*.

In some cases, subspecies can interbreed and, in doing so, develop new morphological and genetic characteristics. This is especially true in more plastic plant populations which, in many cases, do not have the same reproductive or genetic barriers seen in some animals. Genetic exchange between subspecies of plants can lead to genetic diversity, the introduction of adaptive or maladaptive traits, disruption of coadaptive genetic complexes (in addition to polyploidy, aneuploidy), and the creation of new ecotypes, subspecies, or species.[8] For example, bread wheat *Triticum aestivum* is an allohexaploid species (that is, its genome comes from a combination of three interrelated diploid [2N] genomes). This genetic combination resulted from three donor species designated as A, B, and D. Donor species A (*Triticum urartu*), donor species B (*Aegilops speltoides*), and donor species D (*Aegilops tauschii*) gave rise to common bread wheat with 42 chromosomes (AABBDD genomes). The plant most likely occurred 8000–10,000 ya somewhere in the Fertile Crescent.

The problem with classification is that if we look closely enough at morphological, physiological, cellular, ecological, and molecular data, we will find differences between all organisms, whether related or not resulting in population-level variation. Population-level variation can come from a variety of sources, including specific adaptations, environmental differences, or slight differences in genetic background. These variations can lead to morphotypes where the apparent morphology is so different that they appear to be distinct species, but on closer observation they are the same.[9,10] One example of this is the difference in height and skin color between people of the Netherlands and the pygmies of Cameroon. Looking at these individuals from a strictly morphological perspective might suggest that they are different species. However, their ability to interbreed, their use of language, and DNA evidence tell us that they are the same species.

Although there are more than 20 different species concepts, based on the criteria described above, many of these concepts overlap in their ideology. However, many concepts and their associated definitions are incompatible and can lead to different conclusions concerning the boundaries and numbers of species. Most taxonomists have narrowed down the number of species concepts. However, there is still debate over the process of delimitation because of various ways to classify the same organism[1] (Table 1). It should be noted that there have been attempts to unify species concepts

Table 1 Species concepts.

Concept	Property defining concept
Biological	Interbreeding (natural reproduction resulting in viable and fertile offspring)
Isolation	Intrinsic reproductive isolation (absence of interbreeding between hetero-specific organisms based on intrinsic properties, as opposed to extrinsic [geographic] barriers)
Recognition	Shared specific mate recognition or fertilization system (mechanisms by which conspecific organisms, or their gametes, recognize one another for mating and fertilization)
Ecological	Same niche or adaptive zone (all components of the environment with which conspecific organisms interact)
Evolutionary	Unique evolutionary role, tendencies, and historical fate
Cohesion	Phenotypic cohesion (genetic or demographic exchangeability)
Phylogenetic	Heterogeneous (see next four entries)
Hennigian	Ancestor becomes extinct when lineage splits
Monophyletic	Monophyly (consisting of an ancestor and all of its descendants; commonly inferred from possession of shared derived character states)
Genealogical	Exclusive coalescence of alleles (all alleles of a given gene are descended from a common ancestral allele not shared with those of other species)
Diagnosable	Diagnostic (qualitative, fixed difference)
Phenetic	Form a phenetic cluster (quantitative difference)
Genotypic cluster (definition)	Form a genotypic cluster (deficits of genetic intermediates; e.g., heterozygotes)

Species concepts and the properties associated with the concept. All of these ideas conform to a single general concept under which species are equated with separately evolving metapopulation lineages. Many of the properties are more appropriately interpreted as lines of evidence relevant to assessing lineage separation.
Taken from K De Queiroz. Species concepts and species delimitation. *Syst Biol* 2007;56:879–886.

(unified species concept) since defining a species has proven so difficult. But even this definition is not all-encompassing.

Because of the considerable disparity in species concepts, the science philosopher John S. Wilkins has narrowed the species concepts into seven different concept categories. Wilkins recognized that many of the species concepts involved sexually reproducing organisms, and so his first two classifications recognize the difference between sexual and asexual organisms. (1) Since reproductive isolation cannot be used as a criterion for asexual species, the concept of agamospecies is used to identifying certain diagnostic traits to distinguish between closely related asexual lineages, and (2) the concept of biospecies is used for reproductively isolated sexual organisms. This is the most popular mechanism, although reproductive isolation may be a product of speciation and not the cause.

Wilkins also recognized the effects of ecological niches and genetics when identifying species, thereby identifying three other concept classifications: (3) ecospecies based on ecological niches, and the adaptation of organisms to unique resources; (4) evolutionary species based on lineage descent works for both sexual and asexual species; and (5) genetic species based on their gene pool including karyotype and DNA sequence.

Wilkins' last two classifications relate to the morphological interpretations and classifications based on a variety of data: (6) morphospecies based on form or phenotype following the Linnaeus classification scheme; and (7) taxonomic species, a species as determined by a taxonomist that could be based on a variety of different data and analyses.[11]

Today it appears that most taxonomists use several different methods for identifying and classifying species. Here we will only discuss four of these methods, which are most often used when identifying species:

(1) Biological concept, a group of individuals in which any two of the opposite sex can produce fertile offspring but cannot produce viable offspring outside the group (reproductive isolation).

(2) Genealogical concept, the exclusive coalescence of all alleles of a given gene descended from a common ancestor, or karyotype and/or DNA sequence not shared with those of other species.

(3) Ecological concept, a set of organisms adapted to and who exploit a particular set of resources in the environment (niche), where differences in morphology and behavior are related to differences in the ecological resources.

(4) Chronospecies concept, where a species is derived from examining continual and uniform changes in a sequential developmental pattern leading to physical, morphological, and/or genetic changes making it distinct from other ancestors. The chronospecies concept is typically used by paleoanthropologists when examining fossils.[1]

Some of the phylogenetic species concepts, which define a species as a group whose members are descended from a common ancestor and who all possess specific defining (or derived) traits, recognize too many species, whereas some of the reproductive isolationist concepts ignore asexual and allopatric species (species that occur over time when populations of the same species become isolated).[12] The reason for incompatibilities among these different concepts is due to the different biological disciplines, methods of data collection, and principles upon which the concepts are based. The importance placed on morphometric, genetic, regional, and lineage analyses exposes the differences in interest between different subgroups of biologists. Reproductive incompatibilities are important to biologists who study hybrid zones, niche variations are important to ecologists, genetics are key for population geneticists and molecular systematists, while morphological differences are central for paleontologists. It appears that the biologists who adopt a multidisciplinary approach, and who can step back from their research interests, might have the best approach because all of the concepts have some merits.

The species concept debate is complicated by various views such as: (1) only populations are real, and species are artifacts[13]; (2) species are arbitrary constructs of the taxonomist for convenience; (3) species are real entities[14]; and (4) species are the empirical units of evolution and ecology, and if monophyletic groups are real, then so are species.[15]

The argument that species are artifacts stems from the idea that like other artifacts, species are invented by humans and their desire to classify something that in this case is only temporary. In relation to species, the term "artifact" implies that species are artificial because of the complexity and variability within a group of species. On the other hand, those who argue that species are real entities point out that by looking at distinguishing features and characteristics, one can define species, and that the term is useful.

Several taxonomists have advocated for pluralism, or the use of multiple species concepts,[16] although different situations or questions may call for the use of different species concepts. The pluralistic view suggests that any evidence relevant to the species question needs to be considered. Total evidence could include morphological information, genetic characters, behavioral traits, ecological considerations, and more. While perhaps not satisfying multiple species concepts, the requirement for total evidence may be the best approach.[17]

It seems that every taxonomist is trying to find the perfect species concept that works for all scenarios and types of organisms.[18] Because of this, it has become apparent that when defining a species, scientists have created their own problem through their disagreements and by using different forms of data and analysis.

Importance of species identification

When looking at the complexities of evolutionary theory, it is sometimes difficult to justify why we should care about species. This goes back to the debates and disagreements over what constitutes a species and how we identify it. Although it may seem trivial, species comprise one of the fundamental units of biology, and a consensus on what constitutes a species is important for global, political, and financial reasons. This is especially the case among conservation biologists, where species are useful units when describing biodiversity and therefore can help to inform conservation action. Again, we see a discrepancy in species definitions where many conservation biologists use the evolutionary significant unit (ESU) or wildlife species (a distinct organism) as their definition of a species.[19]

Another important reason to study species is that species that are introduced into an exotic environment can endanger natural species or, in extreme cases, cause their extinction, resulting in habitat and economic losses.[20] One classic example of species invaders and the level of destruction that they can cause can be seen in Australia, where the introduction of exotic plants, animals, and microbes has led to habitat destruction. It is estimated that Australia has more than 600,000 species, many of which were introduced by early inhabitants from the Americas and Europe, who deliberately brought many exotic species onto the continent. For example, cane toads and rabbits have caused millions of dollars in damage and have cost millions of dollars in attempts to eradicate them. The cane toad was introduced from South America to help lower the number of cane beetles (*Dermolepida albohirtum*) that destroy sugar cane. The cane toad is now considered a pest and an invasive species that never did help to eliminate cane beetles. Rabbits were introduced to Australia from Europe in the 18th century and are now serious pests and an invasive species that have been blamed for the destruction of crops, the eremophila plant, and various species of trees. Because the rabbits feed on seedlings, many plants and trees are never able to reproduce, leading to regions where certain trees and plants are extinct. Unfortunately, scientists, residents, and visitors continue to bring in invasive plants, animals without natural predators, and microbial pathogens. These invasive species are often brought in purposely or unexpectedly through the postal service, and through international transport of items as a part of global trade. In addition to introduced species, there are still numerous endemic plants and animals that have not been classified. One example is the estimated 200 mosquito species that have not been named or studied. The concern is that some of these undocumented insects may be carrying diseases.[21–23]

Introduced pathogen species and/or their vectors can be especially harmful to human health (HIV, West Nile virus, Coronavirus), agriculture (African Swine Fever, hoof-and-mouth disease), forests (mountain pine beetle, crown gall), aquatic systems (nonnative fish and shellfish pathogens), and wildlife (chronic wasting disease, rabies). Of increased importance are viruses that spread from agricultural animals and wildlife to humans. In these cases, animal viruses or other animal pathogens are transferred to humans either because of increased exposure of humans to infected animals or because the pathogen acquires genetic variations that permit it to overcome transmission barriers, allowing it to infect human hosts. When animal viruses are transmitted to humans, devastating outbreaks—and in severe cases, pandemics—can result. The steps involved in the transfers of animal viruses to new human hosts require infection of an initial individual leading to viral variants that can spread efficiently through human-to-human transmission and thus into the population. The increase in urbanization has resulted in more introduced species, more human-wildlife contact, and more zoonotic disease transmission. Although the process of animal viruses entering human hosts is rare and the possibility of the virus acquiring mutations for the ability of human-to-human transmission is even rarer, if the virus does mutate to this level, the results can be devastating because it produces disease never before seen in the human population. If the disease is not controlled and is highly contagious, it can cause a pandemic.[20,24–26]

Pandemics have been a natural part of human history. Cholera, bubonic plague, smallpox, and influenza are some of the most severe diseases in human history, and outbreaks of these diseases across international borders have killed millions of people. Most recently the Coronavirus SARS-cov-2, which causes the disease Covid-19, has created a pandemic that has moved around the world at a rapid pace. Finding the virus and animal sources of these diseases and their treatments will become ever more important as the human population grows and more people travel internationally.

Understanding how species adapt to new environments and how they evolve is also crucial in cancer research. Recently, investigators have recognized that the spread of cancer cells is dependent on environmental perturbations, akin to the evolution of species, only at the cellular level. Analogous to two groups of organisms competing for resources during evolution, cancer cells and normal tissues compete for resources, and the tissue that is better adapted to the environment succeeds. Usually, normal cells thrive because they are a good fit for the biochemical signals, nutrients, and physical cues received from surrounding healthy tissue. Cancer cells can be initiated by mutations in normal tissue, resulting in cells that are most often poorly adapted for the environment, and normal cells outcompete the cancer for resources. But if the surrounding normal tissues are damaged by inflammation, old age, physiological changes, or extreme environmental conditions the cancer cells outcompete normal cells, and the change in the surroundings leads to the cancer cells' success and growth.[27]

Investigators have found evidence in animal models to support this theory of "cancer adaptive oncogenesis." This evolutionary perspective has resulted in a new approach to cancer therapy. Chemotherapy has been used to shrink tumors, but the tumors often come back resistant to chemotherapeutic drugs. The evolutionary approach, tested in animals and humans with advanced prostate cancer, limits the natural selection of cancer cells through a restrained use of chemotherapy by giving low doses of the chemotherapeutic drug. The goal is to maintain a small population of vulnerable chemo-sensitive cancer cells. The chemo-sensitive population does well enough to prevent chemoresistant cells from taking over. The results of the prevention and therapeutic strategies may ward off cancer before it becomes lethal for patients who do not respond to large doses of chemotherapy. Again, the basis of this new treatment revolves around the idea of how species adapt to new environments and cells acting to their environment.[27,28]

Artificial species (those created by human selective breeding) are also important for global and financial reasons. The production of subspecies by breeders is well established.

Plant breeders have developed subspecies that have higher seed yields and are resistant to pests and some herbicides. Breeders have also produced polyploid fruit to increase the size of strawberries or to create seedless watermelon. Although this breeding has not produced new species and rarely produces plants that show reproductive isolation, there are genetically modified plants with genes that prevent them from cross-pollinating with the original parent.

Artificial species are also created by animal husbandry, which involves breeding, feeding, and the production of a variety of domesticated and wild meat animals and the conversion of some animal commodities to edible and useful products. Probably the best example of artificial animal species is domestic dogs, which display a wide variety of physical and behavioral attributes. Laboratory animals are other examples of organisms that are artificially bred and in many cases could not survive in the wild.

Experiments with *Drosophila* have also shown the potential to produce artificial species. Two investigators, Rice and Salt,[29] exposed *Drosophila* to various conditions in the hope of establishing two species through sympatric speciation (evolution of a new species from a surviving ancestral species while both continue to inhabit the same geographic region). Two groups emerged favoring different environmental conditions, and after 35 generations, the 2 groups would not interbreed. This demonstrated that the first step in speciation (reproductive isolation) could occur under artificial conditions.[29]

The concept of speciation and its role in evolution is also important in everyday life and the realm of politics. The politics of evolution occur both within the scientific community and outside. Creationism is the most powerful force outside of science that refutes evolution, and its persuasiveness comes from the idea that humans must be unique from all other animals and the biological complexity that we see could never occur without a higher force. This is a convincing argument for most people who have a difficult time understanding the forces of evolution. However, ideological factions within science also provide conflicting views. Scientists disagree over methodological and philosophical perspectives. These conflicts within biology, education, religious philosophy, and the confusion within the general public can have far-reaching implications in the belief of science and its value. Some of the public's misunderstanding comes from the interpretations of evolution since the time of Darwin. Darwin's definition of a species as a "kind" similar to those mentioned in the Bible no longer applies. Today biologists see a species as a group of organisms whose members share specific characteristics representing a present manifestation of genes that will constantly change over time. From a biological perspective, there is no "kind." For example, when we see modern-day humans (MDH), we are looking at a snapshot in time that consisted of a varying array of gene assemblies constructed over millions of years, and our ancestors looked very different from MDH. Millions of years from now, those genes will be different, and we will also be different if we survive extinctive forces. But there was never a specific instance in time where a human was born from a nonhuman, i.e., there is no essential human, no kind.[30]

Speciation

When introducing the concept of speciation, Darwin's book title tells it all: *The Origin of Species*. The term "speciation" refers to the formation of new species during the course of evolution. Although Darwin had limited resources to determine exactly how species are formed, he knew that one of the major forces involved was natural selection. Darwin realized that there were distinct species and that natural selection played a role in their origin, but he was not sure why, because he did not see intermediate species forms in nature or understand what other forces might be involved. We know that natural selection plays a significant role in speciation and can influence reproductive isolation, adaptation, predator-prey interactions, host-parasite interactions, competition, mutualism, prezygotic fitness, and coevolution.

Today we recognize that all species arise from preexisting species, and there are several modes of speciation that include the mechanisms of geographical isolation, reproductive isolation, natural selection, hybridization, mutation, and genetic drift. Speciation is both a microevolutionary process (accumulation of genetic changes) and a macroevolutionary process (large-scale changes in organism phenotype, geographic range, and adaptive diversity).

All proposed modes of speciation have occurred, but the contribution of each to biodiversity is still debated. In each mechanism, a subgroup of the original population becomes separated and then is exposed to new environmental and selective pressures. The new pressures may include variation in predator behavior, new resources, competition for resources with other groups, reproductive success, novel mutations, and genetic drift. Then if the subgroup is fortunate enough to reproduce quickly and carry traits that allow them to adapt, the new group will flourish and become more and more phenotypically distinct as gene flow from the original group is blocked or reduced. Those subgroups that do not flourish will be lost to extinction. One macroevolutionary force in selection of species is larger size and larger geographic range, which can influence the rate of speciation and species section. Fossil records indicate that the successes or failures of lineages are not necessarily determined by the same factors. The fossil records show that after a mass extinction, the species that dominated before the event and recovered did not always recover and reassume its domination. After mass extinctions, it appears that species were favored that had a broad diet and wide geographic range.[4]

Geographical modes of speciation: Allopatric, peripatric, parapatric, and sympatric

Allopatric speciation and peripatric speciation

Allopatric speciation, also referred to as geographic speciation, is a mode of speciation that occurs when a group from a population of the same species becomes isolated from the parent group, leading to the absence of mating and gene flow between the group and parent population. It is considered the most common method of speciation. Geographic barriers can place a group of individuals in a situation where the environment is different, leading to different selective pressures and, eventually, different species. This type of speciation is typically divided into vicariance and peripatric speciation (Fig. 1). Vicariant speciation refers to instances where the geographical ranges between organisms do not

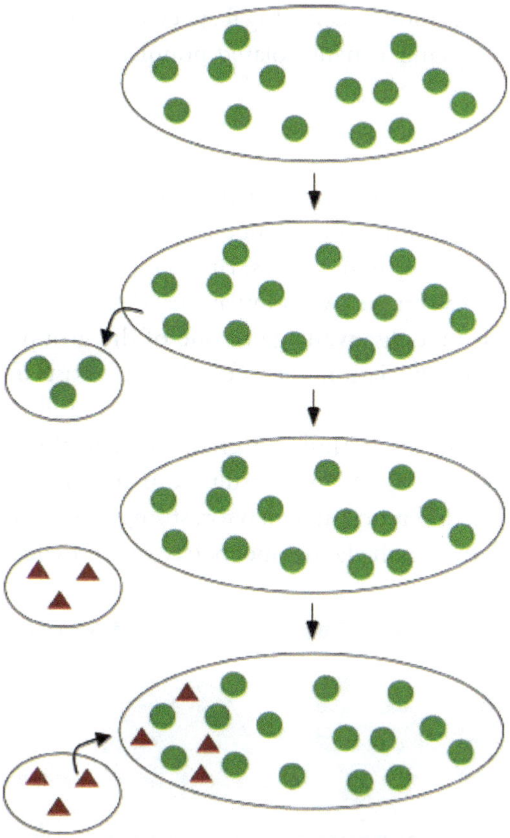

Fig. 1 Allopatric and peripatric speciation. A population of individuals of the same species becomes isolated from the parent group leading to the absence of mating and gene flow between the group and parent population. Isolation can occur by geographical barriers which can place a group of individuals in a situation where the environment is different, leading to different selective pressures and, eventually, different species. This type of speciation is typically divided into allopatric (vicariant) and peripatric speciation. Vicariant speciation refers to instances where the geographical ranges between organisms do not overlap, whereas peripatric speciation refers to instances where a group becomes separated without interaction by a distinct geographical range. If the new species goes back to the original population they cannot interbreed. *(Taken from Marko PB,* Encyclopedia of Ecology. *2008:131–138.)*

overlap, and the groups are split by a major geographic barrier (volcano, island, mountain, earthquake, etc.).

Peripatric speciation is sometimes associated with allopatric speciation because it refers to a similar mechanism with small groups. Peripatric speciation refers to instances where a small group becomes separated by a geographical range that is distinct, but the ranges are adjacent (Fig. 1). The three key features of peripatric speciation are: (1) the small size of the population; (2) the strong selection pressures caused by the colonization of a new environment; and (3) the effects of genetic drift on the small population. An offshoot of peripatric speciation is centrifugal speciation, in which the range of species expands and contracts, leaving a small isolated population.

Parapatric speciation

Parapatric speciation is when two subpopulations form at a border between two parent species. The subspecies continue to exchange genes while evolving reproductive isolation mechanisms to become separate species. Broad reproductive isolation occurs when a hybrid zone (junction between the two populations) develops, which may become eliminated later due to a selective disadvantage of one of the groups (Fig. 2). Three characteristics define parapatric speciation: (1) populations exist in either continuous or discontinuous geographic ranges; (2) mating occurs nonrandomly across a phenotypic or genotypic gradient; and (3) within the gradient gene flow occurs unequally. The unequal distribution of individuals may result from incomplete geographical barriers, environmental gradients, differences in behavior, strength of selection, or social selection. In natural habitats, variation in the landscape, behavior, and selection can generate spatial

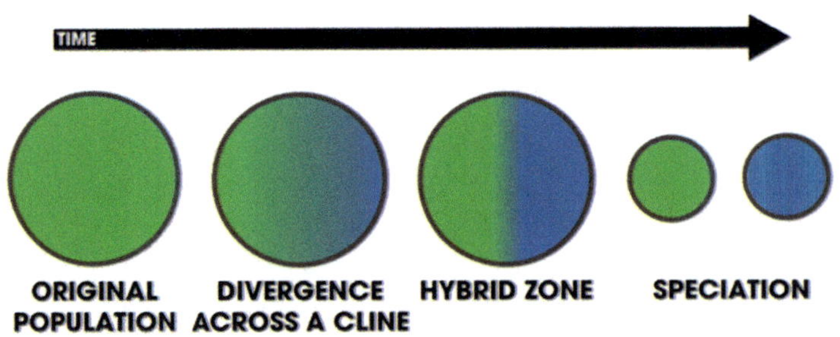

Parapatric speciation is when two subpopulations form at a border between two parent species. The subspecies continue to exchange genes while evolving reproductive isolation mechanisms to become a separate species. The diagram represents a population subject to a selective gradient of phenotypic or genotypic frequencies (a separate but adjacent niche). Each end of the gradient undergoes different selective pressures. Reproductive isolation occurs upon the formation of a hybrid zone which effectively completes speciation.

Fig. 2 Parapatric speciation. *(From Wikipedia, the free encyclopedia.)*

variation that can provide the starting point for genetic divergence, leading to parapatric speciation.[31–33]

Although parapatric speciation is thought to be rare, mathematical models, observational evidence by Coyne and Orr, and laboratory studies by Odeen and Florin have confirmed its existence in nature.[34,35] Examples of situations resulting in parapatric speciation are clines that contain hybrid zones, adjacent populations that have evolved reproductive isolation due to differences in flowering time, pathogens that are the product of host jumping or that adapt to irregularly distributed host species, and endemic species that exist in specialized habitats next to sister species in nonspecialized habitats.[36,37]

Sympatric speciation

Sympatric speciation is the evolution of a new species from a surviving ancestral species. At the same time, both continue to inhabit the same geographic region, and one finds a new niche within the original territory. In this case, organism ranges overlap, and the organisms are often closely related (sister species). Although sympatric speciation falls under one of the modes of geographical speciation, there is no geographical constraint to interbreeding in contrast to allopatric speciation where there is complete spatial segregation (see Fig. 3).

Similar to parapatric speciation, sympatric speciation is rare but does occur in nature, commonly among plants. One contributor to sympatric speciation is polyploidy. Plants can tolerate substantial changes in the number of chromosome sets (polyploidy). Many plants undergo polyploidy through hybridization between two species, but still occupy

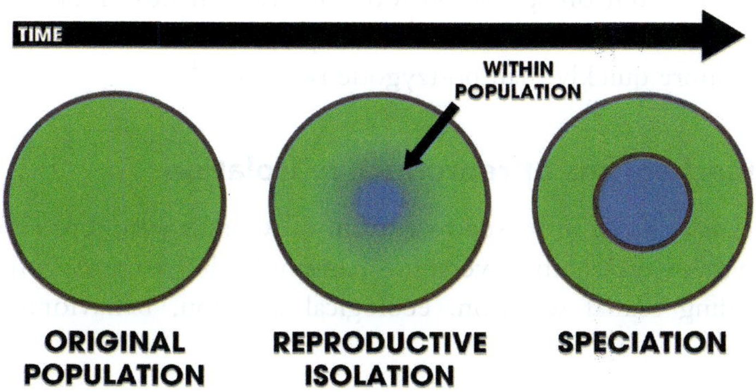

Sympatric speciation is the evolution of a new species from a surviving ancestral species. At the same time, both continue to inhabit the same geographic region, and one finds a new niche within the original territory. In this case, organism ranges overlap, and often times the organisms are closely related (sister species). There is no geographical constraint to interbreeding in contrast to allopatric speciation where there is complete spatial segregation.

Fig. 3 Sympatric speciation.

the same environment. Polyploidy has been recognized as an important evolutionary phenomenon in plants, but less so in animals, where it is not compatible with the complex development and sexual mode of reproduction. Although polyploidy occurs most often in plants, there are some examples from animals, particularly in fish and amphibians.[38]

Polyploidy can eventually lead to reproductive isolation and speciation, and is probably one of the fastest methods of reproductive isolation. Sympatric speciation may also be more common among bacteria because of asexual reproduction and horizontal gene transfer.[39,40]

Homozygous individuals in a population could also have a selective advantage over heterozygotes due to greater fitness because they have accumulated the allelic combinations that are best suited to the environment. Homozygosity could then lead to speciation. This form of disruptive selection could occur due to phenotypic changes caused by genomic alterations in the population.

As was pointed out previously, a major component in the formation of new species is reproductive isolation, which can be caused by prezygotic and postzygotic barriers. Prezygotic barriers prevent animals from mating by temporal isolation (different reproductive seasons), ecological isolation (occupying different habitats or mating environments), behavioral isolation (not recognizing mating rituals or calls), or mechanical isolation (physically unable to mate due to size or negative immune or physiological responses to sperm viability). Postzygotic barriers prevent the formation of viable embryos or viable offspring, or result in the formation of sterile hybrids. This can involve gamete incompatibility (sperm and egg pro-nuclei cannot combine to form a nucleus) or zygotic mortality (zygote dies after fertilization).

Much of the research on speciation today focuses on how reproductive isolation mechanisms evolve. Studies with birds and *Drosophila* have shown that prezygotic isolation can evolve more quickly than postzygotic isolation.[41]

Prezygotic mechanisms of reproductive isolation

Several theories have been proposed to explain how prezygotic reproductive isolation evolved. These theories describe several mechanisms leading to prezygotic reproductive isolation, including sexual selection, ecological isolation, behavioral isolation, and gametic isolation.

Sexual selection

One mechanism that explains prezygotic isolation is sexual selection. Sexual selection is variation in mating success (measured by the number of offspring produced) among individuals with different phenotypes within a population. Selection could occur due to males competing for mates, females preferentially mating with specific males, or individuals from geographically isolated regions no longer finding each other sexually attractive. Sexual selection may explain the number of species within biological groups where some biological orders have hundreds of thousands of organisms within a single order, while others have only a handful.[42]

Sexual selection among animals is typically due to female preferences for certain males, and several mechanisms of sexual selection can generate diversity and reproductive isolation (a component of speciation). The mechanisms include: (1) differences between environments which include sensory drive (how communication signals are recognized and easily detected); (2) good genes, which involves the selection of traits that appear to impart fitness; and (3) sexual dimorphism and interactions between the sexes (Fisher's runaway model and sexual conflict). (See Chapter 3 for a more detailed explanation.)

Sensory drive is a process whereby some aspect of the sensory world predisposes individuals to prefer particular communication signals that could include evolved traits for hearing, sight, and smell. These, in turn, can influence mate preference, whether consciously or not. An example of this in humans may be the preference or tolerance of females for certain types of male perspiration smells[42,43] (see Chapter 3).

The "good genes hypothesis" is a form of sexual selection in which females obtain beneficial genes from mating with particular males who have evolved phenotypes that indicate good genes. These are context-dependent benefits—ultimately leading to reproductive isolation from individuals without the desirable phenotypes. In addition, males could have conditional dependent benefits, which make them desirable for mating due to being well suited to a particular locality or environment.

Fisher's runaway sexual selection is a model of sexual selection that explains the exaggeration of both male display traits and female preferences for those displays in the absence of perceived benefits. It appears that the genetic correlation between male traits and female preference creates continual evolutionary exaggeration of both until the cost of producing traits balances the reproductive benefit. In other words, as female preference becomes stronger, males must evolve to satisfy those preferences and be an acceptable mate without sacrificing beneficial traits.[42]

Interactions between the sexes can include sexual conflict when interest in reproduction and strategies for reproduction conflict between the two sexes. This conflict can arise to antagonistic levels during the coevolution of each sex and can include both behavior and morphology. This conflict creates evolutionary change in each sex in response to the other. It can cause the sexes to take different evolutionary trajectories as a consequence of ecological and/or genetic variation and the strong selection imposed by each sex.[42] Another mechanism that explains reproductive isolation is phenotype. There are three main classes of phenotypic reproductive isolation: ecological, behavioral, and gametic.

Ecological isolation

Adaptation to different environments can lead to reproductive isolation. Different habitats can lead to varying levels of adaptation, such as salt tolerance in plants. Certain species of sunflower can tolerate high levels of salt in the soil while others cannot.

The salt-tolerant hybrid species fills a niche that the parental groups cannot. Genetic analysis of the parental species compared to the salt-tolerant species indicates that 17 chromosomal regions contribute to the salt-tolerant phenotype.[41]

Behavioral isolation

Differences in courtship rituals and courtship demonstrations can cause prezygotic reproductive isolation between species. This can result from misinterpretations of sexual signals or differences in chemical signals. Studies have shown that at least five regions of the third chromosome in *Drosophila melanogaster* contribute to the organism's ability to distinguish pheromones of different species of *Drosophila*. Other *Drosophila* species show epistatic effects between the second and X chromosome in pheromone signaling. Color patterns and songs can also be important in courtship rituals, and variation in these can lead to prezygotic reproductive isolation.[41]

Gametic isolation

Prezygotic isolation can occur after mating but before the formation of the zygote (post-mating prezygotic isolation). For example, in abalones (genus *Haliotis*), successful fertilization requires the seminal fluid protein lysin to interact with the egg protein VERL. The sperm protein lysin shows a high degree of variation between abalone species. It appears that positive selection, for the most effective form of lysin, has driven the divergence of these proteins between species.[41]

The genetic components of speciation occur in stages: (1) the geographical separation of a small population that becomes reproductively isolated; and (2) hybrids become less fertile and eventually become sterile (most likely due to failure of meiosis in the hybrid). In vertebrates, hybrid sterility typically occurs in the male first and is associated with fertility genes on the sex chromosomes. The last stage in speciation, when organisms from the two groups fail to produce offspring, is a product of postzygotic barriers. These stages, along with the principles of population genetics proposed by Haldane, Fisher, and Sewell Wright, have contributed to our current understanding of speciation.[44,45]

Reproductive isolation can occur as a result of major mutations or an accumulation of small-effect mutations, and natural and sexual selection can drive mutations to a high frequency in a population. These mutations can produce both prezygotic and postzygotic barriers by altering individual protein function, or protein networks and their outputs. Researchers have identified several genes associated with reproductive isolation. Examples include *Hmr* (male hybrid rescue), *Lhr* (lethal hybrid rescue), *Nup96* and *Nup160* (nucleoporin 96 and 160[44]), and *Hst* (hybrid sterility[45,46]). Over long periods of time, chromosome rearrangements (duplications, deletions, inversions, and translocations) can occur, resulting in reproductive isolation and speciation. All of these rearrangements can occur through transposons and crossing over (Fig. 4).

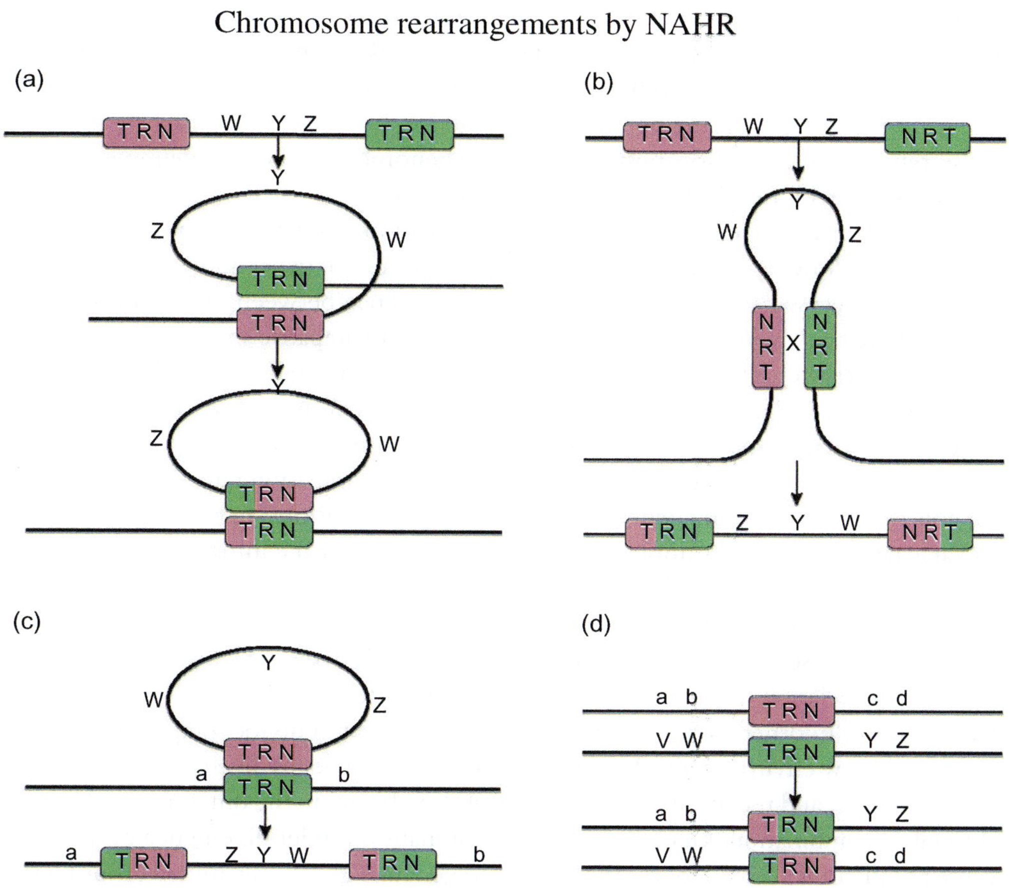

Fig. 4 Nonallelic homologous recombination and its function in chromosome rearrangements. Figure represents the various types of recombination that can occur in a nonallelic homologous recombination (NAHR) resulting in a deletion, inversion, insertion and translocation. The figure shows a transposable element (TRN) that can act as a sight for NAHR. (A) Shows a chromosome loop forming as a result of two homologous TRN sequences existing as direct repeats on the chromosome. Recombination between the elements results in a deletion (ring chromosome) of the sequence between the two elements and a hybrid (red-green) transposon. (B) Shows the recombination between two inverted elements along the same chromosome resulting in an inversion of the intervening sequences. (C) Shows a ring chromosome with a TRN element and a nonhomologous chromosome with a TRN element. Recombination between the two TRN elements results in an insertion of segmem *ZYW* into the linear chromosome. (D) Shows two nonhomologous chromosomes (*abcd* and *vwyz*) lining up as a result of two TRN sequences on the different chromosomes. Recombination between the TRN sequences results in a chromosome translocation, where *a* and *b* are now linked to *y* and *z*, and *v* and *w* are now linked to *c* and *d*.

Comprehending reproductive isolation requires understanding the evolution of the genetic barriers to gene flow between populations. Researchers are just beginning to understand how speciation genes are organized within the genome, and how genes and genomes evolved during population divergence. In other words, how are genes that cause reproductive isolation positioned in the genome, and how is genome structure linked to the evolution of individual speciation genes?

When considering cases where geographic isolation separates groups (allopatric speciation), not allowing for sexual interaction, the position of genes and genome structure may not be necessary to the establishment of divergence between populations. In situations where taxa overlap geographically (sympatric or parapatric speciation), or secondary contact between allopatric species, and exchange of genes takes place, the architecture of the genome can be very important during the process of speciation. This is because in order to form a new species, the groups must overcome gene exchange due to migration and hybridization. In these cases, gene linkage and recombination rates can be important for enhancing selection and genetic drift in differing populations. In the case of sympatric or parapatric speciation, genome architecture is essential for the initial stages of isolation, whereas in the case of secondary contact, genome architecture is important in maintaining reproductive isolation.[47]

The term "selection-recombination antagonism" describes how recombination breaks up associations between selected loci in populations and loci that cause reproductive isolation. In other words, divergent selection will build up loci helping populations to adapt to new environments. In contrast, other loci will be involved in assortative mating, gene flow, and recombination, where each may break up the different combinations through recombination. Gene flow and recombination hinder divergence because they cause populations adapted to specific habitats to generate genetically mixed offspring with locally adapted populations. The only way that genetic mixture could be avoided in this case is if genes were so tightly linked that recombination was limited.

Another mechanism leading to selection-recombination antagonism is when a population of organisms mates in a unique habitat, reducing gene flow with others that use a different habitat. If selection is strong and migration nonrandom, linkage disequilibrium may be established for genes based on survival and habitat without linkage. This "multifarious selection" affecting multiple loci could reduce gene flow enough to maintain divergence across the genome.[47]

Hybrid speciation

Hybridization has often been viewed as a destructive force that can diminish successfully established gene pools, but it is now being recognized as a potential force in evolution because it can lead to novel genotypes, some of which could have the potential for rapid adaptation to new environmental conditions.[48]

Although several modes of speciation invoke the formation of sterile hybrids between the parent group and isolated group, reproductive isolation and speciation episodes of gene flow between species can sometimes lead to new genetic combinations. Natural hybridization between closely related groups is a common phenomenon in both plants and animals, and hybridization is increasingly being recognized as a potential force that could lead to a mixture of novel genotypes that have the potential for rapid adaptation to new environments. This type of hybridization, leading to viable offspring, could generate new species because the hybrids could display novel combinations of traits that induce reproductive isolation from their parental species. Although hybrid speciation is not common, some animal species, such as fruit flies and *Heliconius* butterflies, along with some fish, one marine mammal, one form of dolphin, and a few birds are the result of hybrid speciation.[48,49]

Schumer et al.[50] proposed a new model for hybrid speciation, which does not incorporate inbreeding or assume that hybrids have an ecological or reproductive fitness advantage over parental populations. Their simulated model showed that reproductive isolation between hybrids and parental species could evolve frequently and rapidly, even in the presence of substantial ongoing immigration from parental species and strong selection against hybrids. In their model, both species contribute equally to the hybrid population. In subsequent generations, recombination breaks up genetic ancestry blocks and selection drives the fixation of parental genotypes at incompatibility loci. In some proportion of cases, incompatibility loci pairs become fixed for opposite parental species genotypes.

Schumer et al.'s nonadaptive model generates patterns of species diversity and relatedness resembling adaptive radiation (diversification of organisms into forms filling specialized ecological niches). The authors reveal several known hybrid species (sparrows and sunflowers) that exhibit patterns of reproductive isolation consistent with their model.

In other studies, hybrid populations derived from three species were more likely to evolve reproductive isolation than those derived from two species. This supports arguments that hybridization-supplied genetic diversity can lead to the evolution of novel "adaptive systems" and promote speciation. Researches concluded that whether homoploid hybrid speciation (hybridization without change in chromosome number) is a common mechanism of speciation remains an unanswered question.[51]

Neanderthal human introgression and the concept of speciation

Evidence indicates that anatomically modern humans (AMH) who evolved in Africa most likely did so as a result of hybrid speciation and possibly parapatric speciation as they moved and interacted with a variety of hominins.

Forty years ago, scientists felt they had solved the puzzle of human evolution. The evidence pointed to AMH originating in East Africa and then moving out of Africa to occupy Europe, Eurasia, and East Asia before moving on to inhabit the rest of the world. Some of this evidence revolved around finding Oldowan stone artifacts and hominin-induced stone tool cutmarks dated to 2.6 mya found in East Africa.

However, in 2018 a north African site (Ain Boucherit) yielded Oldowan tools estimated to 2.4 mya ago, suggesting that northern Africa and the Sahara may yield further ancient archeological materials. The distance from East Africa to Ain Boucherit implied either very rapid expansion of stone tool manufacture from East Africa to other parts of the continent or a possible multiple-origin scenario of ancestral hominins and stone technology.[52]

Over the last decade, the timeline of human evolution and the activities leading to the evolution of AMH have shifted from being those of a simple tree as new species evolved and branched off to something more complex. More and more ancient human species have been found to survive to within the last 200,000 years, suggesting that in some places, at least, there may have been significant overlap between our own ancestors and these ancient species. Southeast Asia, for example, was once home to numerous species of human. In an early migration out of Africa, *Homo erectus* reached China and Indonesia, Denisovans reached the vicinity of Papua New Guinea, and Neanderthals occupied Europe and Eurasia.

Fossils of *H. floresiensis*, *H. luzonensis*, and *H. naledi* from Indonesia, the Philippines, and South Africa, respectively, further substantiate the theory that different human species arose as humans moved to different geographical regions. This instinct to travel was probably inherent from the beginning and resulted in hybridization with other individuals whom they encountered. This can be seen, for example, in the discovery of hundreds of *Homo naledi* fossils which display a unique mix of AMH and archaic traits.[53]

As we continue to find more intermediate human fossil forms and identify different human species, our understanding of the origins and diversity of our human lineage continues to suggest significant overlap with other ancient species and introgression. Over the last several years, our introgression with Denisovans and Neanderthals has become evident through our ability to identify DNA from these species in AMH. There is also evidence that another unknown ancient species contributed to the genome of many AMH.

Neanderthals are a human species which persisted in Europe and Eurasia through multiple glacial-interglacial cycles in the Middle Paleolithic (MP), which most scholars place between 300 and 35 kya. Although they were present as far north as southern Siberia and as far south as southern Israel, there is no evidence of them in Africa. The term "Neanderthal" is often used as a derogatory term when describing cognitive ability or poise, but detailed studies of Neanderthal lifestyle and their genetics have begun to reveal that they lived a life similar to that of humans of the time. Neanderthals are the best studied of all extinct hominins and are our closest extinct relatives, and research has revealed they may have been more sophisticated than previously thought. While we are closely related to Neanderthals, they are not our direct ancestors, and fossil evidence suggests that they developed separately in Europe, most likely from *H. heidelbergensis*, who are thought to have descended from *H. erectus*. Similarly, *H. erectus* gave rise to the African form of *H. heidelbergensis* (sometimes referred to as *H. rhodesiensis*), which evolved into *H. sapiens* around 300 kya. Although Neanderthals and humans are morphologically distinct, researchers agree they share a relatively recent common ancestor in *H. heidelbergensis* and were capable of interbreeding.

When determining species, most biologists would turn to the biological concept and the ability of individuals to interbreed, resulting in viable offspring. However, when examining the species-level distinction between Neanderthals and modern humans, the chronospecies concept is applicable. As mentioned previously, this concept determines species by examining continual and uniform changes in a sequential developmental pattern leading to physical, morphological, and/or genetic changes that make it distinct from other ancestors.

The distinction of species from the fossil record is difficult. Deciding if a new fossil is a member of an already described species, a subspecies, or if it should become a new species requires a close examination of morphology, geographical location, and possible DNA analysis. Even after comprehensive analyses, there can be confusion due to variation within species caused by differences in gender, age, nutrition, environmental influences, and allele distribution.

Using the chronospecies concept, the majority of scientists have concluded that Neanderthals were a distinct species from modern humans even though there is evidence of interbreeding between the two. This conclusion is based on the anatomical differences between the divergent developmental patterns of humans and Neanderthals[54] as children[55] and mitochondrial DNA and Y-chromosome analysis.[56] In addition, the argument of large temporal and geographical separation (with Neanderthals evolving in Europe and humans in Africa) adds to the separate species conclusion.

On the other hand, researchers who subscribe to the biological concept of speciation have suggested that humans and Neanderthals are the same species or at least a subspecies. Currently, humans, except those of sub-Saharan Africa descent, contain 1%–4% of Neanderthal DNA. Studies of anatomically modern humans from Europe dating from 38 to 36 kya indicate that those individuals contained even higher levels of Neanderthal DNA than modern Europeans.[57,58] The findings suggest that Neanderthal DNA was mixed and lost over successive generations, most likely through natural selection, leading to the low percentage of Neanderthal DNA found in present-day Europeans.

Studies have indicated that Neanderthal DNA entered the human gene pool during several distinct time periods between 50 and 60 kya.[59] There appear to have been at least two admixture events between Neanderthals and the ancestors of present-day people. The first probably occurred during modern human migration out of Africa, and a second event between the ancestors of present-day Asians.[60] In between those interbreeding episodes, there appears to have been periods where human and Neanderthal populations lived side by side, sharing the same environment and geographical regions, adopting similar tool technologies and even some cultural practices, with little evidence of successful interbreeding.

These intermittent periods of interbreeding suggest that there may have been some reproductive barriers already in place or evolving barriers preventing continuous gene flow between the two populations. The most likely explanation is that genetic compatibility between modern humans and Neanderthals decreased over time, until finally almost all, or possibly all, genetic exchange ceased. If, in fact, the process of reproductive isolation was complete by the time Neanderthals became extinct, they would legitimately be considered a separate species by most investigators; however, others may conclude that they should be classified as a subspecies.

The potential reproductive barriers between humans and Neanderthals have been revealed by sequencing mitochondria and the Y-chromosome DNA of humans and Neanderthals. A study by Mendez et al.[61] examined Y-chromosome sequences of humans and Neanderthals, and estimated that the Y-chromosomes of humans and Neanderthals diverged around 806–447 kya. Their estimate suggested that the Y-chromosome divergence mirrors the population divergence of Neanderthals and modern human ancestors. The data also showed that the Neanderthal Y that Mendez et al. describe has never been observed in modern humans, indicating that the Neanderthal Y-chromosome is most likely extinct. This further suggests that the Y-chromosome of the male Neanderthal proved to be unviable in male hybrids, possibly due to incompatibility with other human genes.

In addition, the study by Mendez et al.[61] suggested a possible reason why the Neanderthal Y chromosome may have created conditions that frequently led to miscarriages when a Neanderthal male and modern human female interbred. They discovered three gene mutations on the Y chromosome of a Neanderthal (*PCDH11Y*, *TMSB4Y*, and *KDM5D*) that can produce minor male-specific histocompatibility antigens; these could have produced damaging changes in protein structure that can trigger immune responses from women during pregnancy, leading to miscarriages.

In other studies, comparing mitochondrial DNA from humans and Neanderthals, investigators determined the divergence time for mtDNA to be around 498–295 kya[56,62,63] which is considerably younger than the time estimates obtained from autosomal and Y-chromosome data. The Neanderthal and modern human sequences differed by approximately 27.2 substitutions, and Neanderthal mtDNA has never been found in modern humans.

The mtDNA data indicate that interbreeding between Neanderthal females and human males must have resulted in sterile female offspring or inviable hybrids, while Y-chromosome data indicate Neanderthal males mating with human females must have resulted in sterile males or spontaneous abortion of fetal males. Therefore, the most likely offspring that resulted from interbreeding were hybrid females that resulted from intercourse between Neanderthal males and human females, i.e., only female hybrids proved to be fertile, so no Neanderthal Y chromosomes were passed on. This outcome highlights a concept known as Haldane's Rule, which states that if one of the genders in a hybrid species turns out to be sterile, that gender is most likely the heterogametic sex (the gender with different sex chromosomes).

Another criterion for distinguishing different species between two groups is the extinction of one group. Because an organism's gene combinations are constantly confronting environmental changes and potential dangers, some organisms will possess characteristics that make them superior while competing for resources (species selection). When considering humans and Neanderthals, heritable variation in traits that affect fitness at the species level will have potential consequences. Adaptation to environmental changes and heritable characteristics, such as intelligence, and the neural development for cooperation have been mentioned as possible advantages that may have led to humans outcompeting Neanderthals for resources; this allowed them to pass on their genes while Neanderthals became extinct. Over time, those humans who represented gene combinations that conferred superior survival and breeding competence passed on the gene combinations leading to forms, behaviors, and ultimately the species we recognize today as anatomically modern humans.

Although natural selection can explain some of the forces responsible for our evolution, natural selection is not predictive of future species because of its reliance on the environment. Species interact with the environment, and nobody can predict what environment our species may encounter over time or what settings Neanderthals might have faced if they had survived.[30]

Neanderthals did not simply constitute a separate species lineage, but formed part of a larger pervasive European and eastern European clade to which distinctive fossil forms belong. This clade occurred in Europe and parts of Asia with the larger and more multiregional species *Homo heidelbergensis*, indicating that hominid history is more complicated than merely that of a single lineage evolving toward the terminal species *Homo neanderthalensis*.[64]

Humans and Neanderthals were not the only humans to interbreed. Evidence for a Denisovan Neanderthal hybrid came from a 90,000-year-old flake of bone found in a Russian cave. DNA evidence showed that the fragment had equal amounts of Denisovan and Neanderthal DNA, and that that it came from an individual who had a Neanderthal mother and a Denisovan father. The mother came from a population that lived later in

Europe, suggesting that migrations of Neanderthals between eastern and western Eurasia occurred sometime after 120 kya. The finding suggests that interbreeding between Late Pleistocene hominin groups was common when the mother and father encountered each other.[65]

In sum, research has shown that modern humans overlapped with Neanderthal and Denisovan populations for a period, and that they interbred. As a result, many people today have small amounts of genetic material from these ancient ancestors. Neanderthals, who interbred with *H. sapiens* in Asia, resulted in a higher admixture of Neanderthal DNA in East Asian populations. Geographical variation in degree of interbreeding between *H. sapiens* and Neanderthals likely contributed to the amount of DNA in AMH and to neurological and behavioral differences in modern humans. Similarly, people with Denisovan genetic admixture were better able to dwell in mountainous regions, allowing them to cross the Himalayas and persist in Southeast Asia and Oceania. In the sub-Saharan region, unaffected by Denisovan or Neanderthal interbreeding, *H. sapiens* most likely interbred with *H. heidelbergensis*.[66]

Conclusions

The concept of species is one of the major questions that occurs when describing evolutionary transitions. The definition of a species has perplexed scientists and philosophers for more than a century. Although the Latin word *species* translates to the word "kind," definitions encompass a wide range: a group of similar or closely related individuals; a specific taxonomic unit; a group of individuals capable of breeding and reproducing viable offspring; or a unit of biodiversity. From a biological perspective, there is no "kind." The difficulty in defining a species comes from the fact that a species is just a snapshot of an individual or group of individuals over evolutionary time so that even members of the same species do not always look alike, act alike, or interbreed.

Although biologists disagree on a specific definition of species, there is an agreement that the key events in evolution are those that create new species. Thus, change within a species, "speciation," is a fundamental aspect of evolution. Speciation is an opportunity to form a new species, an imaginary end point of variation at a particular point in time, with an unknown long-term outcome.

Thus, when humans try to look at populations over long periods, they are observing genetic exchanges between individuals in the population that are continually shifting. Although scientists have been attempting to categorize the idea of species into various concepts, against the background of continual variation, all our classification systems break down. The genetic variation, changes in environmental influences, phenotypic variation, and natural selection make it difficult to conceive of a perfect example of a species.

Even when using the strict biological definition of species (reproductive isolation), looking at our own species, it is hard to imagine during our evolution, that there would never have been a point, where our ancestor of one generation would not have been interfertile with another member of its own species from the previous generation.

It seems that every taxonomist is trying to find the perfect species concept that works for all scenarios and types of organisms. Because of this, it has become apparent that when defining a species, scientists have created their own problem through their disagreements and by using different forms of data and analysis.

When examining Neanderthals and humans, the definition of species is blurred by the variation and similarities we see between the two groups. It appears that if Neanderthals had survived, they would have become more distinct and fit better into our definition of species—but we will never know.

No species definition works for populations over long periods of time, and as a result, species classifications are arbitrary and used for convenience. But to communicate effectively and spread ideas, we need labels for things, and so we divide life into discrete categories, even though we now realize that life is a continuous spectrum, and we are only looking at a snapshot in time when naming a species.

References

1. De Queiroz K. Species concepts and species delimitation. *Syst Biol.* 2007;56:879–886.
2. Wilkins JS. What is a species? Essences and generation. *Theory Biosci.* 2010;129:141–148.
3. Bowler PJ. *Evolution: The History of an Idea.* 3rd ed. Berkeley, CA: University of California Press; 2003:177–223.
4. Schluter D. Speciation and macroevolution. In: Losos JB, et al., eds. *The Princeton Guide to Evolution.* Princeton University Press; 2014:483–485.
5. Mallet J. A species definition for the modern synthesis. *Trends Ecol Evol.* 1995;10:294–299.
6. Beheregaray LB, et al. Ecological speciation in the tropics: insights from comparative genetic studies in Amazonia. *Front Genet.* 2015. https://doi.org/10.3389/fgene.2014.00477.
7. Barraclough TG. Speciation patterns. In: Losos JB, et al., eds. *The Princeton Guide to Evolution.* Princeton University Press; 2014:496.
8. Nagy ES. The selection for native characters in hybrids between two locally adapted plant species. *Evolution.* 1997;51:1469–1480.
9. Zelnio K. Species concepts. *Sci Am.* 2012;(April). https://blogs.scientificamerican.com/evo-eco-lab/species-concepts/.
10. Carney SL, Flores JF, Orobona KM, Butterfield DA, Fisher CR, Schaeffer SW. Environmental differences in hemoglobin gene expression in the hydrothermal vent tubeworm, *Ridgeia piscesae. Comp Biochem Physiol B.* 2007;146:326–337.
11. Zachos FE. Species concepts in biology. In: *Historical Development, Theoretical Foundations and Practical Relevance.* Springer; 2016. ISBN:978-3-319-44964-7.
12. Shapiro BJ, et al. What is speciation? *PLoS Genet.* 2016. https://doi.org/10.1371/journal.pgen.1005860.
13. Brooks DR, McLennan DA. Species: turning a conundrum into a research program. *J Nematol.* 1999;31:117–133.
14. Mayr E. What is a species, and what is not? *Philos Sci.* 1996;63:262–277.
15. Wiley EO, Mayden RL. The evolutionary species concept. In: Debate A, Wheeler QD, Meier R, eds. *Species Concepts and Phylogenetic Theory.* New York: Colombia University Press; 2000:70–89.

16. Mayden RL. Consilience and a hierarchy of species concepts: advance toward closure on the species debate. *J Nematol.* 1999;31:95–116.
17. Fitzhugh K. The 'requirement of total evidence' and its role in phylogenetic inference. *Biol Philos.* 2006;21:309–351.
18. Hey J. The mind of the species problems. *Trends Ecol Evol.* 2001;16:326–329. https://doi.org/10.1016/S0169-5347(01)02145-0.
19. Hey J. Why should we care about species? *Nat Educ.* 2009;2(5):2.
20. Simberloff D. A rising tide of species and literature: a review of some recent books on biological invasions. *Bioscience.* 2004;54:247–254.
21. Baskin Y. *A Plague of Rats and Rubbervines.* Washington, DC: Island Press; 2002.
22. Bossard CC, Randall JM, Hoshovsky MC, eds. *Invasive Plants of California's Wildlands.* Berkeley: University of California Press; 2000.
23. Thiele K. It's funny to name species after celebrities, but there's a serious side too. *The Conversation.* 2018. https://theconversation.com/its-funny-to-name-species-after-celebrities-but-theres-a-serious-side-too-95513?xid=PS_smithsonian.
24. Gozlan RE, et al. Effect of microbial pathogens on the diversity of aquatic populations, notably in Europe. *Microbes Infect.* 2006;8:1358–1364.
25. Jernigan DB. Update: public health response to the coronavirus disease 2019 outbreak—United States, February 24, 2020. *MMWR Morb Mortal Wkly Rep.* 2020. https://doi.org/10.15585/mmwr.mm6908e1. ePub: 25 February 2020.
26. Parrish CR, et al. Cross-species virus transmission and the emergence of new epidemic diseases. *Microbiol Mol Biol Rev.* 2008;72:457–470.
27. DeGregori J, Gatenby RA. Darwin's cancer fix. *Sci Am.* 2019;321:52–57. https://doi.org/10.1038/scientificamerican0819-52.
28. Zhang J, et al. Integrating evolutionary dynamics into treatment of metastatic castrate-resistant prostate cancer. *Nat Commun.* 2017;8:1816.
29. Rice WR, Salt GW. The evolution of reproductive isolation as a correlated character under sympatric conditions: experimental evidence. *Evolution.* 1990;44:1140–1152.
30. Prindle DF. *The Politics of Evolution.* New York, NY: Routledge Press; 2015:1–8.
31. Hochberg ME, et al. Socially mediated speciation. *Evolution.* 2003;57(1):154–158.
32. Mallet J. The speciation revolution. *J Evol Biol.* 2001;14(6):887–888.
33. Turelli M. Theory and speciation. *Trends Ecol Evol.* 2001;16(7):330–343.
34. Ödeen A, Florin A-B. Effective population size may limit the power of laboratory experiments to demonstrate sympatric and parapatric speciation. *Proc R Soc Lond B.* 2000;267(1443):601–606.
35. Coyne JA, Orr HA. *Speciation.* Sinauer Associates; 2004:118–124.
36. Antonovics J. Evolution in closely adjacent plant populations X: long-term persistence of prereproductive isolation at a mine boundary. *Heredity.* 2006;97(1):33–37.
37. Butlin RK, et al. Sympatric, parapatric or allopatric: the most important way to classify speciation? *Philos Trans R Soc Lond B.* 2008;363(1506):2997–3007.
38. Wertheim B, et al. Polyploidy in animals: effects of gene expression on sex determination, evolution and ecology. *Cytogenet Genome Res.* 2013;140:256–269.
39. Otto SP, Whitton J. Polyploid incidence and evolution. *Annu Rev Genet.* 2000;34:401–437.
40. Mable BK, et al. Genome duplication in amphibians and fish: an extended synthesis. *J Zool.* 2011;284:151–182.
41. Orr HA, McNabney D. Genetics of speciation. In: Losos JB, et al., eds. *The Princeton Guide to Evolution.* Princeton University Press; 2014:543.
42. Boughman JW. Speciation and sexual selection. In: Losos JB, et al., eds. *The Princeton Guide to Evolution.* Princeton University Press; 2014:520.
43. Havlicek J, Roberts SC. MHC-correlated mate choice in humans: a review. *Psychoneuroendocrinology.* 2009;34:497–512.
44. Sloan PR. Originating species: Darwin on the species problem. In: Ruse M, Richards RJ, eds. *Cambridge Companion to the Origin of Species.* Cambridge UK; New York: Cambridge University Press; 2009:67–86.

45. Bard J. *Principles of Evolution: Systems, Species and the History of Life*. New York, NY: Garland Science; 2017:237–251.
46. Pilder SH, et al. Hybrid sterility-6: a mouse t complex locus controlling sperm flagellar assembly and movement. *Dev Biol*. 1993;159:631–642.
47. Seehausen O, et al. Genomics and the origin of species. *Nat Rev Genet*. 2014;15:176–192.
48. Nolte AW, Tautz D. Understanding the onset of hybrid speciation. *Trends Genet*. 2010;26:54–58.
49. Mallet J. Hybrid speciation. *Nature*. 2007;446:279–283.
50. Schumer M, et al. Reproductive isolation of hybrid populations driven by genetic incompatibilities. *PLoS Genet*. 2015. https://doi.org/10.1371/journal.pgen.1005041.
51. Comeault AA, Matute DR. Genetic divergence and the number of hybridizing species affect the path to homoploid hybrid speciation. *PNAS*. 2018;115:9761–9766. https://doi.org/10.1073/pnas.1809685115.
52. Sahnouni M, et al. 1.9-million- and 2.4-million-year-old artifacts and stone tool–cutmarked bones from Ain Boucherit, Algeria. *Science*. 2018;362:1297–1301.
53. Berger LR, et al. *Homo naledi*, a new species of the genus *Homo* from the Dinaledi Chamber, South Africa. *elife*. 2015;4. https://doi.org/10.7554/eLife.09560, e09560.
54. Harvati K, et al. Neanderthal taxonomy reconsidered: implications of 3D primate models of intra- and interspecific differences. *Proc Natl Acad Sci USA*. 2004;101:1147–1152.
55. Rosas A, et al. The growth pattern of Neandertals, reconstructed from a juvenile skeleton from El Sidrón (Spain). *Science*. 2017;357:1282–1287. https://doi.org/10.1126/science.aan6463.
56. Rieux A, et al. Improved calibration of the human mitochondrial clock using ancient genomes. *Mol Biol Evol*. 2014;31:2780–2792. https://doi.org/10.1093/molbev/msu222.
57. Seguin-Orlando A, et al. Genomic structure in Europeans dating back at least 36,200 years. *Science*. 2014;6213:1113–1118. https://doi.org/10.1126/science.aaa0114.
58. Lazaridis L, et al. Ancient human genomes suggest three ancestral populations for present-day Europeans. *Nature*. 2014;513:409–413.
59. Green RE, et al. A draft sequence of the Neandertal genome. *Science*. 2010;328:710–722.
60. Fu Q, et al. Genome sequence of a 45,000-year-old modern human from western Siberia. *Nature*. 2014;514:445–449.
61. Mendez FL, et al. The divergence of Neandertal and modern human y chromosomes. *Am J Hum Genet*. 2016;98:728–734.
62. Fu Q, et al. A revised timescale for human evolution based on ancient mitochondrial genomes. *Curr Biol*. 2013;23:553–559.
63. Posth C, et al. Deeply divergent archaic mitochondrial genome provides lower time boundary for African gene flow into Neanderthals. *Nat Commun*. 2017;8:16046.
64. Tatersall I, DeSalle Bones R. *Brains and DNA: The Human Genome and Human Evolution (Wallace and Darwin)*. Bunker Hill Publishing; 2007.
65. Slon V, et al. The genome of the offspring of a Neanderthal mother and a Denisovan father. *Nature*. 2018;561:113.
66. Nosil P, Felder JL. Genomic divergence during speciation: causes and consequences. *Philos Trans R Soc*. 2012;367B;332–342.

CHAPTER 3

The evolution of sex

An intellectual is a person who's found one thing that's more interesting than sex.
Aldous Huxley

In broad terms, reproduction can be either asexual or sexual. Asexual reproduction arises from a single organism giving rise to offspring that inherit the same chromosomes and genes of the parent. The first form of reproduction was asexual resulting from the fission of a single-cell organism into two daughter cells (Fig. 1). Asexual reproduction is the primary form of reproduction for most single-cell organisms; however, with the exception of mammals and birds, there are examples of asexual species in almost every taxonomic group including reptiles, fish, insects, and plants. Asexual reproduction can occur through fission, budding (resulting in a mother and daughter cell), vegetative propagation, spore formation, and parthenogenesis/apomixis (where a haploid cell develops into a new organism).

Asexual organisms can be either diploid (two sets of chromosomes) or haploid (one set of chromosomes). Diploid individuals reproduce asexually without recombination (through the process of mitosis), and there is evidence that some diploid individuals that reproduce asexually may have reproduced sexually in the past.[1] Deleterious or beneficial recessive mutations will be masked by dominant alleles in an asexual diploid organism. Asexual individuals that are haploid will express beneficial or deleterious alleles regardless of their form. At first it may seem that diploid individuals have an advantage in that they can accumulate large numbers of recessive deleterious alleles, whereas haploid individuals with a single deleterious allele will not survive. However, selection works much faster in haploid individuals, whereas selection in diploid individuals does not take place as quickly when deleterious recessive alleles are not expressed immediately (masked in the heterozygote). This leads to diploid sexually reproductive populations accumulating large numbers of recessive deleterious mutations. This occurs because only 50% of copies of a recessive allele for a particular recessive disease are actually producing the disease phenotype, while the other 50% are preserved in the population through unaffected, heterozygous carriers.

The first eukaryotes to engage in sex were single-celled protists that appeared approximately 2 billion years ago, more than 1.3 billion years before the development of the first

Sex and Cohabitation Among Early Humans
https://doi.org/10.1016/B978-0-12-819086-9.00007-5

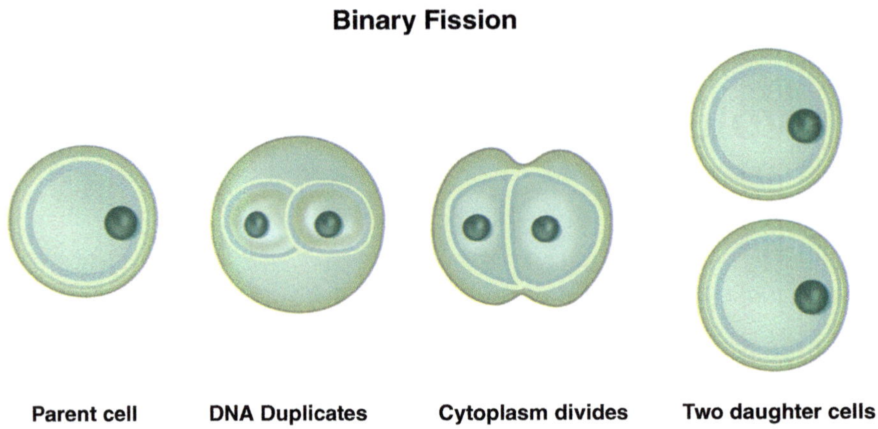

Fig. 1 Asexual reproduction (https://byjus.com/biology/binary-fission/).

animals. The first single-cell organisms that engaged in genetic exchange used conjugation, transformation, and transduction, all of which fall under the umbrella of parasexuality. Parasexuality involves genetic exchange without meiosis and no formation of a zygote. Conjugation is the transfer of genetic material between cells by direct cell-to-cell contact or by a bridge-like connection between two cells (see Fig. 2). In transformation, a cell takes up a piece of DNA from its environment, whereas in transduction, DNA is moved from one cell to another by a virus.[2]

Today most eukaryotic sexual reproduction results from combining genetic information from two individuals of different types (male and female). In many eukaryotes, male and female genders are determined by specific sex chromosomes or by a combination of genes that the individual inherits, depending on the organism in question. In some cases, a single organism can have both male and female reproductive tissues (hermaphrodite). Scientists estimate that today, more than 99% of eukaryotes can reproduce sexually; however, some organisms, such as aphids, and yeast have the ability to be both sexual and asexual. (Gender determination is discussed later in this chapter.)

Although most eukaryotes reproduce sexually, the evolutionary rationale for sexual reproduction is not clear and many hypotheses have been proposed to try to explain its existence.[3] The evolutionary advantage of sexual reproduction, and the complex recombination and rearrangement of chromosomes that accompany sex, beg the question: why donate half of your genetic inheritance to a partner and shuffle (and potentially alter) beneficial combinations of alleles, which allowed you to survive and reproduce?

In asexual species, a female reproduces without the genetic contribution from a male, making identical genetic daughters, with the exception of rare mutations. This reproductive mechanism allows the organism to increase the size of the population

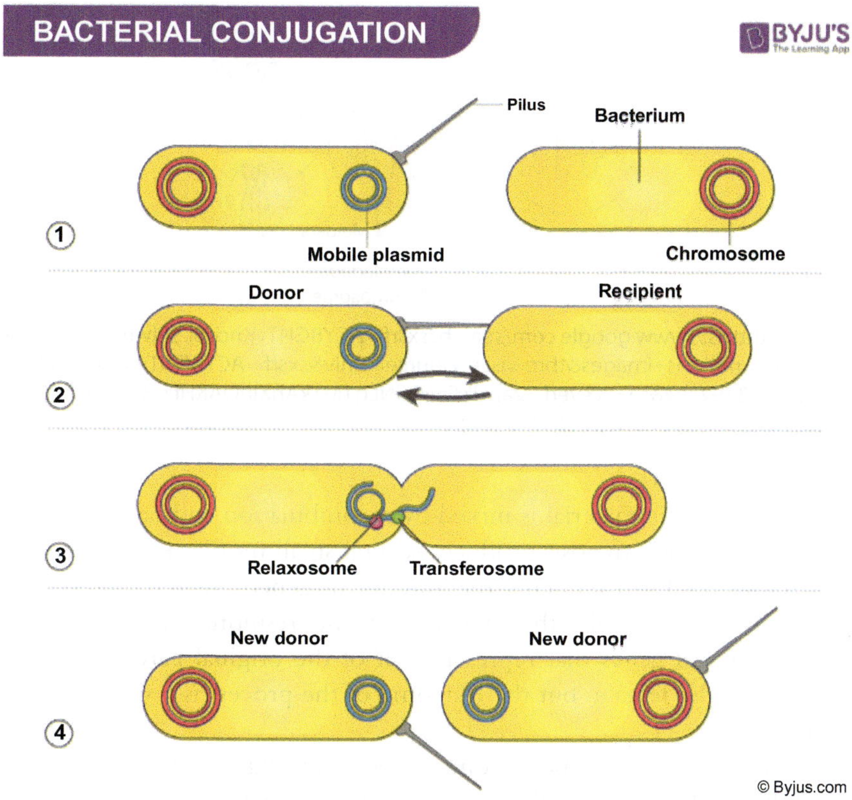

Fig. 2 Parasexuality—conjugation (https://www.google.com/search?sxsrf=ACYBGNS6tPb57cLnsL0lWzl3uKbrN0wSmw:1568034716592&q=conjugation+image&tbm=isch&source=univ&sxsrf=ACYBGNS6tPb57cLnsL0lWzl3uKbrN0wSmw:1568034716592&sa=X&ved=2ahUKEwiYpJzC6MPkAhVrmeAKHR6lA9lQsAR6BAgJEAE&biw=1383&bih=723#imgrc=XdnJZSwByvNzTM).

quickly by continuously making more females—and making females that have the gene combinations to help them survive. Since females are the sex that invests the most in reproduction (making eggs, providing a safe environment for development, nurturing young), there is no need to produce males.

So why did the majority of eukaryotic species evolve some form of sex or genetic exchange, when it required a new energy-consuming mechanism of cell division (meiosis) in order to make sex cells in two different individuals? Why did they develop two different genders? Why mix up a genome that was already sufficient for survival? Why put energy into finding a mate?

Sex at the cellular level involves a process in which sex cells (sperm and egg) are formed through reshuffling of the organism's DNA and reducing the number of chromosomes by half. The process happens through a special type of cell division called meiosis (Fig. 3).

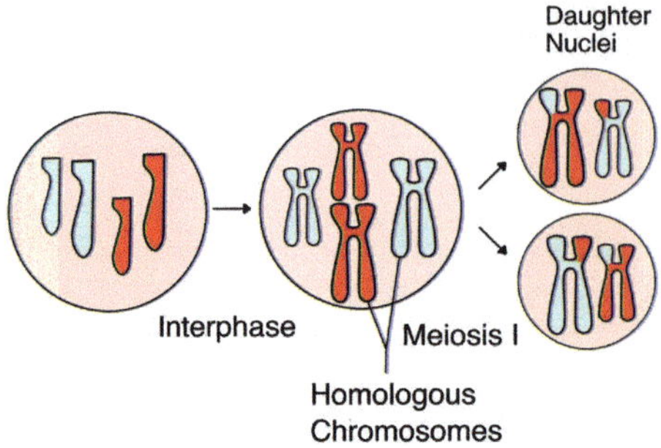

Fig. 3 Meiosis (https://www.google.com/search?sxsrf=ACYBGNTvKndzSLSRlwe7RbXTDjXn4l_pyA:1568063792404&q=meiosis+images&tbm=isch&source=univ&sxsrf=ACYBGNTvKndzSLSRlwe7RbXTDjXn4l_pyA:1568063792404&sa=X&ved=2ahUKEwiQ9NLq1MTkAhXlhOAKHfM3C7kQsAR6BAgHEAE&biw=1383&bih=723#imgrc=swe5CeSV1-U4fM).

During meiosis, genetic material is mixed by recombination and random assortment of chromosomes. In males, meiosis produces several small motile sex cells called gametes (sperm). Meiosis results in reducing the amount of genetic material, found in the rest of the body's cells, by half; this means that the resulting gamete has a mixture of chromosomes and genes that represent half of the original parent's cells. A similar process occurs in the female, but the outcome of the process is a single relatively large gamete: the egg.

When male and female gametes come together during sex, it is referred to as fertilization and the resulting cell is called a zygote. The zygote has a combined genome from the two different parents. This single-cell zygote replicates to form an embryo, and the cells within the embryo divide and eventually develop into an individual. This process is energy-consuming but results in variation of genes, morphology, and chromosomes in the next generation.

From a biological perspective, sex is not necessary and reproduction and variation (through mutation) can go on indefinitely without sex. In a stabilized environment, asexual organisms will outcompete sexual organisms because there is no incentive to adapt quickly. Although sex and recombination during meiosis can increase the amount of variability, this increased variability is not always favorable, and natural selection alone can build up variation in a population. Selection remains the main evolutionary force providing variation, and sexual reproduction and recombination can break down genetic combinations built up by selection. In other words, sex can reduce fitness by mixing together genes from different parents after natural selection has already established the most fit genetic combination. For example, in cases involving multiple genes whenever those genes interact in such a way that intermediate genotypes (+ − + − + −) have lower fitness than the extreme genotypes (+ + + + + + or − − − − − −), mixing genes will lower fitness.[2]

In general, mathematical models have confirmed that selection builds more variation than expected from randomly combined genes whenever intermediate genotypes have lower than expected fitness. In such cases, sexual reproduction and recombination could destroy the genetic associations that selection has built, and therefore could result in decreased (rather than increased) variation among offspring. Even when sexual reproduction does restore genetic variation, producing more variable offspring, it does not necessarily support the evolution of sexual reproduction. This refutes one of the arguments often raised in an attempt to explain the relationship between sex and evolution. For example, if heterozygotes are more fit than homozygotes, heterozygotes will increase in the population; however, sexual reproduction regenerates homozygotes from crosses among heterozygotes. Although sexual reproduction results in greater genetic variation among offspring, the variation will still result in the increase of homozygotes with low fitness (through simple Mendelian genetics).[2]

The advantages of sexual reproduction do not explain how sexual reproduction arose in the first place. The problems with the evolution of sexual reproduction (cost of finding a mate, cost of meiosis as a new cell division process, random variability from each parent, development of two different sex types, potential for sexual disease, loss of beneficial genes through recombination, etc.) should have curtailed the evolution of sexual reproduction. In addition, sexual reproduction is so complex that it is extremely unlikely to have evolved by mutation on a frequent enough basis that it would be expected to become fixed by a few organisms, survived obvious disadvantages, and spread into different kingdoms. Sexual reproduction would only benefit future generations, and not the population that gave rise to the first sexual reproduction. Therefore, it appears that the selective forces that gave rise to sexual reproduction must have been very different than those responsible for its maintenance and spread once it evolved. This would have required natural selection to retain beneficial biological mutations that were not used immediately, for future generations that had potential future utility. Once those mutations had accumulated to evolve into sexual reproduction, they were initiated and spread.

Sex evolves when selection pressures change over time and space

So why sex? Sex is highly desirable because although sexless organisms may be at a temporary advantage, they cannot evolve at the same rate when environmental readjustments and adaptations are necessary.[4] One instance where increased variability due to sex could be favorable is in increasing the rate of adaptation to parasites, disease, and predators. Continual adaptation to new challenges is known as Red Queen Evolution (if an organism wants to stay where they are in evolutionary terms and survive, they need to stay one step ahead of their competition or challenges).[5] This hypothesis states that organisms must constantly adapt, evolve, and proliferate in order to survive in a changing environment and overcome predators, parasites, etc.

This coevolution, particularly between hosts and parasites, can lead to sustained alterations in genotype frequencies, and forms the core for one of the leading hypotheses for the persistence of sexual reproduction.[6,7] In asexual host species, coevolutionary interactions with predators or parasites may form a strong selection pressure to evolve sexual reproduction as a way to reduce the risk of infection in offspring.[8]

It has been established that recombination frequency varies in many organisms in response to maternal age, temperature, nutritional status, and stress. In addition, it appears that sex enables a lineage to adapt to changing conditions by varying recombination frequency (genetic mixing) in response to parasitic challenges, resulting in offspring that are very different than their parents. Theoretical models indicate that increased recombination can occur if an organism's fitness and recombination frequency are negatively correlated. A negative correlation enables beneficial combinations of alleles on linked haplotypes while providing less fit allele combinations to be disrupted and reassembled into potentially more fit haplotypes.[9,10]

For example, recent studies by Singh et al.[3] observed that in *Drosophila*, offspring showed increased recombination when parents were exposed to bacterial infection or predation by a parasitic wasp. However, the increase in the offspring's genetic diversity was due not to an overall increased rate of recombination during meiosis but to an unequal allocation of gametes (specific gametes selected to form offspring) that had undergone recombination. This suggested that the response was driven by transmission distortion (nonMendelian transmission of genes or chromosomes) rather than an increased recombination rate.

The response occurred in different genetic backgrounds, developmental stages, and challenges by different parasites. In other words, selection occurred for offspring that showed an increased amount of recombination over those that had less recombination. The study by Singh et al extends the Red Queen model to include the increased production of recombinant offspring and revealed the remarkable ability of hosts to distort their recombination fraction in rapid response to environmental cues.

In addition, evidence continues to show that recombination frequency in a variety of species is flexible in response to factors such as maternal age, temperature, nutritional status, and social stress.[11–13] Several theoretical models suggest that flexible recombination can evolve if organismal fitness and recombination frequency are negatively correlated.[9,10] This negative correlation would enable the organism to maintain beneficial combinations of alleles on linked haplotypes while also providing opportunities for less fit combinations of alleles to be recombined into potentially more fit haplotypes.

Yet in principle, asexual reproduction should be more efficient: it avoids the costs of mating and allows individuals to pass all (rather than half) of their genetic material on to their offspring. Extensive theoretical work has sought to understand why sex is pervasive despite these substantial costs. Examining the costs of asexual reproduction, however, may shed some light on how sex may be worthwhile. One example of a high cost in asexual organisms is the phenomenon of clonal selection, which explains why beneficial mutations can take a long time to get fixed or even disappear in asexually reproducing populations. Normally, in asexual organisms a beneficial mutation would be likely to get fixed if it occurred alone, but it may fail to be fixed, or even be lost, if another beneficial mutation arises in a different individual in the same population—i.e., the two individuals with their beneficial mutations and their clones can interfere by competing with each other in the population. For example, if a beneficial mutation *A* arises in an asexual individual within a population, it can only be passed on by clonal replication and the frequency of *A* will only increase slowly in the population. If during this time a different beneficial mutation *B* occurs in the same population, organisms with that mutation will compete with the *A* mutants through natural selection. When two asexual organisms compete with each other in a population, by each attempting to fix different beneficial traits, the beneficial trait of one of these organisms should interfere with that of the other in the population. Eventually one of the beneficial traits is completely lost and one becomes ubiquitous in the population via competition.

This process of competition among asexual organisms must take place in the population to allow the population to adapt through selection, but it introduces inefficiency into the pace of natural selection. Instead of one population changing over time, the two subpopulations (*A* and *B*) are being selected. This is essentially a waste of time compounded by the hitch-hiking of deleterious alleles with the advantageous allele, which ultimately results in fixation of deleterious alleles in the successful subpopulation.

This same scenario in a sexually reproducing population would produce individuals with beneficial mutation *A* and individuals with beneficial mutation *B* and either (*A* or *B*) would be more fit individuals than those in the population without the mutations. When carriers of mutation *A* and *B* mate and produce offspring, the more fit genotype AB can arise. Individuals with genotype AB would have a higher fitness than others in the population and therefore would be likely to produce more offspring, assuming *A* and *B* mutations were compatible and did not produce negative outcomes. In other words, in a sexually reproducing population, the *A* and *B* mutants would not necessarily compete but could end up producing a more fit organism.

To examine the effects of recombination on fitness, McDonald et al.[14] examined the evolutionary dynamics that drive the effects of sex at the genomic level. They compared the DNA sequence-level changes during adaptation in sexual and asexual *Saccharomyces cerevisiae* populations. Using the same organism with different modes of reproduction allowed them to identify the specific molecular mechanisms by which sex speeds adaptation. They found that sex alters the variety of mutations that get fixed, and confirmed that it does so by alleviating clonal interference. Sex provides this relief to clonal interference by bringing together beneficial mutations that arise on different genetic backgrounds that would otherwise compete against one another. They also showed that sexual recombination prevented deleterious mutations from hitch-hiking and becoming fixed in the population, by allowing individuals which possessed one of the adaptive alleles (under selection) to recombine and replace deleterious alleles with less harmful alleles. Their results demonstrated that sex both speeds adaptation and allows natural selection to sort beneficial mutations from deleterious mutations more efficiently.

Another argument for the evolution of sex is that sex inevitably leads to the evolution of cooperation and the formation of groups. Group selection is a proposed mechanism of evolution in which natural selection acts at the level of the group, instead of at the more conventional level of the individual. Group selection occurs when the benefits of cooperation or altruism between subpopulations are greater than the individual benefits of egotism/selfishness within a subpopulation. Group selection does this because those individuals that took on the practice of sexual reproduction could not have taken the risk of mating with individuals that demonstrated unusual physical or behavioral traits that could have resulted in an unhealthy environment for offspring. Group selection would eventually stabilize certain phenotypes and create stable groups with particular strategies and advantages, especially the advantage of cooperation. Selfish individuals and introverts would not be desirable mates and would not interact at the group level, which would make them incompatible with sexual behavior. Once groups were formed, they might become isolated and take on different behaviors. Groups that happened, by chance, to reach a cooperative strategy would be fitter than those that disbanded. Establishment of cooperative behavior would then allow the group to defend against alternative mating strategies that might be brought into the population through mutation or immigration, with the ultimate result being the establishment of cooperative groups. Social behaviors such as altruism and group relationships can influence the outcomes of intraspecific competition and interspecific interactions by making the group stronger via looking out for others in the group or forming relationships between cooperating groups. This type of alliance between individuals would result in greater efficiency in hunting, caring for offspring, resource management, division of labor, and passing down information and technology to future generations.[15,16]

This level of cooperation may have also had an influence on brain size and intellect. Gonzalez-Forero and Gardner[17] developed a computer model to examine the evolution of the brain. The data suggested that the evolution of adult *Homo sapiens* brains and bodies occurred when individuals faced a combination of 60% ecological, 30% cooperative, and 10% between-group competitive challenges. They suggest that between-individual competition was unimportant for driving the evolution of human brain size. They contend that human intelligence and the ability to store and share cultural knowledge would have likely carried a high survival value, and that a cooperative culture was part of sexual selection (selecting individuals who worked well within the group). This supports the idea that some of the evolutionary forces involved in brain development and intellect were sexually selected, as an increase in brain size alone would not confer enough fitness relative to its high maintenance cost. An example of this might be the large brains of Neanderthals compared to those of humans, and the fossil evidence suggesting that Neanderthals lived not in a cooperative society but instead in small interbreeding groups. The larger brains of the Neanderthals did not confer a greater level of fitness than those of humans who did live in cooperative groups.[17,18]

Gender determination

Gender determination is fundamental to the life history of sexual species. Yet the extent to which evolutionary changes in the master sex-determining genes trigger changes to maintain proper development of males and females is still not fully understood. Thus, understanding the central gene regulators and environmental influences on sexual development could help us to understand the mechanisms and the key contributors to the evolution of gender determination. Today there is a clear distinction between sex determination (physical attributes of a particular sex) and gender, with discussions involving physical and psychological gender identity, transgender, and gender-neutral individuals. In this chapter we will use "sex determination" and "gender" as equivalent terms, although we understand there is a cultural distinction.[19]

In some eukaryotes, sex chromosome combinations and their gene products direct sex determination, while in other eukaryotes, specific genes (not chromosome combinations) play a role in gender determination. Sex-determining genes (or chromosomes carrying those genes) are usually present in one sex while absent in the other, or alternatively gene protein products exist in different doses between males and females.

Among vertebrates, various sex-determining mechanisms direct male or female development through genes and/or environmental cues (sensitivity to temperature or other environmental conditions). In humans, sex-determining gene products determine the differentiation of gonads into testes or ovaries. Similar genetic processes in other mammals, birds, snakes, and amphibians determine their gonadal development.

In other species, including some fishes, most turtles, and all known crocodilians, sex determination is not limited to specific genes or chromosomes, but instead is most commonly determined by environmental temperature experienced during fetal development. Most investigators agree that this form of sex determination renders species vulnerable to extinction if they are unable to counter the effect of rapid climate change by adapting, or having alternative methods of reproduction such as parthenogenesis.

Although specific genes, chromosome combinations, or environmental influences are the typical methods used for gender determination, other species have mixed mechanisms of sex determination, i.e., components of both the genotype and environmental conditions that work to determine sex. In these cases, temperature may affect gene expression or the activity of gene products. In addition to gender determined by a mixed gene-environment process, there are other variations in how gender is determined. One example is in honeybees, where the fertilization or non-fertilization of eggs determines gender. Fertilized eggs develop into females while male offspring develop from haploid unfertilized eggs[20]; the fertilized eggs, which have two sets of chromosomes, differentiate into queens and female worker bees. This process of sex determination is known as the haplodiploid system. We now know that about 20% of animal species use a haplodiploid mode of reproduction.[21]

Although it is not clear how different sex-determining mechanisms evolved, there is speculation that some genes evolved to attain a master regulatory role within the network and pathways leading to gender determination. Gene evolution may have included alterations in the timing and/or conditions of gene expression, or genes shifting positions within or between chromosomes. Changes in *cis* and *trans* regulatory regions of genes could have had significant effects leading to changes in timing and or conditions under which genes were expressed or silenced, or could have changed the environmental conditions influencing gene expression. Gene activation[22] or genes shifting positions (cytogenetic reshuffling) in the regulatory pathways between roles as master regulators or downstream actors could also influence gender.[23] In addition, changes in the protein coding sequences of genes could lead to alterations in enzymatic rates, cofactor binding, and/or effects on protein structure influencing regulatory pathways.

Transitions leading to sex chromosome evolution could have been characterized by translocations of a master sex-determining gene onto a proto-sex chromosome, whose overall effect on the sex determination network would change the way gender was determined in a species.

Most vertebrates have genes that are essential for gender development and are classified as male-promoting or female-promoting, based on their role on the developing gonads and whether these roles are fairly conserved among species. For example, in humans the SRY gene and gene product (the testis-determining factor) are fundamental in human sex determination, acting as the switch that initiates testis development and thereby male sex determination. In addition to SRY, the SRY-box-related (SOX) genes have been identified on autosomes, and mutations in the SRY or SOX genes have been implicated in sex reversal.[24] There are other sex-determining genes involved in phenotypic sex, and a few of these genes are mentioned below.

The *SOX9* gene plays a critical role during embryonic development and is especially important for development of the skeleton and determination of sex before birth. The CYP19A1 gene provides instructions for making an enzyme called aromatase. This converts a class of hormones called androgens, which are involved in male sexual development, to different forms of the female sex hormone estrogen. The expression of aromatase varies among different cell types, but in females it is most active in the ovaries, acting to guide sexual development. In males, aromatase is most active in fat tissue. In both sexes, estrogen plays a role in bone development and blood sugar levels. The *Dmrt1 gene* plays a key role in male sex determination and differentiation by controlling testis development and sperm proliferation, leading to abundant and continuous production of sperm. Dmrt1 is also involved postnatally by maintaining testis determination and preventing feminization through repression of female-promoting genes and activating male-specific genes.[22,23]

In humans, biological sex (phenotypic sex) is typically determined by: (1) the presence or absence of the Y chromosome; (2) external and internal genitalia; (3) level of sex hormones; and (4) behavior. Sexual dimorphism also plays a role in human sex determination with males typically being taller, hairier, heavier, and having more muscle mass than females.[19] Regardless of how sex determination evolved and how we define it, it has become obvious over the last several years that gender is not a binary trait.

The sex chromosome system in human males typically involves two different sex chromosomes (X and Y), causing males to be known as the heterogametic sex, while females are defined as the homogametic sex (two similar XX chromosomes). The Sex-determining Region on the Y chromosome is referred to as SRY. As mentioned above, the SRY gene and gene product (the testis-determining factor) are fundamental in human sex determination, acting as the switch that initiates testis development and thereby male sex determination during embryo development. In other words, individuals with SRY should develop into males and certain mutations in SRY or the absence of SRY should develop as females; however, there are other sex-determining genes involved in phenotypic sex. Mutations of the SRY gene or structural anomalies of the SRY region on the Y chromosome can result in a variety of conditions including XY gonadal dysgenesis (defective gonad development) or XY females resulting from loss of SRY gene function on the Y chromosome. Recombination of the SRY gene onto an X chromosome can lead to males with two X chromosomes (XX males). In addition to gene or sex chromosome abnormalities, defects of androgen biosynthesis can also lead to sex reversal or intersex. Another complication during sex development is the activation of cell receptors. In order for cells to be activated by hormones, they must have the proper membrane receptors that act as transmitters of information from outside the cell. In some cases, the genes responsible for hormone production are active; however, the cells that should respond to these hormones have defective receptor genes and do not have the proper membrane receptors, and therefore do not respond to the hormones. One example is the response to androgens where androgens are produced but are not received by cell receptors, which can lead to sex reversal or intersex.

Today there is a greater understanding of some of the genes involved in gender determination and the complex metabolic pathways that lead to gender during fetal development. Because there are a large number of genes involved in gender determination, mutations in any of these genes can result in partial or complete failure of sex organ differentiation. In addition, the level of those genes' expression, the timing of gene expression, and the failure of certain developmental cell receptors to recognize sex-determining gene products can lead to variation in the expression of gender and to individuals with intermediate sexual traits, sterility, and gender ambiguity. Today there are distinctions between biological sex and gender. Such a discrepancy is highlighted by the existence of intersex individuals whose psychosexual development and gender sometimes do not match their assigned biological sex. Intersex is defined as a congenital anomaly of the reproductive and sexual system. The Intersex Initiative, a North American-based organization, estimates that one in 2000 children in the United States are born visibly intersex. Although the physical manifestation of intersex (variations of sex chromosomes, gonads, sex hormones, or genitals not typically associated with male and female bodies) may not be directly related to psychosexual development and gender determination, it helps to illustrate the complexity of sexual development and to discredit the idea that sex and sexual identity are binary.

There are several other biological conditions leading to variations in sex determination. Some of the more common conditions are listed here.

Hermaphroditism

This is a genetic autosomal condition in which affected individuals have both mature ovarian and testicular tissue.

Congenital adrenal hyperplasia

This is an inherited autosomal recessive condition that can affect both boys and girls. It is the most common cause of intersexuality in females with XX, where untreated girls develop an outwardly male appearance. The condition results from a genetically caused deficiency of cortisol, produced by the adrenal cortex, and increased androgenic effects due to a compensatory increase in adrenocortical hormone.

Androgen insensitivity syndrome

This syndrome is an X-linked recessive disorder in which affected individuals have external female genitalia and breast development despite being genetically male with both XY. In this case, tissues do not have androgen receptors so do not respond to androgens yet do respond to estrogens.

Turner syndrome

Turner syndrome occurs in individuals who have one X chromosome, no Y chromosome, and are phenotypically female. Affected individuals are short in stature, generally lack prominent female secondary sexual characteristics, are sterile, and occasionally show slight mental retardation.

XXX females

Women with three X chromosomes experience normal development of sexual traits and are fertile. Affected individuals are usually taller than average, have slender builds, and may have slight learning difficulties.

Klinefelter syndrome

Individuals with XXY or XY/XXY mosaics express the male phenotype. Affected individuals are often tall and produce relatively small amounts of testosterone. Males express abnormal testis development, leading to infertility, and have incompletely developed secondary male sex characteristics.

XYY males (Jacobs syndrome)

These individuals are usually taller than average, fertile, and are prone to acne because they produce higher than average levels of testosterone.

Homosexuality

Although homosexuality has been documented in many mammals, it remains an evolutionary enigma because of the need for reproduction in order to sustain the species. Normally any orientation that lowers reproduction in a species will be removed by natural selection. It has been reported that 2%–10% of individuals of both sexes engage in sex with same-sex partners, either exclusively or in addition to sex with opposite-sex partners. Twin and other studies have shown that same-sex behavior has a genetic component, but these studies have not been widely recognized because of the small sample sizes that could realistically determine the genetic contributions for complex behavioral traits. In general, behaviors have both genetic and environmental components with most genetically complex (polygenic), and contributions from the environment (multifactorial). Recent genome-wide association studies have shown that there is no gay gene, but as expected, both heterosexual and homosexual behaviors are polygenic.[25] Recent identical twin studies have also indicated that environmental influences may have an effect on homosexual behavior. Studies by Watts et al.[26] used physiological indicators to determine if identical twins who reported differences in sexual orientation showed differences in arousal when presented with same-sex or different sex genitalia. Their data provided physiological evidence confirming twins' discordant sexual orientations, thereby supporting the importance of the nonshared environment for the development of sexual orientation and sexual arousal.

Social and environmental conditions can also influence sexual orientation and behavior. Androphilia and gynephilia are terms used in behavioral science to describe sexual orientation, as alternatives to the binary homosexual and heterosexual descriptors.

Androphilia describes sexual attraction to men or masculinity; gynephilia describes the sexual attraction to women or femininity. In some cases, androphilia can be recognized in young children before being expressed in adults. Studies by Bartlett and Vasey[27] showed that atypical behavior (male attraction to males and males showing female-typical behaviors) could be detected in childhood and subsequently expressed in adults. They concluded that adult androphilia may be a cross-cultural universal pattern of psychosexual development shared by most males who are predominantly androphilic.

Environmental factors can also influence sexual development. Estrogens are a group of sex hormones that promote the development and maintenance of female characteristics including the growth and development of female secondary sexual characteristics, such as breasts, pubic and armpit hair, and the regulation of the menstrual cycle and reproductive system. Natural and manmade chemicals can mimic the effects of estrogen and influence development. Phytoestrogens, xenoestrogens, and antiandrogens induce estrogen hormone imbalances, can impact sexual development, and can lead to a range of health problems and unwanted physical changes.

Phytoestrogens are natural compounds found in plants and plant-based foods. Xenoestrogens (herbicides, pesticides, PCBs, plasticizers, and polystyrenes) or antiandrogens (polyaromatic hydrocarbons, linuron, and vinclozolin) are found in the environment. All of these compounds can mimic the effects of estrogen and may act as endocrine disruptors that can upset normal hormonal function by binding to cell estrogen receptors.

Phytoestrogens have been linked to some positive effects in women undergoing menopause, but have also shown to increased risk of breast cancer, endometriosis, and endometrial cancer in some women. Results from Sultan et al.[28] indicated that increased exposure to environmental xenoestrogens and/or antiandrogens may be responsible for some demasculinizing effects during male fetal development by disturbing male sexual differentiation and leading to abnormal penis and testicular growth.

Sexual orientation (homosexual, heterosexual, bisexual, or asexual), gender identity (intersex, transgender), and sexual behavior are complex issues showing various gradations that are only now becoming a focus of more biological, psychological, and social research. As investigators find out more about the difference between sexual attraction, sexual identity, and sexual behavior, and the wide spectrum of expression among humans, it has become obvious that no single theory can explain human sexuality.

Menstruation and fertility

Menstruation is essential to human reproduction. It is also one of the biological processes that makes us special because humans, chimpanzees, bats, and elephant shrews are the only animals on earth that go through it. The vast majority of mammals signal fertility through estrus, the period when females are ovulating and display their sexual receptivity via genital swelling, behavioral changes, or pronounced alterations in body odor. The female human body, however, conceals this critical window of reproduction; thus monitoring ovulation, endometrial receptivity, fertilization, implantation, and gestation requires sampling biological and biochemical markers.[29]

The most visible sign of potential fertility is a woman's first production of menstrual blood, which appears at puberty. However, the production of menstrual blood at puberty appears after the first fertile period has closed. The endometrial lining of the uterus thickens over the course of a woman's cycle as her estrogen level rises, and if none of the eggs she releases form an embryo that implants in her uterus, then levels of estrogen and progesterone drop, triggering the uterine lining to shed (menstruation).

Unlike female chimpanzees, which undergo estrus to signal fertility and seek mates, human females are not always aware of when they are most fertile. Ancestral men and women would have had a hard time determining when a female was ovulating and most fertile. Ancestral men and women must have realized that menstruation was somehow associated with fertility or lack of fertility. This recognition of fertility may have led to men's preference for a mate that had been through puberty, knowing that they would be the most fertile.

With all the knowledge of what causes menstruation, scientists continue to ask: why do we share this process with a limited number of species? And why is menstrual blood different from regular blood in that it cannot clot and is full of immune agents? Why do human females shed their endometrial lining when most mammals that experience estrus appear to reabsorb their endometrial linings at the end of each cycle? And why do up to 80% of women experience cramps, bloating, fatigue, anger, or other symptoms just before the onset of menstruation? Very little is known about menstruation, and what scientists do know is often badly communicated with the public, making attitudes toward menstruation overwhelmingly negative.[30]

In the 1980s, Beverly Strassmann proposed that concealing ovulation could entice more paternal partners. She speculated that because a woman's fertile window is more or less invisible, it might encourage pair-bonding—where human males invest in fewer sexual relationships and protect and care for the resulting offspring as a way to ensure their paternity. Strassmann hypothesized that indigenous taboos associated with menstruation in certain groups were cultural practices based on religious beliefs that served to enhance reproduction. Her research on modern indigenous communities offered clues about how humans lived thousands of years ago. Strassmann hoped to show that long-standing cultural taboos around menstruation had developed to support reproductive goals.[30,31]

Strassmann studied the Mali, an indigenous community of West Africa, and the women's use of menstrual huts. She collected urine samples from 93 women to test hormone levels and showed that their use of the huts correlated with menstruation patterns. She observed that most of the women who used the huts became pregnant soon after their visits to the huts and concluded that the huts made a woman's fertility status obvious to her husband and family. Since women enter their fertile window after menstruation, males could track female fertility and schedule sexual activity accordingly.[32]

In addition, Strassmann determined that in the Dogon aboriginal population, women had far fewer periods than American women. This is because women tended to get pregnant earlier, have more babies, and spend more time breastfeeding in Dogon communities. (A Dogon woman menstruated around 100 times in her life, compared with the average American woman's experience of 400 periods in her lifetime.) Strassmann concluded that because Dogon women nurse consistently after birth (and do not pump or go extended periods without nursing), their experience is closer to what all women would have experienced before the development of contraception, which shuts down a woman's natural cycle.[30,33]

The experience of lower menstruation in Dogon women and women not on contraception is because prolactin (produced in the pituitary gland) is the primary hormone responsible for breast milk production and also helps to prevent menstruation. In general, breastfeeding keeps hormone levels high, so the longer a woman nurses, the more likely she will be to experience a light menstruation or no period at all. As the baby is weaned, periods return relatively quickly because the pituitary gland will sense this feeding change and produce less prolactin. However, the range of breastfeeding effects is enormous and some women resume their menstrual cycles soon after giving birth, while others do not resume menstruating until the baby is weaned (which can be months or years later). This depends on how sensitive the mother's body is, and how frequently the baby nurses. Infrequent nursing (only several times a day) can lead to ovulation and resumption of menstruation. Most nonlactating women will not ovulate until 6 weeks postpartum. A small number of women will ovulate earlier, potentially putting them at risk for pregnancy sooner, although the fertility of these early ovulations is not well-established. The potential risk of pregnancy soon after delivery underscores the importance of initiating postpartum contraception.[34]

Another area of interest in female reproduction is premenstrual dysphoric disorder (PMDD), which involves severe irritability, depression, or anxiety 1–2 weeks prior to menstruation, with symptoms easing 2–3 days after menstruation begins (also referred to as premenstrual symptoms or PMS). Up to 80% of women report these symptoms, so investigators have begun to consider if there is any evolutionary advantage to having these symptoms.

Michael Gillings, Professor of Molecular Evolution at Macquarie University in Australia, became interested in women's experiences with PMS and argued that it offered a selective advantage by causing tension between pair-bonds, and therefore it might have helped women to dissolve relationships with infertile men.[35] Not only was Gillings' hypothesis controversial, but also some researchers disagree that PMS is a product of evolution; instead they contend that PMS and its symptoms are more cultural than biological because it manifests differently in women around the world, with women in Western cultures experiencing PMS at a higher frequency than in other cultures.[36,37]

Scientists are also divided over whether the act of bleeding itself serves an evolutionary purpose since other animals reabsorb the uterine lining. Many evolutionary biologists think that the essential feature of women's cycles is the ability of the uterus to thicken its lining in preparation for embryo implantation. They argue that once the endometrium is established, it requires constant blood flow and metabolic support, requiring energy consumption, so it is more efficient for the female to dispose of it and rebuild it each cycle than it is to maintain it until needed. Human physiology does not allow for reabsorption, so bleeding may be an insignificant by-product of evolution.[30]

Mating

To understand human sexual strategies that led to mating, it is important to recognize the recurrent selection pressures or adaptive challenges of the past, and the psychological adaptations and/or strategic solutions that they created.

The evolution of characteristics that are focused on mating benefits, rather than survival benefits, is known as sexual selection. Sexual selection in mammals takes two forms. The first is intrasexual competition among male competitors for a mate, and the second is where one sex (usually females) chooses a mate with particular desirable characteristics. In both cases, the individuals who mate pass on specific genetic traits to their offspring, while the genetic traits of those excluded from mating perish.

Both of these forms of sexual selection can be dependent on the operational sex ratio (OSR), which is the ratio of males sexually competing to mate, compared to sexually competing females. This differs from the typical sex ratio, which includes all males and females regardless of their sexual activity or competition for mates. The idea is that the OSR affects the mating competition of males and females in a population, and acts as a measure of how intense sexual competition is in a species. It can also be useful in determining the potential rate of reproduction. The OSR can reveal bias toward a particular sex when sexually ready members of one gender are more abundant than the other, i.e., a male-biased OSR indicates more sexually competing males than sexually competing females. Equal numbers of males and females versus variation in the ratio of males and females can shift sexual practices. As the OSR shifts, sexual strategies among men and women may change, resulting in variations in selection pressures and adaptations to new conditions.

Today the OSR still plays a role in mating. Studies using data from eight Makushi communities of southern Guyana, characterized by varying OSRs in the region, showed that even within a single ethnic group, male mating effort varies in predictable ways with the OSR. At male-biased sex ratios, men's and women's investment in mating effort were indistinguishable, and only when men were in the minority were they more inclined toward short-term, low investment relationships than women. Their results support the behavioral ecological tenet that reproductive strategies are predictable and contingent on varying situational factors.[38]

The sex ratio has had different effects on human sexual selection and mate choice over time and in different cultures. In general, the sex ratio at birth and the population sex ratio are remarkably constant in human populations (104–107 males for every 100 females).[39] The discrepancy between male and female births is thought to be the result of sperm carrying the Y chromosome having increased mobility over sperm carrying the X chromosome due to the size difference of the X and Y chromosomes (the X chromosome is about three times the size of the Y in terms of base pairs).

The tradition of son preference in parts of North Africa and Asia has distorted natural sex ratios.[39] In 2018, for example, there was a male bias in India with 1070 males per 1000 females. In China, the ratio was 1016 males to 1000 females. In both cases the ratio was higher than the 2018 global sex ratio of 104 males:100 females. There are various reasons for the male bias in Asia, including selective abortion of females, neglect of female children, high maternal mortality, female infanticide, and families stopping reproductive efforts after having a male child. Differential gender mortality and underreporting have not contributed significantly to these abnormal sex ratios. This male bias has caused problems with men seeking long-term relationships because of the lower numbers of females in the populations.[39–41]

Sexual strategies

Although the OSR plays a role in sex selection, other factors contribute significantly to sexual selection. Sexual strategies are adaptive solutions to ensure that desirable traits continue in the population through reproductive success. In ancient cultures, physical contests most likely overrode other mechanisms of sexual selection because individuals could exclude their competitors by force or threat of force, making mate choice, sperm competition, and other mechanisms less likely. The human mating environment, the spatial and temporal clustering of mates and competitors, and physical characteristics predict that contests should have been the primary mechanism of sexual selection in men. Size, strength, aggression, and the use of weapons probably helped ancestral males win contests. Male mate choice was probably central because ancestral females could not constrain the choices of larger and more aggressive males, and attractive women would have had greater male investment. However, over time, male dominance became less of a factor in mating, and other aspects of mate selection eventually became a dominant force over strict competition. Sexual strategies, female mate choice, sperm competition, and sexual coercion became factors in sexual selection.[42]

Sexual strategies include psychological adaptations like love, desire for sex, jealousy, perception of attractiveness, signs of promiscuity, and self-value (confidence in ability to attract desirable mates). David Buss argues that although the term "sexual strategies" is a useful phrase, it is misleading in the sense that it implies conscious intent, because sexual strategies do not necessarily require conscious planning or awareness of the actor. Sexual strategies evolved in a similar way to other survival mechanisms by looking for mates who could provide resources, protection, and commitment, which result in survival and reproductive advantages. For early human evolution, females who chose these qualities in a male had a better chance of their own and their children's survival.[43]

From an evolutionary perspective, men who were indifferent to the promiscuity of their mates risked compromising their paternity. They risked investing time, energy, and resources in children who might not be their own. Ancestral women, in contrast, did not risk the loss of parenthood from male infidelity, because maternity was always certain. However, women who allowed their mates to be promiscuous risked losing their mate's commitment, protection, and resources.[43,44]

Cultural differences have had an effect on mating. Some cultures allow polygamy (multiple female partners), while others are polyandrous (females with multiple male partners), and others polyamorous (many partners). These all have an effect on mating and offspring outcomes. Our ancient ancestors most likely practiced a variety of sexual customs, including monogamy, depending on circumstances. One factor involved in these sexual customs is the operational sex ratio, as mentioned above. Equal numbers of males and females versus variation in the ratio of males and females could shift sexual practices. As sex ratios shift, sexual strategies among men and women may change, resulting in variations in selection pressures and adaptations to new conditions, leading to relationships with more than one wife or husband at the same time.

Why men commit to a monogamous relationship is an evolutionary puzzle. If the goal is to spread your genetic characteristics, men only need to inseminate a woman and then move on to the next mate. Promiscuity would ensure that a man could spread his genes to a variety of offspring. The incentive for a man to stay loyal may have come from women who required signs that a male would commit to a relationship before engaging in sex. The male benefits would come from paternity, influence, and knowledge imparted to the child, repeated and exclusive sex from their mate, and ability to influence mate selection for their children. These benefits work both ways because men who are willing to commit to a long-term relationship most likely have committed to a woman with similar ideals and goals for herself and their children.[45] This, of course, is in contrast to some contemporary practices where partners can choose short-term mating with little to no commitment.

Today, factors that contribute significantly to long-term sexual selection and relationships are characteristics such as health, intelligence, attractiveness, athletic ability, socioeconomic status, musicality, ethnicity, and resourcefulness. These are desirable attributes that are selected for and passed on to the next generation. If these characteristics continue to be desirable in the group, they will increase in frequency within the population.[46] Today, culture, consciousness, and free will also affect selection, but it appears that these may have a relatively smaller influence on mating choice than desirable characteristics.

There are various strategies to retaining mates and also getting rid of mates that may no longer provide sex, protection, and support, or who are infertile or promiscuous. Humans today are the result of strategies that kept good mates and got rid of bad mates.

Our ancestors most likely weighed the cost-benefits of keeping partners or discarding them.[47] Today divorce occurs in most cultures, leading to mate replacement. Humans that separate often attract new mates by reentering the cycle of selection, attraction, and retention. Occasionally mates who separate at an older age carry with them advanced social status and increased economic security, which can be beneficial when looking for a new mate.

Sexual conflict is when interest and strategies for reproduction conflict between the two sexes. This conflict can arise to antagonistic levels during the coevolution of each sex, and includes both behavior and morphology. This conflict creates evolutionary change in each sex in response to the other, and can cause the sexes to take different evolutionary trajectories as a consequence of ecological and/or genetic variation and the strong selection imposed by each sex. An example in humans is when men and women clash over resources and sexual access. In the evolutionary psychology of human mating, the sexual strategy adopted by one sex can conflict with the strategy adopted by the other sex. David Buss calls these phenomena "strategic interference." Consider the differences in men's and women's proclivities to seek casual short-term sex. Although there are exceptions, men usually have a lower limit for engaging in casual and random sex while women do not usually engage in random sexual encounters.[48]

Townsend and Wasserman[49] showed that contemporary men and women possess different emotional mechanisms that motivate and evaluate casual sexual activities. Consequently, even when women expressed indifference to emotional involvement and commitment, and voluntarily engage in casual sex, their goals, feelings, and the associations between their sexual behavior and prospects for long-term investment differed significantly from those of men. Women associated their sexual behavior with investment potential: long-term, short-term, and their partner's ability and willingness to invest. For men, these associations are weaker or inversed. Women who engaged in casual sex expressed worry and vulnerability, and for females, a greater number of partners was associated with increased worry and vulnerability, whereas for males the trend was the opposite.

Studies by Haselton et al.[50] indicated that negative emotions have evolved or have been coopted by selection, in part, to defend against deception and reduce its negative consequences. These studies supported the hypothesis that emotions impact sex-linked forms of strategic interference. They describe three clusters of sex differences: emotional upset about resource deception, commitment deception, and sexual deception.

Competition for mates does not just occur in males. Among primates, female rhesus monkeys, gelada baboons, and savanna baboons use physical aggression against other females to secure sex and win males for sexual access, long-term relationships, and physical protection.[51] However, in humans, female aggression is very different than male aggression and usually consists of undermining their rivals and other deceptive tactics.

Research on female aggression consistently reports that women use indirect aggression (harm to another while masking aggressive intent) to an equivalent or greater extent than men.[52,53] Specific examples of indirect aggression include spreading false rumors, gossiping, exclusion from a social group, insinuations without direct accusation, criticizing appearance or personality, or denigrating other women for their perceived loose sexual mores.[54]

Girls' use of indirect aggression begins around age 11 and persists into adulthood, using more indirect forms of aggression across various stages of life.[55,56] A cross-cultural survey of female aggression among 317 societies by Burbank[57] found that female aggression was mostly indirect and rarely inflicted physical injury. However, more recent studies have shown that physical sexual aggression by women exists in higher frequency than previously thought. Stemple et al.[58] analyzed data from four large-scale federal agency surveys conducted by the Centers for Disease Control and Prevention and the Bureau of Justice Statistics from 2008 through 2013. They determined that the data contradicted the common belief that female sexual perpetration is rare. They then reviewed the literature to identify patterns and provide context, including high-risk populations among college students and inmates. Although they are a small minority, women do commit acts of sexual aggression against men, women, and children. This type of sexual aggression includes numerous sexual activities forced upon a victim without the victim's consent. Female acts of sexual aggression may include coerced sex, anal or vaginal penetration, oral sex, kissing, exposing genitals, and using objects to cause harm.[54]

In addition to selective strategies, the exhilaration of love dominates society in the form of music, literature, film, and imagination. People of all cultures experience love, and because love (deep empathy and overwhelming passion for someone) is so pervasive and part of the human experience, it is clear it was a significant part of evolution. However, love remains a scientific puzzle and it is not clear why people spend their lives and energy trying to find love and a fruitful relationship. Evolutionary strategies for love seem to be as critical as those of survival, and to explain the existence of love and mating we can only look at some of the current practices and benefits. One of these benefits appears to be monogamy and the insurance of commitment, child-rearing by two parents, and hope for a lifelong friendship. However, love can be complex, can exist between more than two people, can change over time, and can lead to hatred or lack of empathy and passion for someone. Studies have shown that 85% of people in Western cultures have experienced at least one romantic breakup and its sometimes devastating emotional outcomes. However, humans are remarkably resilient and even with high rates of divorce or breakups, individuals will still seek other relationships, indicating that love is still a driving force behind finding a mate.[59]

Sexual dimorphism

Sexual dimorphism also plays a role in selection, and dimorphism may have benefits that evolved as a result of selective pressures. Males and females differing in size can specialize in, and exploit, different food sources without competing, and variation in body size can help when trying to avoid predators. Differences in body size might also have been useful in securing a mate where females would sum up the cost benefit of mating with males of larger or smaller body size. Various levels of sexual dimorphism are seen as markers of sexual selection. Studies have shown the earliest hominids showed strong levels of dimorphism, while human dimorphism has diminished over time. Other great apes continue to exhibit strong sexual dimorphism.[60]

Sexual selection has been used to explain a number of dimorphic human anatomical features, including differences in sexual behavior, breast size, facial hair, pubic hair, and penis size. Most primates have flat breasts while human female breasts are rounded and filled with fatty tissue. Some biologists have suggested that rounded female breasts serve as signals of fertility. Morris and Dawkins[61,62] hypothesized that the loss of the penis bone in humans as opposed to other primates was a result of sexual selection by females looking for a clear sign of good health in prospective mates. They assumed that erection failure may have served as a warning of health problems and a lack of mental wellness. Because *Homo sapiens* have a thicker penis than other great apes do, some investigators have suggested that the larger size was the result of female selection, while others have suggested it was the result of natural selection. Gallup and Burch[63] suggested that one reason for natural selection resulting in a larger penis might have been the efficiency of a larger penis in displacing sperm of rival males during intercourse, and that semen placement was directly proportional to the depth of pelvic thrusting, increasing the chances of fertilization.[62–65]

Human females' preferences and selectivity in a partner are more complex and puzzling than the mate preferences of any other species. In many species, female selectivity is predictable and based on pheromones, male size, weight, male coloring, results of male competition, male behavior, and cognitive differences. In humans there are no obvious pheromones, and male size, weight, competition, etc. differ significantly in mate choice by females. The evolutionary roots of women's desires and mate selectivity requires going back in time, to before primates emerged from their mammalian ancestors. Women begin to produce a lifetime supply of oocytes (large female reproductive cells) during their fetal development. Although females invest the energy to produce millions of these cells (primary oocytes), only about 400–500 of these will mature into eggs and be capable of being fertilized, and only a few of these mature eggs, once fertilized, will result in a pregnancy. However, one sexual encounter can result in a 9-month energy-consuming investment that foregoes other fruitful mating opportunities. Once a child is born, the mother is likely to be invested in lactation, nursing, nurturing, and protection. These investments and time consumption mean that women's reproductive opportunities are limited and that they can only reproduce a few times during their lifetime. Males, on the other hand, can reproduce as often as they can find a willing and fertile female. Male investment may have only been a few minutes or hours during a casual sex encounter, while a female's investment may involve years caring for offspring.

Because women invest so much in reproduction, they are highly selective when it comes to mating. Female selectivity has an evolutionary explanation. Ancestral women would have suffered the consequences of poor mate selection. These could have involved unhealthy offspring, lack of protection, lack of resources, and offspring with undesirable genes. Further, these women would be denied the additional, potentially successful mating opportunities through their poor choices. Ancestral women who preferred intelligent mates would have raised their odds of securing material, and economic resources, and since there is a genetic component to intelligence, these favorable qualities would have been passed on genetically to their sons and daughters.[47] Ancestral women would also have focused on men's most valuable characteristics (hunting skills, ability for protection, commitment) and potential for resource acquisition when choosing a mate(s). These traits may have been recognizable by a man's health, physical abilities, position within a group, and economic resources. These ancestral women may have been best served by a monogamous relationship where a single male may have provided more resources for offspring than several temporary and unreliable partners would have. Several primate and mammalian studies have concluded that there is a strong benefit to monogamy. Primate studies have shown that the primary male benefit of monogamy is that it can increase the male's chances of siring many offspring who survive long enough to reproduce. Although it seems counterintuitive that mating with only one female would be less adaptive for a male than spreading his seed widely, the conclusions were that children who live without a father in the household are more likely to die in childhood; a similar conclusion is drawn in human government statistics. Other mammalian studies have indicated that monogamy evolved in nonhuman mammals where breeding females are intolerant of each other and female density is low, suggesting that it represents a mating strategy that has developed where males are unable to defend access to multiple females, and therefore guarding a single female was a male's best reproductive strategy.[66,67]

High selectivity is manifest in today's females who may be selective based on men's athletic ability, intelligence, physical prowess, kindness, empathy, emotional stability, sense of humor, ambition, and social skills and status. Studies have shown that high mate value (attractive) females have more discerning mate preferences relative to lower mate value women. Arnocky found that men who perceive themselves as high in overall mate value are selective in their mate choice in a manner similar to high mate value women. Men who thought they were of high mate value tended to marry earlier in life and have a partner who was sociable, ambitious, and high in social status, and who had a desire for children.[68,69]

Mate selection is not only the choice of desirable physical characteristics, social characteristics, and the ability to provide resources. Current studies suggest that molecular signals may be involved that may contribute to mate choice. Chemical signals known as pheromones are used to mark out territories, attract mates, and promote social cohesion and coordination in insects, and some mammals. Pheromones are volatile organic molecules, which are effective at very low concentrations. While pheromones are not known to exist in humans, there is increasing evidence that human body odor is involved in interpersonal communication. Odor conveys cues related to mate fitness, genetic relatedness, age, gender, and sexual orientation. In addition, perspiration from highly emotional individuals can modify physiological activity in individuals smelling the odors. Each individual has a unique odor signature, which is influenced by genetic factors and a number of environmental factors including diet and reproductive status, particularly for women.[70]

Some of the first evidence for chemical attractants in humans came from experiments for the detection of major histocompatibility complex (MHC) molecules in humans. The MHC (MHC is known as HLA in humans) is a source of unique individual odors that influence individual recognition, mating preferences, nesting behavior, and selective block of pregnancy in animals.[71]

In 1995, Wedekind set up a test of women's sensitivity to male odors by recruiting 49 women and 44 men. The men were selected for their variety of HLA gene types. HLA genes code for cell surface proteins that are essential for acquired immunity and a person's ability to recognize foreign molecules. Wedekind gave the men clean T-shirts to wear for 2 nights without showering or use of deodorants. The researchers put each T-shirt in a box equipped with a smelling hole. The women were asked to smell the boxes and describe each odor as to intensity, pleasantness, and sexiness. Overall, the women preferred the scents of T-shirts worn by men whose HLA genes were different than their own. The results suggested that evolution has provided humans, as well as mice and insects, with a transmitter and olfactory receiver for genetic information that could influence mate choice.[71]

One interesting finding of Wedekind's experiments was that women's preference reversed (they chose T-shirts with similar MHC genes) if the women were on oral contraceptives. Wedekind speculated that this might have been because contraceptives stop ovulation and put the body into a hormonally pregnant state, so that during that time there would be no reason to seek out a mate whose MHC complemented that of the female. Several investigators have expressed concern over Wedekind's results of women on contraception, indicating that MHC-similarity couples could lead to fertility problems and might lead to the breakdown of relationships when the women stop using contraceptives, because odor perception appears to play a role in maintaining attraction.[72,73]

In 2002, Jacob et al.[74] set up more sophisticated experiments that could eliminate the high variability in background odors encoded by the rest of the genome and influenced by cultural practices. Their experiments eliminated a low signal-to-noise ratio that could mask HLA-based olfactory cues. They showed that women can detect differences of one HLA allele among male odor donors with different HLA genotypes. Jacob et al. established that the mechanism for a woman's ability to discriminate and choose odors was based on HLA alleles inherited from her father but not her mother. The parents'

HLA alleles that she did not inherit showed no relationship with odor choice, despite exposure to these HLA-encoded odors (from her parents) throughout her life. The authors' data indicate that paternally inherited HLA-associated odors influence odor preference and may serve as social cues.[74]

Mating behavior changes with age. Older females and males may be less desirable due to age or health. Women who are older or have undergone the menopause become less desirable for mating if the objective is childbirth, and men who are older are sometimes less desirable for health reasons. There can be a cultural component to this behavior. In societies where men choose their brides, young women are most desirable and demand the highest dowry, and likewise women prefer wealthier men. Mulder[75] studied the Kipsigis people of Kenya and found that Kipsigis women, or their parents on their behalf, preferentially chose men offering high-quality breeding opportunities determined by the number of acres available on which to settle. In addition, there was a preference for bachelors over monogamists or polygynists due to the perceived reproductive costs associated with having a large number of cowives—costs which women attempt to minimize through their marital choice. Older women in this case were of lower value and had fewer choices than younger women.

Ancient humans, whether men or women, and Neanderthals most likely shared similar psychology as well as similar biology. Although women and men are psychologically different and have faced different evolutionary pressures, there was most likely still variability in preference when choosing a mate. Both species probably had some concerns about loyalty, jealousy, and ability to provide resources when choosing a mate, but attractiveness would have been subjective.

Today we know that there were sexual interactions among Neanderthal and humans; however, it is not clear if these interactions were the result of mutual attraction, if consensual and long-term relationships were formed, or if love was involved. The evidence is clear that the two species mated, resulting in viable offspring and that traces of Neanderthal genomic DNA still exist in modern-day humans. However, it is impossible to determine the extent of relationships and the reasons for mating.

It is most likely that not all sexual encounters between Neanderthals and humans resulted in offspring or were consensual. Throughout human history there have been conflicts leading to rape and the birth of offspring. In these cases, the care of offspring and the resources provided by consensual mating were not available, and women were left to raise the child on their own. This type of mating may have been frequent in Neanderthal-human encounters where women may have been taken and raped by both groups, in many cases leading to hybrid children who were shunned or disparaged within the group. Molecular evidence indicates that Neanderthal females did not pass on their mitochondrial DNA to modern humans. This finding suggests that female children born from male humans and female Neanderthals: (1) did not survive to reproduce; (2) were not desirable for reproduction; or (3) were sterile and therefore did not pass on their mitochondria to human-Neanderthal hybrids[76] (Tables 1 and 2).

Table 1 Techniques for determining absolute age (https://www.geocraft.com/WVFossils/DatingMethods.html).

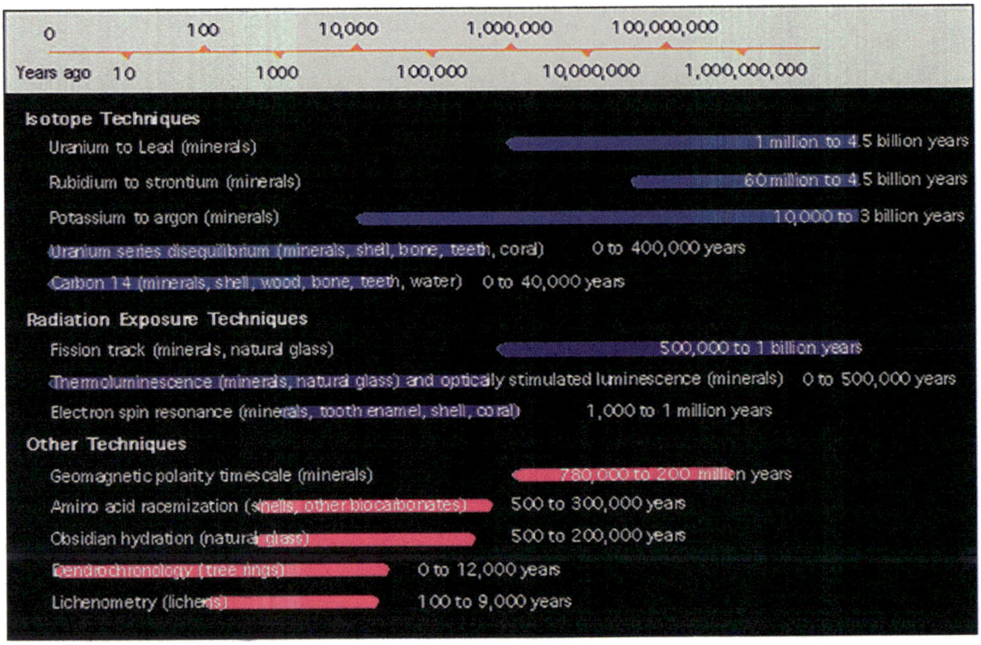

Table 2 Family of U-series methods (https://en.wikipedia.org/wiki/Uranium%E2%80%93thorium_dating).

U-series dating methods			
Isotope ratio measured	**Analytical method**	**Time range (ka)**	**Materials**
$^{230}Th/^{234}U$	Alpha spec. mass spec.	1–350	Carbonates, phosphates, organic matter
$^{231}Pa/^{235}U$	Alpha spec.	1–300	Carbonates, phosphates
$^{234}U/^{238}U$	Alpha spec. mass spec.	100–1000	Carbonates, phosphates
U-trend	Alpha spec.	10–1000(?)	Detrital sediment
^{226}Ra	Alpha spec.	0.5–10	Carbonates
$^{230}Th/^{232}Th$	Alpha spec.	5–300	Marinesediment
$^{231}Pa/^{230}Th$	Alpha spec.	5–300	Marinesediment
$^{4}He/U$	mass spec. (gas)	20–400(?)	Coral

Conclusion

In evolutionary terms, sex is important and fuels evolutionary change by adding variation to the gene pool. It appears that increased genetic variability, due to sexual reproduction, is especially important during adaptation. When choosing a mate, both men and women value dependability, intelligence, long-term commitment, and the possibility that their child will be brought up in a nurturing environment. Although there are significant costs to human sexual reproduction, including increased energy consumption for mate selection, sex cell production, and child-rearing, it appears that love, passion, and mate choice drive human sexuality leading to new generations with the ability to adapt to new environmental conditions. Because of the complexities involved in human sexual selection and sexual reproduction, the urge to pass our genes to the next generation has potentially changed human culture. Original reproductive practices and sex selection were most likely based on physical prowess and competition (similar to what we see in many animal species); however, present-day selection among humans is much more complex and mate choice can depend on a variety of characteristics (ethnicity, income, educational background, socio-economic factors, musical ability, etc.) that were not available or as valuable to our ancestors. Theories about how and why sexual reproduction evolved include: ability to adapt quickly to changing environments; ability to mask deleterious recessive mutations; ability to store unused beneficial mutations for future generations; sex leading to the evolution of cooperation and the formation of groups; and the development of a more complex brain as a result of group interactions. Molecular evidence indicates that there were sexual interactions among Neanderthals and humans; however, it is not clear if these interactions were the result of mutual attraction, if they resulted in long-term relationships, or even if they were consensual.

References

1. Lam F, et al. On the genealogy of asexual diploids. *J Comp Biol.* 2011;18:415–428. https://doi.org/10.1089/cmb.2010.0270 2011.
2. Otto S. Sexual reproduction and the evolution of sex. *Nat Educ.* 2008;1:182–183.
3. Singh ND, et al. Fruit flies diversify their offspring in response to parasite infection. *Science.* 2015;349:747–750. https://doi.org/10.1126/science.aab1768.
4. Becks L, Agrawal AF. The evolution of sex is favored during adaptation to new environments. *PLoS Biol.* 2012;10:e1001317.
5. Lythgoe KA, Read AF. Catching the red queen? The advice of the rose. *Trends Ecol Evol.* 1998;13:473–474.
6. Van Valen L. A new evolutionary law. *Evol Theory.* 1973;1:1–30.
7. Bell G. *The Masterpiece of Nature: The Evolution and Genetics of Sexuality.* Berkeley: University of California Press; 1982.
8. Nash DR, et al. A mosaic of chemical coevolution in a large blue butterfly. *Science.* 2008;319:88–90.
9. Agrawal AF, et al. The evolution of plastic recombination. *Genetics.* 2005;171:803–812. https://doi.org/10.1534/genetics.105.041301. 16020791.

10. Hadany L, Beker T. On the evolutionary advantage of fitness-associated recombination. *Genetics*. 2003;165:2167–2179. 14704195.
11. Belyaev DK, Borodin PM. The influence of stress on variation and its role in evolution. *Biol Zentbl*. 1982;101:705–714.
12. Lloyd A, et al. Plasticity of meiotic recombination rates in response to temperature in *Arabidopsis*. *Genetics*. 2018;208:1409–1420.
13. Stevison LS, et al. Recombination rate plasticity: revealing mechanisms by design. *Philos Trans R Soc Lond B Biol Sci*. 2017;372:20160459.
14. McDonald MJ, et al. Sex speeds adaptation by altering the dynamics of molecular evolution. *Nature*. 2016;531:233–236.
15. Koeslag JH. The prisoner's dilemma game, and the evolutionary inevitability of cooperation. *J Theor Biol*. 1997;189:53–61.
16. Simon C. Why norwegians don't have their pigs in the forest: illuminating nordic 'co-operation'. *Behav Soc Issues*. 2017;26:172–186.
17. Gonzalez-Forero M. Inference of ecological and social drivers of human brain-size evolution. *Nature*. 2018;557:554–557.
18. Schillaci MA. Sexual selection and the evolution of brain size in primates. *PLoS One*. 2006;1:e62.
19. Arnold AP. The end of gonad-centric sex determination in mammals. *Trends Genet*. 2012;28:55–61.
20. Mackensen O. Viability and sex determination in the honey bee *Apis mellifera* L. *Genetics*. 1951;36:500–509.
21. Gempe T, Beye M. Sex determination in honeybees. *Nat Educ*. 2009;2:1.
22. Valenzuela N, et al. Extreme thermal fluctuations from climate change unexpectedly accelerate demographic collapse of vertebrates with temperature-dependent sex determination. *Sci Rep*. 2019;9:4254. https://doi.org/10.1038/s41598-019-40597-4.
23. Beukeboom LW, Perrin N. *The evolution of sex determination*. Oxford Scholarship Online; 2014. https://doi.org/10.1093/acprof:oso/9780199657148.001.0001.
24. Jiang L, et al. Systematic identification and evolution analysis of sox genes in Coturnix japonica based on comparative genomics. *Genes*. 2019;10:314. https://doi.org/10.3390/genes10040314.
25. Ganna A, et al. Large-scale GWAS reveals insights into the genetic architecture of same-sex sexual behavior. *Science*. 2019;365:eaat7693.
26. Watts TM, et al. Sexual arousal patterns of identical twins with discordant sexual orientations. *Sci Rep*. 2018;8:14970.
27. Bartlett NH, Vasey PL. A retrospective study of childhood gender-atypical behavior in samoan *Fa'afafine*. *Arch Sex Behav*. 2006;2006(35):659–666. https://doi.org/10.1007/s10508-006-9055-1.
28. Sultan C, et al. Environmental xenoestrogens, antiandrogens and disorders of male sexual differentiation. *Mol Cell Endocrinol*. 2001;178:99–105.
29. Campbell K, Rockett J. Biomarkers of ovulation, endometrial receptivity, fertilization, implantation and early pregnancy progression. *Paediatr Perinat Epidemiol*. 2006;(Suppl 1):13–25. https://doi.org/10.1111/j.1365-3016.2006.00766.x.
30. Sole-Smith V. The point of a period. *Sci Am*. 2019;320:32–40. https://doi.org/10.1038/scientificamerican0519-32.
31. Strassmann BI. Menstrual hut visits by Dogon women: a hormonal test distinguishes deceit from honest signaling. *Behav Ecol*. 1993;3:304–315.
32. Strassmann BI. The biology of menstruation in *Homo sapiens*: total lifetime menses, fecundity, and nonsynchrony in a natural-fertility population. *Curr Anthropol*. 1997;38:123–129.
33. Strassmann BI. The function of menstrual taboos among the dogon. *Hum Nat*. 1992;3:89–131.
34. Jackson E, Glasier A. Return of ovulation and menses in postpartum nonlactating women: a systematic review. *Obstet Gynecol*. 2011;3:657–662. https://doi.org/10.1097/AOG.0b013e31820ce18c.
35. Gillings MR. Were there evolutionary advantages to premenstrual syndrome? *Evol Appl*. 2014. https://doi.org/10.1111/eva.12190.
36. Johnson TM. Premenstrual syndrome as a western culture-specific disorder. *Cult Med Psychiatry*. 1987;11:337–356.
37. Browne TK. Is premenstrual dysphoric disorder really a disorder? *J Bioeth Inq*. 2015;12:313–330. https://doi.org/10.1007/s11673-014-9567-7.

38. Schacht R, Mulder BM. Sex ratio effects on reproductive strategies in humans. *R Soc Open Sci.* 2015;2:140402. https://doi.org/10.1098/rsos.140402.
39. Hesketh T, Xing ZW. Abnormal sex ratios in human populations: causes and consequences. *Proc Natl Acad Sci USA.* 2006;103:13271–13275. https://doi.org/10.1073/pnas.0602203103.
40. Srivastava P. *Women and Men in India (A Statistical Compilation of Gender Related Indicators in India)*; 2018. http://www.mospi.gov.in/sites/default/files/publication_reports/Women%20and%20Men%20%20in%20India%202018.pdf.
41. Gu B, Roy K. Sex ratio at birth in China, with reference to other areas in East Asia: what we know. *Asia-Pac Popul J.* 1995;10:17–42.
42. Puts DA. Beauty and the beast: mechanisms of sexual selection in humans. *Evol Hum Behav.* 2010;31:157–175.
43. Buss DM. The evolution of human intrasexual competition: tactics of mate attraction. *J Pers Soc Psychol.* 1988;54:616–628.
44. Buss DM. Sex differences in human mate preferences: evolutionary hypotheses tested in 37 cultures. *Behav Brain Sci.* 1998;12:1–49.
45. Apostolous M. Parent–offspring conflict over mating: domains of agreement and disagreement. *Evol Psychol.* 2015;13. https://doi.org/10.1177/1474704915604561.
46. Hanevik HI, et al. Can IVF influence human evolution? *Hum Reprod.* 2016;31:1397–1402. https://doi.org/10.1093/humrep/dew089.
47. Buss DM, Strategies of human mating. *Phsychol Topics.* 2006;15:239–260.
48. Buss DM. Conflict between the sexes: strategic interference and the evocation of anger and upset. *J Pers Soc Psychol.* 1989;56:735–747.
49. Townsend JM, Wasserman TH. Sexual hookups among college students: sex differences in emotional reactions. *Arch Sex Behav.* 2011;40:1173–1181.
50. Haselton MG, et al. Sex, lies, and strategic interference: the psychology of deception between the sexes. *Soc Personal Soc Psychol.* 2005;31:3–23.
51. Smuts BB. *Sex and Friendship in Baboons.* New York: Aldine Publishing; 2009. ISBN:978-0-202-02027-3. First printing 1985).
52. Archer J, Coyne SM. An integrated review of indirect, relational and social aggression. *Personal Soc Psychol Rev.* 2005;9:212–230. https://doi.org/10.1207/s15327957pspr0903_2.
53. Arnocky S, et al. Jealousy mediates the relationship between attractiveness comparison and females' indirect aggression. *Pers Relat.* 2011;19:290–303. https://doi.org/10.1111/j.1475-6811.2011.01362.x.
54. Denson TF, et al. Aggression in women: behavior, brain and hormones. *Front Behav Neurosci.* 2018. https://doi.org/10.3389/fnbeh.2018.00081.
55. Österman K, et al. Cross-cultural evidence of female indirect aggression. *Aggress Behav.* 1998;24:1–8. https://doi.org/10.1002/(sici)1098-2337(1998)24:1<1::aid-ab1>3.3.co;2-f.
56. Archer J. Sex differences in aggression in real-world settings: a meta-analytic review. *Rev Gen Psychol.* 2004;2004(8):291–322. https://doi.org/10.1037/1089-2680.8.4.291.
57. Burbank VK. Female aggression in cross-cultural perspective. *Behav Sci Res.* 1987;21:70–100. https://doi.org/10.1177/106939718702100103.
58. Stemple L, et al. Sexual victimization perpetrated by women: federal data reveal surprising prevalence. *Aggress Violent Behav.* 2017;34:302–311.
59. Buss DM. The evolution of love in humans. In: Sternberg RJ, Sternberg K, eds. *The New Psychology of Love. New York*: Cambridge University Press; 2018:42–63. https://doi.org/10.1017/9781108658225.004 [chapter 3].
60. Foley RA, Lewin R. *Principles of Human Evolution.* 2nd ed. Wiley-Blackwell; 2013. ISBN:978-1-118-68799-4.
61. Morris D. *The Naked Woman: A Study of the Female Body.* Thomas Dunne Books: St. Martin's Press, NY; 2004.
62. Dawkins R. *The Selfish Gene.* 30th ed. UK: Oxford University Press; 2006.
63. Gallup G, Burch R. *The Semen-Displacement Hypothesis: Semen Hydraulics and the Intra-Pair Copulation Proclivity Model of Female Infidelity.* Cambridge University Press; 2006. https://doi.org/10.1017/CBO9780511617812.007.

64. Miller GF. How mate choice shaped human nature: a review of sexual selection and human evolution. In: Crawford CB, Krebs DL, eds. *Handbook of Evolutionary Psychology: Ideas, Issues, and Applications*. Lawrence Erlbaum Associates Publishers; 1998:87–129.
65. Dixson AF. *Sexual Selection and the Origins of Human Mating*. UK: Oxford University Press; 2009:61–65.
66. Opie C, et al. Male infanticide leads to social monogamy in primates. *PNAS*. 2013;110:13328–13332. https://doi.org/10.1073/pnas.1307903110.
67. Lukas D, Clutton-Brock TH. The evolution of social monogamy in mammals. *Science*. 2013;341:526–530. https://doi.org/10.1126/science.1238677.
68. Arnocky S. Self-perceived mate value, facial attractiveness, and mate preferences: do desirable men want it all? *Evol Psychol*. 2018. https://doi.org/10.1177/1474704918763271.
69. Schmitt DP, Weekes-Shackelford VA, Shackelford TK, eds. The evolution of culturally-variable sex differences: Men and women are not always different, but when they are…it appears not to result from patriarchy or sex role socialization. In: *The Evolution of Sexuality*. New York, NY: Springer; 2015:221–256.
70. Gaby JM, Dalton P. Discrimination between individual body odors is unaffected by perfume. *Perception*. 2019;48:1104–1123. https://doi.org/10.1177/0301006619872055 [Epub 2019 Aug 31].
71. Allen C, et al. Preparation for fatherhood: a role for olfactory communication during human pregnancy? *Physiol Behav*. 2019;206:175–180. https://doi.org/10.1016/j.physbeh.2019.03.030.
72. Allen C, et al. Evidence for odor-mediated assortative mating in humans: the impact of hormonal contraception and artificial fragrances. *Physiology & Behavior*. 2019. https://doi.org/10.1016/j.physbeh.2019.05.002.
73. Roberts SC, et al. MHC-correlated odor preferences in humans and the use of oral contraceptives. *Proc Biol Sci*. 2008;275:2715–2722. https://doi.org/10.1098/rspb.2008.0825.
74. Jacob S, et al. Paternally inherited HLA alleles are associated with women's choice of male odor. *Nat Genet*. 2002;30:175–179. https://doi.org/10.1038/ng830.
75. Mulder BM. Kipsigis women's preferences for wealthy men: evidence for female choice in mammals? *Behav Ecol Soc Biol*. 1990;27:255–264.
76. Wang CC, et al. Neanderthal DNA and modern human origins. *Quat Int*. 2013;295:126–129.

CHAPTER 4

Archeological tools used to study hominin evolution

Archaeology is about facts.... Forget any ideas you've got about lost cities, exotic travel and digging up the world. We do not follow maps to buried treasure, and X never, ever marks the spot.

Indiana Jones and the Last Crusade

An integrative approach

Today the study of hominin evolution brings together a number of approaches and techniques from various seemingly unrelated disciplines, including history, geology, physical anthropology, archeology, linguistics, genetics, molecular biology, and statistics. Traditionally these fields have operated rather independently from each other, with researchers often going about their fact-finding missions as if knowledge exists compartmentalized in well-defined entities unrelated to information from other disciplines. There are various reasons for this isolationist condition occurring among the sciences. One is the mode that specialists have been traditionally educated and trained at times in isolation from other disciplines. Even today our educational systems partition curricula into departments, divisions, and colleges, often teaching students in isolation from other syllabi and at times duplicating the subject matter. One example is departments of biology and chemistry both teaching biochemistry.

A consequence of this nonholistic approach to education is an education system that trains students on narrow bodies of knowledge incapable of integrating science. These conditions often take the form of a general lack of knowledge and appreciation for information derived from other related fields. These gaps among disciplines often manifest themselves in lack of collaboration, which promotes confusion and impacts negatively on progress. At times this lack of communication and collaboration takes the form of scientific animosity, as experts express their dislike for the approach and quality of the data generated by researchers from other fields. Limited understanding and rapidly increasing complexity of the methodology used by different disciplines also comprise an important factor for this schism. For example, at times the methodology and procedures employed by specific fields are unintelligible to workers in different but pertinent areas due to the complexity of the technology and jargon used.

Sex and Cohabitation Among Early Humans
https://doi.org/10.1016/B978-0-12-819086-9.00009-9

In spite of the increasing sophistication in procedures experienced by scientific disciplines in recent decades, which one might think should require increased specialization, there is a general tendency in recent years for an integrative approach to scientific research. In topics such as hominin evolution, for example, we are beginning to see how archeological studies are benefiting from recent data emanating from linguistics, molecular biology, and genetics. This change in approach and philosophy is largely due to a more open-minded attitude to scientific research and an understanding that information from one field can often benefit another area, enabling confirmation of conclusions. And where discrepancies are observed among data from different fields, this alerts researchers that models are incorrect or experimental faults exist. Nowadays, empirical results from ancient and contemporary DNA, for example, are routinely compared with archeological data to assess the validity of human dispersions theories. And in other cases, linguistic phylogenies are contrasted with population relationships based on anatomical characters. Although there is still some degree of resistance by some sectors of the scientific community to this type of integrative approach, an increasing number of researchers believe that this type of holistic style tends to reinforce conclusions and extend the significance of the data.

In this spirit, this and the next two chapters present in an integrative manner a number of important methodologies from various disciplines routinely used in the study of hominin evolution.

Archeology

By definition, archeology is the study of the material and cultural history of people. It usually takes the form of research into the culture of populations by analyzing their artifacts, writings, symbols, inscriptions, architecture, monuments, and excavated utensils. Generally speaking, archeology is the study of the by-products of hominin ingenuity as well as imprints left behind by domesticates (plants and animals). Although these signals from the past are frequently material items, such as utensils or artwork, they could also include folklore, music, and myths.

To appreciate fully the contributions of archeology to the overall body of human knowledge and to frame the evolution of archeology as a discipline, one should first be introduced to the topic by way of a brief history of the discipline, highlighting key milestones. The personalities described below were selected for review because, due to their contributions, they represent important stepping-stones in the progress of the discipline.

Origins and evolution of archeology

The roots of archeology can be traced back to ancient times. The discipline has its genesis in human interest in historical events. Generally, individuals, such as monarchs and emperors, having a vested interest in the glory of their ancestors and their past, developed a preoccupation and passion for history and archeology.

Fig. 1 Image of Nabonidus (https://www.tripadvisor.com/LocationPhotoDirectLink-g652373-d2651160-i195555615-Sanliurfa_Archeology_and_Mosaic_Museum-Sanliurfa_Sanliurfa_Province.html).

A number of scholars trace the origins of archeology to a remarkable Neo-Babylonian monarch named Nabonidus, who lived from 556 to 539 BCE[1] (Fig. 1). Nabonidus was the last king of the Neo-Babylonian Empire and is acknowledged to have been responsible for a number of excavations of foundations and restorations of artifacts in Southwest Asia. Although a number of experts question Nabonidus' true motives, indicating that he was as driven by personal interests, politics, and/or religious fervor as he was by curiosity about the unknown, his activities of exploration were clearly ahead of his time. In his efforts, he is described as being persistent to the point of obsession.

Nabonidus was involved in excavations in a number of ancient cities of Mesopotamia, such as Ur, Larsa, Sippar, and Harran.[2] His main objective was to restore and rebuild

Fig. 2 Bust of Herodotus *(left)* and Thucydides *(right)* (https://www.pinterest.co.kr/pin/560346378608389879).

temples. In fact, during a 10-year period, he delegated the governing of the empire to his son Belshazzar to devote all of his time to the restoration and building of places of worship. In addition to rebuilding, Nabonidus was also keen on dating the ruins and restoration of artifacts.[3] In one instance, for example, he instructed skilled artisans to restore the head and face of a statuette of Sargon, a previous ruler. This event represents one of the oldest accounts of reparation of an archeological artifact by experts. Although other monarchs before Nabonidus identified previous structures and built on their foundations for convenience, his efforts went beyond by providing an accurate and thorough account of the ruins. Some of Nabonidus' reports of excavations are so complete that they could have been extracted from a recent archeological report from the 1800s.

Herodotus (484–425 BCE) was another prominent figure of ancient times, and is considered by many the "father of comparative anthropology"[4] (Fig. 2). He lived and worked about a century after Nabonidus.[5] Herodotus was born in Halicarnassus, Asia Minor, in what is today Eastern Turkey, in the coastal town of Bodrum. Born to a well-to-do family, he traveled considerably within the known world, systematically recording information that he later used in his single book *Inquiries*.[6] Although Herodotus cited the work of others, the material in his book was complemented thoroughly by first-hand experiences accumulated during his travels. Today the nine-part book is known as *The Histories*.[7] In the first six sections, Herodotus provides descriptions of the people of the world, including their customs, folklore, and traditions. In this sense, his *Histories* represent a wide-ranging anthropological treatise demonstrating an encyclopedic level of knowledge. This masterpiece is considered the number one prose work from the great intellectual era of 5th-century Greece. His contemporary and scolding critic, Thucydides (Fig. 2), wrote the second, *The History of the Peloponnesian War*. In *The Histories*, Herodotus examined and recorded everything of interest to him, all aspects of ancient life. It is a wealth of information that encompasses areas such as natural sciences, ethnography, geography, and history. His approach could be described as holistic and comprehensive, total anthropology. In his study of nature, he examined his subjects using a systematic and objective approach by gathering the data and then critically arranging them into historical accounts. In addition, he uses his observations to draw moral lessons, thus initiating the Greco-Roman historiographical tradition.

Herodotus also devoted a considerable amount of space in his *Histories* to the Greco-Persian conflicts. He makes his intentions clear when he states at the beginning of his book:

> *Here are presented the results of the enquiry carried out by Herodotus of Halicarnassus. The purpose is to prevent the traces of human events from being erased by time, and to preserve the fame of the important and remarkable achievements produced by both Greeks and non-Greeks; among the matters covered is, in particular, the cause of the hostilities between Greeks and non-Greeks.*
>
> **—Herodotus, The Histories**

Although critics have, at times, been displeased when he included a large number of what seemed to be legends and fanciful accounts, an increasing amount of information that he provided has since been confirmed by archeologists.

One of Herodotus' main critics, Thucydides (460–400 BCE), was in his own right another outstanding historian during the Greek Golden Age.[8] In *The History of the Peloponnesian War*, he narrates the 5th-century BCE conflict between the city-states of Sparta and Athens that ended in 411 BCE. He is credited for adopting an objective and strict approach to the gathering and reporting of information. No gods are involved in his narratives. His accounts of the Peloponnesian War are impartial in analyzing causes and effects.

More recently, a number of personalities have contributed to the methodology and approach in the field of archeology. Ciriaco de' Pizzicolli (1391–1453) is of particular interest (Fig. 3). A pre-Renaissance man, de' Pizzicolli has been described as the "Father of Modern Classical Archaeology" due to his contributions to the discipline.[9] Born in Ancona, a port-city on the Italian east coast (Adriatic Sea), to a wealthy merchant

Fig. 3 Portrait of Ciriaco de' Pizzicolli (https://en.wikipedia.org/wiki/Ciriaco_de%27_Pizzicolli).

family, Ciriaco traveled throughout the Mediterranean basin, initially as a merchant, and then increasingly to satisfy his curiosity about the past. He systematically identified and cataloged archeological sites for decades throughout the Eastern Mediterranean, accurately recording Greek and Roman centers and inscriptions, and collecting a number of artifacts. He was relentless and prolific, and in his notebook entitled *Commentaria* he included depictions of ancient structures with detailed annotations. Unfortunately, his six-volume notebook was never published, and only a fraction of his work reached manuscript form. His *Commentaria* perished in a fire in 1514. Knowledge of his work reaches us indirectly from the writing of more recent scholars. De' Pizzicolli particularly made a mark in the field of archeology by contributing to the field of epigraphy as he interpreted ancient writings, dating and placing them in the context of their times and authors. Today we know of long-destroyed monuments and scriptures thanks to de' Pizzicolli's work.

Antiquarianism

During the 15th and 16th centuries, Europe experienced a revival of the arts and sciences, including a renewed interest in the past. This process was part of an intellectual movement known as the Renaissance. This epoch is characterized by a resurrection of knowledge from classical Greece and Rome that had been largely forgotten during the Middle Ages after the collapse of the Roman Empire. Significantly, this period featured a general reawakening of a sense of wonder. As part of this new outlook on life, a group of scholars developed a passion for the past and specifically for collecting artifacts. These individuals, known as antiquarians, began to search for and collect ancient objects. Although much of this activity lacked organization, was not done systematically, and was at times even destructive, there was an incipient interest in answering questions related to how ancient people lived and died by examining their possessions and culture.[10] During this time, archeology was performed, for the most part, rather rudimentally by excavating, not necessarily carefully, without preserving the strata, in a vacuum of theory.

This antiquarian movement was driven for the most part by independently wealthy individuals who had the time and resources to undertake travel, searches, and excavations. Today this type of activity would be considered a hobby. These antiquarians were driven individuals keen on physical and empirical evidence. They were primarily in search of facts, and not necessarily concerned with exploring theory.

The age of enlightenment

Europe experienced another dramatic cultural transformation during the 17th and 18th centuries. This revolution in philosophical thought is known today as the Age of Enlightenment or Reason.[11] Included in this philosophical movement was a scientific revolution that promulgated the principles of reasoning, logic, and the scientific method as the basis for legitimacy and sound decision-making. As expected, all the sciences, including archeology, underwent a dramatic transformation during this period. This fresh way to view the universe advocated not only for a separation of church and state but also for a rejection of religious dogmas dictated by faith; it underscored the power of the scientific method and rational thinking as a tool to arrive at sound solutions. This new philosophy also promoted the concept of reductionism as a mechanism to provide explanations in terms of ever-smaller units—the essence of things. This method **is still** employed by many scholars **for** scientific inquiry.

Nationalistic archeology

It was during the Age of Enlightenment that the practice of antiquarianism gradually morphed into a more systematic and organized process.[12] Starting in the late 18th century and continuing during the 19th century, the practice of archeology changed from individual efforts focused on personal collections to larger amalgamations of artifacts that were made available to the public for viewing and appreciation. These collections took the form of private or state-run exhibitions. These institutions became a source of national pride, and many Western European nations and universities competed to procure and possess these precious antiquities. Nationalistic fervor was so extreme at times that the process of acquiring the artifacts was more like stealing or illegal trafficking. These precious relics have rarely been returned to their country of origin, although their extradition is constantly being demanded. Today visits to any of a number of leading

Fig. 4 Giovanni Battista Belzoni (https://upload.wikimedia.org/wikipedia/commons/9/96/Giovanni_Battista_Belzoni_by_William_Brockedon.jpg).

museums, such as the Louvre and the British Museum, among others, provides ample evidence of the systematic sacking that occurred.

During the rise of the nationalistic movement, it was common practice for prominent museums and institutions to employ individuals to travel, collect, and bring back valuable artifacts and, in some instances, even parts of monuments. One notorious effort to acquire precious relics dates back to the early 1800s when Henry Salt, the British Consul to Egypt, acting on behalf of the British Museum, hired the well-known Italian archeologist/explorer Giovanni Battista Belzoni (1778–1823) (Fig. 4) to collect and send to England large number of artifacts.[13] Belzoni's efforts single-handedly contributed to the growth and importance of the British Museum. Belzoni, nicknamed "the Great," was responsible for bringing to the British Museum from Thebes the famed statue of Young Ramesses II (the Younger Memnon), a 7-ton bust that is currently exhibited conspicuously as part of the Egyptian Halls (Fig. 5). Curiously, the shipment to England of this precious statue was authorized by Muhammad Ali, at the time the Pasha (high-ranking officer commissioned by the Ottoman Empire) of Egypt.

Fig. 5 Rameses II, at the British Museum (https://en.wikipedia.org/wiki/Giovanni_Battista_Belzoni#/media/File:Egypt-statue.jpg).

Another well-known case of archeological vandalism during this epoch was the removal and relocation to Britain of low and high reliefs as well as metopes (rectangular sculptures occupying a space as part of the architecture) from the Parthenon in Athens (Fig. 6) by Thomas Bruce, seventh Earl of Elgin[14] (Fig. 7). This incident has been the source of extraordinary consternation among scholars, heads of state, and the public in general.

The Parthenon was built on the Acropolis during the Golden Age of classical Greece between 447 and 438 BCE under the direction of the architect, sculpture, and painter Phidias. The exquisite marble sculptures portrayed scenes commemorating the Athenian Cult and mythology (Fig. 6). The friezes carved in low relief decorated the building high around all four sides of the building. The architrave (molding) around the columns is in high relief while the pediment (triangular gable) sculptures were carved in the round.

Fig. 6 Marble sculpture from the Parthenon (https://en.wikipedia.org/wiki/Elgin_Marbles#/media/File:Dionysos_pediment_Parthenon_BM.jpg).

The story of how the marbles ended up in the British Museum begins in 1798 when Thomas Bruce was named ambassador and Minister Plenipotentiary of the British Majesty to the court of the Sultan of Turkey, Selim III. At the time, Greece was under the control of the Ottoman Empire. Prior to his departure for Greece, the Earl of Elgin contacted British government officials and propositioned them to cast and draw the sculptures and friezes of the Parthenon. The project was turned down, but Thomas Bruce decided to do it himself with the assistance of a Neapolitan painter named Giovani Lusieri. Somehow the original intent of the project was radically transformed, and instead of making casts and drawings, Lusieri in 1801 began to dismantle and remove sculptures from the Parthenon and other temples within the Acropolis. This was done under the direction of Thomas Bruce. The process of removing the art from the Acropolis lasted until 1812, and during those 11 years the Earl removed about half of the available sculptures from the Parthenon, Propylaea, and Erechtheum.[15] Bruce paid for the dismantling and shipment to Britain. At that time his intentions were to use the priceless art to decorate his mansion in Scotland. However, his ex-wife destroyed his plans when she sued him and ruined him financially during their divorce. In 1816, Thomas Bruce sold the loot to the British government for the equivalent today of about 6 million pounds (approximately 7.8 million dollars)—a bargain considering that the art was and still is priceless.

As might be expected, this series of events generated turmoil in international circles. In the age of nationalistic archeology, part of the British populace supported the acquisition, alleging that the relocation was done to protect the precious art considered to be

Fig. 7 Seventh Earl of Elgin (https://en.wikipedia.org/wiki/Thomas_Bruce,_7th_Earl_of_Elgin#/media/File:7th_Earl_of_Elgin_by_Anton_Graff_around_1788.jpg).

the heritage of humanity. However, others characterized the incident as an act of vandalism and robbery. A public hearing was conducted in Parliament, resulting in the exoneration of Thomas Bruce. In his defense, he stated that he was provided with a permit by the Ottoman Empire to remove and transport the sculptures out of Greece. This document has never been found, and it is doubtful that it ever existed. Today the sculptures that were not removed from the Parthenon by Thomas Bruce are in the Acropolis Museum. Empty wall spaces are left as part of the exhibition in the positions of the missing freezes, as if Greece is still waiting for the art to be returned to occupy its rightful place.

The start of excavations

In the first half of the 17th century, a number of ancient sites started attracting the attention of pioneering archeologists interested in conducting larger-scale excavations, as opposed to minor digs in search for artifacts. However, at the time the methodology for excavating was still poorly developed and the procedures were not systematic in nature. The excavations were largely conducted in a disorganized manner. These early enterprises had a tendency to be hazardous. Early archeologists were not interested in the context in which the artifacts were found, and the concept of stratification was nonexistent.

In this environment, a number of excavations were initiated. In Italy, for example, works on the ruins of Herculaneum and Pompeii were initiated in 1738 and 1748, respectively.[16] These efforts were funded by the King of Naples, who had a keen interest in recovering the treasures buried under the ash generated by Vesuvius in CE 79. He was convinced that since the populations had been unable to evacuate in time, considerable wealth would be found within the ruins.

In Britain during this period, a number of well-known sites were studied, including Stonehenge and other megalithic locations. William Stukeley was one of the most important early archeologists who attempted to date Stonehenge, suggesting that the structures were pre-Roman Druidic in origin.[17] Other major excavations during the first half of the 19th century included the Assyrian palace of Sargon II unearthed by Paul Emile Botta, as well as the ruins of Babylon and the Library of Ashurbanipal discovered by Austen Henry Layard. In addition, in the 1870s, the ruins of Troy were identified and stratigraphically excavated. This finding generated considerable popular interest since many had considered the city a fable. Accounts of the existence of the site were derived from Homer's epic poem *The Iliad*, which describes the city as the location of the Trojan War.

Development of methodological archeology

In the early 1800s, some novel ideas began to emerge in the field of archeology. One was the principle of stratigraphy and another was the concept of "deep time" or geological time. These two notions have since become essential concepts of archeology.

Stratigraphy, as it applies to archeology, states that layers of overlapping sediments can be employed to trace successive periods of geological time. The notion of stratigraphy was actually borrowed from the field of geology. It was during the late 18th century and early 19th century that, while performing routine surveying of English mines, the self-taught English geologist William Smith perceived layers of rocks aligned in a predictable order.[18] Furthermore, the same sediments were observed in the same position relative to each other in different locations. Smith also noticed that each sediment layer was characterized by specific groups of fossils that were found in corresponding layers in various parts of England (Fig. 8). This keen observation prompted him to formulate the "principle of faunal succession," which states that fossils are arranged vertically, with older ones in deeper sediments, and that their positions can be used to identify equivalent layers in various regions of the country. For his efforts, he was nicknamed "Strata Smith" by scholars of his time. In an effort to publish his work, he sold his fossil collection to the

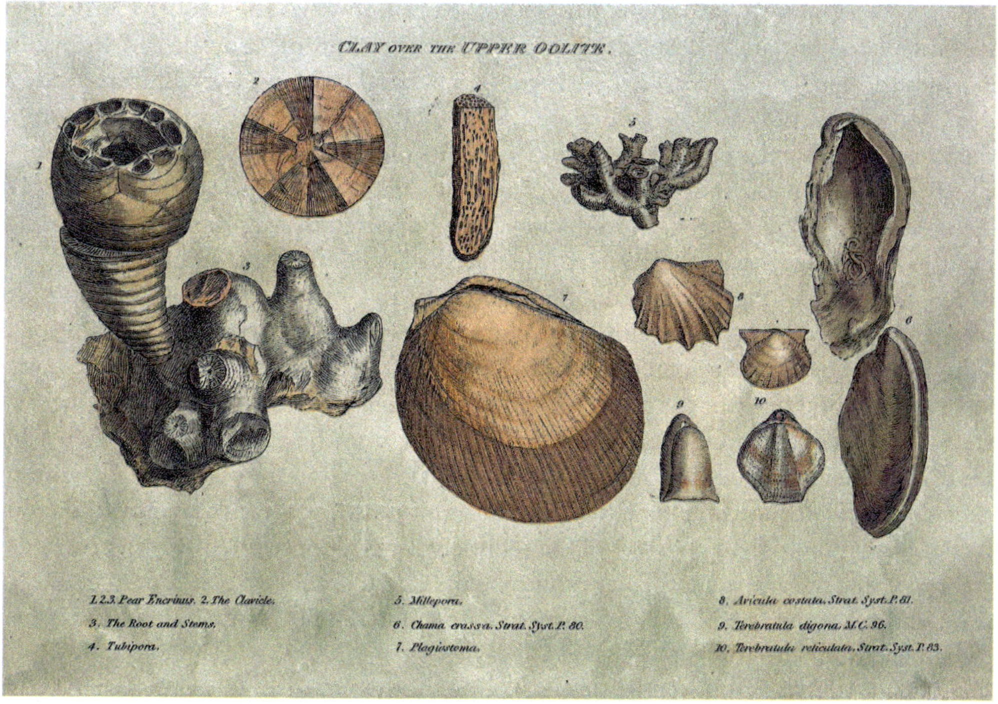

Fig. 8 Engraving by William Smith illustrating fossil deposit (https://upload.wikimedia.org/wikipedia/commons/b/b0/Smith_fossils2.jpg).

British Museum for £700 (equivalent to £56,000 or $72,000 today), but went to prison for failing to clear his debts.[19]

The sense of chronology inherent in Smith's principle of faunal succession was later applied by archeologists and antiquarians, such as the Danish scholar Christian Jürgensen Thomsen, to organize hominin artifacts in time based on the three-age system (stone, bronze, and iron) that he helped to refine.[20] After being appointed director of the antiquarian collections in 1816, which later evolved into the National Museum of Denmark, Thomsen undertook the task of characterizing all the artifacts in the exhibits in the context of the three-age system. By grouping artifacts that are found together in equivalent time periods, he was able to establish an empirical-based system for organizing items into well-defined prehistoric epochs. Independent of the scientific merit, this provided for a system that facilitated communication among scholars using the same nomenclature.

The concept of deep time also emerged as new methodologies were being introduced into the field of archeology in the late 18th century. The origins of the Earth have preoccupied scholars and theologians for many centuries. Prior to the realization of geological time, the Earth was envisioned as being created relatively recently, on October 23, 4004 BCE. This remarkably exact date was extrapolated from specific events recorded in the Old Testament. Interestingly, this genesis (later known as the Ussher Chronology) was calculated by James Ussher (1581–1656), Archbishop of Armagh and Primate of All Ireland, by reading the Old Testament literally and working back in time.[21] However, some keen observations of exposed sediments at the Siccar Point promontory on the east coast of Scotland prompted geologist James Hutton (1726–97) to start thinking of the Earth's age on a much different timescale that was much older than the Old Testament-based calculations would allow. While contemplating the layers in the cliff, Hutton and other scholars began to realize that the large number of strata observed seem to have been laid long ago in a discontinuous process, in which periods of deposition were interrupted by lack of sediment accumulation. These observations led Hutton to formulate the notion of unconformity of sediments, which made him realize that it must have taken a long time for the layers to form. Hutton and others felt that while looking at the endless layers of sediments, they were "looking into the abyss of time".[22] In 1788, Hutton further wrote about the Siccar Point Strata, "we find no vestige of a beginning, no prospect of an end," referring to the seemingly infinite number of layers and the slow rate of formation of the sequences of the sedimentary and nonsedimentary layers. Today, we know that the age of our planet cannot be measured in biblical time but in geological time, and it is roughly 4.55 billion years old[23] or about 750,000 times older than Ussher's calculations.

This idea of deep time as opposed to a biblical timescale was also advocated by the human archeologist Jacques Boucher de Crèvecœur de Perthes (1788–1868), who proposed that hominins existed in Europe as early as the Pleistocene [2.6 million years ago (mya)]. In his book *Antiquités celtiques et antédiluviennes* (1847), he includes calculations and drawings of lithics to support the ancient existence of humans in Europe.[24] Initially his colleagues at the French Academy of Sciences disregarded his findings and ideas, not because the evidence was lacking but because of their novelty. In addition, the fact that Boucher de Perthes was an eccentric provincial bureaucrat with a reputation of being a romantic and possessing a vivid imagination did not help.[25] Many felt that he was overinterpreting the data. In fact, members of the French Academy of Sciences were so incredulous of Boucher de Perthes' findings that they requested the English geologist Joseph Prestwich and the archeologist John Evans to verify the authenticity and the antiquity of his discoveries. Years later, as additional similar artifacts were discovered throughout Europe, Boucher de Perthes' ideas started gaining acceptance. At the International Exposition of 1867 (French Exposition Universelle d'Art et d'Industrie), the second world's fair held in Paris, Boucher de Perthes presented a number of bifaced flints (Fig. 9) that he found in association with elephant and rhinoceros bones. These findings were well received by the attendees.

Another fundamental concept that was ushered in during the age of methodological archeology was the realization that all artifacts could be of potential informative value. In other words, archeologists began to realize that both deformed and "ugly" as well as the

Fig. 9 Biface of Menchecourt-les-Abbeville, exhibited at the Universal Exhibition of 1867 (https://upload.wikimedia.org/wikipedia/commons/thumb/5/56/Biface_de_Boucher_de_Perthes_MHNT.jpg/1280px-Biface_de_Boucher_de_Perthes_MHNT.jpg).

beautiful artifacts should be collected, carefully examined, and studied. It is likely that this dramatic change in procedure came about as scholars began to see the field not as a collection process driven by a desire to exhibit the curious, but as a rigorous scientific discipline aiming at investigating the past. This new philosophical approach to conducting archeology reflects a fundamental change in the field. Scholars began inquiring about the culture and ideas of the people who made the artifacts and how they lived; they no longer focused only on the inherent esthetics and monetary value of the items. Now practitioners of archeology were primarily interested in understanding the past. This procedural change represented a quantum leap from the collection-driven outlook of antiquarianism, which was often destructive and seemed more like a treasure hunt.

This new outlook, for example, was reflected in the new practice of meticulously collecting broken pieces of pottery with little esthetic value. Distinct attempts began to be made not to leave anything of informative value behind. In this new era, fragmented pieces of pots and ceramics began to be examined carefully and systematically put together. Items, large and small, were gathered from the sites and final judgments on their informative value were not reached until they had been examined and analyzed in the context of the rest of the collection and theories at hand.

Contemporary archeology

Survey

Today archeology tends to follow specific sets of guidelines and methods designed to minimize damage to the sites and artifacts and maximize the amount of information extracted from the search or dig. Usually the first step prior to starting an excavation is to perform a survey, which serves different purposes. Primarily, it locates the site(s) of interest, delineates the boundaries of the site, and allows the archeologists a good opportunity to sample the site. Surveys also may be the end or aim of a project, providing crucial information about past human activities in a regional context.

Preliminary search

A site can be aperçu in a number of ways. Information on a site may be gathered prior to arriving at the site by probing into church, synagogue or mosque records, or searching library records for previous investigations reported in the published literature and/or archives based on similar expeditions from the past. It is likely that fieldwork is preceded by a phase of computer work in which formal and informal written records, photographs, drawings, and internet search engines are consulted.

In addition, interviews with members of contemporary populations may provide useful information prior to fieldwork. The history of a group of people is often imprinted in the folklore and oral tradition of present-day populations. Stories, legends, songs, and dances often provide a treasure trove of information. Experiencing the music, art, and beliefs of people usually provides clues on their origins and relationships with other populations. In addition, by interviewing individuals, archeologists can determine where high concentrations of artifacts have been found. This type of information is extremely useful for the future success of the excavation, in addition to saving time, effort, and funds.

Another method of gathering preliminary information is the assessment of the environmental resources that are and were available in the area of interest. Historic and prehistoric people were selective in deciding where to settle. Given the choice, populations inhabit regions that provide appropriate shelter and food for survival. These are conscious decisions that were made based, for example, on the expected migration of herbs, proximity to streams or lakes, and the fertility of the land. By ascertaining the resources provided by the surroundings, an archeologist might be able to delineate the location of past settlements. However, to make these determinations, investigators need to be familiar with the conditions of the habitat at the putative time of habitation. In other words, if the population in question dates to the Early Neolithic, then it is necessary to be familiar with the climate and available resources at that time. The conditions of a locale may change in a matter of centuries. Wetlands, for example, may be transformed to deserts or vice versa in a relatively short period of time.

Walking survey

A surface or walking survey usually follows the preliminary survey. In this process, archeologists systematically comb through the area of interest step by step, carefully going over every space, identifying artifacts, and recording their locations. This type of survey is inexpensive and fast, but suffers from failing to detect items below the surface; it may also be misleading regarding the location of areas of interest. Nevertheless, it is a logical starting point prior to a more extensive and time-consuming excavation.

Remote reconnaissance

Remote reconnaissance is gaining acceptance as technology improves and the cost of instrumentation diminishes. In remote reconnaissance, information on specific items is obtained without making physical or visual contact with the object. This type of approach is usually not practical for the initial stages of the survey, which tend to encompass extensive surface areas, but may be very informative once specific locations of interest have been delineated.

Examples of remote reconnaissance include technologies such as metal detection, electrical resistivity, ground-penetrating radar, and magnetometers. In the case of electrical resistivity, the resistance between negative and positive poles is determined as electricity is applied. This technique operates based on the principle that objects underground conduct electricity differently. For example, an underground stone coffin is an inferior medium to transmit a current, thus if the electrical poles are placed on either side of the crate and a closed electrical circuit is applied, greater resistance would be registered compared to an area made up of just soil.

Another example of indirect survey is ground-penetrating radar. This method is based on the same principle as conventional radar, whereby a radar pulse in the form of an electromagnetic pulse in the microwave band of the radio spectrum is directed toward the medium and a returning signal is detected. Suitable strata for this application include rocks, dirt, ice, fresh water, and cement. Artifacts, features, and layers reflect the emitted wave differently and the sensor picks up the return signal accordingly. The travel time of the reflected signal is directly proportional to the depth of the structure.

Magnetometers are also employed to identify spatial variation in the Earth's magnetic fields generated by certain metals. However, these sensitive instruments are limited to detecting only ferrous metals. In the field of archeology, magnetometers are regularly employed to locate buried object in marine and terrestrial environments, including shipwrecks, fire hearths, and walls of bricks or stones made up of basalt and granite that contain magnetic material. Walking tracks and roadways can sometimes be mapped due to differential compaction in magnetic soils or disturbances in the clays. In all of these indirect methods, the data are graphed to generate a three-dimensional map.

Although metal detectors have been traditionally used by hobbyists or amateurs in search of treasure, the instrument is gradually gaining greater acceptance among archeologists. In its simplest form, a metal detector consists of an oscillator (electronic instrument used for the purpose of generating an electric signal) that generates an electrical current. The current then passes through a coil producing an alternating field. Fluctuations in the magnetic field resulting from the presence of an object are detected by a magnetometer, which often produces a sound and a change in a meter. With this type of device, metals such as iron, nickel, copper, brass, aluminum, tin, lead, gold, silver, and bronze are detected. Some instruments are even capable of discriminating among different metals and alloys by the different sounds emitted by the various metals.

The negative reputation experienced by nonprofessional detectorists is not totally unfounded, since these instruments are easily accessible to the uninformed layperson, being relatively inexpensive and small. Furthermore, in the hands of the public, metal detectors are not necessarily designed to examine an archeological site and to study its historical context, but simply to find the object and dig it out. Therefore, in the wrong hands, metal detectors can be destructive by altering or obliterating very valuable evidence. Yet the instrument possesses some inherent qualities for the archeologist. For one thing, it provides fast and easy identification of locations of interest for further careful examination. In addition, due to the huge number of amateurs searching, it allows for the exploration of many areas not frequented by professionals. Of course, this scenario assumes that the layperson would eventually notify the professionals and allow experts to intervene and conduct appropriate studies.

Aerial and satellite reconnaissance

Aerial and satellite reconnaissance provides archeologists with a powerful new investigative tool. Not only does it enable researchers, for the first time in human history, to see things that were previously undetected, but also it affords a wide-ranging survey of an area. These extensive scans provide both the identification of structures and information on the context of the surrounding regions. In other words, scholars are able to assess, for example, if the nearby lands are or were capable of sustain human and animal habitation as well as crops. Archeologists are no longer restricted to limited surface areas afforded during a walking survey; they can also visualize extensive regions and examine entire landscapes in one view. In addition, surveys from airplanes, drones, and satellites are noninvasive and can pinpoint locations of interests. These methods are very useful in setting the stage for further detailed ground inspections and excavations.

The relatively new technology employed in satellite imaging systems relies, for example, on thermal or infrared emissions capable of penetrating about 1 m in depth. Infrared, with its longer wavelengths compared to visible light, is able to identify objects beneath the Earth's surface and identify items not visible during a walking survey. Using the various wavelengths available to a satellite source, the system allows for the identification of different types of soil, plant life, topology, and edifices, since they all provide specific spectral signatures. This type of methodology has been very useful in finding manmade structures buried in the ground and/or concealed by vegetation. The application of these

Fig. 10 LiDAR view of the Copan Acropolis (https://www.researchgate.net/figure/LiDAR-Point-cloud-of-the-Copan-Acropolis-classified-by-elevation_fig9_283852300).

state-of-the-art physical science tools to archeology is known as archaeometry, and this is now a major part of archeology.

Today, specific aerial and space sensor technologies are routinely used to detect and classify objects on the ground and under water (e.g., lakes and oceans), and are capable of penetrating through thick vegetation. These systems employ signals, such as visual, lasers (LIDAR), electromagnetic, thermal, and infrared radiation emitted from airplanes, balloons, remote-controlled drones, and satellites. They all generate high-resolution images and 3D topologies. LIDAR utilizes laser return signals and wavelengths to reconstruct 3D pictures of targets (Fig. 10).

In addition to architectural structures, such as temples and habitation, humans tend to leave behind residues of their existence. We throw away and abandon material not needed, including garbage and other by-products that signal previous human occupation. This debris may create differences in ground texture, structure, and/or elevation. Much of the resulting topological variation can only be observed from a distance, preferably from above. Inspection at close range does not allow effective detection due to a limited perspective. Soil marks, crop marks, and topological variations near ground level could easily go undetected during a regular walking survey.

Considering the rapid progress in this type of technology driven by unrelated fields, such as defense, it is likely that the capabilities of such instrumentation will increase. Progress in satellite imagery may in future allow visualizing of smaller items and identification of the chemical nature of objects underground.

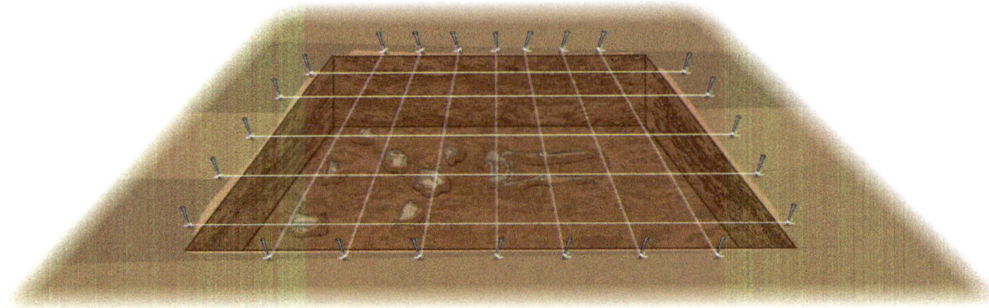

Fig. 11 Grid (http://www.nysm.nysed.gov/programs/archaeology-van-schaick-mansion).

Excavation

Subsequent to surveys, a decision is made whether to continue on to the next step of the project—whether to excavate or not. The decision is not a trivial one since it takes into account a number of issues, including how promising the site is, the size of the area, the potential cost of the project, the difficulty presented by the terrain, the expertise required, etc.

If a decision is made to continue with the project, the area of interest is delineated and the archeologists in the field partition the future dig into grids. Grids are subdivisions approximately 1 ft or so above the ground, usually made of strings (Fig. 11). They divide the site of interest into square areas of equal sizes. A grid is often identified by a datum (in the form of a plaque or slab) or a fixed reference point. In the United States, a datum may be provided by USGS (United States Geological Survey), a government agency that studies natural resources and the hazards that impact them.

Although, in retrospect, the method of partitioning an area into grids prior to the actual digging is a logical step, it was not until the 1920s and 1930s that two field archeologists, Mortimer Wheeler and Kathleen Kenyon, introduced the practice. The grid serves a number of functions. It portions the site to allow for the precise location and documentation of artifacts. It also provides orientation and guidance to organize the dig. Wheeler and Kenyon also initiated a number of other systematic practices that forever changed the discipline, making it more standardized and analytical. One was the excavation of trenches, leaving balks or strips of earth in between the excavation dugouts. This practice allows for the retention of a record of the strata and the stratigraphy of the site (Fig. 12). These alternating series of grooves and soil allow archeologists to keep an intact record of the strata in the various areas of the excavation. Another practice that was introduced at the time was the rapid publication of the excavation, including specific details of the dig such as location of items and relationships to other artifacts and/or fossils. Although this is a logical approach, often specific accounts of fieldworks have vanished or

Fig. 12 Excavation trenches (https://commons.wikimedia.org/wiki/File:Aerial_photograph_of_Maiden_Castle_from_the_west,_1937.jpg).

been distorted as investigators were slow to publish the data formally, and in some cases this never happened. Thus, at the time, a conscious effort was made to archive the information permanently before it was lost. Details are often compromised when left to memory.

These innovations are part of what is known today as modern archeology, and these two archeologists were so influential in this systematic scientific approach that the process is referred to as the Wheeler-Kenyon method. These procedures were later employed in the excavations of sites such as Jericho (1952–58) by Kenyon.

Recording the events

Keeping a detailed record of the excavation is essential. People often think of archeology as a digging activity, but in its proper form, the discipline is mostly recording the findings. Meticulous record-keeping is especially important in archeology because the mere process of conducting the activity is bound to alter or destroy the evidence and data. The soil that is removed cannot be placed back as it was originally. The items retrieved cannot be put back in their exact original positions. Therefore, keeping the best annotation possible is thus essential. Records in the form of notes, written material, maps, graphs, and photos are all that will remain of the original condition of the site. Evidence of the authentic condition of the site only remains in the testimony of the archeologists.

Annotations are made in the form of field notes. This type of record-keeping takes place continuously, during the actual removal of soil from the dig, the careful inspection of the sediments for artifacts, and at night after the physical labor is finished. Directors of excavation routinely carry notebooks, small enough to fit in a pocket and retrieve immediately when required to annotate something important. The information recorded usually includes measurements, quality, and relationships among strata, artifacts identified, and overall work done that day. The notes also include the tests performed on the soils and strata as well as what happened in each section of the grid. This information is accompanied by drawings and hand-generated graphs. Some researchers nowadays annotate with electronic devices. Whether in traditional notebooks or personal computerized devices, the notes read as a diary with many details. The material written in these notes provides a basis for generating final reports and eventually manuscripts for publication.

In addition to notes, a number of other records about the excavation are kept including detailed descriptions of each layer and their relationships to other strata nearby, an account of the features (items such as pits, walls, and ditches) found, drawings, photographs, and a separate document itemizing and illustrating all the artifacts found. Information regarding in which specific part of the grid the sediment was located, its depth, type of soil (e.g., sandy, clay, silty), and possible age is recorded. A list of itemized drawings and photographs pertaining to the layer is generated. The sketches are done from different angles with a key for scale and including distances and relationships within the grid and other nearby features. The drawings highlight the stratigraphy and artifacts discovered.

The recording of artifacts should be done while they are in situ, and subsequently they should be cleaned and cataloged individually, or as a group if they are small. Nowadays the cleaning process should avoid the use of water since porous items such as bones, wood, and fossils tend to absorb fluids that carry with them extraneous contaminants. In the era of DNA profiling and isotope dating, care must be taken to prevent contamination of the interior of artifacts with foreign substances carried in solution by water or other fluids. These impurities or foreign matter could potentially alter the true date of an artifact if the substance introduced is younger or older than the authentic object. Alternatively, it could prevent genetic analysis by interfering with the DNA amplification (polymerase chain reaction or PCR) and/or sequencing reactions. Each item is labeled with a code, which refers to a description in a log, including the specific site of origin. The findings' designations are written in ink on the items themselves over a layer of nail polish or shellac. After drying, the codes are also covered with nail polish or shellac for protection.

Postexcavation analysis

Dating and ancient DNA sequencing

Once artifacts and items have been excavated and collected from the site, the arduous process of studying them begins. This is meticulous work that often takes years. In some cases, the nature of the study is such that the lifetimes of several individuals are devoted to answering specific questions. Thus, it is not unusual for manuscripts based on extensive projects to take years to appear in the published literature.

Initially the artifacts are inspected structurally to assess their general shape, texture, style and patterns of engravings and colors. This coarse examination aims to establish similarities with previously classified collections. These comparisons, in their most basic form, are purely typological, meaning that artifacts are classified according to their morphological characteristics. Invariably, archeologists try to identify other sites with similar artifact assemblages in order to establish common origin and cultural connections.

The postexcavation analysis work is usually allocated to various laboratories, each equipped with different types of instruments and expertise. This multidisciplinary approach is necessary since archeology has evolved from a descriptive to a highly technical science influenced and entwined with a number of other different disciplines from the natural and physical sciences. This new, highly integrative, and multifaceted archeology is known as archaeometry.

In addition, as new technologies are developed, specific artifacts are revisited and reanalyzed to gather information that was previously unattainable. Isotopic dating, material analysis, and DNA sequencing of very ancient fossils are a few technical innovations that allow scientists nowadays to reevaluate items previously stored on museums' storage shelves. The reassessment of the ages of fossils and sequencing of ancient DNA are areas within archeology that have experienced dramatic advances in recent years. This was demonstrated in a landmark publication in 2016 reporting on the nuclear DNA sequencing of 430,000-year-old hominin fossils from Sima de los Huesos, Spain.[26] Prior to this research, the oldest hominin remains sequenced were in the range of 40,000–45,000 years old. This new methodology led to a quantum leap in DNA sequencing by a factor of 10.

Rigorous age determination is paramount in the study of hominin evolution. Accurate dating of items is essential to ascertain issues such as the tempo and complexity of hominin evolution. The resolution of a number of highly controversial questions including the geographical origin(s) of hominins and specifically AMHs, the migration patterns of the various hominin groups, the coexistence and interactions of hominin species, the anatomical, cultural/technological and genetic continuity patterns, and the replacement of populations is dependent on the correct assessment of the age of archeological material.

Relative dating

Relative dating of artifacts is routinely performed in archeology. It is fast and inexpensive and often times it can be done in situ, as part of the dig. Although scientifically it is always desirable to generate absolute dates, at times it is practical and highly beneficial to have a rough idea of the age of a site and/or a certain layer. Occasionally, the circumstances of the terrain are such that none of the absolute methods is applicable, or the funds to perform absolute dating are just not available. In general, this is how relative dating works: archeologists uncover certain remains as part of a stratum. Almost instinctually, they try to establish correspondence between the flora and fauna of known ages and the newly discovered ones. This is done by finding morphological/topological parallelisms using knowledge from zoology, botany, taxonomy, and geology. The idea is that if the ages of specific remains from different sites are known, it is possible that the newly discovered material is of comparable chronology. This approach is a good start, but it is subject to a number of problems. One problem is that biological (species) or geological (types of sediments) markers from different sites do not always correspond in a timescale since, for example, similar organisms, migrate and inhabit regions at different times. In addition, morphology can be misleading due to forces of convergent evolution, as similar anatomy can evolve independently from unique ancestries.

Absolute dating

Isotope ratios

Archaeometry encompasses technological advances in dating artifacts, including radioisotope ratios. The rate of isotopic decay is a commonly employed method to assess the age of artifacts and fossils. Table 1 itemizes a list of isotope techniques, including effective time ranges, currently in use to ascertain absolute dates. All of these systems aim to measure the ratio of isotopes (different forms of chemical elements having different numbers of neutrons) in the substance. Isotopes of the same element possess almost identical properties, but scientists are able to discriminate among them and determine their abundance. Isotope ratio systems function as atomic clocks that start ticking and registering time when, for example, an animal or a plant dies, in the case of $^{14}C/^{12}C$, or when molten rock or lava is created, as in the case of $^{40}K/^{40}Ar$. In the case carbon dating, the ratios of ^{14}C and ^{12}C are measured *directly* in organic matter while in the $^{40}K/^{40}Ar$ system, the amounts of argon in rocks that melted are used to assess *indirectly* the ages of nearby objects of interest. Each system has limitations, specifically in their reliable time range of the estimates and the corresponding degree of error or variation in the dates.

$^{14}C/^{12}C$

The radiocarbon method is the most well-known isotope-type dating system.[27] This methodology was developed by Willard Libby in the 1940s, who subsequently received

Table 1 Techniques for determining absolute age (https://www.geocraft.com/WVFossils/DatingMethods.html).

Years ago: 0, 10, 100, 1000, 10,000, 100,000, 1,000,000, 10,000,000, 100,000,000, 1,000,000,000	
Isotope Techniques	
Uranium to Lead (minerals)	1 million to 4.5 billion years
Rubidium to strontium (minerals)	60 million to 4.5 billion years
Potassium to argon (minerals)	10,000 to 3 billion years
Uranium series disequilibrium (minerals, shell, bone, teeth, coral)	0 to 400,000 years
Carbon 14 (minerals, shell, wood, bone, teeth, water)	0 to 40,000 years
Radiation Exposure Techniques	
Fission track (minerals, natural glass)	500,000 to 1 billion years
Thermoluminescence (minerals, natural glass) and optically stimulated luminescence (minerals)	0 to 500,000 years
Electron spin resonance (minerals, tooth enamel, shell, coral)	1,000 to 1 million years
Other Techniques	
Geomagnetic polarity timescale (minerals)	780,000 to 200 million years
Amino acid racemization (shells, other biocarbonates)	500 to 300,000 years
Obsidian hydration (natural glass)	500 to 200,000 years
Dendrochronology (tree rings)	0 to 12,000 years
Lichenometry (lichens)	100 to 9,000 years

the Nobel Prize in Chemistry in 1960 for his achievement.[28] Libby and collaborators initially tested their radiocarbon dating system using samples of known ages. In their 1949 articles published in the journal *Science*, Libby and colleagues reported on the age of two ancient Egyptian mummies, kings Zoser and Sneferu. Previously dated at 2625 BCE $\pm$ 75 years by historical and archeological records, the radiocarbon methodology yielded estimates of 2800 BCE $\pm$ 250 years.[29] Very soon after this breakthrough, archeologists around the world started benefiting from this procedure.

In retrospect, it is fair to say that Libby's discovery revolutionized the field of archeology. And although the technique is limited in terms of being able to test only organic matter and in its accuracy extending only to the last 50,000 years, radiocarbon age estimation is considered by many as the most important archeological innovation of the last century. This technology allowed archeologists, for the first time, to ascertain relatively accurately the absolute ages of artifacts made up of plant and animal material. In many instances, radiocarbon dating changed relative or inaccurate dating to specific reliable chronologies.

The radiocarbon method relies on the fact that ^{14}C is constantly being generated in the Earth's atmosphere by cosmic rays as they impact atoms of nitrogen, changing them to ^{14}C. Initially, plants incorporate ^{14}C as carbon dioxide, which they use to make carbohydrates through photosynthesis; animals then assimilate this into their bodies secondarily when they eat the plants and other animals. However, as soon as plants or animals die, the intake of ^{14}C from the environment by the organic matter abruptly stops as the radioactive carbon already inside the bodies continue to decay into its nonradioactive form, ^{12}C. This conversion is paramount for this dating system because it sets a clock to determine time. From that point on, the lifeless bodies begin to experience a gradual decrease in ^{14}C and a concomitant increase in ^{12}C. So, the older the fossil or artifact is, the less ^{14}C remains in the matter. In other words, this radioactive decay sets in motion a change in the $^{14}C/^{12}C$ ratio as a function of time that can be measured and translated to the number of years since death.

These calculations require calibration curves since the concentration of ^{14}C changes with time. These adjustments are needed to account, for example, for the impact of carbon emission from fossil oil and gas since the carbon in fossil fuel is entirely ^{12}C due to its very old age. Thus, release of carbon emission would impact date assessments by generating inflated time estimates. Conversely, the atmospheric nuclear testing that occurred during the 1950s and 1960s has increased the concentration of ^{14}C in the biosphere and within living tissue. If this is not compensated for during the time estimations, it will make fossils and organic artifacts appear younger than they actually are.

$^{40}K/^{40}Ar$

The ^{40}K-^{40}Ar dating method relies on the radioactive decay of radioactive potassium 40 to radioactive argon 40 in minerals and rocks subsequent to melting. It is ideal for dating material such as lava, which solidifies rapidly. Due to the abundance of potassium on our planet, this technique can also be applied to date other substances such as mica, clay, minerals, and tephra.

As rocks are heated to high temperatures and melt, any ^{40}Ar contained in them escapes into the atmosphere in gas form. This sets the atomic clock to zero, since there will be no ^{40}Ar left behind in the recrystallized rock. When the rock resolidifies, it starts accumulating ^{40}Ar again as the newly generated Argon remains trapped in the rock. Thus, any ^{40}Ar found after the melting event must have been produced subsequent to the solidification of the rock. Investigators only need to measure accurately the ratio of $^{40}K/^{40}Ar$, consider the rate of decay of ^{40}K, and, based on those values, estimate the time elapsed since the rock resolidified. ^{40}Ar is measured using a mass spectrometer, an instrument that identifies and quantifies isotopes of different atoms according to mass.

However, unlike the $^{14}C/^{12}C$ system, $^{40}K/^{40}Ar$ does not probe the fossil or artifact itself, but the rock or lava encapsulating or surrounding it. In other words, the ratio of $^{40}K/^{40}Ar$ acts as a proxy to estimate the age of the item nearby. Therefore, it is an indirect method and may introduce errors if the death of the organism or creation of the artifact does not coincide with the melting of the rock. For this reason, it is essential that the relationship between the rocks being dated and the fossil or artifact is well-established. For archeologists, the ideal scenario is for lava flow to cover the strata containing the fossils or artifacts, immediately after death or creation, respectively.

In contrast to the $^{14}C/^{12}C$ system, which is particularly useful for relatively recent dates, $^{40}K/^{40}Ar$, due to its extended half-life of 1.3 billion years, is effective for very old archeological material. In addition, since ^{40}Ar takes a while to accumulate after the rock melts, there is insufficient amount of the isotope present in the resolidified rock to allow for reliable readings for periods of less than 100,000 years. However, even rocks 4 billion years old—close to the age of the Earth—can be dated with this system. Geological events, such as the impact of a meteor about 65 mya that created the Chicxulub crater in the Yucatan Peninsula, Mexico and likely prompted the extinction of most of the dinosaurs, have been dated with $^{40}K/^{40}Ar$.[30] More recent chronologies established with this methodology include the notorious occupations by *Homo habilis*, *Paranthropus boisei* (an australopithecine), *Homo erectus*, and *Homo sapiens* 1.9 mya, 1.8 mya, 1.2 mya, and 17,000 ya, respectively, in Olduvai Gorge, Tanzania.[31]

Uranium series

A number of other isotope techniques for dating have been developed over the 20th century. Among them is the uranium disequilibrium series capable of determining the age of substances such as shells, bone, teeth, corals, and marine sediments.[32] As the name indicates, this system encompasses a number of different isotope combinations that can be measured independently, each providing different time ranges (Table 2). The uranium series differs from other radioisotope methods in that it does not access the accumulation of a stable end isotope produced from decay, but it ascertains the time it takes to re-establish equilibrium between the daughter and parent isotopes (e.g., ^{230}Th and ^{234}U, respectively, in Table 2). Table 2 illustrates eight different daughter/parent isotope combinations that are part of this uranium disequilibrium series. Each of the modes

Table 2 Family of U-series methods (https://en.wikipedia.org/wiki/Uranium%E2%80%93thorium_dating).

U-series dating methods[2]			
Isotope ratio measured	**Analytical method**	**Time range (ka)**	**Materials**
$^{230}Th/^{234}U$	Alpha spec. mass spec.	1–350	Carbonates, phosphates, organic matter
$^{231}Pa/^{235}U$	Alpha spec.	1–300	Carbonates, phosphates
$^{234}U/^{238}U$	Alpha spec. mass spec.	100–1000	Carbonates, phosphates
U-trend	Alpha spec.	10–1000(?)	Detrital sediment
^{226}Ra	Alpha spec.	0.5–10	Carbonates
$^{230}Th/^{232}Th$	Alpha spec.	5–300	Marine sediment
$^{231}Pa/^{230}Th$	Alpha spec.	5–300	Marine sediment

The question mark denotes uncertainty.

described in Table 2 is named after the parent and daughter isotope measured, indicating the materials that are usually tested.

The uranium to lead (^{238}U to ^{206}Pb and ^{235}U to ^{207}Pb) system, for example, determines the ratios of two different sets of radioisotopes. These are used to date the formation of minerals and provide ways to find out the ages of associated artifacts and fossils. The ^{238}U to ^{206}Pb chain exhibits a half-life of 4.47 billion years, while the ^{235}U to ^{207}Pb decay is 710 million years. The system has been reported to be sensitive enough to date rocks as young as 1 million years.[33] Clair Cameron Patterson developed this isotope system and in 1953 dated the age of the Earth at 4.55 billion years, $\pm$70 million years, using material from the Canyon Diablo (Northern Arizona) meteorite. This estimation of the Earth's age is still considered the most reliable.

^{87}Rb/^{87}Sr

The decay of rubidium to strontium (^{87}Rb to ^{87}Sr) has also been widely used to date minerals, especially samples that have not been subject to repeated alterations such as serial melting. It has been a useful tool to estimate the age of meteorites and lunar rocks, as these types of samples have not been subject to weathering since their initial formation. In addition, because rubidium is an abundant trace element in the Earth's crust, the ratio of ^{87}Rb to ^{87}Sr can be conveniently employed to date the last melting event of various types of rocks. As with the other radioisotope systems described above, the principle is the same: the initial amount of the parent ^{87}Rb is ascertained and the age of the sample is determined by measuring the Rb and Sr concentrations and the ^{87}Sr/^{86}Sr ratios. This isotope system was used to date Precambrian microfossils from Sinai to 934 $\pm$ 80 mya.[34]

Radiation exposure techniques

Radiation exposure methodologies take advantage of the impact of radiation on materials. Radiation may cause spontaneous split or decay of an element, leading to a release of energy. The splitting or decay occurs at a rate directly proportional to the dose of the radiation received. Thus, if the radiation dosage for a particular location is known, it is possible to calculate the age of artifacts. The examples described below represent radiation-based dating systems routinely used.

Fission-track

Fission-track is a procedure used for dating minerals and glass. It takes advantage of the damage produced by the spontaneous fission (radioactive decay of an element as it splits into smaller atoms) of ^{238}uranium on minerals.[35] Although a number of isotopes undergo spontaneous fission, only ^{238}U splits at a rapid enough rate to measure its effect over geologic time and hence be useful in dating. The damage produced by the energy released from the fission of ^{238}U is evident in the tracks left in the mineral or glass (Fig. 13). These indentations are formed as the energy is given out and travels through the substance. The

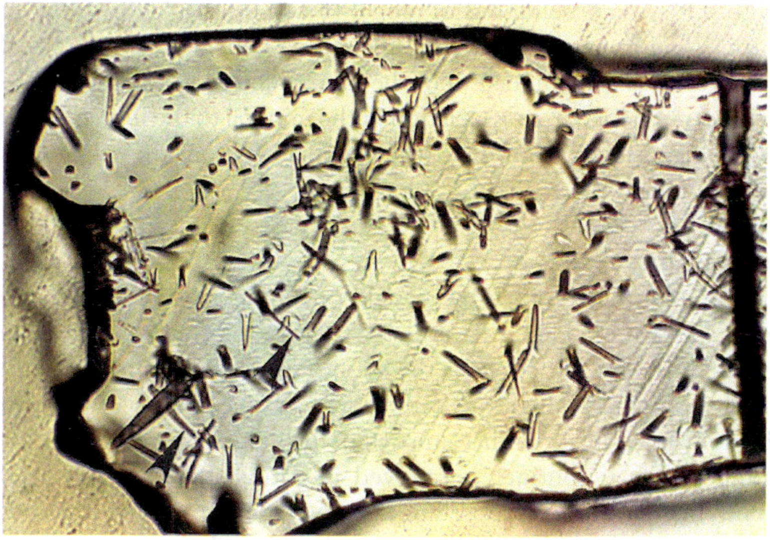

Fig. 13 Damage trails or tracks left by fission fragments in uranium-containing materials (https://pangea.stanford.edu/research/groups/thermochronology).

damage looks like tube-shaped holes. These holes in the matter are counted using an instrument as basic as a conventional light microscope.

In brief, fission-track dating works as follows. The amount of naturally present fission-tracks in the mineral is ascertained by microscopy. The total concentration of uranium in the matter is then assessed by irradiating it with neutrons to produce a second population of tracks (alterations). The actual age of the mineral is the ratio of the naturally present spontaneous fission-tracks to neutron-induced fission-tracks. Since spontaneous fission occurs at a known rate, it is possible to calculate age.

The formula used to calculate the age of the mineral is A (age) $= N \times Ds/Di \times 6 \times 10^{-8}$, in which N is the dose of neutron administered during the procedure to elicit fission expressed as neutrons per square centimeter, Ds is the original (spontaneous fission) track density, and Di stands for the number of induced fission-tracks while 6×10^{-8} is the constant rate of fission.

Fission-track exhibits certain advantages as well as shortcomings for dating. The procedure is relatively simple, inexpensive, and fast. Other convenient characteristics include the abundance of ^{238}U in many types of minerals, and the damage generated by fission is easily quantifiable with a small sample. In addition, is not subject to contamination, which could otherwise compromise the accuracy of the dating. However, its time-range limitation is 500,000 to 1 billion years. Samples younger than 100,000 years contain too few tracks to be reliably counted to generate age estimates. Therefore, fossils and artifacts pertinent to the evolution and migration of anatomically modern humans (AMHs) are not dateable with this system, being out of range. In addition, as with all the methods that associated rocks and minerals to artifacts, the dating generated is indirect; the age estimates are derived from the minerals, not the item of interest. As previously stated, indirect dating relies on the accurate association of the mineral to the item of interest. In other words, if rocks had moved from their true original location, they would provide dates unrelated to the item.

Thermoluminescence

The thermoluminescence emitted by crystal-containing substances is routinely used to date materials. This methodology assesses the release of energy from exited electrons trapped within the crystal lattices since the last heating and reformation of the crystal lattices within the substance. This sequestered energy that accumulates over time is a measure of the ionizing radiation that impacted the lattice since it was last formed.[36] Thermoluminescence is routinely employed in dating items, such as ceramics, lava, and sediments, that have been exposed to ionizing radiation.[37] The methodology is also used to determine the age of buried archeological artifacts that at some time were exposed to radiation mainly from sunlight. Radioactive elements in the soil also produced radiation that impacts minerals, contributing to the age determination. For ceramics, this is the time elapsed since the item was last fired. This technology has been employed to date a number of important findings including artifacts belonging to the Valdivia culture of coastal Ecuador, one of the oldest of pre-Columbian America,[38] and Egyptian pottery from the Giza plateau dating back to the Fourth Dynasty in the Old Kingdom.[39]

As previously stated for other indirect methods, this system relies on the accurate assessment of the physical relationship of the artifact(s) of interest and the thermoluminescence-emitting material. Its useful range for dating extents to 500,000 ya and is particularly useful in instances where $^{14}C/^{12}C$ dating is not possible due to the type of substrate of interest such as sediments that are not carbon-based. Although its accuracy is not as precise as carbon dating, experiments have shown close correspondence between the two techniques (Fig. 14).

Thermoluminescence dating is based on the fact that solids-containing crystals absorb energy from ionizing radiation (alpha, beta, gamma, cosmic rays, high-energy particles). Sunlight is a major source of ionizing radiation, and materials are constantly being bombarded with it. This energy frees electrons from the rigid lattice structure of the crystal, allowing them to move around, some eventually becoming trapped in imperfections or electron-holes within the lattice. This occurs at relatively low ambient temperatures. This phenomenon is accumulative in nature and directly proportional to the radiation received from the sun and the immediate surroundings. An external stimulus of energy, such as heat (about 450°C/842°F) or intense light, causes the electrons to be released from the traps within the lattice. As the electrons are free from the holes, they return to a lower energy level (energy state), and in the process they emit light or photons.

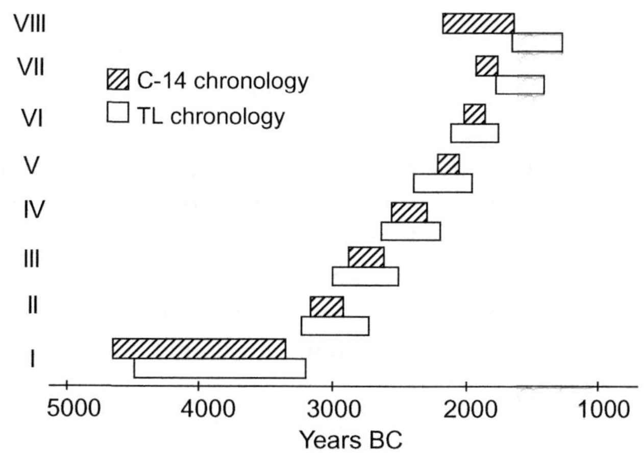

Fig. 14 Correspondence between 14C chronology and thermoluminescence dating. The *X*-axis represents age while the *Y*-axis denotes the eight main cultural stages of the Valdivia culture of coastal Ecuador (http://hbar.phys.msu.ru/gorm/dating/tldating.pdf).

The amount of light produced can be quantified. Thus, by estimating the rate of ionizing radiation received in a given locale where the archeological artifact of interest is found, the amount of thermoluminescence released after the stimulus (e.g., heat or light) is used to calculate the time elapsed since the formation of the crystals within the mineral. The longer the time radiation impacts crystals, the larger the amount of trapped electrons and the greater the energy (photons) released when the stimulus is provided. This relationship is mathematically expressed in the formula: age (date) = accumulated dose of ambient radiation/dose accumulated per year. Every time the matter is heated or exposed to sunlight, it undergoes a zeroing or resetting event in the history of the material by removing the preexisting trapped electrons and allowing the electrons to start accumulating again.

However, there are a number of drawbacks associated with this method. One shortcoming is the potential error in estimating the amount of ionizing radiation impacting the sample, considering that certain types of terrains experience geological changes with time. A second flaw is that the procedure is destructive in the sense that the material being tested is consumed.

Electron spin resonance

In electron spin resonance (ESR), the dosage of radiation on matter since its formation is used to assess age. This method determines time by measuring the unpaired electrons sequestered within substances such as quartz, flints, carbonate crystals, and fossil remains. ESR determines age by quantifying the amount of high-energy electrons that accumulate since the matter was formed or previously heated.[40]

This dating system has been used to determine the age of human remains such as dentition as well as artifacts like charred flints and ceramics.[41] The methodology has been particularly useful in identifying the age of ancient teeth. Together with the related thermoluminescence technique, also based on radiation exposure (see previous section), ESR provides information on the critical time period during the late middle to early late Pleistocene (150,000–50,000 ya) where $^{14}C/^{12}C$ dating is out of its usable range. This technology was first applied in the field of archeology in 1975 by Motoji Ikeya to date stalactites and stalagmites.[42] More recently, ESR was employed to date accurately the early AMH samples from the Qafzeh Cave in Israel to approximately 90–120 ka.[43] ESR was also instrumental in dating hominin fossils and burned flints from the Skhul and Kebara Caves in Israel.

In dating fossils by ESR, archeologists take advantage of the fact that electrons exist in two states: a ground state and excited state. In human remains such as dentition, when teeth are formed all the electrons are in the ground state. However, as time passes after tooth formation, increasing amounts of electrons in the ground state start transforming into an excited state. These changes into higher energy state occur when radioactive radiation impacts the tissues. As previously discussed, the sun is the main source of radioactive radiation. In these high-energy states, these electrons travel through the matter, eventually reaching areas within the substance capable of trapping them. These sites are the impurities, defects, or holes found within the lattices of minerals. These captured electrons become paramagnetic regions that emit detectable characteristic resonance signals. The strength of these signals can be measured. Furthermore, the intensity of the signals is directly proportional to the number of traps and energized electrons in the matter. The age of an artifact derived from ESR can be expressed mathematically as follows:

$$\text{Age} = \frac{\text{Accumulated dose}}{\text{Dose rate}}$$

With this system, two variables need to be ascertained to determine age: the accumulated dose and the dose rate. The accumulated dose is the amount of radioactivity that a tissue (such as a tooth) has received since it was buried. This value is the total resonance signal emitted by the object. The dose rate is obtained directly by determining the amount of radioactivity to which the sample and its surroundings are exposed. This value is a function of the types and amounts of radioactive elements in the sample and in the immediate area as well as the irradiation from the sun. Since the amounts of radioactivity generated from internal sources within the tissue (e.g., enamel) and its surroundings tend to differ, at least two different measurements are required to compute the total dose rate (the denominator in the age equation above). ESR has the advantage of not being destructive, and repeated measurements can be performed on the same sample.

Geomagnetic polarity

Our planet possesses a magnetic field. This field is produced by our planet's molten core, which exhibits moving electric currents that generate dynamo-type effects that periodically change polarity (Fig. 15). This field extends from the interior of our planet into space (Fig. 16). It prevents the solar wind from the sun with its ionizing radiation from reaching the Earth and obliterating the ozone layer, which insulates us from damaging ultraviolet radiation.

Remarkably, this magnetic field inverts (from north to south and from south to north) as a function of time in a random manner. During the course of Earth's existence, these inversions have happened probably more than 100 times.[44] This phenomenon is a global one, affecting the entire planet equally. The magnetic poles produced by the fields are not

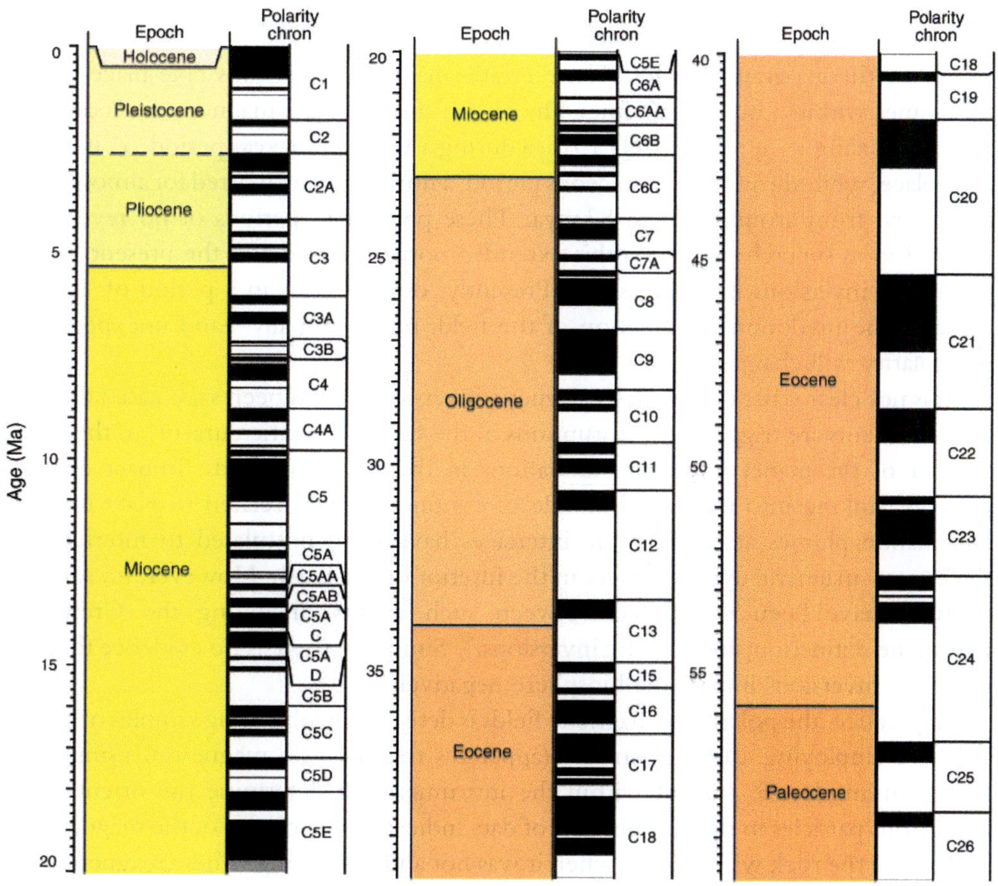

Fig. 15 Geomagnetic timescale dating. *Dark areas* denote periods where the polarity matches today's normal polarity; *light areas* denote periods where that polarity is reversed (http://sp.lyellcollection.org/content/373/1/1).

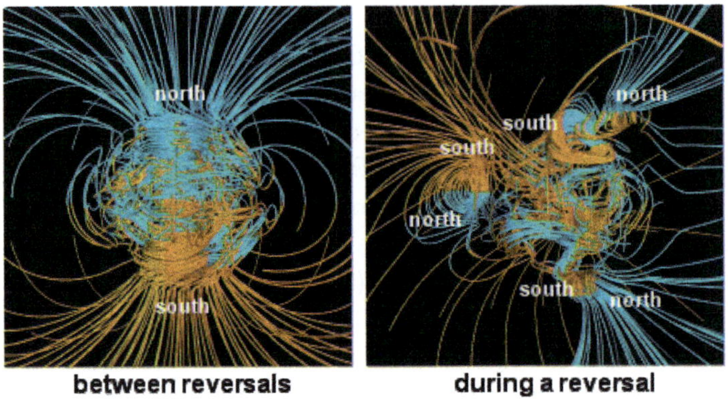

Fig. 16 Simulation of Earth's magnetic fields (https://commons.wikimedia.org/wiki/File:NASA_54559main_comparison1_strip.gif).

related to the geographic poles. These erratic magnetic inversions take place quickly, sometimes within a human lifetime. The rate of the recorded magnetic reversals is variable. For example, approximately 15 mya during a 12-million-year period, 51 inversions took place, while during the Cretaceous period, a normal polarity lasted for almost 40 million years, from around 120 to 83 mya. These prolonged periods of no reversals are referred to as superchron. From this extended normal period and the present, the frequency of inversions has increased.[45] Presently, our planet is in a period of "normal" polarity, the predominant direction of the field, but eventually—and unexpectedly—the polarity will change.

It is not clear what causes these magnetic inversions. It has been speculated that these reversal events are triggered by disruptions in the flow of electric currents in the molten interior of the planet leading to alterations in the magnetic fields. Impact events of meteors, sinking into the liquid mantle of continental slabs related to plate tectonics, and mantle plumes at core-mantle interfaces have been postulated to interfere with the electro-magnetic dynamo effect in the interior of the Earth. However, no solid correlations have been established between such events, including the Cretaceous-Paleogene extinction, and specific inversions.[46] Similarly, there is no evidence that these magnetic inversions impact the biosphere negatively.

In practice, the polarity of magnetic fields is detected by obtaining samples of rocks or lava and employing a magnetometer (apparatus that accesses magnetism using disturbances in an electric current within the instrument) to determine the orientation of the ferrous particles in them. This type of data indicates the polarity of the magnetic pole at the time the rock was formed, when it was hot and solidifying. These magnetic values are used to construct a timeline of the polarity of stratified rocks in the area of interest. This type of geomagnetic data would provide the age of items such as articles made of clay, which are rich in iron or rocks associated with artifacts. In addition to clay, geomagnetic chronology is amendable to dating adobes, ceramics, bricks, tiles, combustion, kilns and burn strata (soils, floors, walls), volcanic eruptions, and sediments from lakes, caves, and ditches. The concept was first described in the 1930s by E. Thellier.[47]

In a typical scenario involving an excavation, archeologists first generate a standard curve of polarity orientations versus age in core samples from the site. These polarity standards can then be used to determine the chronology of samples of unknown age. In other words, the age of material, such as clay, can be determined by correlating the magnetic orientation of the utensil with the values in the standard curve. An example of such a standard curve was described by Stacey Lengyel in 2010.[48] The magnetic polarity (age) of an iron-containing item such as a clay pot is fixed or set at the time it was last heated. After the clay cools, the iron-containing particles retain their polarity until the substance is heated again. Ovens made out of clay are also ideal for polarity reversal dating since they are repeatedly heated every time they are used, resetting the magnetic clock. Some age estimations are done indirectly since some of the artifacts of interest lack iron. In such cases, since most Earth sediments contain iron, the dates of items are extrapolated from closely associated soil layers.

The geomagnetic polarity chronology system has been used routinely to date a number of archeological items including the timing of the end-Triassic mass extinction and the postextinction recovery period in Southern Africa.[49] Establishing the chronology of this mass extinction is essential for answering questions relating to the diversity of dinosaurs during a critical period in their evolution. Specifically, in relation to hominin evolution, magnetic polarity was recently used to date human tools discovered in China (Shangchen Province) to more than 2 mya.[50] The age of these items is very significant because it shows that early humans left Africa for East Asia much earlier than we thought. In fact, polarity dating demonstrated that Archaic humans were in Asia about 1.3 my before our species, *Homo sapiens*, evolved. The dating of these East Asian Archaic humans has revolutionized the field of archeology by forcing scholars to reevaluate the African origin of *sapiens* and the genetic continuity of early hominins (e.g., *Homo erectus*) and AMHs.

In spite of its power, geomagnetic polarity dating has its shortcomings. For example, the system does not provide absolute dates. And since it depends on the switch between two polarities, it is subject to miscorrelations. Therefore, investigators usually employ an additional absolute method, such as $^{40}K/^{40}Ar$ dating, to locate the variation within the time range. In addition, the useful timescale is in the hundreds of thousands to millions of years. Thus, it is only applicable to early hominin fossils and not more recent ones.

Amino acid racemization

Racemization dating is based on changes in the configuration of amino acids. Amino acids are the constituents of proteins. There are around 20 different types of amino acids in proteins, and they are linked head to tail in a chemical conga line to form proteins.

There are around 500 naturally occurring amino acids, though only 20 are present in tissues. Proteins are made of combinations of these 20 different amino acids in chains of various sizes.

Amino acid racemization, as a dating technique, takes advantage of the fact that all amino acids, except glycine, exist in one of two forms: "D" or "L".[51] These two versions of amino acids have the same atoms, but they exhibit a different arrangement of the atoms in space. In other words, the "D" and "L" forms are mirror images of each other with different orientations of atoms around the alpha carbon (the first carbon atom that attaches to the functional groups). This phenomenon is known as stereoisomerism of amino acids (Fig. 17).

Curiously, living organisms maintain all amino acids in the "L" form. This is due to the fact that the enzymes (biological catalysts) in animal and plants only recognize the "L" version of amino acids during protein synthesis. The biological reason for this phenomenon is not known. However, as soon as an organism dies, the organic matter loses control over this constraint and the "L" amino acids start converting into the "D" forms; the older the fossil, the greater the amount of "D" amino acids. The conversion from "L" to "D" or vice versa is referred to as racemization. In other words, while the plant or animal is alive, the ratio of D to L is close to 0, but after death, it gradually begins to drift in the direction of 1. Thus, the ratio of D to L amino acids (D/L value) represents a mechanism to access relative age. The concept was conceived in 1967 by P.E. Hare and R.M. Mitterer as a method to date fossilized shells.[52]

In brief, this is how the procedure works. The first step is to isolate the specific amino acid of interest from the sample (e.g., fossil bone). The fossil is processed by grinding it, followed by extraction of proteins and then the separation of its constituent amino acids. Aspartic acid has been the most widely used amino acid in fossil bone dating. The breaking down of proteins into amino acids is done by acid hydrolysis. Then, the amino acid mixture is tagged differentially with fluorescent compounds specific for the "L" and "D" amino acids. Differential fluorescent emission is used to discriminate between the two forms. The final steps are to separate the specific amino acid of interest (e.g., aspartic) by subjecting the mixture to chromatography or electrophoresis (procedures capable of separating molecules based on their size, shape, and charge), visualize the relative

Fig. 17 Stereoisomerism of amino acids (https://www.chemguide.co.uk/basicorg/isomerism/optical.html).

amounts of the "L" and "D" forms by determining the intensity of the two types of fluorescence, and calculate the D:L ratio.

Applicable to the chronology of hominin evolution, the racemization method is effective for dating within the Quaternary Period (Table 1); it is suitable for different organic matter including bone and shells as well as sediments. Specific applications of the technique include dating of amber-entombed insects ranging in age from 40 million to 130 million years[53] and Paleo-Indians skeletons from North America to 50,000 years before the present.[54]

However, although in principle the method of ascertaining the age of fossils and artifacts by amino acid racemization is potentially promising, currently a number of variables plague the system.[55] For example, parameters such as temperature, humidity, acidity (pH), and other characteristics in the immediate vicinity of the object impact the rate of conversion from L to D. In addition, changes in D/L ratios seem to take place suddenly. These issues underscore the need for additional investigation on the effects of environmental conditions on racemization. At this time, it is clear that it is necessary to restrict amino acid chronologies to fossils with known environmental histories. Furthermore, independent dating with other methodologies should be performed to verify racemization results.

Other dating methodologies

Various other technologies are currently employed to assess age, including obsidian hydration (determines age in either absolute or relative terms of an artifact made of obsidian), dendrochronology (dating tree rings which help calibrate radiocarbon age estimations), palynology (study of pollen, spores and dust, organic and inorganic), and lichenometry (examines lichens to determine the age of exposed rocks using the rate of growth of the organism or radial size over time), among others. These methodologies, although scientifically valid, have limited applications due to their restricted effective time range and/or narrow spectrum of applications due to the finite distribution of the material analyzed.

Composition, texture, and style of artifacts

In addition to dating, analyses to determine the chemical composition of artifacts are paramount to assess the origin and relationships of hominin groups. The discipline that encompasses the study of chemical composition is known as chemometrics, and this addresses some of the most fundamental questions in archeology research. The chemical constitution of an artifact provides important information on its function, societal meaning, manufacturing techniques, dispersals, technology transfer, trade, distribution, and interactions among hominin groups.

Chemometrics falls within the realm of archaeometry and as such this analysis suite usually requires the input and expertise of scholars from diverse fields. Furthermore, due to the diversity of the scientific aims and chemical compositions, an educated decision by archeologists is needed in deciding which methodology to implement, especially if the material is limited and/or precious. For example, for the purpose of substance analysis, items are usually partitioned into two categories: organic and inorganic matter. Because organic archeological materials are perishable, most of the preserved artifacts available for measurements are inorganic such as lithics, pottery, metals, and glass. Among the inorganic materials frequently found among archeological remains, pottery and stone tools are usually the most abundant and informative.

Pottery and ceramic

Ceramics predates farming by at least 20,000 years. Dating back to the Upper Paleolithic (50,000–10,000 ya), pottery is considered one of the oldest human inventions.[56] Most of the innovations and evolution in pottery are prehistoric. Many prehistoric cultures are known for the pottery they made, and they are named after the type of clay vessels they produced. In some instances, little is known about the populations other than their pottery. The Gravettian (European culture dating back to the Upper Paleolithic about 37,000–12,000 ya) Venus of Vestonicka from what is today the Czech Republic may be as old as 31,000 years (Fig. 18). The ubiquitous nature of pottery is illustrated by the worldwide distribution of earliest examples, including Europe, China, Japan, and Russia. This extensive distribution may be indicative of independent origins.

Pottery is usually created from clay. The process begins when the potter identifies a source of clay exhibiting the desired properties. The clay is made soft by adding water, which allows manipulation of the mass. Additives are included as filler material to prevent the vessel from breaking during the heating process and to provide specific characteristics (esthetic and/or practical). The clay is then prepared by kneading to generate an equal distribution of moisture. Wedging helps the removal of air from the matrix. The potter gives the desired shape to the clay, allows it to air dry, and then bakes it. Decoration and painting may be done prior to and/or after baking.

Pottery and ceramics are superb indicators and markers of cultural activity and traditions. Populations used pottery as a medium of expression. These items are not only utilitarian in nature, but they also provide a mechanism for identification—for populations to express their social identity and ancestry. In addition, they are objects of beauty as decorative pottery and a source of pleasure and pride.

The composition, texture, design, and overall style of utilitarian pottery and ceramics of esthetic value are excellent factors to ascertain provenance and establish connections among populations. Similarities and parallelisms among utensils from different regions are often interpreted as technological transfer, trade, and contact with the possibility even of gene flow.

In order to facilitate classification and comparative studies, pottery is described using a number of parameters. For example, the shape and color of the vessel are two easily noted

Fig. 18 Venus of Vestonicka. The earliest known ceramics are the Gravettian figurines, which date to 29,000–25,000 BCE (https://en.wikipedia.org/wiki/Pottery#/media/File:Vestonicka_venuse_edit.jg; https://commons.wikimedia.org/wiki/File:Vestonicka_venuse_edit.jpg).

characteristics. Color is usually examined using the standardized Munsell System (after Albert H. Munsell, who conceived this system early in the 1900s, which was adopted by the United States Department of Agriculture as the official color system for soil research in the 1930s). Another apparent trait is the presence or absence of decorations as well as the style. Decorations can be present inside or outside the pot and include markings, lines, curves, and drawings as well as painting, slips, glazes, wheel marks, incisions, and excisions. Ornamentation usually possesses high cultural value and the styles are diagnostic of specific cultures and origin. Clay type is assessed in terms of grain size (e.g., fine, medium, or coarse) and hardness. This provides information on the origin of the clay, and in some instances on the specific clay bed from which the raw material was extracted.

Charts and tables listing clay types are employed as points of reference for assignment. In addition, sizeable computerized databases of pottery types from different ages, locations, and cultures from around the world are available to identify the dates and sources of objects. This information is also useful to establish function. Another source of information is the temper in the clay. Temper is the material added to clay prior to firing to add strength and prevent the vessel from cracking during the heating process. In this type of analysis, the size and type of filler are noted. The temper may be gravel or crushed crystalline rock, crushed pottery, shells, and even sticks and grasses. The diversity and size of the temper material contribute to the texture of the vessel. The type of filler is often diagnostic of the population responsible for the pottery as well as the geographic origin.

The type of topological analysis described in the previous paragraph represents the essential parameters regularly studied in traditional archeology. Several of the morphological parameters described above are evaluated simply with the naked eye or a regular $10\times$ light microscope. However, with technological advances in different fields of analytical sciences, various approaches to access the fine composition of artifacts have become available. The data provided by these recent technological approaches of modern archeology allow scholars to search deeper into questions relating to composition, provenance, and population interactions. Not only is exact determination of main matrices possible, but also appraisal and measurements of highly informative minor components that usually relate to origin may be ascertained.

Some of the technologies that contributes to the advances of this modern archeology of clay fingerprinting include the following:

- X-ray fluorescence (XRF) employs secondary fluorescent emission to chemically analyze and characterize materials subsequent to being bombarded with high-energy X-rays or gamma rays.
- Inductively coupled plasma mass spectrometry (ICP-MS) quantifies small amount of metals and nonmetals by ionizing the artifacts and measuring its components' mass to charge ratios.
- Neutron activation analysis (NAA) uses high-speed neutrons to strike the item of interest and generate radioactive isotopes that give off element-specific radioactive emissions and decay profiles that identify the substance and its quantity.
- Laser-induced breakdown spectroscopy (LIBS) utilizes high-energy laser pulses to differentially excite the atoms that make up artifacts; each element then emits a signature wavelength light that identifies the substance and is quantified to access the abundance of the element.
- Transmission electron microscopy (TEM) works by directing a beam of electrons into a thin slide of a substance. The differential density of the structures within the matter reflects the electrons differentially at various angles, striking a screen and producing an image.
- Scanning electron microscopy (SEM) is a type of electron microscopy that delineates the topological surface of an object by scanning it with a stream of electrons.

Lithics

Stone artifacts are, in many ways, like pottery and ceramics. Stone tools are extensions and reflections of the hominins who created them. Just like pottery, the characteristics of lithics, such as style and composition, change and evolve with time and space, and therefore can be used as cultural markers to assign source and determine relationships among populations. Furthermore, stone tools are useful in dating sites and hominin fossils. The similarities between these two types of artifacts extend to the affinities in methodology employed to study pottery and stone tools. For example, traits such as shape, style, and color are morphological characteristics routinely scored in both types of remains. The parallelisms also include the analytical procedures used to ascertain the chemical nature of the matrix making up the bulk of the substance and the identification of minor components, impurities, and contaminants as well as the examination of its fine structure. For example, NAA and LIBS are strategies to identify the chemical makeup of stone tools, while TEM and SEM often help visualize substance homogeneity and cutting edge fine structure of rocks, respectively. On the other hand, lithics differs from pottery in its greater durability. Pottery tends to have a rather limited life due to its propensity to break, while most rocks do not fracture easily. Another important difference between the two types of artifacts is the time period during which they were used by hominins. Pottery and ceramics encompass a time range extending from the Upper Paleolithic to the present, while the study of stone tool manufacturing is informative, starting with the earliest known hominins to the Neolithic, including some contemporary populations who are still living in the Stone Age. These time range differences limit the use of these two types of inorganic artifacts to study hominin interactions. On one hand, pottery is incapable of addressing question involving early hominins; on the other, lithics loses resolution when investigating more recent hominin evolution. Ideally, a combination of pottery and lithic artifacts would provide an entire timeline of settlement in cases of continuous hominin occupation of specific sites.

In the study of lithics, the most fundamental initial determination is whether the finding is in fact the product of hominin manufacturing or the result of natural forces. This assessment is not a trivial one and is often challenging to settle. In practice, scholars frequently argue for years or indefinitely for one interpretation over the other. Items that look like they were made or modified to serve some purpose for hominins but were in fact created by erosion or natural fractionation of rocks are referred to as geofacts. These objects often look like tools with sharp edges suitable for cutting and dismembering of prey, yet their appearance is coincidental and thus misleading.

Authentic artifacts have been modeled into knives, blades, and projectile points since the origin of hominins. Certain characteristics of stone tools such as size, shape, and modes of manufacturing are indicative of function. Points of different sizes, for example, are associated with hunting or combat, axes with dismembering prey, and scrapers for removing flesh from hides. In addition, the wear at the edges may provide information on the function and authenticity of the artifact. At times, signs of use are evident in their dulled edges and residual animal matter or debris left on the blade. These remnants permit dating and identification of the material that was cut or manipulated using the tool as well as the identity of the prey. Species prey identification may be attempted by DNA fingerprinting.

Tradition- or culture-specific standard techniques were developed at different stages of hominin evolution to produce tools.[57] The methodologies employed and the resulting shape and styles of tools are diagnostic markers for assessing time and the people responsible for making them. For example, five main tool-making traditions appear sequentially in hominin evolution:

- Oldowan (named after Olduvai Gorge, Tanzania, dates to about 2.6 mya, and is characterized by simple construction based on cores with an edge or sharp tip);
- Acheulean (first appears in Kenya around 1.7 mya and exhibits bifaced cores shaped into axes);
- Mousterian (dated to 160,000–40,000 ya, this industry evolved from the Acheulean and is represented by retouched smaller and sharper knife-like artifacts and scrapers made from prepared cores by Neanderthals in Europe and hominins in Africa);
- Aurignacian (characterized by long blades dating to the Upper Paleolithic, 50,000–10,000 ya, were made by early modern *sapiens* in Europe); and
- Microlithic (these tools date back to 35,000–3,000 ya and were made primarily into composite instruments by AMHs in Europe, Africa, Asia, and Australia by mounting small flakes onto a piece of wood or a bone handle).

The general approach in stone tool making is to strike a core or rock at certain specific angles with another hard rock that functions as a hammer. By carefully manipulating the point of impact and chipping away flakes from the flint block, the core is shaped into various tools. These tools include axes, choppers, points, and scrapers (Fig. 19). In addition, a certain amount of debitage (waste material) is left behind, which was used by hominins to create arrowheads.

Ecofacts

In addition to pottery and stone tools, information about hominins is gathered from organic material, such as the plants and animals that were part of the diet, habitat, or ecosystem. The flora associated with a region is indicative of the environment as well as the animals inhabiting the region and the diet of the hominin populations as well as their behavior.

The examination of pollen species, for example, is particularly informative because plants are ubiquitous, tend to be region-specific, and produce substantial amounts of pollen. In addition, pollen morphology is species-specific and easily discernible with a regular light microscope. Furthermore, pollen is very hardy and survives prolonged periods of time intact. In a typical project involving pollen identification, an archeologist secures a core from the excavation by drilling a hollow pipe into the soil, marsh, or peat bog. The

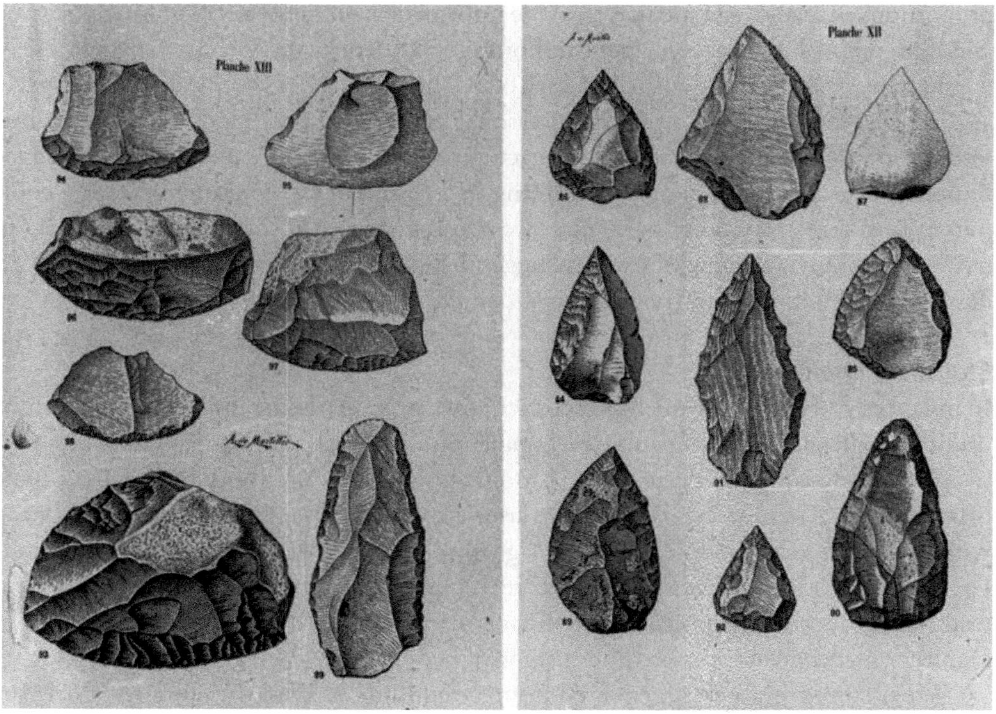

Fig. 19 Middle Paleolithic points and scrapers (https://www.ancient.eu/image/6185/drawings-of-middle-palaeolithic-tools-points–scra).

extracted layers of sediment are taken to the laboratory where experts identify the pollen species and their concentrations.

This information is then tabulated and employed to create pollen sequences indicating floral changes as a function of time. This, in turn, can be related to dates and provide inferences about climatic conditions as well as the hominins' mode of survival. Archeologists also identify plant species using the method of flotation or water sieving. This process involves placing excavated material in a container with water. Material denser than water will sink to the bottom, while less dense matter will float. The material on the surface and at the bottom of the container is collected and assorted into categories and saved for further examination by experts.

Examination of the fauna in a site is equally important since the species identified are diagnostic of diet, climate, and way of living. Changes in the proportion of various small animal species such as rodents, birds, lizards, and insects among others may provide information on habitat and climatic fluctuations, since some of these species are well-known environmental indicators, being very sensitive to ecological changes. Megafauna, on the other hand, are indicative of the hominin's diet and habits. For example, the number and proportions of species at a site may provide information on the type of hunting practiced, and if predatorial marks are evident in bones (e.g., teeth marks), it may suggest a scavenging subsistence rather than a predatory one. The presence of specific migratory species at a site may indicate the temporary or seasonal use of a camp as opposed to a permanent settlement. Bones also provide data on cultural and religious practices. For instance, bones of animals may show signs of burning (e.g., change in coloration indicating cooking) or breakage as a clue of bone marrow consumption. On the other hand, marks on bones of their own clan members may suggest habitual cannibalism, ritualistic activity, or desperation resulting from harsh survival conditions and hunger.

Statistical analyses

In instances where numerous characteristics or variables are being evaluated to determine origin and affiliations among cultures, populations, raw materials, artifacts, or traits, multivariable analysis methods are needed to establish statistically the significant compositional groups. These statistical tools are used in various disciplines to delineate groups with similar characteristics (e.g., assemblage of populations or tools) and understand differences among them. Confirmation of relationships by statistically significant associations of groups of artifacts from distant regions may help establish migrational connections between locations and groups of people.

When comparing items recovered from a multitude of sites, the amount of data that needs to be compared is huge. In those instances, the archeologist is usually dealing with dozens of variables (traits) and hundreds or thousands of samples (e.g., points or arrowheads) from different digs. When confronted with such a daunting task, computerized analysis is clearly a necessity, and these types of large multivariable datasets require multivariate analysis (MVA) computer software. Without these analytical programs, it is not possible to identify patterns and trends from the large amount of data. As a prelude for this type of work, the data are arranged into a matrix in which the samples are listed in rows and the characteristics (e.g., thickness of points at the base) in columns. Subsequent to running the program, the results are visualized with routines that represent the outcomes in the form of plots, graphs, and tables. These illustrations are routinely included in summaries, reports, and publications.

A number of MVA programs have been adapted from the field of statistics. Each of these programs was designed with a specific set of algorithms based on explicit assumptions.[58] Thus, expert knowledge of the available applications is essential to determine the proper program(s) for the analysis of the data. For this purpose, a detailed consideration of the software assumptions is highly recommended. Expert advice and assistance in MVA and statistical interpretations should be obtained prior to starting a project. These programs are not uniquely employed in archeology, but are routinely used to analyze data from different disciplines including fields pertinent to the study of hominin evolution such as physical anthropology, linguistics, and molecular biology. Among the most frequently used programs used in archeology, Principal Component Analysis (PCA) and Hierarchical Clusters Analysis (HCA) are two of the most relevant.[59]

In the case of PCAs, a large number of variables are reduced to a small set of variables or principal components (PCs). PCA takes an amalgamation approach to examine variables. The intent is to summarize as much of the original variation as possible using the minimum number of variables or principal components. The first principal component (PC1) generated by the software accounts for the largest amount of the variability of the original data, while succeeding components represent decreasing percentages of the remaining variability. Fig. 20 illustrates a PCA graph based on 26 chemical elements (calcium, iron, potassium, etc.) detected in Bronze Age pottery from a site in Turkey.[60] PC1 appears on the *X*-axis while the PC2 is on the *Y*-axis. Notice the four groupings of pottery delineated by the PCA plot. This type of partitioning provides evidence of the existence of at least four types of clay used to make this Bronze Age pottery.

Cluster analysis works differently. HCA is a program that uses a dissimilarity matrix. The results are visualized in the form of trees in which the samples are grouped according to similarities to each other. In other words, HCA places together artifacts that are more

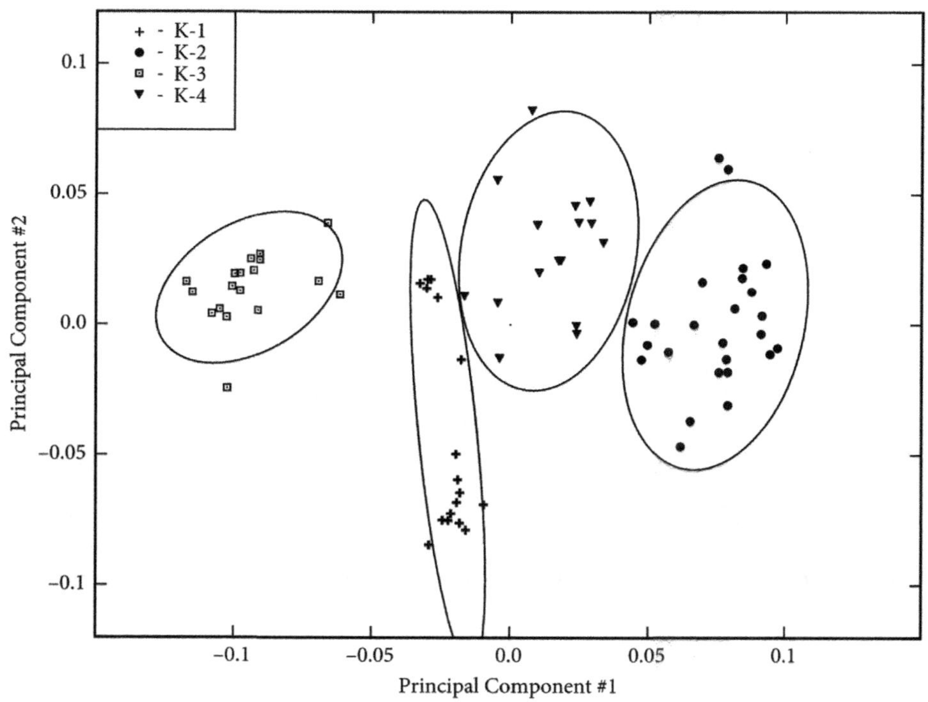

Fig. 20 Principal component analysis (PCA) based on 26 chemical elements in pottery from a Bronze Age site in Turkey (http://www.oxfordhandbooks.com/view/10.1093/oxfordhb/9780199935413.001.0001/oxfordhb-9780199935413-e-8).

similar to each other than to those that are not. Samples that have characteristics (variables) that make them very similar to each other are separated by small distances from each other and partitioned in the same cluster or branch. Conversely, samples that possess very different traits are very distant from each other in far-off groupings (Fig. 21). The HCA program converts the degree or extent of differences among artifacts to linear distance in a tree or dendrogram. The program is capable of calculating distances (likeness) between all pairs of artifacts using one of a number of measures.

The dendrogram or tree in Fig. 21 represents the same 80 Bronze Age pottery samples tested for the 26 chemical elements and analyzed with the PCA program illustrated in Fig. 20.[60] Notice how the numerous bifurcations in cascading hierarchical format illustrate the different degrees of similarities among the pottery samples.

Provenance analysis

In addition to topological characteristics, the ratio of isotopes of certain elements such as lead, strontium, and oxygen provides information on the geographical location of origin of the raw material used to make artifacts as well as migration routes and systems of commerce. In hominin fossils, these elements and isotopes may also be indicative of dietary

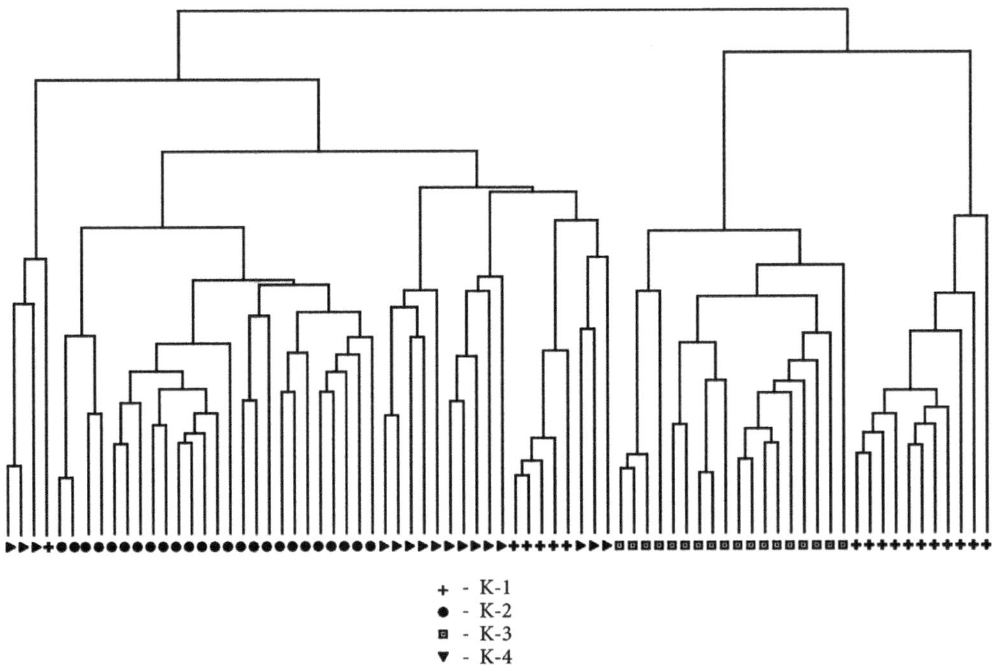

Fig. 21 Hierarchical clusters analysis (HCA) based on 26 chemical elements in pottery from a Bronze Age site in Turkey (http://www.oxfordhandbooks.com/view/10.1093/oxfordhb/9780199935413.001.0001/oxfordhb-9780199935413-e-8).

characteristics and the place where individuals were born and/or raised. The premise of this type of analysis is that raw materials, animals, and plants are distributed differentially in our planet with specific locations exhibiting characteristic presence and concentration profiles. In addition, in the case of animal and plants, their presence and distribution fluctuate with time, thus this type of data may provide temporal information as well.

Provenance may also refer to the location of the place of the workshops or production sites where the artifacts were made. Additional resolution in this type of approach is provided by the signature of minor components which, although often only accidentally included, tend to be very region-specific and excellent markers to ascertain source and relationships.

Since the chemicals present in the stones used for making lithic artifacts are not altered in the process of creating the tool, standard methodologies used in geological research are employed to analyze composition. Obsidian is a good example of how provenance of lithic is done. Using methodologies such as XRF, ICP, and/or NAA, the presence and concentration of major and trace elements are determined. The amounts of major elements provide information on the specific volcanic region to which the obsidian belongs, while information on trace elements may identify precisely the flow or quarry that provided for the raw material that led to the artifact.

References

1. Beaulieu P-A. *The Reign of Nabonidus, King of Babylon 556–539 B.C.E.* New Haven, CT, USA: Yale University Press; 1989.
2. Grassi GF. Nabonidus, king of Babylon. *Middle East Top Argum.* 2014;3:125–135.
3. Dougherty RP. *Nabonidus and Belshazzar; A Study of the Closing Events of the Neo-Babylonian Empire.* New Haven, CT, USA: Yale University Press; 1929.
4. Burn AR. *Herodotus: The Histories.* London, UK: Penguin Classics; 1972.
5. Blanco W. Herodotus. In: Roberts JT, ed. *Herodotus: The Histories: The Complete Translation, Backgrounds, Commentaries.* New York, USA: W. W. Norton and Company; 2013. Translated by Walter Blanco.
6. Waterfield R. *Herodotus: The Histories.* A New Translation by Robin Waterfield, Oxford, UK: Oxford University Press; 2008.
7. Marincola JM, De Selincourt A. *The Histories of Herodotus.* New York, NY, USA: Penguin Classics; 2006.
8. Perez Z. *Thucydides.* Princeton, NJ, USA: Princeton University Press; 2015.
9. Ashmole B. Ciriac of Ancona. *Proc Br Acad.* 1959;45:25–41.
10. Sweet R. *Antiquaries: The Discovery of the Past in Eighteenth-Century Britain.* London, UK: Bloomsbury Group; 2004.
11. Outram D. *Panorama of the Enlightenment.* Los Angeles, CA, USA: Getty Publications; 2006.
12. Shnirel'man VA. Nationalism and archeology. *Anthropol Archeol Eurasia.* 2013;52:13–32.
13. Noël HI. *Belzoni: The Giant Archaeologists Love to Hate.* Charlotteville, VA, USA: University of Virginia Press; 2011.
14. St Clair W. *Lord Elgin and the Marbles.* Oxford, UK: Oxford University Press; 1967.
15. Hunt P, Smith AH. Lord Elgin and his collection. *J Hell Stud.* 1916;36:163–372.
16. Mastrolorenzo G, et al. Herculaneum victims of Vesuvius in AD 79. *Nature.* 2001;410:769–770.
17. Hutton R. *Blood and Mistletoe: The History of the Druids in Britain.* New Haven, CT, USA: Yale University Press; 2009.

18. Torrens H. William Smith (1769-1839): his struggles as a consultant, in both geology and engineering, to simultaneously earn a living and finance his scientific projects to 1820. *Earth Sci Hist.* 2016;35:1–46.
19. Eyles J. William Smith: the sale of his geological collection to the British museum. *Ann Sci.* 1967;23:177–212.
20. Rowley-Conwy P. The concept of prehistory and the invention of the terms 'prehistoric' and 'prehistorian': the Scandinavian origin, 1833–1850. *Eur J Archaeol.* 2006;9:103–130.
21. Ussher J. In: Larry, Pierce M, eds. *Annals of the World: James Ussher's Classic Survey of World History.* Green Forest, AR, USA: Master Books; 1650.
22. McPhee J. *Book 1: Basin and Range, in Annals of the Former World.* New York, NY, USA: Farrar, Straus and Giroux; 1981.
23. Dalrymple GB. The age of the Earth in the twentieth century: a problem (mostly) solved. *Geol Soc Spec Publ.* 2001;190:205–221.
24. Boucher de Perthes J. *Antiquités Celtiques et Antédiluviennes.* Paris: Treuttel et Würtz; 1849. 3 vol.
25. Sackett J. Boucher de Perthes and the discovery of human antiquity. *Bullet Hist Archaeol.* 2014;24:Art. 2.
26. Meyer M, et al. Nuclear DNA sequences from the Middle Pleistocene Sima de los Huesos hominins. *Nature.* 2016;531:504–507.
27. Bowman S. *Radiocarbon Dating.* London, UK: British Museum Press; 1995.
28. Anderson EC, et al. Radiocarbon from cosmic radiation. *Science.* 1947;105:576–577.
29. Arnold JR, Libby WF. Age determinations by radiocarbon content: checks with samples of known age. *Science.* 1949;110:678–680.
30. Schulte P, et al. The Chicxulub asteroid impact and mass extinction at the Cretaceous-Paleogene boundary. *Science.* 2010;327:1214–1218.
31. Tattersall I. *The Fossil Trail: How We Know What We Think We Know About Human Evolution.* Oxford, UK: Oxford University Press; 1995.
32. Schwarcz HP. Uranium series dating of quaternary deposits. *Quat Int.* 1989;1:7–17.
33. Schoene B. *U–Th–Pb Geochronology.* Princeton, NJ, USA: Princeton University; 2014.
34. Shimron AE, Brookins DG. b-Sr radiometric age of late Precambrian fossil-bearing and associated rocks from Sinai. *Earth Planet Sci Lett.* 1974;24:136–140.
35. Naeser ND, Naeser CW. Fission track dating. *Dev Palaeontol Stratigr.* 1984;7:87–100.
36. Kurz MD, Brook EJ. Surface exposure dating with cosmogenic nuclides. In: Beck C, ed. *Dating in Exposed and Surface Contexts.* Albuquerque, NM, USA: University of New Mexico Press; 1994:139–159.
37. Michels J. *Dating Methods in Archaeology.* New York, NY, USA: Seminar Press; 1973.
38. Martini M, Sibilia E. Radiation in archaeometry: archaeological dating. *Radiat Phys Chem.* 2001;61:241–246.
39. Sekkina MA, et al. Thermoluminescence archaeological dating of pottery in the Egyptian pyramids zone. *Ceramics-Silikáty.* 2003;47:94–99.
40. Rink WJ. Electron spin resonance (ESR) dating and ESR applications in quaternary science and archaeometry. *Radiat Meas.* 1997;27:975–1025.
41. Grun R, Stringer CB. Electron spin resonance dating and the evolution of modern humans. *Archaeometry.* 1991;33:153–199.
42. Motoji I. *New Applications of Electron Spin Resonance.* Pulau Ujong, Singapore: World Scientific; 1993.
43. Bar-Yosef O. The chronology of the Middle Paleolithic of the Levant. In: Akazawa T, Aoki K, Bar-Yosef O, eds. *Neandertals and Modern Humans in Western Asia.* New York, NY, USA: Plenum Press; 1998.
44. Sagnotti L, et al. Extremely rapid directional change during Matuyama-Brunhes geomagnetic polarity reversal. *Geophys J Int.* 2014;199:1110–1124.
45. Merrill RT, et al. *The Magnetic Field of the Earth: Paleomagnetism, the Core, and the Deep Mantle.* Cambridge, MA, USA: Academic Press; 1998.
46. Merrill RT, McFadden PL. Paleomagnetism and the nature of the geodynamo. *Science.* 1990;248:345–350.
47. Thellier E. Sur l'aimantation des terres cuites et ses applications geophysiques. *Ann l'Instit Phys Globe.* 1938;16:157–302.
48. Lengyel S. The pre–A.D. 585 extension of the U.S. Southwest Archaeomagnetic reference curve. *J Archaeol Sci.* 2010;37:3081–3090.

49. Sciscio L, et al. Magnetostratigraphy across the Triassic-Jurassic boundary in the main Karoo Basin. *Gondwana Res*. 2017;51:177–192.
50. Zhaoyu Z, et al. Hominin occupation of the Chinese Loess Plateau since about 2.1 million years ago. *Nature*. 2018;559:608–612.
51. Candida L, et al. Kinetics of amino acid racemization (epimerization) in the dentine of fossil and modern bear teeth. *Int J Chem Kinet*. 2003;35:576–591.
52. Hare PE, Mitterer RM. Nonprotein amino acids in fossil shells. *Year B Carnegie Inst Wash*. 1967;65:362–364.
53. Bada JL. Amino acid racemization in amber-entombed insects: implications for DNA preservation. *Geochim Cosmochim Acta*. 1994;58:3131–3135.
54. Bada JL, et al. New evidence for the antiquity of man in North America deduced from aspartic acid racemization. *Science*. 1974;184:791–793.
55. Bender ML. Reliability of amino acid racemization dating and paleotemperature analysis on bones. *Nature*. 1974;252:378–379.
56. Vandiver PB, et al. The origins of ceramic technology at Dolni Vestonice, Czechoslovakia. *Science*. 1989;246:1002–1008.
57. Clarke D. *Analytical Archaeology*. New York, NY, USA: Columbia University Press; 1978.
58. Buxeda i Garrigos J, Kilikoglou V. Total variation as a measure of variability in chemical datasets. In: Van Zelst L, ed. *Patterns and Process: A Festschrift in Honor of Dr. Edward V. Sayre*. Washington, DC, USA: Smithsonian Press; 2003.
59. Hall M. Quantifying trends in site location with multidimensional scaling. *World Archaeol*. 1982;14:131–152.
60. Glascock MD. *Archaeology, Archaeological Methodology and Techniques*. Oxford, UK: Oxford Handbook on Line; 2016.

CHAPTER 5

Hominin fossils: Anatomical trends

No subject has lately excited more curiosity and general interest among geologists and the public than the question of the antiquity of the human race.

Charles Lyell[1]

The new physical anthropology

In 1949, a concerted effort was made by American physical anthropologist (primatologist) Sherwood Washburn (1911–2000) (Fig. 1) to put forward what he called "the new physical anthropology." Washburn teamed up with Ukrainian-American geneticist and evolutionary biologist Theodosius Grygorovych Dobzhansky (1900–75) in 1950 to promote a fundamental change in the way physical anthropology was practiced. In the formulation of his ideas, Washburn was heavily influenced by Sir Arthur Keith's *A New Theory of Human Evolution*.[2] Both Washburn and Dobzhansky were attentive to the advances in the field of evolutionary biology as well as genetics and the increasing drift in relation to the way traditional physical anthropology was practiced. At the time, the sanctioned, long-established practice of physical anthropology was primarily descriptive and taxonomic in nature while other scientific disciplines were moving to a more systematic cause-effect approach directed at testing theories and establishing connections. In his vision, Washburn was interested in facilitating a transformation in the field of physical anthropology from a basic and at times anecdotal approach to one in which hypotheses and theories are tested and the relationships between structure and function explored. These ideas were eloquently expressed by Washburn in the following passage:

> *The old physical anthropology was primarily a technique. The common core of the science was measurement of external form with calipers. The new physical anthropology is primarily an area of interest, the desire to understand the process of primate evolution and human variation by the most efficient techniques available.*[3]

At this crossroads in the field of physical anthropology, Washburn and Dobzhansky acted as advisors to Croatian-American geneticist Milislav Demerec (1895–1966) in organizing the XV Cold Spring Harbor Laboratory Conference on the topic of the Evolution of Man in 1950 in Huntington, New York. At the time, Demerec was interested in promoting better understanding between geneticists and anthropologists while Washburn and Dobzhansky were motivated by a desire to incorporate evolutionary theory into the discipline of physical anthropology.

Sex and Cohabitation Among Early Humans
https://doi.org/10.1016/B978-0-12-819086-9.00010-5

Fig. 1 Sherwood Washburn (http://www.nasonline.org/publications/biographical-memoirs/memoir-pdfs/washburn-sherwood.pdf).

In addition to the distinctive way scholars approached the study of the two disciplines, geneticists and anthropologists were divided on a number of fundamental ideas. The contrasting positions were very evident in the types of presentations delivered during the CSHL meeting and in the argumentative nature of the discussions that ensued during the presentations. In fact, it turned out that the conference was an ideological battleground encompassing several topics. One point of contention was how the two groups viewed the concept of race. For instance, prominent geneticists, such as the British-American anthropologist Montague Francis Ashley-Montagu (1905–99) (Fig. 2), indicated that the term "race" should be avoided and the term "ethnic group" should be used instead. Later in his career, Ashley-Montagu further stated that the term "race was fine for scientists to use in talking to scientists, but in communicating with laymen they ought to say ethnic group if only to make them think."[4] Clearly, Montagu was very concerned about the negative impact that the notion of population differences would have on society and how the public would perceive the existence of human diversity. It is likely that Montagu's early life growing up in the racist society of the time in London and the events of the Jewish genocide by the Third Reich influenced some of his views. Montagu's original name was Israel Ehrenberg and he often described the antiSemitic abuse he endured when he ventured out of his Jewish neighborhood in London. Although at the time of the CSHL meeting in 1950 many anthropologists shared the idea that all human races originated from different geographic groups, a growing number of geneticists agreed

Fig. 2 Ashley Montagu in 1958 (https://en.wikipedia.org/wiki/Ashley_Montagu#/media/File:Ashley-Montagu-1958.jpg).

with Montagu's ideas that the concept of race is meaninglessness and strongly motivated by racism. Montagu's opinions are still popular among many human biologists today. The idea that races had their genesis in unique ancient populations was problematic, since that implied that races were old.

Other issues were intensely argued at this landmark meeting that underscored the changing tides in physical anthropology. For example, in a controversial seminar, German-American leading evolutionary biologist Ernst Walter Mayr (1904–2005) (Fig. 3) expressed consternation about the liberal assignment of names to new, poorly characterized fossil findings. Mayr's criticism centered on the common practice of coining different genus and species names to hominin fossils with minimal numbers of remains and justification. In an interview given to CSHL, Mayr remembered saying, "Let's cut all this mass of names. Let's cut it down to 2 or 3 genera, and maybe 5 or 6 species." He further recollected: "Everybody was totally outraged [at me]. And part of it was, of course, nonsense." He then added: "*Pithecanthropus* is nonsense! That's just *Homo*. That's not a different genus!" (http://library.cshl.edu/oralhistory/interview/cshl/symposia/1950-CSHL-symposium-origin-and-evolution). Furthermore, Mayr was particularly astonished at the common application of scientific names with no regard for the International Code of Zoological Nomenclature.

In addition, along these lines, Washburn expressed his dismay at the situation when he argued against the practice of assigning specific behaviors to populations based on little or

Fig. 3 Ernst Walter Mayr (http://www.nasonline.org/publications/biographical-memoirs/memoir-pdfs/mayr-ernst.pdf).

no objective evidence. It is worth mentioning that, at the end of Washburn's seminar, the chair of the session Earnest Albert Hooton (1887–1954), a classical anthropologist well known for his topological views, got up from his chair and loudly said, "I hate the word population." Hooton was well-known for his views on races and exemplified some of the racism of the time.[5]

Over the years since the 1950 pivotal conference at CSHL, the field of physical anthropology has received numerous infusion of novel ideas based on evolutionary theory. These concepts include individual versus group selection,[6] the theory of senescence,[7] parental investment and reproductive effort,[8] the theory of cooperation among nonrelatives and reciprocal altruism,[9] punctuated equilibrium as an alternative to gradualism,[10] epigenetic mechanisms,[11] hominin introgression or interbreeding,[12] and multilevel selection,[13] among others. Undoubtedly, this influx of ideas into the discipline of physical anthropology has enriched and helped move the field into the mainstream of biological theory, and yet, despite all this activity and input of new thoughts contributing to the expansion of evolutionary theory, some of the arguments are still as intensely debated now as in 1950. Today, fueled by discoveries in molecular biology and advances in human population genetics, the biological basis of races is still a prominent topic of discussion in biological anthropology.

Nevertheless, today, more than half a century after the CSHL meeting, Washburn's views on the concept of the new physical anthropology have been, for the most part, realized. In addition to the philosophical change to a theory-based science, the application of new technology from a number of disciplines has contributed to this transition. Today this transformation is evident, as many scholars use the term "biological anthropology" instead of "physical anthropology."

Technical advances

The fundamental reformation experienced by the field of physical anthropology that transformed the discipline from a descriptive and topological cataloging of fossils to a more hypothesis-driven and systematic cause-effect approach during the second half of the 20th century was also influenced by a technical revolution in the biological and physical sciences. These technical innovations nourished the fields of archeology and physical anthropology and facilitated the metamorphosis that Washburn envisioned. Chapter 4 describes a number of technical discoveries that impacted not only the field of archeology, but anthropology as well. These advances include the relative and absolute dating of fossils and the molecular biology revolution, which included in vitro DNA amplification or PCR (**P**olymerase **C**hain **R**eaction) as well as ancient DNA sequencing. During this period, methods such as isotope ratios, radiation exposure, amino acid racemization, and geomagnetic polarity were introduced to date hominin remains (Chapter 4). In addition, a number of procedures employed to study the chemical nature of matter were applied to the field of physical anthropology to assess the composition of remains. Mechanisms to analyze large amounts of data were incorporated from the field of statistics. These multivariable procedures, described in Chapters 4 and 6, allow for the comparison of a large number of variables and their presentation in the form of graphs and plots.

In more recent times, a number of methodologies and practices have been incorporated into the study of hominin fossils. Two of the most promising technological advances are computed tomography (CT) and synchrotron imaging. In CT, large numbers of X-ray images from various angles are combined to generate a tomographic view (3D picture based on the differential passage of energy waves on structures) of a specific object. Computers then combine the individual images, resulting in a virtual cross-sectional picture of the anatomy of interest. This procedure is nondisruptive and allows the investigator to see inside the structure without damaging precious fossils. The final product is a three-dimensional image of the object. Manipulation of the data also allows for the visualization of specific internal structures from different angles based on their differential absorption of the X-rays. Access to the internal anatomy of a fossil would be out of reach during routine external examination. The nondisruptive nature of the procedure and the storage of the data in electronic form also provides for verification by repetition of the measurements and ease of data comparison and sharing with the rest of the scientific community. Variations of the original CT technique, such as spiral CT, have allowed superior resolution, processing of cross-sectional images, and 3D surface reconstruction.[14] Other extensions of the CT procedure utilize 3D data and stereo lithography with UV lasers and photo-hardening resin to create 3D models of structures, layer after layer.[15]

A number of notable remains have been investigated employing CT. Among them is the Ötzi corpse of an AMH (anatomical modern human) who lived about 5500 years ago during the Chalcolithic epoch (period in between the first farming communities of the Neolithic and the urban societies of the Bronze Age) discovered in the Alps in what is today the Austrian-Italian border[16] and the *Homo naledi* fossils, an extinct hominin species from South Africa.[17]

Yet of particular significance are the CT-guided stereo lithography studies of the Middle Pleistocene (781,000–126,000 years ago) skulls from Petralona (Chalkidiki Peninsula, Greece) and Broken Hill (Zambia). These investigations demonstrated the capacity of the technology in assessing the internal morphology of hominin fossils.[18] These skulls are of particular interest because they exhibit a mosaic of traits associated with three distinct hominin species. Specifically, typical of *Homo erectus*, these fossils possess an angled occipital bone (trapezoidal-shaped bone found at the lower-back area of the cranium), massive supra-orbital torus (brow ridge), and thick cranium, while they also include Neanderthal characteristics, such as a zygomaxillary (mid-facial skeleton complex) morphology and the characteristic brow ridge. Remarkably, these skulls also possess a number of archetypal traits of much more recent late archaic humans, including parietal expansion (parietal lobe behind the frontal lobe), enlarged braincase, and reduced total facial prognathism (projection of jaws forward beyond the plane of the skull). This collage of traits has generated considerable confusion and controversy among scholars, contributing to the classification of these fossils into various groups, including *Homo neanderthalensis*, *H. erectus*, archaic *H. sapiens*, and *H. heidelbergensis*. In the content of this discussion, it should be emphasized that the confusing diverse diagnostic characteristics mentioned above are based on external features that do not reflect the internal structure of the skulls. The examination provided by computer generated three-dimensional stereo lithographic models, on the other hand, afforded internal endocranial information, which revealed additional structures incompatible with the external resemblance of skulls to the previously mentioned hominins.[18] Specifically, the internal scrutiny uncovered, for the first time, extensive pneumatization (presence of air cavities or sinuses) of the skull, which in turn was correlated with dramatic rearrangements of the skulls related to derived (*sapiens*-like) brain morphology. Fig. 4 illustrates one of the sinuses revealed by this technology.

Other physical anthropology applications have benefited considerably from CT, including investigations involving the developmental biology of early hominins. This field of work has recently attracted interest since the limited information available suggests little parallelism among hominins. Specifically, recent data with CT technology have demonstrated unique brain and teeth developmental patterns among hominins.

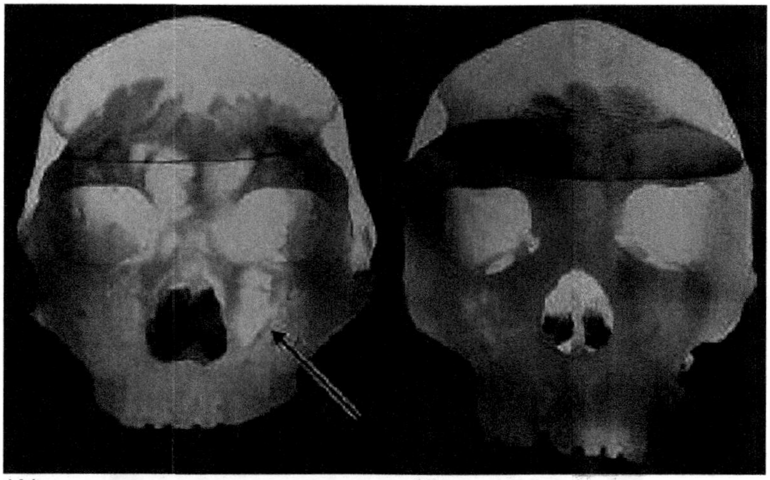

Fig. 4 Frontal views of the stereo lithographic models of (A) Petralona and (B) Broken Hill fossil skulls. The *arrow* indicates the position of a strong crest protruding into the maxillary sinus (file:///Users/ReneJHerrera/Downloads/A_comparative_study_of_stereolithographi%20(1).pdf).

Traditionally, the developmental biology of extinct hominins has been studied based on patterns exhibited by living great apes and humans. In retrospect, this approach was somewhat naïve since modeling extinct hominin development based on living species assumes that environmental conditions and selection pressures on early and recent hominin evolution are comparable. Today we know than hominin evolution has been complex, with a multiplicity of groups living in diverse habitats—factors that likely have contributed to unique anatomical specializations and developmental histories. Limited studies on *Homo erectus* have demonstrated, for example, that its developmental pattern was unique, not resembling that of extant species of apes or humans, and could provide misleading developmental information.[19]

To address these issues, CT technology is being employed to allow delineation of the limited number of juvenile hominin fossils. Fig. 5, for example, illustrates the 3D reconstruction from CT images of the Selam fossilized skull of a 3-year-old *Australopithecus afarensis* female found in Dikika, Ethiopia. This image clearly shows the endocast in brown, inner ear in red, deciduous teeth in yellow, and permanent dentition in white.[20]

Synchrotron imaging is the other technology at the forefront of physical anthropology. It relies on an accelerator, which propels particles at high velocities around a fixed, closed-loop path. The concept was invented in 1944 by Ukrainian-Russian experimental physicist Vladimir Iosifovich Veksler (1907–66), while the first functional electron synchrotron was built by American chemist Edwin Mattison McMillan (1907–91) in 1945. Today, the synchrotron X-ray technique, a variation of the original principle, is used to

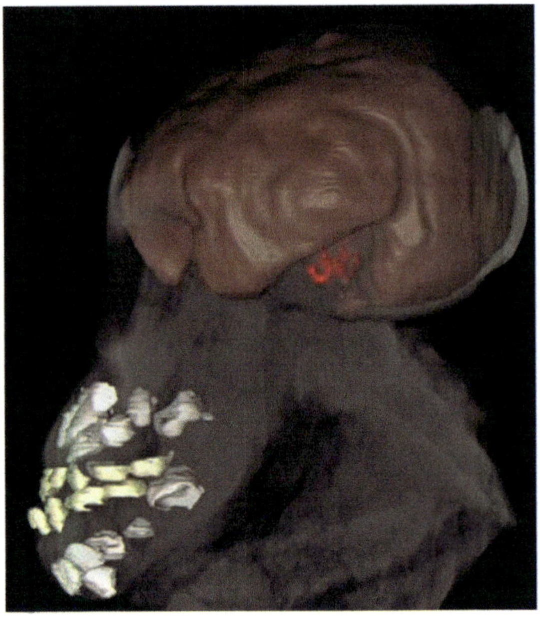

Fig. 5 Computed tomographic 3D reconstruction of the Selam fossilized skull of a 3-year-old *Australopithecus afarensis* female hominin found in Dikika, Ethiopia (https://www.nature.com/scitable/knowledge/library/reconstructing-hominin-life-history-96635644).

detect small density differences within fossil matter, which are then employed to generate high-resolution images.

Like CT, synchrotron microtomography uses X-rays to create cross-sections of a structure, which are then used to produce 3D images without destroying the fossil. Yet in comparison to CT technology, synchrotron images exhibit fewer blockage artifacts and the images tend to be sharper with features that are more distinct with less background noise.[21] A potential negative side effect of synchrotron technology, however, is the color changes observed in teeth subsequent to exposure to synchrotron radiation.[22] Although exposure to daylight or UV light usually regenerates the original color of the enamel, it is important to recognize potential DNA degradation if the object is also intended for molecular biology analysis. These concerns also apply to CT procedures, since X-rays are also employed.

Synchrotron technology has been used to see inside objects such as teeth of AMHs and Neanderthals (Fig. 6) to reveal growth lines and assess potential developmental differences between the two species.[23] Synchrotron images have uncovered that Neanderthal tooth crowns grew more rapidly than AMHs', allowing for faster dental maturation. Middle Paleolithic juvenile AMHs, on the other hand, exhibit greater developmental similarity to contemporary humans. Furthermore, both Neanderthals and AMHs have

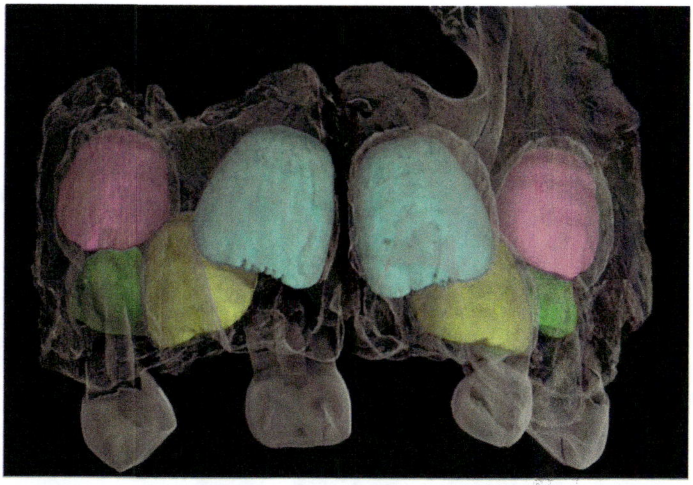

Fig. 6 Phase contrast synchrotron scan of dentition of a 3-year old Neanderthal (https://www.pnas.org/content/107/49/20923; https://www.researchgate.net/publication/47794338_Dental_Evidence_for_Ontogenetic_Differences_Between_Modern_Humans_and_Neanderthals/download).

longer dental development relative to earlier hominins. This period of dental maturation is especially long in AMHs.[24]

When the synchrotron methodology was applied to the Selam remains discussed above, the images indicated that the 3.3 million-year-old partial skeleton includes all seven cervical (neck) and 12 thoracic vertebrae and belonged to a 6- to 8-year-old juvenile.[25] The Selam skeleton is the only evidence for the presence of 12 thoracic (rib-bearing) vertebrae in hominins prior to 60,000 years ago (Fig. 7). Prior to these synchrotron images, it was not clear whether *Australopithecus* had 12 thoracic vertebrae, as in most humans, or 13 as in most African apes, and where the transition from thoracic to lumbar vertebrae took place along the spine. The resolution and definition provided by synchrotron scanning provided evidence that Australopithecines, or at least *afarensis*, possessed 12 thoracic vertebrae with a transition from thoracic to lumbar at the 11th thoracic level. This location is also seen in all subsequent early hominins and is higher than in AMHs and extant apes. The Selam skeleton is the earliest hominin specimen exhibiting this unique type of segmented spine.

In addition to their contribution to the visualization of the internal structure of fossils, CT and synchrotron technologies have impacted the field of physical anthropology by providing greater accessibility of the images to the scientific community. The mere fact that the end product of the analysis is electronic reproductions of fossils facilitates the dissemination of the data. Furthermore, in recent times, it has been common practice to publish in open access articles 3D data (i.e., surface scans). Increasingly, the publication of digitized fossil images provides for transparency and dissemination of data and scientific

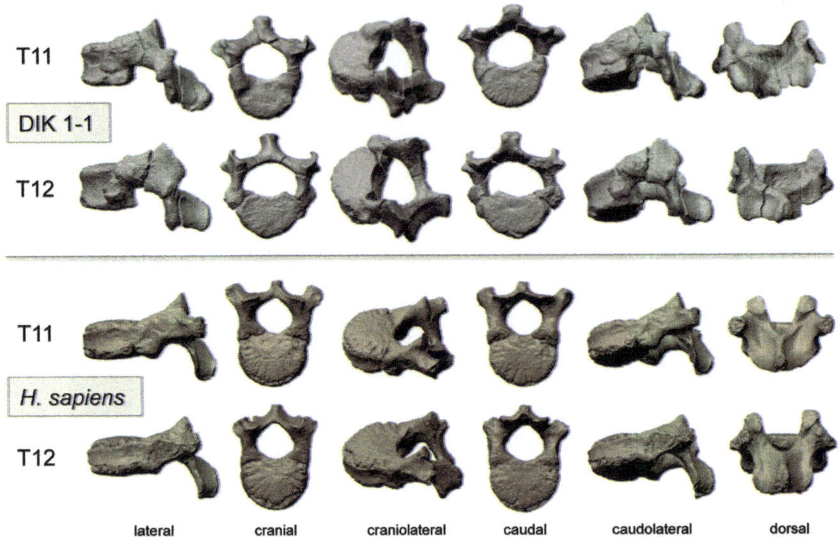

Fig. 7 Reconstructions of the 11th and 12th thoracic vertebrae of DIK-1-1 and a modern *H. sapiens* juvenile (age estimated at roughly 6–8 years). Reconstruction is based on synchrotron data (https://www.pnas.org/content/early/2017/05/09/1702229114).

progress. Specific examples of this practice are the skeletal remains of *Homo naledi* that were made available on the MorphoSource website (https://www.morphosource.org), a databank for 3D fossil images. This databank, created and managed by Duke University, provides immediate exchange of information among the research community. This new philosophical approach in the field of physical anthropology allows access to collections of real bones with restricted or no available access.

Although usually overlooked, there are the advantages to using CT and synchrotron technologies in the process of cleaning fossils of soil or contaminants. The 3D images generated are of great value in guiding researchers during restoration and preventing accidents in which fossils are physically damaged while trying to free the remains from earth.

Misinterpretation with a twist of imagination: The first hominin fossils found

Although Feldhofer 1 (Neanderthal 1) is often regarded as the first hominin fossil unearthed, at least three previous findings that went initially unreported were made. The Red Lady of Paviland, Engis 2, and Gibraltar 1 were found during the first half of the 19th century. In general, scholars of the time and even those extending into the early 20th century tended to identify hominin remains as representative of early "races" of modern man or abnormal (deformed) representatives of contemporary humans. During these early years of physical anthropology, scientists and people in general perceived humanity's relationship to nature differently than today. Scholars were still roped-in with creationist thoughts and dealing with how to accommodate data generated by the scientific method with dogmatic religious beliefs. In addition, the early hominin discoveries described below preceded three key publications that forever changed the way we think about our relationship to other living things. These books are Darwin's *On the Origin of Species* (1859),[26] Huxley's *Man's Place in Nature* (1863),[27] and Lyell's *The Antiquity of Man* (1863).[28] Following these publications, scholars increasingly began to see *sapiens* just as part of a biological continuum within the animal kingdom and not as a special and unique creation.

The Red Lady

The Red Lady of Paviland was unearthed on January 18, 1823 from a coastal cave/rock shelter and grave site on the Gower Peninsula of South Wales in Great Britain. At the time of their discovery, these remains were misinterpreted and misclassified.[29] William Buckland (1784–1856), a well-known theologian as well as paleontologist, walked into Paviland Cave, locally known as Goat's Hole Cave, and discovered a partial human skeleton as well as bones of extinct mammals, including a mammoth. At the time, he was Professor of Geology at Oxford University. Although he was aware of extinct animals—in fact, Buckland drafted the first description of a fossil dinosaur, coining the taxa *Megalosaurus*—Buckland was a creationist and believed that no humans had coexisted with any extinct animals.

During the actual excavation, Buckland thought the bones were the remains of a customs officer murdered by smugglers. Yet by the time of his initial publication, he had changed the gender to female with a more colorful story to go along. He explained that the remains were of a local prostitute or witch and associated them with a burial that was dug in historical times, possibly by the builders of nearby preRoman fortifications.[30] He was convinced that skeleton postdated the biblical Great Deluge. Buckland deduced the sex of the individual from the various personal ornamental objects found at the location, such as ivory rods, ivory rings, bracelet fragments, and jewelry carved from a mammoth tusk, as well as perforated periwinkle shells. All of these items were heavily covered with red ocher (iron oxide), indicative of a ceremonial interment. At the head of the skeleton was a mammoth skull, complete with both tusks. Marker stones were placed nearby. It is possible that Buckland's conclusions were fueled by a vivid imagination and eccentric personality. He was known to give dramatic/theatrical lectures on horseback and to mimic the movement of animals such as dinosaurs during his lectures, and he regularly performed fieldwork in his academic gown (http://www.oum.ox.ac.uk/learning/pdfs/buckland.pdf).

Today, we know that the skeleton discovered by Buckland was of a young male from the Early Upper Paleolithic and it has been dated to 33,000 years before the present by carbon decay.[31] Therefore, the Red Lady's ancestors arrived in Europe approximately 10,000 years prior as part of the dispersal of AMHs that first populated Europe.

At the time the Red Lady was alive, Paviland was about 100 km inland and part of the now-underwater Channel River valley. Conforming to other Paleolithic burials, the corpse was headless, resting in a supine position (flat on its back), likely a ritualistic practice. The long bones of the right side and the spine were missing. The male was 25–30 years old at the time of his death, about 5′ 8″ in height, and weighed around 73 kg. The mtDNA haplogroup of the remains was reported to be H, one of the most common extant lineages in Europe,[32] although the reliability of the mtDNA typing has been questioned on the basis of the possibility of contamination from contemporary sources.

Clearly, the Red Lady discovery took place at a time when the practice of archeology was primitive. Furthermore, the Red Lady was not only the first grave of a human fossil found, but also the first hominin fossil discovered anywhere in the world. No carbon dating was available. All of this made the discovery a novelty. Buckland performed no stratigraphy, nor did he collect any spatial data during the excavations.

Engis 2

The remains of Engis 2 were discovered in 1829 by the Dutch-Belgian physician, naturalist, and paleontologist Philippe-Charles Schmerling (1790–1836) in the Flémalle region of central Belgium.[33] The site is named after a group of caves known as Grottes d'Engis, meaning Engis Caves. One fossil, named Engis 2, was found in one of these, referred to as lower or Awirs cave (Fig. 8). Engis 2 is the second oldest hominin fossil finding after the discovery of the Red Lady of Paviland in Wales in 1823.

In this lower cave, the remains of two individuals were found. In addition to Engis 2, which was eventually classified as a Neanderthal, a second fossil of a Neolithic AMH was disinterred. The AMH remains were of an older person and were designated Engis 1. In addition to Engis 1 and 2, the lower cave contained faunal bones (bear, hyena, rhinoceros) and an assortment of human bones, including one human incisor, an upper jaw bone with teeth, two vertebrae, a collarbone, fragments of a radial bone and ulna, a metacarpal bone, and metatarsal bones. This lower cave has since collapsed.

The remains of Engis 2 consist of a calvarium (skullcap that covers the cranial cavity and brain), pieces of upper and lower jaw, one maxillary bone, and one upper incisor. When first discovered, Schmerling thought that he had found an assemblage of three different individuals. He realized that the fossils were ancient and postulated that they belonged to an "Ethiopian Type" individual of the Diluvian period. Later, in 1863, the famous English biologist/anthropologist Thomas Henry Huxley (1825–95) declared the remains were of a man of a "low degree of civilization." Huxley was one of the most prominent proponents of Darwin's theory of natural selection.

Fig. 8 Awirs Cave (https://en.wikipedia.org/wiki/Philippe-Charles_Schmerling#/media/File:Schmerling_Caves05.JPG).

Today, it is known that the fossil belonged to a 3- to 6-year-old Neanderthal child who lived about 40 kya.[24] Other estimates suggest dates as old as 70 kya. The child's skull was not recognized as a Neanderthal until a century later in 1936.

Gibraltar 1

On March 3, 1848, 23-year-old Navy Lieutenant Edmund Henry Réné Flint, at the time secretary of the Gibraltar Scientific Society, presented to the rest of the committee a human-like skull during a regular meeting. In the minutes of the gathering, it was recorded "Presented a human skull from Forbes' Quarry, North Front by the Secretary." It is not evident if Flint himself found the skull or someone else did. Furthermore, the exact location of the find was not noted, nor was any information relating to it. Apparently, the fossil was found during routine mining operations at the quarry. To the people at the meeting, it was just an old skull. It is possible that the fossil was found in a small limestone cave 40 feet long by 8 feet wide on the north side of the Rock. After the meeting, the skull was regarded as being the remains of an ancient human who died before the universal flood. It was placed, unceremoniously accumulating dust, on a cupboard in the Garrison Library and ignored for years.

Not knowing exactly what Gibraltar 1 represented, it was sent to London in 1864. The transfer to London was facilitated by the well-known Russian-born British zoologist and paleontologist George Busk (1807–86) who visited Gibraltar. Busk was familiar with a similar skull from Germany named Feldhofer 1 (Neanderthal 1) since he translated the original report on the specimen from German to English in 1861. He was now pondering on the identity of two fossils that were quite different from AMHs and strange looking in different ways, and yet he was unclear what exactly they were. At the time, the opinion of scholars about Gibraltar 1 ranged from being the remains of a deformed individual to a member of an ancient human race, while others suggested it could be an entirely new species.[34]

After examining both skulls, Busk was convinced that they belonged to a previously uncharacterized type of ancient human and they were not diseased individuals. He indicated that the fossils were not "a mere individual oddity, but instead a group of creatures that once roamed from the Rhine to the Pillars of Hercules."[35] Charles Darwin and Thomas Huxley also examined Gibraltar 1 and indicated that the skull was of an extinct human type. It was about that time that Busk started considering an appropriate name (taxa) for the specimen from Gibraltar, and his friend, the Scottish paleoanthropologist Hugh Falconer (1808–65), suggested *Homo calpicus*, *calpicus* being the Rock of Gibraltar in Latin.[36] Yet Busk took too long pondering about the name and, in 1863, British-Irish geologist William King (1809–86) proposed the taxa *Homo neanderthalensis* to the British Science Association, which appeared in print a year later in a detailed study of the Neander Valley specimen. Since Busk had already committed himself to the view that the Gibraltar and Neander specimens belong to the same group, he was bound to the name coined by King and lost the opportunity to name the Gibraltar skull Calpican. It seems that if Busk would have been less hesitant, we would be talking today about our Calpican instead of our Neanderthal ancestral DNA. Other than the name difference, one may wonder about the impact of having a specimen such as Gibraltar 1, with softer and less robust features compared to Neanderthal 1, first in the public eye. It is likely that the early views of Neanderthals as being brutish, idiotic cavemen would have been different, and the fossils would have been accepted as a *sapiens*' sister species more easily. It is also perplexing that a year after these events, the Gibraltar skull dropped off the radar again. Busk never performed the full analysis of the skull that he promised in his preliminary report,[35] and the find ended up once again forgotten in an obscure room of the Royal College of Surgeons in London, becoming the "twice forgotten skull."

It was not until 1886, subsequent to the discovery of two near-complete Neanderthal skeletons at the Spy Cave in Belgium,[37] that Gibraltar 1 was finally recognized as a distinct species unique from AMHs. Gibraltar 1 is now part of an exhibition on human origins at the Natural History Museum in London.

No radiocarbon dates have been reported for Gibraltar 1 and no stratigraphic data exist for secondary assessment of age from the soil. A different fossil from another site in Gibraltar produced dates ranging from 30 to 50 kya.[24] Today it is known that Gibraltar 1 is the remains of an adult woman with features, morphology, and metrics characteristic of Neanderthals. Considering that the Rock of Gibraltar was one of the last refuges of a species in the last stages of extinction, Gibraltar 1 may have been one of the last of its kind to survive.

Feldhofer 1 or Neanderthal 1

Neanderthal 1 was discovered in 1856 in a limestone cave (Kleine Feldhofer Grotte) in the Neandertal valley, a location about 13 km east of the city of Düsseldorf in western Germany. The fossil was unearthed 8 years after Gibraltar 1. However, unlike the Gibraltar specimen, from the very beginning the discovery the fossil generated controversy internationally over what it could represent. Most of the arguments were based on religious grounds. It was contended that since God's creation of the universe and the world (according to Christians) was only a few thousand years old, the bones' age could not exceed Genesis and must belong to an ill person who had died recently. Nevertheless, the reactions to the fossil were so vigorous and widespread that it is credited as being the dawn of paleoanthropology. It was reported, described, and named in 1864.[38] The original find included a skullcap, two femora, the three right arm bones, two of the left arm bones, an ilium, and fragments of a scapula and ribs. In 1999, archeologists made an additional discovery in sediments of the now-destroyed cave consisting of fragments of bones that include one that fits exactly the original femur.[39]

It is reported that two workers in the process of mining a limestone quarry made the discovery. The laborers were enlarging the entrance of what seemed to be a cave by removing the hard clay when at a depth of about a couple of feet a number of bones were unearthed. The workers initially failed to notice the remains and they were dispersed with the excavated debris some distance away from the cave. Somehow, the fossils were subsequently spotted, collected, and taken to the mine owners, who thought they were the remains of a cave bear. Eventually the bones were given to German paleoanthropologist Johann Carl Fuhlrott (1803–77) (Fig. 9). As Fuhlrott writes in a letter to Thomas Henry Huxley, "the workers were not careful in gathering the remains and only collected the larger ones, thus it is likely that they got only a portion of the complete skeleton."[40] Fuhlrott recalls returning to the quarry to look for the rest of the specimen, but failed to find anything more.

To his credit, Fuhlrott initially recognized the bones for what they were: a previously unknown type of human, not a deformed specimen.[41] Shortly after, Fuhlrott shared the findings with German anatomist, anthropologist, and professor of Anatomy at the University of Bonn Hermann Schaaffhausen (1816–93). They both agreed that the bones represented the remnants of an ancient human race, different from contemporary humans. Yet this interpretation was not initially welcomed by the public, since it was

Fig. 9 Johann Carl Fuhlrott (https://en.wikipedia.org/wiki/Neanderthal_1#/media/File:Fuhlrott.jpg).

in direct conflict with the scriptures. As with the previously discovered Engis 1 and Gibraltar 1, the initial reaction was that the skull, with its oval shape, low receding front, prominent brow ridges, and thick bones, belonged to a deformed contemporary human. Even Thomas Henry Huxley, a firm believer of the theory of natural selection, thought that Neanderthal 1 was within the range of variation of modern humans.[42]

Other opinions were expressed by the local press, which on September 4, 1856 published an announcement that read:

> *Examination of the skeleton, namely the skull, revealed the individual belonged to the tribe of the Flat Heads, which still live in the American West and of which several skulls have been found in recent years on the upper Danube in Sigmaringen. Maybe the find can help to settle the issue of whether the skeleton belonged to an early central European original inhabitant people or simply to one of Attila's roaming horde's men.*

On June 2, 1857 Schaaffhausen and Fuhlrott delivered the results of their investigations to the members of the Natural History Society of the Prussian Rhineland and Westphalen.[43] In their reports, they indicated that the remains were of "a representative of a native tribe who had inhabited Germany before the arrival of the ancestors of modern humans." And that the bone dated to "prehistoric times, probably from the Diluvial period and therefore belong to an archetypal individual of our race." Then Fuhlrott modified his opinion and indicated, "these bones are ante-Diluvial, forms of fossil human remains." The typical reaction by contemporary scholars was expressed by the editors of one of Fuhlott's articles, who in 1859 wrote in a postscript at the end of one of his manuscripts "[the manuscript] puts forward views, that cannot be shared."[42]

In 1863, when William King described the specimen and named it *Homo neanderthalensis*, he initially indicated that he felt "strongly inclined to believe that it is not only specifically but generically distinct from Man." Yet he later changed his mind and went further, suggesting that the differences between the fossil and contemporary humans went beyond unique species, that in fact they were different genera, indicating that Neanderthal 1 was not a *Homo*. Other authorities such as prominent zoologist and evolutionist Ernst Haeckel (1834–1919) suggested the taxon *Homo stupidus*. This name clearly indicates the perception of the public and scientific community toward the fossils as a brute subhuman.

A bizarre twist to the history of Neanderthal 1 came in 1864 when creationist and anatomist August Franz Josef Karl Mayer (1787–1865) concluded that the bones belonged to a Cossack, a Russian military man who came to Germany some decades earlier to battle, and somehow ended up alone inside the limestone cave. Mayer elaborated further indicating that the soldier was under the command of Prussian general Chernichef, who camped just east of the Rhine on January 14, 1814, during the Napoleonic Wars. He went on to explain that the Cossack was a soldier in the cavalry and thus spent much of his life on horseback and that the constant seating position on a horse brought about the bowed legs that resembled rickets. Furthermore, the man was in so much pain from his condition that he decided to crawl into the cave, which is 30 meters up a cliff side, and wait to die. The pain he suffered was so strong that he was constantly furrowing his eyebrows, which in turn created the massive brow ridges observed in the skull.[44] From the hands of Mayer, Neanderthal 1 went to German anthropologist and pathologist Rudolf Carl Virchow (1821–1902) who provided a different version altogether. Virchow indicated that the bones were recent and belonged to an old man that suffered from childhood rickets, had experienced severe impacts to the head, and suffered arthritis in his old age. At the time, Virchow's opinion was highly regarded in Germany and his thoughts on the issue affected the progress of research in human evolution for several decades. These accounts were so absurd that they rival in imagination the descriptions of the Red Lady from Wales. In retrospect, it is difficult to comprehend how post-Enlightenment/Reason scholars in the mid-19th century felt justified in coming forth with such outlandish stories.

Trends in hominin evolution

What triggered the split between the apes and hominins?

Subsequent to the split of the human and great ape lineages, about 7–6 million years ago, the human evolutionary line experienced a number of genetic, morphological, and physiological changes that characterize hominins. Many experts are of the opinion that the

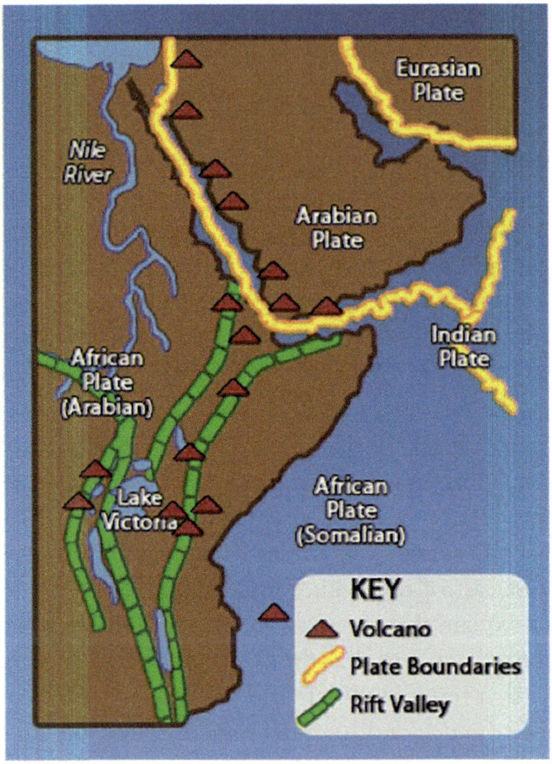

Fig. 10 The East African Rift Zone (http://www.eclp.com.na/demo/geography/mod1page18.html).

separation of the two evolutionary lines that eventually led to the creation of the great apes and human lineages occurred in East SubSaharan Africa and was triggered by a number of climatic/environmental changes connected with the creation of the East African Ridge (EAR) (Fig. 10). It is postulated that the creation of the EAR resulted in the transformation of forest to dry savannah to the east of the mountain chain. The EAR is the largest seismically active rift system on Earth today, and its formation is thought to be connected to a number of geological forces, including tectonic activity, earth crust density, mantle plumes, and volcanic activity.[45]

Although volcanism may have commenced as early as 45 to 33 million years ago (mya) and the initial uplifting in the region may have started as early as 38 to 35 mya[46] the high relief characteristic of the EAR developed between 18 and 13 mya.[47] These geological changes over the span of the last 10 million years have been rather dramatic as plate tectonic (continental drift) movements are separating Africa east of the Great Lakes (Lakes Victoria, Tanganyka, and Malawi, among others) from the rest of the continent. It has been postulated that the uplifting and the creation of the EAR and the Great Lakes transformed the habitat east of the mountain range and prompted intense selection pressure for traits conducive to terrestrial instead of arboreal existence.[48,49]

In geological time scale, the habitat east of the EAR was transformed rapidly due to the sequestration of moisture by the mountain chain, which blocked and limited rain to the east. An actual increase in aridity east of the EAR is supported by a number of studies.[50] West of the massifs, tropical forests persisted, providing suitable habitats for arboreal forest dwellers including great apes and various other primates, while to the east, it is theorized that hominins evolved in a different direction due to distinct environmental pressures and selection. This outcome is intuitive, and over the years since the 1980s, investigators have been compiling data correlating past speciation episodes with climatic changes.[51] Specifically as it relates to hominin evolution, the increase in aridity has been linked to vegetation shift and habitat change.[52]

In addition to the environmental changes created by the mountains, the process of continental split driven by the separation of the African Plate will eventually rupture the Somali Plate, creating a new continent in the Indian Ocean. This separation is expected to occur gradually during the next 10 million years.[53] It is known that the African Plate is separating into two tectonic plates, the Somali and Nubian Plates, at a rate of 6–7 mm per year. Fig. 10 illustrates the locations of the various plates and fault lines in East Africa.

Another factor that is thought to have impacted the inhabitants of East Africa are the numerous large and deep lakes created by this local continental drift. The areas in the vicinity of these lakes, which run north to south along the rift zone, are rather sensitive to changes in rainfall and evaporation episodes promoting periodic cycles of fluctuating environmental conditions. These alternating climatic and habitat conditions are the basis of the turnover-pulse hypothesis. Formulated by paleontologist Elisabeth Vrba, this hypothesis postulates that ecological changes can result in rapid extinction and high turnover of new species across multiple different lineages.[54] Fig. 11 shows graphically how the presence of lakes in the EAR Valley tended to correlate with hominin diversity and cranial capacity. There is increasing evidence indicating that in addition to the long-range diminution of rain created by the mountains of the ridge zone, the rapidly alternating dry-wet conditions generated by the East African Great Lakes triggered important events in hominin evolution.[49] In relation with these correlations, it is noteworthy that most of the available evidence points to East Africa as the region where all four major stages of human evolution have occurred.[55] These four stages include the emergence of the earliest proto-hominins (the genera *Sahelanthropus*, *Orrorin*, and *Ardipithecus*) between 7 and 4 mya, the appearance of the *Australopithecus* around 4 mya and the robust *Paranthropus* approximately 2.7 mya, the genus *Homo* about 2 mya, and the genesis of AMHs around 315,000 ya.[56]

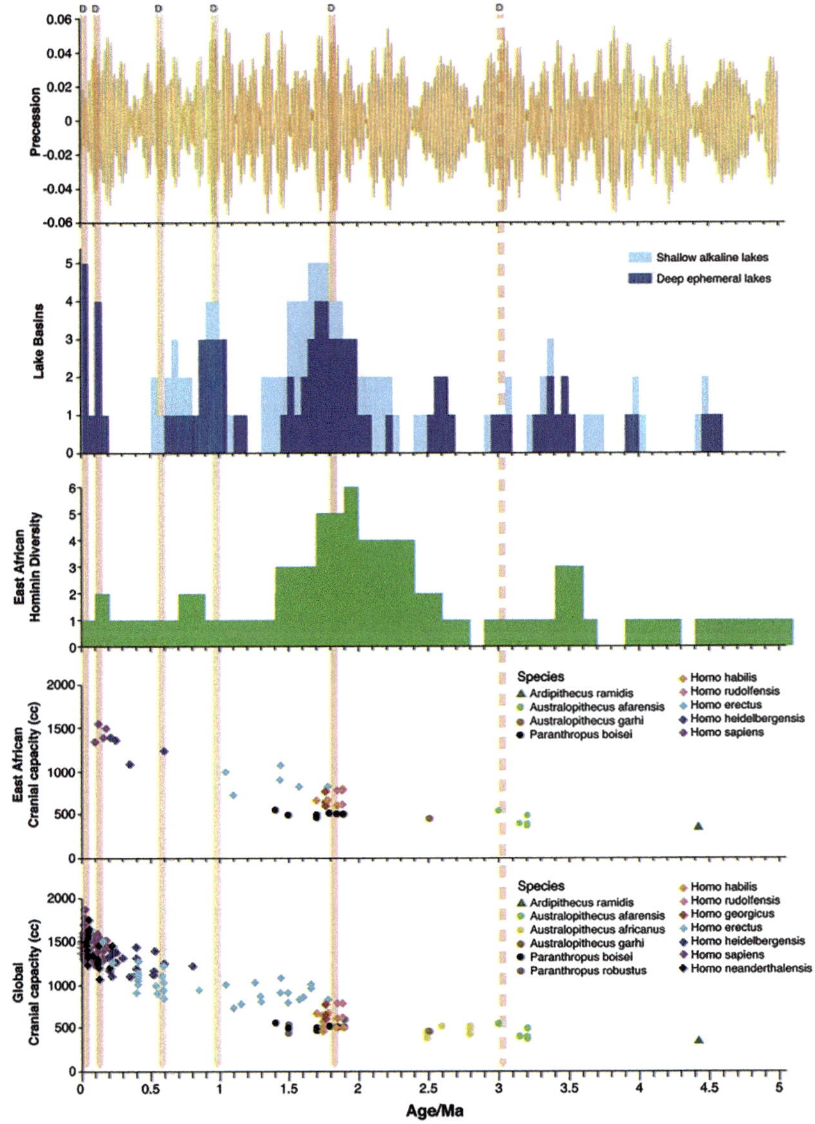

Fig. 11 Correlation between the presence of lakes and hominin diversity and cranial capacity (https://www.sciencedirect.com/science/article/pii/S0277379114002418).

Hallmarks of hominin evolution

The traditional view of hominin evolution in which a limited number of lineages incurred changes sequentially in geological time is currently no longer tenable. Scholars used to envision hominin evolution as one taxon gradually metamorphosing into another. This perception was partially due to the fragmentary nature of the data available and the limited technology to study the remains. This simplistic model is being replaced by a much more complex scenario in which a multiplicity of hominin groups and lineages lived in temporal and physical proximity to each other, possibly interacting culturally and biologically. A number of factors are forcing these changes in perspective among the scientific community. These include the number of archeological and physical anthropological specimens now available, which have provided information on the complexity of evolutionary events previously unknown. Furthermore, the amount of information derived from techniques from different fields, new, more precise dating methods, the application of statistical approaches designed to integrate and correlate large amount of data, and the realization of previously unknown evolutionary processes such as punctuated evolution, have provided novel ways to examine hominin evolution. In addition, an overall philosophical transformation in the way science is conducted from isolated discipline-specific investigations to integrative, theory-driven approaches has contributed to more complete evaluations of specimens and species. Using this holistic approach, investigators have collaborated and combined evidence from different disciplines, such as paleontology, molecular genetics, chemistry, geology, paleo-biogeography, paleoecology, and social studies, to answer these questions. Yet in spite of these advances and new discoveries, certain aspects of hominin phylogeny remain unclear and at times puzzling, and as the number of new groups increases, the number of kinship questions and blind spots grows, providing for a complex hominin tree.

Mosaicism

In addition to the greater evolutionary complexity that is now substantiated by the number of different hominin species that shared habitats, other characteristics help define hominin evolution. One particularly striking theme that occurs throughout hominin phylogeny is mosaicism. This phenomenon is apparent when a mixture of ancestral and derived traits is present in an organ or part of the anatomy of an organism. It is clear that certain parts of the anatomy evolve faster than others, resulting in body parts that retain the ancestral condition while others experience change. In addition, different hominin species exhibit different patterns of mosaicism. In other words, different combinations of ancestral and derived traits are observed from species to species. It seems as if certain characters that are part of structures or organs have changed at different rates, responding to unique selection pressures, possibly reflecting an underlying hierarchical organization based on differential genetic and developmental origins. Is as if specific structures are semiautonomously controlled by inheritance, reflecting subsets or modules, which correspond to developmental origins. Below, a number of specific species examples are discussed.

- *Homo naledi* (315–236 kya)

The *hominin* fossil record is permeated with examples of mosaicism. In some instances, the collage of ancestral and derived characteristics makes sense, considering the overall evolutionary trend affecting the anatomy in question. In other instances, mosaicism is manifested in bizarre combinations that do not seem to follow any evolutionary progression. Take, for instance, the fossil assemblage from the Rising Star caves of South Africa, near the city of Johannesburg,[57] where the remains of a new species was discovered in October 2013. This unusual hominin species was named *Homo naledi* and it lived as recently as 250,000 ya. Therefore, this species was contemporary with the larger-brained AMHs and they probably lived side by side as competitors. In addition, it is likely that *naledi* was not a direct ancestor of AMHs, but an offshoot within the genus *Homo*. The quantity and quality of the fossils collected are exceptional. The fossils included every body part of old and young adults as well as infants. It has been reported that a good number of additional fossil remains are still waiting to be unearthed from the chamber. In the material already recovered, important finds include an almost complete hand and foot. *Homo naledi* possessed a very small brain, 560 cc in males and 465 cc for the females, suggesting some degree of sexual dimorphism. Overall, the remains exhibit some body parts reminiscent of ancient groups, such as Australopithecines, other morphologies are in line with more recent groups, such as *Homo erectus* or even AMHs, while other characteristics are unique to *H. naledi*. In general, the anatomy of the fossils above the hip reflects an ancestral body plan, while below the hip it resembles advanced human-like characteristics.

The skulls are a collage of ancestral and derived traits. The combination of cranial characteristics is unique, sharing traits with species across the genus *Homo* including *H. habilis*, *H. rudolfensis*, *H. erectus*, and Middle Pleistocene *Homo*.[58] For example, since the brain size of australopithecines ranged from 420 to 500 cm^3 and *Homo erectus*' averaged about 900 cm^3, *H. naledi* cranial capacity resembled the former, with dimensions that overlap. Yet in spite of its very small brain, *H. naledi* possessed a massive skull and temporal and occipital bulges, characteristics associated with *Homo erectus*. In addition, the postorbital constriction or narrowing of the skull behind the eye sockets is reduced in *H. naledi*. This reduction of the postorbital constriction is considered a derivative trend in hominin evolution. In early *Homo sapiens*, this feature is barely visible, and in contemporary humans, it has disappeared altogether. The teeth of *naledi* are smaller than in Australopithecines, with crowns and cusps reminiscent of AMHs. The muscles of the mandible were small, indicating a softer diet requiring less mastication compared to the ancestral condition.

Postcranially, the rib cage is conical in shape, as in *Australopithecus* and great apes, yet the hands are a combination of ape-like and AMH-like traits never seen in any other hominin. The aggregate includes prominently curved proximate and intermediate phalanges and a wrist, thumb, and palm designed for object manipulation. The thumb, wrist, and palm bones are human-like, whereas the fingers are still curved—an ancestral morphology designed for holding on to branches during arboreal locomotion. For unknown reasons, curved fingers persisted longer than other ancestral traits. This ancestral characteristic is counterintuitive, since *naledi* was bipedal. A mosaic assemblage is also seen in the morphology of the foot, which is predominantly AMH-like and adapted for bipedalism, yet still different from modern humans in having more curved proximal pedal phalanges. This combination of traits suggests that although *naledi* walked upright, its gait differed from AMHs. Remarkably, the feet look like modern humans while the shoulders are ape-like. The pelvis also represents a combination of ancestral and derived characteristics. The base of the pelvis looks very modern human, while the blades spread outward as in Australopithecines.

- *Australopithecus sediba* (~ 2 mya)

The first specimen of *Australopithecus sediba* was discovered in 2008 in Malapa, South Africa, 45 km north-northwest of Johannesburg. This species dates back to about 2 mya and is another good example of mosaicism. *A. sediba* exhibits a dentition similar to *Australopithecus africanus* and yet it possesses derived traits resembling the genus *Homo*.[59] The anatomy of the upper limbs is ancestral, reflecting an arboreal and climbing existence, possibly involving suspension—a characteristic it has in common with other australopithecines.[60] Yet *sediba*'s hand exhibits a diversity of ancestral and derived characteristics, including a number of *Homo*-like features, such as a long thumb and short fingers, traits linked with dexterity.[61] The thorax possesses a mixture of derived and ancestral conditions with a highly flexible vertebral column reminiscent of the Nariokotome *H. erectus*.[62] The anatomy of *sediba*'s lower extremities is unique to the group. This includes the heel, midfoot, knee, and hip.[63]

- *Apidima* 1 (~ 210,000 mya)

A recent reevaluation of the Apidima 1 skull discovered in 1978 in a cave (Apidima) in southeast Greece and now dated at 210,000 ya indicates a unique mixture of ancestral and derived AMH traits.[64] The Apidima 1 skull exhibits several characteristics that are diagnostically Neanderthal and ancestral while its rounded shape is characteristically and uniquely modern human. Specifically, Apidima 1 possesses a rounded posterior cranium, which is a trait that has evolved relatively recently. Furthermore, morphometric analyses including neurocranial shape index and PCA reflect a mosaic of ancestral and derived traits.

This finding is remarkable because it pushes back the presence of *sapiens* in Eurasia by 160,000 years. In addition, alongside the *sapiens* cranium, a second skull dated at 170,000 was classified as an early Neanderthal (Apidima 2). Since contemporary modern humans are thought to derive from a more recent African ancestor lineage that migrated out of Africa about 70,000 ya, this fossil suggests that AMHs dispersed out of Africa in various waves and some became extinct. The presence of both species in close proximity within the cave also points to possible contact between Neanderthals and *sapiens* much earlier

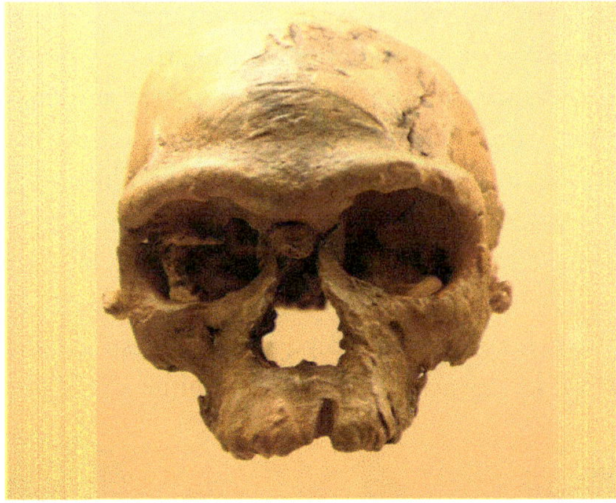

Fig. 12 Skull of *sapiens* Jebel Irhoud-1 (https://en.wikipedia.org/wiki/Jebel_Irhoud#/media/File:Jebel_Irhoud_1._Homo_Sapiens.jpg).

than it was previous thought, as suggested by ancient DNA evidence of introgression of African AMH DNA into Neanderthals.[65]

- Archaic *sapiens*

Although the *Homo naledi* fossils postdate, or at least were contemporary with, early *sapiens* such as Jebel-Irhoud, it is thought that their lineage did not survive to the present, except possibly their DNA through introgression. The ancient AMH remains from Jebel-Irhoud, Morocco date to about 315–334 kya and illustrate that mosaicism is observed in our species as well. The remains were discovered in 1960 approximately 50 km southeast of the city of Safi in Morocco. The mosaicism in the Jebel-Irhoud samples is so pronounced and surprising that initially investigators thought they were dealing with Neanderthal fossils. The assemblage of derived traits include facial, dental, and mandibular AMH anatomy, and the ancestral characteristics comprise endocranial and neurocranial morphology.[56] Specifically, the facial shape is reminiscent of *sapiens* but the morphology of the braincase shows ancestral traits including very prominent and thick brow ridges and an elongated skull (nonglobular) (Fig. 12). Derived characteristics include lack of prognathism (Fig. 13) and a dental developmental pattern similar to AMHs, but with teeth roots that develop faster and crowns that form later than in modern humans.[66]

The examples of mosaicism illustrated above are representative of the phenomenon at large and underscore the previously unappreciated complexity of hominin evolution and the difficulty in deriving conclusions about the phylogenetic relationships of fossils hominins based on incomplete skeletal evidence. Furthermore, the mosaic patterns illustrated above have been correlated with the type of ecological instability experience by hominins in the EAR Valley and in the vicinity of the East African Great Lakes described above.[67]

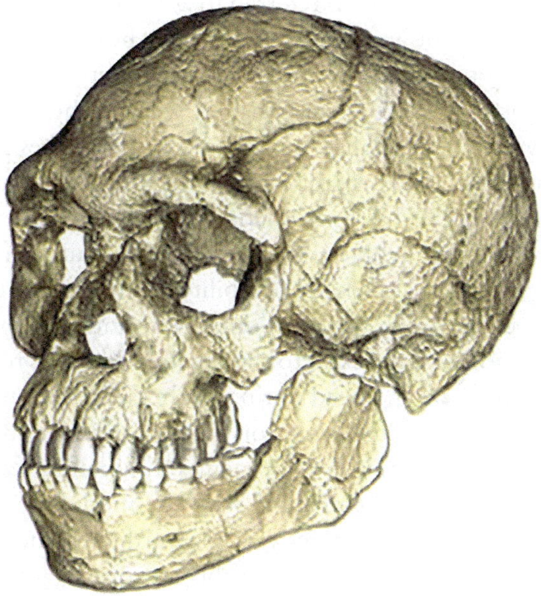

Fig. 13 Computed tomographic (TC) scan of the Jebel Irhoud skull (https://en.wikipedia.org/wiki/Jebel_Irhoud#/media/File:Homo_sapiens_from_Jebel_Irhoud.jpg).

Independence and interconnection of traits

Even though individual anatomical assemblages or modules evolve independently from each other in a nonharmonious way and at different speeds, they eventually need to be interconnected to allow survival in specific environments and habitats. Since natural selection operates on the principle of survival of the fittest at the level of organisms, anatomical modules, even if they evolve independently, ultimately need to function together. What is seen in hominin evolution is that individual organs or anatomical modules exhibit independent phyletic histories, yet these organs end up functioning together as part of the organism. Specific evolutionary developments, such as bipedalism, dexterity, encephalization, language, use of fire, among others, did not evolve at the same time, but independently within lineages with the constraints imposed by natural selection. In addition, since the combination of ancestral and derived traits vary from species to species, it is likely that some these evolutionary trends evolved in different species or lineages independently at different rates.

An example of this phenomenon is illustrated in the evolution of three hominin evolutionary modules: bipedalism, the female reproductive system, and encephalization. Studies suggest that the encephalization process may have been impacted by changes in the female reproductive system, which in turn occurred as our ancestors became bipedal.[68] Scholars envision that as hominins were switching from quadrupedal to bipedal locomotion, the changes led to what is referred to as the obstetric dilemma. The premise of this hypothesis is that as the upright posture of the hominin evolved, it was accompanied by slimming and tightening of the pelvis, which in turn resulted in an increase in risks during childbirth, as large-brained babies were pushed through ever-narrower birth canals. The switch to bipedalism involved major changes in the architecture of the pelvis, making the birth canal significantly narrower. At about the same time, the brain had begun to expand. It has been theorized that as these evolutionary processes were taking place, hominins evolved openings in the skull known as fontanelles in the form of metopic sutures that provide flexibility at the time of birth, as the newborn head is forced to squeeze through the birth canal. The anterior fontanelle, for example, allows the two frontal bones of the skull to slide past each other, reducing the size of the infant's head. This compresses the head during birth, facilitating its passage through the birth canal. It is likely that the initial evolution of bipedalism led to pelvic modifications, which subsequently impacted brain development and organization, providing for brain expansion. Thus, this scenario is compatible with the modular evolution of three evolutionary trends that evolved independently, but nevertheless, they generated constraints on each other as they became integrated into the organism.

Continuity of traits

Continuity of traits is also a phenomenon observed across various extinct species and AMHs. For example, traits such as the shovel-shaped incisors and the flatness at the upper and middle part of the face resulting from blunt nasio-frontal and zygomaxillary angles, respectively, are seen in *erectus* and in most contemporary Asian and Native American populations. In addition, the field of genetics has provided evidence of continuity. Specifically, the *RRM2P4* sequence exhibits a large stretch (a region of 2400-nucleotides) of relatively undisturbed (conserved) DNA between archaic hominins and modern humans.[69] Other genes exhibiting continuity across extinct and extant hominin species include the sequences for 2′-5′-oligoadenylate synthetase 1 (*OAS1*), a protein required for the innate immune response to viral infection, and certain portions of the gene for hemoglobin. Furthermore, relatively large pieces of DNA exhibiting conserved sequences dating back to around 700,000 ya have been detected in SubSahara African populations.[70] This continuity aspect of hominin evolution will be discussed throughout this book in connection with introgression.

Bipedal locomotion

Several evolutionary trends are evident during hominin evolution. These tendencies reflect themselves in the anatomy and function of organs and organ systems. They involve sets of tissues and organs that evolve independently from each other, but nevertheless are linked, due to the constrains imposed by natural selection for integration of organismal

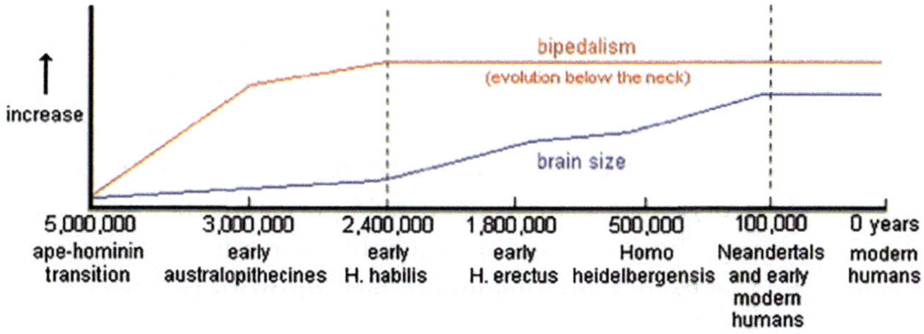

Fig. 14 Timelines of bipedalism and brain size in hominins (http://anthro.palomar.edu/homo2/mod_homo_2.htm).

functions and survival of the individual and the population. Examination of hominin fossils indicates trends that include posture, cranial capacity (brain size), and facial angle, among others. Examinations of these trends provide a useful picture of the evolutionary history and selective forces that affected hominins.

Historically, the expectations of scholars in the field of anthropology were that the brain and intellect were the first characteristics to evolve and, therefore, early hominins possessed large brains and ancestral ape-like postcranial anatomy and locomotion. Yet this conception has been demonstrated to be incorrect. The fact is that upright posture and walking on two legs appeared very early in hominin evolution. Bipedalism preceded encephalization (Fig. 14), and the beginning of the trend dates back to the time hominins separated from the ape lineage, at about 7 mya.

Evolutionary forces that may have influenced upright posture and bipedalism

Human walking is about 75% less energetically costly than either quadrupedal or bipedal walking in chimpanzees. Some hypotheses have supported that bipedalism increased the energetic efficiency of travel and that this was an important factor in the evolution of bipedal locomotion.

In addition, a number of theories have linked the adoption of upright posture and bipedalism to the open savannah of East Africa subsequent to its transformation into an arid region. This thesis stipulates that arboreal hominins were under selection pressure to evolve anatomical characteristics that would allow them to survive in the new dry plains of East Africa.

It has been proposed that in an open savannah early hominins found it advantageous to evolve an upright posture to ward off and intimidate predators. This model is based on the premise that by being visible and appearing big as well as vociferous as possible, attackers would be scared away. In addition, it is possible that an erect posture would be selected for in a grassland environment as a mechanism of vigilance against aggressors or as sexual display to attract mates. Bipedalism also provides for freedom to use the arms and hands for a number of activities that are associated with hominin evolution, including the making of tools and creation of art, carrying food and offspring, as well as taking care of the needy.[71] In a savannah environment, anatomical traits that facilitated these behaviors may have been selected. Even thermoregulation has been proposed as a selection pressure for upright posture, since in an open-field grassland, standing vertical reduces body surface area, minimizing exposure to sunlight, increasing the distance above the ground to contact cooler winds, and helping in heat dissipation.[72]

Anatomical traits impacted by bipedalism

The transition from a tree-dwelling existence to bipedalism is seen in the changing anatomy of hominins. An examination of the anatomy of a quadruped species in the ape lineage, such as the chimpanzee *Pan troglodytes*, and AMHs illustrates that the process of adaptation to bipedalism affected the entire skeleton. In fact, many of the anatomical differences observed between *Homo sapiens* as a biped and the quadruped ancestors are directly or indirectly related to bipedalism. All of the anatomical adaptations necessary for habitual bipedalism can be found in the fossil record and include the following features: (1) the presence of a bicondylar angle (angle between the axis of the shaft of the femur and the perpendicular to the joint axis of the femur); (2) a progressive inferior relocation of the foramen magnum; (3) the presence of a smaller nonopposable big toe; (4) a progressive trend for a pronounced arch of the foot; (5) a more posterior orientation of the anterior portion of the iliac blade (uppermost and largest part of the hip bone); (6) a relatively larger femoral head diameter; (7) an increased femoral neck length; and (8) a slightly larger and antero-posteriorly elongated condyles (rounded protuberance) at the end of the femur.

Bipedalism as a function of hominin species and evolutionary time

Below are some of the landmarks related to upright posture and bipedalism as illustrated by specific extinct hominin species. In the review of the subject, australopithecines are highlighted as a point of reference and comparison due to their intermediate position in the hominin evolutionary tree, the number of complete cranial and postcranial skeletons available for analysis, and the sizable amount of comparative data that exists.

- *Sahelanthropus tchadensis* (7–6 mya)

The very early stages of the evolution of bipedalism are exemplified by the position of the foramen magnum (oval opening beneath the cranium that serves as passage for the central nervous system through the skull connecting the brain with the spinal cord) and relatively small nuchal crest of *Sahelanthropus tchadensis*, possibly the oldest known hominin.[73] This species is dated to around 7 mya and morphological traits clearly indicate that *Sahelanthropus* was bipedal. Fig. 15 illustrates the position of the foramen magnum in this species in

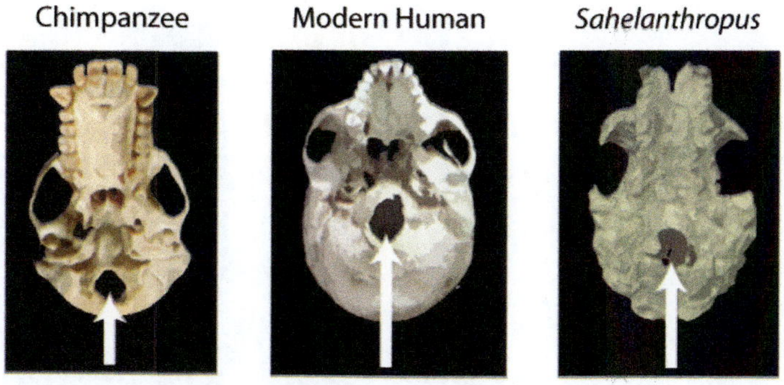

Fig. 15 Location of the foramen magnum (indicated by the *white arrow*) in chimpanzee, *Sahelanthropus*, and AMH (https://www.nature.com/scitable/knowledge/library/the-earliest-hominins-sahelanthropus-orrorin-and-ardipithecus-67648286).

relation to the chimpanzee and shows a shift (intermediate between the chimpanzee and AMH) of the opening to the bottom of the skull. The recovered partial skull from the Djurad desert of Chad represents the most complete evidence of this species. The skull exhibits a collage of ancestral and derived traits. The brain size was equivalent to that of a chimpanzee, about 320–380 cm^3 in volume, yet it possessed small canine teeth, a derived characteristic.

- *Orrorin tugenensis* (6 mya)

Orrorin tugenensis, discovered in the Tungen Hills of Kenya, is likely a descendent of *Sahelanthropus* who lived approximately 6.1–5.7 mya. *Orrorin* is also thought to have been bipedal, based on the morphology of its femur, which possesses a spherical frontal head, an elongated and oval femur neck, and a lesser trochanter (conical protrusion) projecting medially. The thighbone, although similar in size to that of a great ape, is angled and forms a sturdy connection with the pelvic joint to sustain the body's weight as in AMHs. In these discussions, it needs to be emphasized that the features suggestive of bipedalism exhibited by *Sahelanthropus* and *Orrorin* do not necessarily mean that these hominins were fully upright or constantly bipedal. It is likely that the evolution of bipedalism was accompanied by conditions of occasional quadrupedal locomotion and intermittent/intermediate forms of bipedalism. In other words, it is likely that early hominins' bipedalism differed from *sapiens*' in that they moved differently as they walked and may have switched between bipedal and quadrupedal locomotion.

- Australopithecines (3.85–2.95 mya)

Australopithecus afarensis lived from 3.9 to 3 mya. *Afarensis* is the species well known for the famous Laetoli foot imprints on volcanic ash discovered in Tanzania and dated to 3.6 mya. Fig. 16A and B illustrate the relative forces on an ash substrate generated by human, Laetoli hominin, and chimpanzee footprints. These illustrations show how the points of pressure during the stride are dramatically altered as the anatomy of the foot

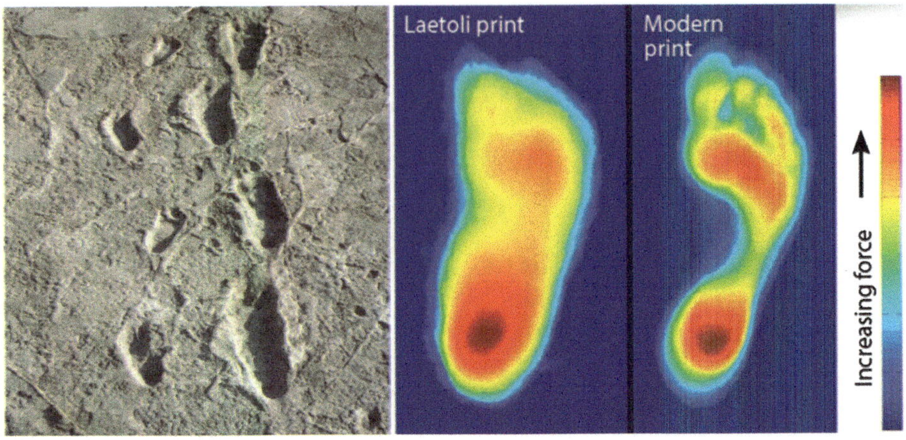

(A)

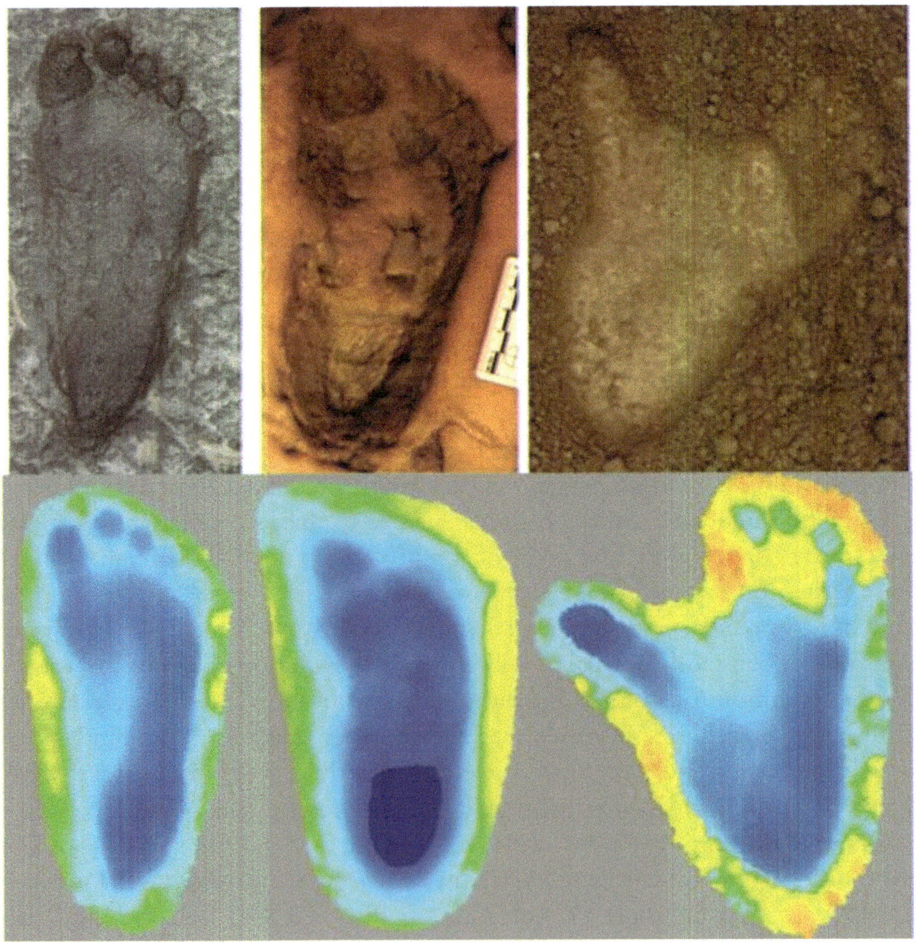

(B)

Fig. 16 See figure legend on next page

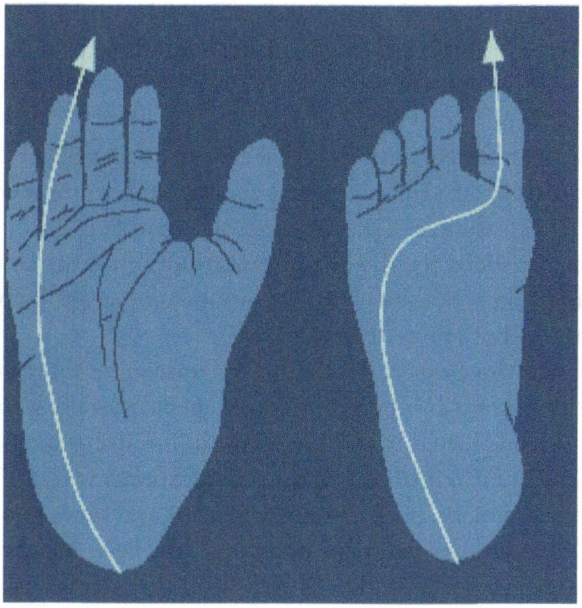

(C)

Fig. 16 (A) Laetoli tracks. Hominin tracks at Laetoli, in Tanzania, made about 3.6 million years ago (left). Computer models of the ancient footprints can be compared with that of a modern human to study gait (right) (http://discovermagazine.com/2015/oct/19-first-impressions; https://www.icr.org/article/human-evolution-story-stumbles-over). (B) Examples of human, Laetoli hominin, and chimpanzee footprints (https://royalsocietypublishing.org/doi/full/10.1098/rspb.2016.0235). (C) Transmission of weight in the foot of an ape and AMH (https://www.pbs.org/wgbh/aso/tryit/evolution/footprints.html).

of hominins changes. Fig. 16C illustrates how the quadrupeds' feet transmit weight from the heel along the outside of the foot and then through the middle toes, while bipeds transfer the weight from the heel along the outside of the foot and then across the ball of the foot and finally through the big toe, which is a more energy-efficient process.

Australopithecus afarensis exhibits a number of anatomical features in transition toward full bipedalism. For example, while the pelvic, leg, and foot bones of *Australopithecus afarensis* clearly show that this species was bipedal, the species was not well adapted for running.[20] In addition, its curved and long finger and toe bones as well as the anatomy of the shoulder blade and the fact that the arms could be swung above the head indicate some degree of arboreal existence. Furthermore, *afarensis*' wrist bones still show some adaptations for knuckle walking. Although the position of the foramen magnum in *Australopithecus afarensis* is located more inferiorly than in *Sahelanthropus tchadensis*, *afarensis* still did not hold its head fully erect.[74]

The transformation of the lumbar vertebra represents another anatomical site of adaptation to bipedalism. Australopithecines, for example, possessed five or six wide lumbar vertebrae that functioned in weight transmission from the upper body to the pelvis, similar to AMHs.[75] A larger number of broader lumbar vertebrae provide for a more efficient transfer of weight and the lumbar curvature allows for a more flexible lower back and permits the hips and trunk to swivel forward when walking. On the other hand, apes, as quadrupeds, have four lumbar vertebrae that are relatively smaller than in australopiths. In addition, these anatomical adaptations that evolved in hominins contributed to a progressive shift in the body's center of gravity. In australopithecines, for example, the center of gravity is intermediate between the quadruped condition at the center on the torso and the AMH state at the center of the pelvis. The lumbar curvature helps transfer the center of gravity within the pelvis above the feet.

The sacrum, a triangular bone at the base of the lumbar vertebrae that connects to and stabilizes the pelvis represents another location of bipedal modification. The evolutionary changes in the sacrum associated with bipedalism in hominins include an incrementally broader surface of the bone and larger sacroiliac joint surfaces (Fig. 17). In contemporary chimpanzees, as an example of quadruped, the sacrum exhibits a relatively smaller sacroiliac joint surface compared to australopithecines. In *Homo sapiens* the sacrum is even larger and wider than in australopiths. In addition, the sacrum increases its curvature as a function of hominin evolution. These size differences and curvature of the sacrum reflect the different schemes of weight transmission through the pelvis during the transition from quadrupedal to bipedal locomotion.[76]

The pelvis also presents a number of critical modifications in the evolution of bipedalism in hominins. One is the progressive enlargement and shift of the iliac ala (wing of the ilium) from a parallel orientation with the plane of the back in quadrupeds to a lateral position in bipeds.[77] In *Australopithecus aphaeresis*, for example, the iliac wings protrude out to the sides giving the pelvis a flat and plate-like appearance. In the pelvic fossils of other hominins, the iliac wings project out to various degrees between *aphaeresis*' extreme

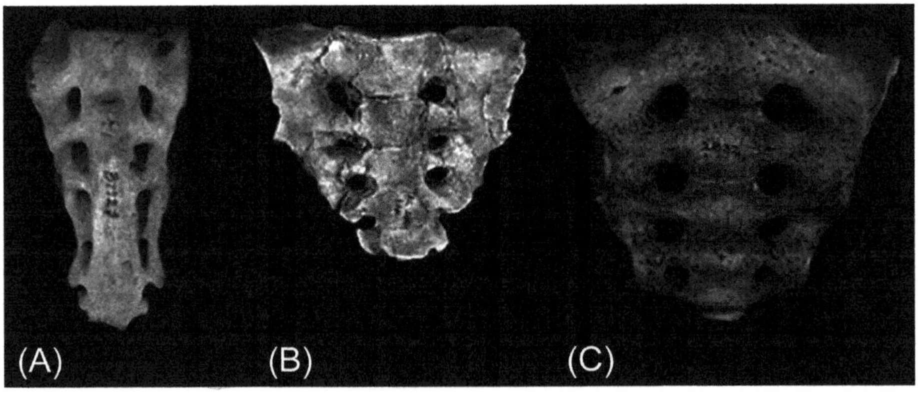

Fig. 17 Sacra. (A) Sacra of a chimpanzee, (B) A.L. 288-1 Australopithecus afarensis, and (C) a modern human (https://royalsocietypublishing.org/doi/full/10.1098/rstb.2010.0112).

and the AMH condition. This lateral location of the pelvis wings moves the hip joints away from the body. This modification also places the weight of the torso on the pelvis and its joints during walking. To deal with the increase in weight on the hip, bipeds possess larger pelvic joints and femoral heads compared to quadrupeds. These structural adaptations work to absorb and accommodate the greater stress as well. This modification of the iliac wings also affects the structure, attachment, and function of the medius and minimus gluteal. These muscles attach the iliac wings to the femur. In bipeds, these muscles act to stabilize the area by preventing the hip from toppling toward the unsupported side. It is interesting that in australopithecines the lateral wings of the pelvis are pointing laterally even more than in AMHs (Fig. 18). This clearly indicates that *Australopithecus* was bipedal, but its gait was different from AMHs.

In bipeds, the femur plays an essential role in the transfer of weight from the upper part of the body to the feet while the organism is standing, walking, or running. The femur possesses a rounded femoral head that articulates with the pelvis at the acetabulum or socket. The femoral head and neck are larger in bipeds, an adaptation for absorbing more weight and stress at the hip joint. The femoral shaft is straighter than in quadropeds with two bulbous condyles (protrusion) at the other end. These condyles are larger and more elliptical in bipeds as compared to the smaller and rounder protuberances seen in quadrupeds. This end of the femur joins with the tibia (lower leg). Australopithecines exhibit a smaller femoral head and longer femoral neck compared to members of the genus *Homo*.[78]

As the size of the pelvis incrementally increased in australopiths and in members of the genus *Homo*, there was still a need to keep the center of gravity balanced. This is very

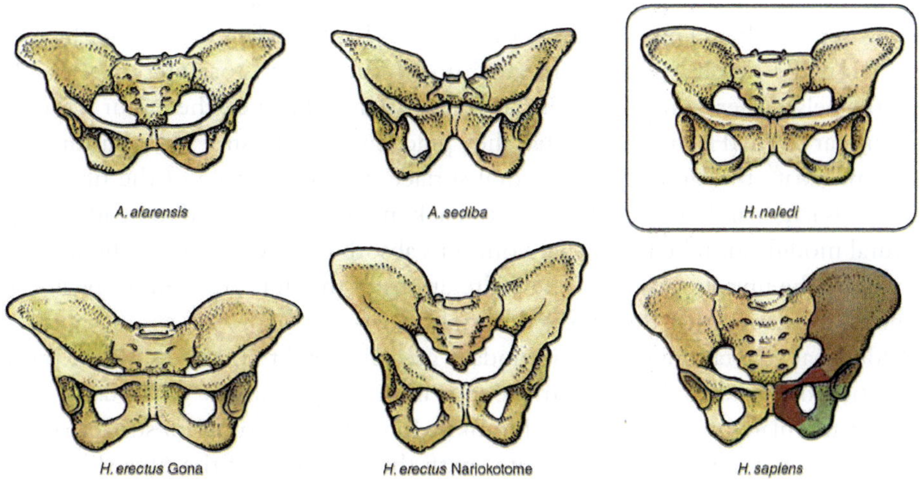

Fig. 18 The hominin pelvis (https://www.americanscientist.org/article/an-updated-prehistory-of-the-human-pelvis).

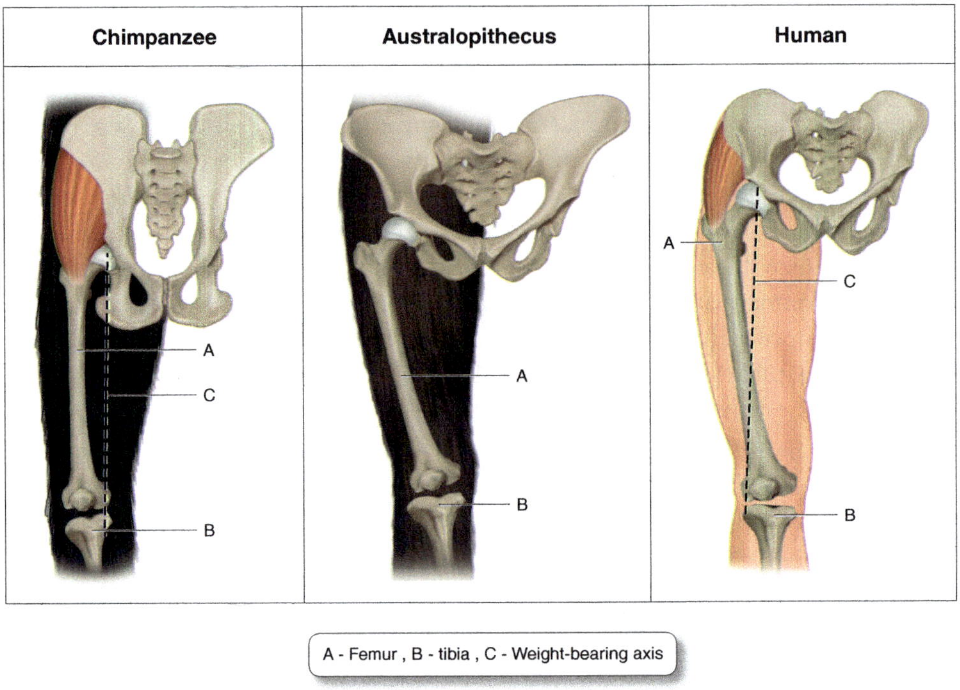

Fig. 19 Angle of the femoral shaft in three taxa (https://answersingenesis.org/human-evolution/lucy/lucy-the-knuckle-walking-abomination).

critical in bipeds, since the center of gravity is sustained by only one leg during walking or running. In the hominin lineage, this paramount necessity was achieved by bringing the knees closer together. This placement of the knees has the effect of moving the feet precisely below the center of gravity. In quadrupeds, the femur is oriented in a straight line under the pelvis; there is no angle (Fig. 19). In australopithecines, the femur is found at an angle similar to AMHs, indicating that they practiced bipedalism.[79] In addition, in order to accommodate the larger distal femoral surface, the top surface of the tibia (articular condyles) is larger and longer from front to back in bipeds compared to quadrupeds. This structural modification helps bipedal hominins absorb the increased weight and pressure imposed by the upright posture. These alterations of the tibia are seen in species of australopiths, indicating that they moved upright.

Additional adaptations to accommodate greater weight and pressure on the lower extremities to provide for bipedalism are seen in the shape and size of the talus. The tibia distal end articulates with the talus at the ankle. In AMHs, the superior surface of the talus is not angular but parallel as it articulates with the distal end of the tibia (Fig. 20). This condition is also observed in *Australopithecus afarensis*.[80] In general, *afarensis* possessed rigid feet relative to quadrupeds, such as apes, which require flexibility, as they use their

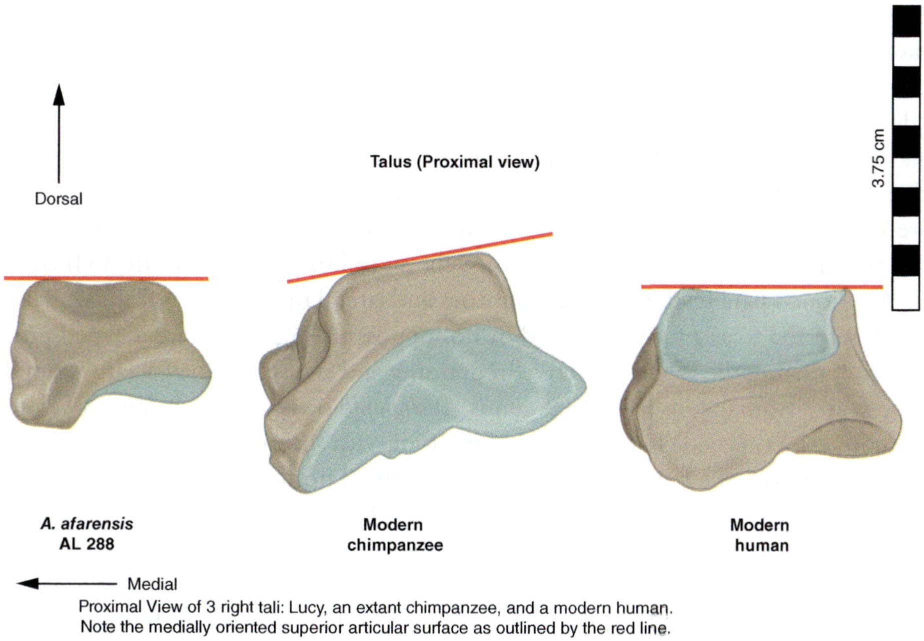

Fig. 20 Articular superior surface *(red line)* of Talus (http://efossils.org/book/tibia-talus-ankle).

extremities to swing among the branches. The greater stability of the australopithecine foot is a logical adaption for bipedalism considering the necessity to sustain the mass of the entire organism and the stress imposed on the feet during locomotion.

The arm-to-leg ratio among hominins is also diagnostic of mode of locomotion. Species exhibiting longer or equal arm lengths compared to legs tend to be quadrupeds. Conversely, organisms with longer legs relative to arm lengths tend to move bipedally. This relationship is determined by calculating the humero-femoral index, which is expressed mathematically as humerus length/femur length $\times$ 100. Tree-dwelling primates have ratios close to 100. Chimpanzees, for example, have a length/femur length ratio of 97.8 and AMHs have a ratio of approximately 71.8, while *afarensis* exhibits a ratio of about 84.6.[74]

The curvature of the hand and foot bones is another trait impacted by bipedal locomotion. Primates that spend time in climbing and swinging from branches possess curved phalanges. This is an adaptation for holding firmly to curved branches. On the other hand, finger and toe bones of bipedal hominins lack such curvature and are flat. Flat phalanges, in turn, provide for better manipulation of objects and dexterity, as curved phalanges hinder precise handling of objects, such as tools. Australopithecines exhibit an intermediate degree of curvature of the hand and foot bones, in between those of AMHs and great apes. This condition is indicative of a hominin still capable of an arboreal existence.[81]

Bipedalism imposes a number of unique requirements, specifically on the foot as the organism walks and runs. A single foot, as the leg transitions from the swing to the stance stage of walking, supports the entire weight of the body. This means that the bipedal foot needs to absorb a much greater force and stress compared to the quadruped foot. In addition, the biped toe is required to exert a thrusting force to initiate the propulsion of the body forward. These requirements are reflected in a number of adaptations, including a more robust and shorter big toe in relation to the other four toes of the foot. Furthermore, the big toe is highly restricted in movement in hominin bipeds. In AMHs, the big toe is positioned parallel to the other four toes and lateral movement is limited. This condition is referred to as a fully adducted or nonopposable big toe. In quadruped primates, on the other hand, the big toe is active in grasping and swinging; thus, a larger opposable big toe is an advantageous trait. The genus *Australopithecus* exhibits an intermediate state in the above-mentioned characteristics of the foot with shorter, moderately adducted big toes.[82] In addition, the heel bone (calcaneus) in biped hominins is larger and more massive than in quadrupeds, an adaptation reflecting the need to accommodate the weight and stress during heel strike. The larger heel bone also provides attachment for strong ligaments that bridge the arch of the foot and tibia. As seen in Fig. 21, *Australopithecus* lacked the foot-arch that is prominently seen in AMHs. The foot-arch is likely an adaptation for absorbing the pressure imposed by the weight of the entire body as it is laid on one foot.

Another indication of bipedal locomotion is the presence of the Achilles tendon, which in the genus *Homo* extends from the calf muscles to the tarsal bones of the heel. In quadrupeds, such as apes, the calf muscles extend right down to the tarsal bones. During running, the Achilles tendon functions as a coil: when extended, it stores energy, releasing it when it contracts as the foot pushes off for the next stride. *Australopithecus*

Fig. 21 Human and *Australopithecus afarensis* foot anatomy (https://www.eleanorvisuals.com/dikika; AMH *Australopithecus afarensis).*

afarensis possessed an Achilles tendon. Taking all of these traits into consideration, it is likely that *Australopithecus afarensis* was a facultative biped.

- *Homo habilis* (2.4–1.4 mya)

Homo habilis represents one of the oldest groups of the genus *Homo*. In fact, some scholars consider this group as part of the genus *Australopithecus*. This lack of consensus in the assignment of the taxa is due to the combinations of ancestral and derived traits it exhibits, which include a slightly larger brain, and smaller face and teeth compared to australopithecines. Furthermore, unlike *Australopithecus*, the foot's arch and general shape are similar to AMHs, providing evidence that this species' foot was more derived than in australopithecines and its locomotion was very similar to *sapiens*'. In terms of ancestral traits, *habilis* retained ape-like features such as long arms and short legs as well as some degree of prognathism (protrusive jaws having a gnathic index over 103). In addition, *habilis* was found in the context of numerous lithics (stone tools), suggesting that the group had the dexterity to manipulate stones and design them to make tools. This is supported by *habilis*' finger bones, which are only slightly curved and intermediate in shape between the curved finger bones of quadruped apes and the straight finger bones of AMHs. This type of finger bone morphology indicates that *habilis* could form a precision grip.

- *Homo erectus* (1.8–0.2 mya)

Homo erectus, or "upright man," as a species consisted of individuals exhibiting a postskeletal anatomy very similar to AMHs, although overall more robust. In many ways, *Homo erectus* represents a landmark in hominin evolution. It was the first hominin group with a body plan that resembled modern humans (Fig. 22). It possessed a slim body frame with lengthy arms and legs, and every detail about its anatomy indicates that it was an agile organism. They were taller than australopithecines and earlier members of the genus *Homo* such as *H. naledi* (only 5 feet tall). The average African *H. erectus* was about 146–185 cm tall, a size that falls within the tallest 17% of the contemporary human male population.

Erectus was a very successful group with a geographical range that included every continent as well as Oceania. Over its long tenure and extensive geographical range, *erectus* exhibited considerable anatomical heterogeneity and a number of important AMH-like characters are seen in this group for the first time. Many of these advanced traits are related to anatomical changes associated with upright posture and bipedalism. *Homo erectus* was not only a walker in an AMH sort of way, and habitually upright, but also an able runner.

Anatomical differences are apparent within the geographical range of *H. erectus* and pronounced variability has been detected even among residents of the same location, such as Dmanisi in the Republic of Georgia in West Asia. If *Homo floresiensis* is indeed a form of *H. erectus* that experienced island dwarfism, as many scholars contest, this pygmy population would add to the impressive anatomical diversity of *erectus*. It is highly likely that the upright posture and bipedal locomotion were paramount in the remarkably fast dispersal and versatility of this group worldwide.

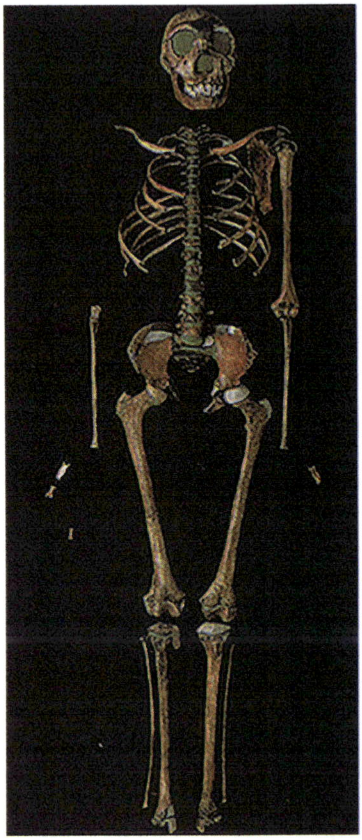

Fig. 22 Skeleton of *Homo erectus* (http://www.sbs.utexas.edu/levin/bio213/humanevol/humanevol.html).

Homo erectus was probably the first hominin that practiced a hunter-gatherer existence, running after game with their long legs and slender physique. This mode of survival contrasted with the way australopithecines lived, mostly as a slower-moving scavenger. *Erectus* successfully settled various habitats and climatic zones and possibly made clothing for protection and built huts. Primitive shelters, such as the ones in Terra Amata, near Nice in southern France, have been attributed to the group. These shelters were fabricated out of hide extended over a wooden frame with foundations made up of stones, and animal skins for a floor. Considering that this species reached Eastern Indonesia (e.g., the island of Flores) in island South East Asia, it is likely that somehow they navigated through one or several bodies of water to reach it. This suggests that *erectus* was the first hominin that build rafts to cross the open ocean. During the Pleistocene, from 2,588,000 to 11,700 ya, two critical deep treacherous bodies of water persisted during the putative time that *erectus* migrated into Eastern Indonesia. One was a 35 km body of water located just east of the island of Bali and the other was a 100 km stretch of ocean (the Makassar Strait) between the islands of Borneo and Sulawesi. *H. erectus* had to cross one of these two bodies of water to reach the island of Flores. During their 2-million-year tenure, *erectus* was able to start and control fire for cooking and probably for protection against predators.

To accomplish all of this, *Homo erectus* must have enjoyed considerable agility, dexterity, and intellect. It is noteworthy that its limbs and extremities are basically the same in morphology and proportions as those of AMHs (Fig. 22); the ratio of hind to forelimb is almost identical to modern humans. Specific anatomical features, such as the styloid process or bone projection at the end of the hand phalanges, is a distinctive, advanced feature that is unique to AMHs, Neanderthals, and *erectus*. The styloid process helps the phalanges lock into the wrist bones, allowing greater amounts of pressure to be applied to the wrist and hand from a grasping thumb and fingers. This advanced trait suggests that *erectus* was using its hands to perform complex tasks and make precise tools, such as the Acheulian utensils usually associated with their fossils.

Other derived traits included arched metatarsals, strong muscles on the backs of their necks, and a highly developed Achilles tendon. All of these derived traits point to generally modern human-like weight transfer and energy-saving mechanisms important for bipedal locomotion and running. It is of interest that *erectus* seems to be one of the first hominins capable of sustained long-distance running.[83] A number of derived traits specifically designed for running are first seen in this group, including strong nuchal ligaments, the rugosity of the nuchal muscle insertions, the large-size gluteus maximus (main extensor muscle of the hip), and the well-developed Achilles tendon. It is theorized that the larger gluteus maximus functioned in stabilizing the trunk while running. It is noteworthy that in AMHs, the nuchal ligaments are not as developed as in *erectus* or Neanderthals, suggesting a reduced dependency on running. Along these lines, it is also notable that the pelvis and pelvis canal of *erectus* were narrower than in AMHs. This suggests that *erectus* may have been more efficient walkers than modern humans.

Yet a number of ancestral traits, such as wide, laterally flaring iliac blades of the pelvis, suggest some differences in locomotion compared to AMHs. In addition, examination of the Ileret foot prints (Fig. 23) from Kenya, dated to 1.5 mya, indicates a foot anatomy retaining some ancestral features.[84] Although the Ileret footprints are generally similar to modern human footprints, indicating a heel-to-toe transfer of weight and pressure, the pattern is slightly different. For example, the footprints lack a clear medial-to-lateral gradient of pressure and yet they exhibit deeper impressions beneath the medial parts of the forefoot as in AMHs. In addition, the angle of the big toe in relation to the other toes and the extent of medial weight transfer indicate a foot anatomy still retaining some ancestral features.

The structure of the clavicle in *Homo erectus* has attracted interest in a number of studies, especially in connection with throwing performance.[85] The capacity to hunt and

Fig. 23 *Homo erectus'* footprint in the sand (https://www.economist.com/science-and-technology/2009/02/26/a-footprint-in-the-sands-of-time).

defend the clan from a distance would have been an advantageous adaptation facilitated by upright posture, bipedalism, and dexterity. As derived characters, *Homo erectus* possessed not only clavicles within the range of AMHs in size and orientation, but also a number of related derived anatomical traits, including inferiorly rotating shoulders, a mobile waist, and low humeral torsion for effective long-distance throwing. Thus, the high-speed thrust of a spear can likely be traced back to at least to 2 mya.

Encephalization

Although the hominin brain has the same overall morphology as the brains of other mammals, its cerebral cortex is far more developed. A number of other animals possess large brains, but when the body size is taken into consideration, contemporary humans exhibit a relative brain size two times and three times larger than the bottlenose dolphin and the chimpanzee, respectively. Much of the human brain, which encompasses about 86 billion neurons, is part of the cortex, an area known for a number of higher-level functions, such as reasoning, abstract ideas, and consciousness.[86] The cortex is responsible not only for self-awareness but also for mindfulness, creative thoughts, and cosmic awareness. This remarkable organ also provides for emotions, empathy, and altruism. The cerebral cortex is divided into four sections: frontal lobe, parietal lobe, occipital lobe, and temporal lobe. Specifically, the frontal lobe is involved with reasoning, organizing, language, locomotion, and emotions. The parietal lobe is important in language processing, movement, orientation, recognition, and perception of stimuli. The occipital lobe is the visual processing center of the mammalian brain, while the temporal lobe, at the bottom of the brain below the lateral fissure on each side of the brain and in close proximity to the ears, is responsible for interpreting sounds, memory, and speech.

The brain cortex of AMHs is highly convoluted and is composed of about 10 billion neurons, each linked to other neurons by 100 trillion synapses or junctions. These neurons communicate using about 100 neurotransmitters that help conduct electrochemical signals at a rate of about 10 billion times per second. Although investigations are constantly discovering facts about neurobiology, it is likely that the human brain will remain a unique organ among both extinct and current species.

The increase in size and complexity of brain tissue as a function of evolutionary time is referred to as encephalization. Often, encephalization is expressed relative to body size, because brain size is generally dependent on the size of the organism. In addition to upright posture and bipedalism, encephalization is undoubtedly a salient characteristic of hominin evolution.[87] Although encephalization is a trend exhibited by a number of animal lineages, the process has been particularly dramatic in hominins. In general, within the class mammalia, high levels of encephalization have been found to correlate with sociability and particularly stable groups.[88]

Encephalization as a trend started evolving later than upright posture and bipedalism. In fact, the first 4.5 million years of hominin evolution is characterized by a very modest increase in brain size (Fig. 14). Throughout this period, hominins began to walk upright and even make simple stone tools. Then, starting with *Homo erectus*, within a period of only 2 million years, remarkably, the hominin brain actually tripled in size (from approximately 400 to 1400 cc). It is estimated that during this time period, about 100,000 neurons and supporting cells increased per generation.[89] This impressive surge in the number of neurons since *erectus* likely translated to a substantial increase in intellectual capacity. Unlike bipedalism, which is generally characterized by a gradual trend, encephalization evolved sporadically with intermittent periods of hiatus or limited increase punctuated with epochs of rapid increases in relative size and complexity. The changes in the hominin brain also involved shifts of function from noncortical parts of the brain to the cortex. In addition to the increase in the number of neurons, the brain experienced a number of significant morphological changes, including expansion and convolutions of the cerebral cortex and greater myelination of neurons, the latter generally implicated with faster neuronal transmission.

Encephalization as a function of evolutionary time

Starting with *Homo erectus*, cranial capacity data indicates that the hominin brain expanded by a combination of sporadic periods of rapid growth interspersed with periods of modest enlargement. Although it is likely that some of this increase is the result of the larger body size of *erectus*, there is also evidence of encephalization unrelated to body size.[90,91] Brain size increases over time were also noted within the *sapiens* superspecies group. When the *sapiens* superspecies group was partitioned into *Homo heidelbergensis*, *Homo neanderthalensis*, and AMHs to assess which species was driving the increase, *heidelbergensis* exhibited no brain size increase, while lineage encephalization is observed for both AMH and Neanderthals.

Specifically, the time intervals delineating statistically significant step changes in hominins include the periods between 0–100 kya, 200–400 kya, 1–1.2 mya, and 1.6–1.8 mya, and 1.8–2.6 mya.[92] The periods of 1.6–1.8 mya and 1.8–2.6 mya correspond to the emergence of early *Homo* (*Homo habilis* approximately 1.9 mya and *Homo erectus* about 1.8–2.0 mya), while the final two statistical significant increments at 200–400 kya and 0–100 kya are concomitant with the advent of the AMHs about 195–334 kya (these dates include the ancient AMH remains from Jebel-Irhoud, Morocco), while the increase that occurred between 0 and 100 kya does not coincide with a species arrival.

Encephalization in Africa versus Eurasia

Overall temporal changes in hominin brain size are observed in Africa and Eurasia (Fig. 24). Yet although some parallelisms are noted, marked specific differences in time and putative causations are apparent between the two continents. In Africa, for example, statistically significant changes in brain size were detected between 1.6 and 1.8 mya as well as between 1.8 mya and earlier, while in Eurasia, statistically significant differences are seen between 1.6 and 1.8 mya, 200 and 400 kya, as well as between the present and 100 kya.

The rapid changes in brain size in Eurasia coincide with migration events involving AMHs, *Homo heidelbergensis*, and *Homo erectus*. In addition, gradual encephalization is also evident within *erectus* at large throughout its geographical range, but not in *heidelbergensis* and *sapiens*. In Africa, on the other hand, no intraspecific brain enlargement is observed. In Africa, encephalization is associated with the genesis of new species or groups.[92] The noted intercontinental differences in brain size dynamics argue for fundamental unique selection pressures acting on hominin populations. It is likely that changes in habitat and environmental fluctuations associated with migration drove encephalization in Eurasia while speciation and possibly population expansion were the determining factors in Africa. Nevertheless, considering that migration of individuals between the two continents is not believed to have been substantial enough to account for similar

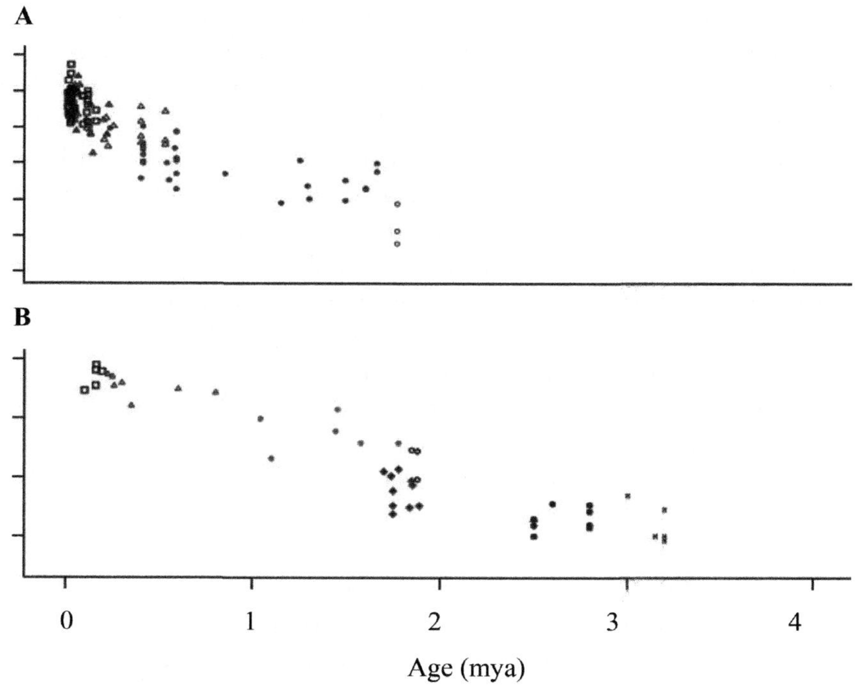

Fig. 24 Continental and species trends in hominin brain size. (A) Depicts brain size in European hominins and (B) depicts brain size change in Africa (*open squares, H. sapiens; filled triangles, H. neanderthalensis; open triangles, H. heidelbergensis; gray filled circles, H. erectus/ergaster; open diamonds, H. rudolfensis; filled diamonds, H. habilis; open circles, H. georgicus; black circles, Australopithecus africanus; crosses, A. afarensis*) (https://www.ncbi.nlm.nih.gov/pmc/articles/PMC3385680/ https://ora.ox.ac.uk/objects/uuid:b12bc11a-f029-4038-baa1-04aa0053b590).

encephalization patterns, any similarity in temporal changes in brain size is likely to be attributed to similar selection pressures on populations in both continents.

Brain reorganization during hominin evolution

In addition to an overall increase in brain size, the process of encephalization in hominin evolution also included reorganization. A number of specific modifications have been described. One of these modifications involves an overall progression toward globularization of the skull. Contemporary AMHs have larger and more globular brains that distinguish them from their extinct *Homo* ancestors. By 300,000 ya, early *Homo sapiens* were already within the range of current-day brain size, yet additional modifications in brain shape occurred gradually subsequently. In fact, it was not until 100,000 to 35,000 ya that AMHs reached present-day brain shape.

Globularization of the skull started sometime after Jebel Irhoud and Omo 2, dated at 315,000 and 195,000 ya, respectively. Thus, the shape of the current human brain was not established at the time *sapiens* originated, but much later, and it is likely that it is still changing, although it is not clear in which direction. Parietal and cerebellar bulging, including shape modifications of the anterior and middle cranial fossae (large depressions in the anterior and middle portion of the floor of the cranial cavity), that occur during prenatal development have driven globularization. Since parietal bulging does not involve surface areas of the brain, it is likely to result from enlargement of internal tissue not visible on the surface of the brain. It is likely that these morphological modifications have resulted in functional changes. A number of critical functions are associated with the parietal region, including orientation, attention, perception of stimuli, sensory-motor transformations underlying planning, visual and spatial integration, imagery, self-awareness, working and long-term memory, numerical processing, and tool use, while the cerebellum is involved in motor-related functions, such as coordination of movements and balance, spatial processing, working memory, language, social cognition, and affective processing.[93] Considering the large spectrum of physiological and cognitive functions controlled by these tissues, it is likely that a number of activities were impacted as a function of the alterations of these brain tissues. For example, certain characteristics that signal behavioral modernity and are uniquely modern human, such as manufacturing ornaments and sophisticated lithic instruments, manipulations of symbols, abstract thought, and art, all appeared recently during AMH's tenure after the onset of brain globularization.

It is of interest that as AMHs started gradually exhibiting brain globularity, they began accumulating signs of more advanced behavioral modernity. By the time skull globularity fell within the range of present-day AMH variation, the full repertoire of modern behavior was in place. This coincided with the transition from the Middle to the Later Stone Age in Africa and from the Middle to the Upper Paleolithic in Europe, approximately 50,000–40,000 ya.

Also noteworthy is the parallelism observed among brain globularity, behavioral modernity, and the appearance and fixation (carried by everyone in the population) of specific genes involved in axonal and dendritic growth or synaptic transmission. It has been demonstrated that these genes were subject to positive selection contributing to their rapid fixation into the AMH gene pool prior to the Upper Paleolithic or last subdivision of the Old Stone Age about 50,000–10,000 ya.[94] These DNA sequences became fixed in the *sapiens* gene pool subsequent to the separation of the AMH lineage from the Neanderthal and Denisovan clade more than 500,000 ya. Genes such as NOVA1, SLITRK1, KATNA1, LUZP1, ARHGAP32, ADSL, HTR2B, and CNTNAP2 that arose during this time period in the AMH lineage are all involved in nervous system functions.[95] In addition, although FOXP2—a gene needed for normal development of speech and language[96]—is present in AMHs as well as in Neanderthals and Denisovans, an internal substitution mutation that affects FOXP2 expression is present only in *sapiens*. This suggests that this AMH-specific mutation may have impacted the evolution of speech in the AMH lineage. This mutation in the FOXP2 gene occurred after the separation of the *sapiens* lineage from the Neanderthals and Denisovans.

In addition to globularization, other anatomical modifications have been noticed within the genus *Homo* starting about 130 kya. These recent rearrangements are localized in the base of the brain and involve the size of the cribriform plate (flat bone above the nasal cavity).[97] This bony plate at the roof of the nasal cavities is permeated with numerous openings for the passage of the olfactory nerves from the nasal cavity. The large size of the cribriform plate is a reflection of the large olfactory bulbs and presumably more acute sense of smell in AMHs. These changes in the cribriform plate morphology were accompanied by other modifications including a wider orbitofrontal cortex (OFC) as well as a size increase and forward projection of the temporal lobe poles (anterior end of the temporal lobe) in AMHs. The OFC is a prefrontal cortex region in the frontal lobes of the brain (Fig. 25) that is involved with emotions, reward, and the cognitive processes of decision-making. The name of the region derives from its location above the orbits of the eyes. It is reasonable that these modifications at the base of the skull led to changes in neurological functions. Furthermore, since olfactory/gustatory functions are link to reward, memory, and learning, modifications in these brain centers likely impacted the processing of emotions and cognitive decision-making. Thus, beyond the basic sense of smell, higher olfactory functions affected various aspects of human behavior. Moreover, it is likely that this remodeling of specific regions of the brain involved in olfactory

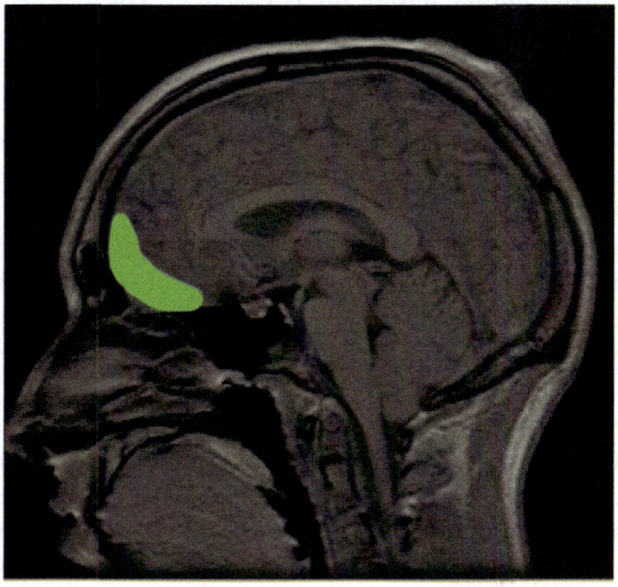

Fig. 25 Region in *green* indicates location of the orbitofrontal cortex (https://en.wikipedia.org/wiki/Orbitofrontal_cortex).

functions, and implicated in cognitive/neurological functions, affected the course of hominin evolution by impacting the intellectual capacity, social behavior, and cultural repertoire of AMHs.

In relation to hominin evolution, it is noteworthy to underscore that considerable anatomically variability is observed within AMHs.[98] Comparative analyses of brain structure have demonstrated statistically significant differences among populations of different racial or ethnic backgrounds.[98] For example, the OFC and amygdala (set of neurons deep within the medial temporal lobe involved in processing emotions) exhibit race-related differences. These two brain regions were shown to occupy a significantly greater proportion of the total cerebral volume in an African-American cohort. Furthermore, population genetic studies indicate that allelic frequencies of genes that are involved in central nervous system function differ substantially across ethnic and racial populations.[99] For instance, variants of the *5HTTLPR* polymorphism have been implicated with differences in frontotemporal structure and function. One of the *5HTTLPR* variants exhibits 25% frequency among African-Americans as compared to 40%–45% in Caucasian Americans, while the other variant is quite common among some Asian populations.

Recent anatomical changes in the AMH brain

A number of studies have documented a number of very recent changes in the size and organization of the AMH brain since the Upper Paleolithic. These modifications occurred in the AMH lineage during the last 200,000 years.[100] One of the observed changes is a reduction in brain size. This decrease in overall brain size has been attributed, in part, to the allometric (proportional to overall body size) decrease of body size experienced by *sapiens* since its origin.[101] Since their genesis, AMHs have experienced a steady loss of bony/muscular robustness and overall size. Further studies comparing *sapiens* fossilized skulls dated to a time period from 190,000 to 25,000 ya to extant populations worldwide show that a statistically significant decrease in absolute endocranial size has occurred.[102] A mean volume of 1350 cm^3 was obtained from a sample of 20,000 contemporary skulls from worldwide populations compared to 1478 cm^3 derived from 28 fossil endocasts (190,000–25,000 ya).[103] In a separate study, based on 15 fossil skulls from 92,000 to 25,000 years in age, the brain volume was estimated to be 1514 cm^3 ($SD = 137.9$).[102]

In addition to differences in volume, the length of the endocasts was found to be statistically ($P < .001$) larger in the fossil sample (15.9 cm) than in the contemporary AMH group (14.9 cm), a difference of 7%.[102] In addition, the width of the cerebellar lobes were found to be statistically significantly greater in the fossil group (9.7 versus 9.2, $P < .05$, 5% different), but the endocranial widths assessed in various areas of the skulls were not different. The endocranial heights are greater in the fossil samples by 3%–16% ($P < .001$), depending on the place of measurement in the skull. Furthermore, the frontal lobes have

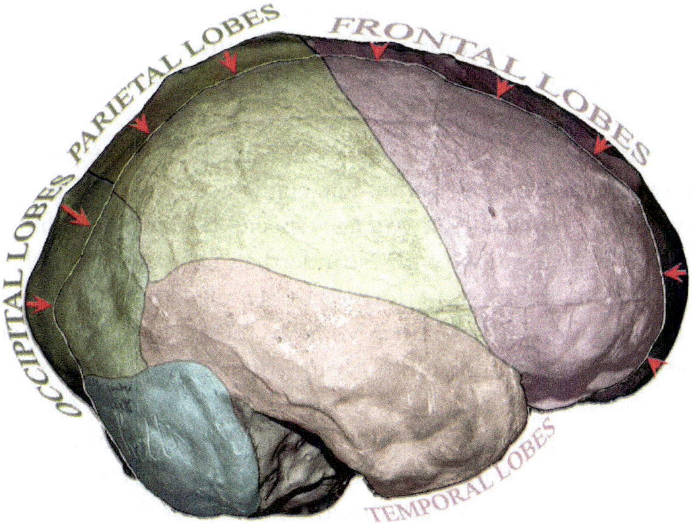

Fig. 26 Schematic representation in the right lateral view of the variations in endocranial anatomy between a fossil (external outline corresponding to Cro-Magnon 1) and a "mean" of contemporary AMH (internal outline corresponding to an endocast based on mean values for the extant AMH samples) (https://link.springer.com/article/10.1007/s13219-012-0069-z#citeas).

become relatively shorter antero-posteriorly in contemporary humans, and their surface area has decreased. The parietal bones have become longer in current AMHs while the surface of the parietal and temporal bones has not changed. Vertically, the occipital lobes are shorter in present-day *sapiens* and their surface has decreased. All of these modification put together contribute to a front to back and top to bottom compression of the skull in contemporary AMHs relative to the human fossils. Fig. 26 provides an illustration summarizing these modifications. No functional differences have been correlated to the abovementioned anatomically differences.

Independent encephalization processes in Neanderthals and AMHs

The Neanderthal lineage originated in Africa as a branch of the genus *Homo* sometime between 800,000 and 400,000 years ago. This split separated the Neanderthal lineage from the early *Homo-H. sapiens* line of descent. It is thought that the common ancestor to both Neanderthals and *sapiens* was *Homo heidelbergensis*. Neanderthals reached Europe before AMHs about 500,000 years ago, likely migrating from Africa across the Near East, eventually reaching East and West Europe.[104]

The skulls of AMHs and Neanderthals brains are about the same size, yet their shape differs. In general, *sapiens* possesses flatter skulls in the back, straight foreheads, and no brow ridges. A more detailed examination of the skulls reveals that certain features at the base of the brain differ between AMHs and Neanderthals. Although both species exhibit larger cribriform plates and olfactory bulbs compared to earlier hominins, the size and spatial position of these structures within their brains differ significantly. AMHs possess larger cribriform plates (12% larger) and olfactory bulbs compared to Neanderthals, suggesting greater olfactory functions in the former. In addition, in contrast to AMHs, the Neanderthal's temporal lobe is less elongated and differs in shape. Furthermore, Neanderthals' skulls exhibit vertically lower middle cranial fossae poles, a primitive condition. These differences suggest that these two *Homo* species had unique evolutionary trajectories in relation to encephalization, with Neanderthals following an archaic evolutionary pattern.[105] Based on the functions performed by the olfactory system, it is tantalizing to consider that the modifications seen in AMHs are related to a greater spectrum of food items consumed by *sapiens* compared to Neanderthals.[106] Furthermore, these olfactory differences may impact a number of AMH-specific behaviors that could set *sapiens* apart from Neanderthals. In connection with the observed morphological differences between Neanderthals and AMHs, it is interesting that the former possesses the ancestral form of certain DNA sequences,[107] suggesting that the fixation of the derive form of genes in *sapiens* is somehow connected with the structural differences mentioned above.

What drove hominin encephalization?

As with upright posture and bipedal locomotion, the process of encephalization has been linked to a number of specific environmental factors. For instance, it has been proposed that environmental diversity, fluctuations, and unpredictability have been important driving forces selecting for larger brain size and higher cognitive functions in hominins over time.[108] It has been proposed from African paleo-climate data that pulsed changes may be more important than long-term climatic trends in directing the course of hominin evolution.[109] Considering the climatic environmental changes and variability observed in East Africa as forest turned into dry savannah, flexibility and innovation in utilizing habitat, resources, and space would have had an adaptive value. Similarly, variable conditions, such as different diets and unpredictable resources encountered by hominins as they migrated into diverse habitats, would have benefited from tool making and use. In turn, this provided for adaptive selection for higher brain functions.

In addition to these environmental changes, increased predation as a result of living in an open savannah has been implicated as a factor driving cognitive capacity and brain size within the genera *Australopithecus* and *Homo*. In most instances, the likelihood of becoming prey to a carnivore would increase in an open field. Along these lines, it has been observed that terrestrial primates inhabiting open field habitants tend to live in larger clans, possibly as a mechanism to counteract predation by collaboration and surveillance.[110] Furthermore, it has been shown that predators exhibit a bias for mammalian pray with small brain size, likely resulting from the greater ease of hunting prey with limited cognitive capacity and less sophisticated escape or defense strategies.[111] If indeed predation is a driving force for clans with larger numbers of individuals, then selection for higher cognitive capacity would provide for a greater repertoire of social behavior. A larger number of members in a group would necessitate greater demands for effective communication, including various types of group interactions and social sophistication. A larger number of individuals within a group would require understanding, flexibility, and altruism to be able to function efficiently as a group and avoid self-destruction. All of these functions would benefit from grater cognitive functions in hominins.

Language, specifically, has been suggested as a factor in directing hominin brain evolution. It has been postulated that the functional/cognitive demands imposed by the development of language played a part in driving hominin brain evolution as it provided an adaptive advantage in a number of activities, including communication, hunting, and defense, among others. If language complexity is the result of brain size and efficiency, as expected, encephalization as a trend may reflect greater sophistication in the development of language. Therefore, the development of language could have been a gradual process of increasing complexity as a function of time. This can be envisioned as a feedback process, as language itself was the advantageous trait selected for that, in turn, led to increases in brain size.

A number of anatomical traits associated with language development are first seen in *Homo* groups dating back to 600,000 ya. These include the expansion of the thoracic nerve and the hypoglossal canal (opening in the occipital bone of the skull through which the twelfth cranial nerve that controls tongue movement passes). It is interesting that the last phase in language complexity, reached about 100,000 ya, coincided with the last dramatic increase in brain size. Linguistic research points to a period about 80,000 ya when complex language evolved in Africa and spread rapidly worldwide. In a study, the number of phonemes (units of sound that distinguishes one word from another) in a global sample of 504 contemporary languages follows a gradual diminution that reflects a serial founder-effect model of expansion from an origin in Africa.[112] These results not only indicate a parallelism involving the fossil record, linguistic, and genetic data, but also link anatomical changes in the brain with language development. Furthermore, they provide linguistic evidence for the African origin of AMHs and the Out of Africa dispersal.

References

1. Lyell C. *The Geological Evidences of the Antiquity of Man, With Remarks on Theories of the Origin of Species by Variation*. London, UK: Murray; 1863.
2. Washburn SL. Review of Sir Arthur Keith's a new theory of human evolution. *Am J Phys Anthropol*. 1949;7:264–266.
3. Washburn SL. The new physical anthropology. *Trans N Y Acad Sci*. 1951;13:298–304.
4. Montagu A, ed. *The Concept of Race*. New York: The Free Press of Glencoe; 1964.
5. Shapiro HL. Earnest A. Hooton: 1887-1954. *Science*. 1954;119(3103):861–862.
6. Wynne-Edwards VC. *Animal Dispersion in Relation to Social Behavior*. Edinburgh: Oliver & Boyd; 1962.
7. Hamilton WD. The moulding of senescence by natural selection. *J Theor Biol*. 1966;12:12–45.

8. Gadgil M, Bosset WH. Life historical consequences of natural selection. *Am Nat.* 1970;104:1–24.
9. Trivers RL. The evolution of reciprocal altruism. *Q Rev Biol.* 1971;46:35–57.
10. Eldridge N, Gould SJ. Punctuated equilibria: an alternative to phyletic gradualism. In: Schopf TJM, ed. *Models in Paleobiology*. San Francisco: Freeman Cooper; 1972:82–115.
11. Gluckman PD, Hanson M. *The Fetal Matrix*. New York, NY: Cambridge University Press; 2004.
12. Sankararaman S, Mallick S, Patterson N, Reich D. The combined landscape of Denisovan and Neanderthal ancestry in present-day humans. *Curr Biol.* 2016;26:1241–1247.
13. Henrich J. *The Secret of Our Success*. Princeton, NJ: Princeton University Press; 2017.
14. Kalender WA, Seissler W, Klotz E, et al. Spiral volumetric CT with single-breath-hold technique, continuous transport, and continuous scanner rotation. *Radiology.* 1990;176:181–183.
15. Zur Nedden D, Knapp R, Wicke K, et al. Skull of a 5,300-year-old mummy: reproduction and investigation with CT-guided stereolithography. *Radiology.* 1994;193:269–272.
16. Seidler H, Bernhard W, Teschler-Nicola M, et al. Some anthropological aspects of the prehistoric Tyrolean ice man. *Science.* 1992;258:455–457.
17. Berger LR, Hawks J, de Ruiter DJ, et al. Homo naledi, a new species of the genus Homo from the Dinaledi Chamber, South Africa. *elife.* 2015;4:e09560.
18. Seidler H, Falk D, Stringer C, et al. A comparative study of stereo lithographically modeled skulls of Petralona and Broken Hill: implications for future studies of middle Pleistocene hominid evolution. *J Hum Evol.* 1997;33:691–703.
19. Dean MC, Smith BH. Growth and development in the Nariokotome youth, KNM-WT 15000. In: Grine FE, Fleagle JG, Leakey RE, eds. *The First Humans: Origin of the Genus Homo*. New York, NY: Springer; 2009:101–120.
20. Alemseged Z, et al. A juvenile early hominin skeleton from Dikika, Ethiopia. *Nature.* 2006;443:296–301.
21. Olejniczak AJ, Tafforeau P, Smith TM, Temming H, Hublin J-J. Technical note: compatibility of microtomographic imaging systems for dental measurements. *Am J Phys Anthropol.* 2007;134:130–134.
22. Richards GD, Jabbour RS, Horton CF, Ibarra CL, MacDowell AA. Color changes in modern and fossil teeth induced by synchrotron microtomography AJPA. 2012;149(October):172–180.
23. Le Cabec A, Tang N, Tafforeau P. Accessing developmental information of fossil hominin teeth using new synchrotron microtomography-based visualization techniques of dental surfaces and interfaces. *PLoS One.* 2015;10(4):e0123019).
24. Smith TM, Tafforeau P, Reid DJ, et al. Dental evidence for ontogenetic differences between modern humans and Neanderthals. *Proc Natl Acad Sci USA.* 2010;107(49):20923–20928.
25. Ward CV, Nalley TK, Spoor F, Tafforeau P, Alemseged Z. Thoracic vertebral count and thoracolumbar transition in *Australopithecus afarensis*. *PNAS.* 2017;114(23):6000–6004.
26. Darwin C. *On the Origin of Species*; 1859. London.
27. *Huxley's Man's Place in Nature*; 1863 [Huxley TH. *Evidence as to Man's Place in Nature*. London; 1863].
28. Lyell C. *The Geological Evidences of the Antiquity of Man*. London: Geological Society Publications; 1863.
29. Jacobi RM, Higham TFG. The 'Red Lady' ages gracefully: new ultrafiltration AMS determinations from Paviland. *J Hum Evol.* 2008;55:898–907.
30. Callaway E. Archaeology: date with history. *Nature.* 2012;485(7396):27–29.
31. Trinkaus E, Richards E. Out of Africa: modern human origins special feature: isotopic evidence for the diets of European Neanderthals and early modern humans. *Proc Natl Acad Sci USA.* 2009;106(38):16034–16039.
32. Sykes BC. Report on DNA recovered from the Red Lay of Paviland. In: Aldhouse-Green S, ed. *Paviland Cave and the "Red Lady": A Definitive Report*. Bristol: SCARAB Research Centre of the University of Wales College, Newort and the Friends of the National Museum of Wales by the Western Academic and Specialist Press; 2000:75–77.
33. Russell MD, LeMort F. Cutmarks on the Engis 2 calvaria? *Am J Phys Anthropol.* 1986;69(3):317–323.
34. King W. The Neanderthal skull. *Anthropol Rev.* 1863;2:393–394.
35. Busk G. Pithecoid priscan man from Gibraltar. *Reader.* 1864;110.
36. Keith A. The early history of the Gibraltar cranium. *Nature.* 1911;87(2184):314).
37. Semal P, Rougier H, Crevecoeur I, et al. New data on the late Neandertals: direct dating of the Belgian Spy fossils. *Am J Phys Anthropol.* 2009;138(4):421–428.
38. King W. The reputed fossil man of the Neanderthal. *Quart J Sci.* 1864;1:88–97.

39. Schmitz RW, Serre D, Bonani G, et al. The Neandertal type site revisited: interdisciplinary investigations of skeletal remains from the Neander Valley, Germany. *PNAS*. 2002;99(20): 13342–13347.
40. Huxley TH. On some fossil remains of man. In: *Evidence as to Man's Place in Nature*. New York: D. Appleton and Company; 1863 [chapter 3].
41. Tattersall I. *The Last Neanderthal: The Rise, Success, and Mysterious Extinction of Our Closest Human Relatives*. Macmillan Publishing Company; 1996.
42. Fuhlrott JC. Menschliche Überreste aus einer Felsengrotte des Düsselthals. Ein Beitrag zur Frage über die Existenz fossiler Menschen. *Verhandlungen des Naturhistorischen Vereins der preußischen Rheinlande und Westphalens*. 1859;16:131–153.
43. Fuhlrott JC. Über die Kalksteinschichten im Neanderthale, Worin 1856 der Homo Neanderthalensis gefunden Wurde. *Correspondenzblatt des Naturhistori-schen Vereins fur Rheinland und Westphalen*. 1868;**25**:62–70.
44. Mayer FJC. Zur Frage über das Alter und die Abstammung des Menschengeschlechtes. *Arch Anat Physiol wiss Med (Müller's Archiv)*. 1864;S. 696–728.
45. Chorowicz J. The East African rift system. *J Afr Earth Sci*. 2005;43(1):379–410.
46. Underwood CJ, King C, E. Steurbaut Eocene initiation of Nile Drainage due to East African uplift. *Palaeogeogr Palaeoclimatol Palaeoecol*. 2013;392:138–145.
47. Wichura H, Bousquet R, Oberhänsli R, Strecker MR, M.H. Trauth evidence for mid-Miocene uplift of the east African plateau. *Geology*. 2010;38:543–546.
48. De Menocal P. African climate change and faunal evolution during the Pliocene-Pleistocene. *Earth Planet Sci Lett*. 2004;220:3–24.
49. Maslin MA, Brierley CM, Milner AM, Shultz S, Trauth MH, Wilson KE. East African climate pulses and early human evolution. *Quat Sci Rev*. 2014;101:1–17.
50. Prömmel K, Cubasch U, F. Kasper A regional climate model study of the impact of tectonic and orbital forcing on African precipitation and vegetation. *Palaeogeogr Palaeoclimatol Palaeoecol*. 2013;369:154–162.
51. Kingston JD. Shifting adaptive landscapes: progress and challenges in reconstructing early hominid environments. *Am J Phys Anthropol Suppl Yearb Phys Anthropol*. 2007;134(Suppl. 45):20–58.
52. de Menocal P. Plio-Pleistocene African climate. *Science*. 1995;270:53–59.
53. Fernandes RMS, Ambrosius BAC, Noomen R, et al. Angular velocities of Nubia and Somalia from continuous GPS data: implications on present-day relative kinematics. *Earth Planet Sci Lett*. 2004;222 (1):197–208.
54. Vrba ES. Environment and evolution: alternative causes of the temporal distribution of evolutionary events. *S Afr J Sci*. 1985;81:229–236.
55. Wood B. Fifty years after *Homo habilis*. *Nature*. 2014;508:31–33.
56. Hublin J-J, Ben-Ncer A, Bailey SE. New fossils from Jebel Irhoud, Morocco and the pan-African origin of *Homo sapiens*. *Nature*. 2017;546:289–292.
57. Dirks PHGM, Roberts EM, Hilbert-Wolf H, et al. The age of *Homo naledi* and associated sediments in the Rising Star Cave, South Africa. *elife*. 2017;6:e24231.
58. Laird MF, Schroeder L, Garvin HM, et al. The skull of *Homo naledi*. *J Hum Evol*. 2017;104:100–123.
59. Irish JD, Guatelli-Steinberg D, Legge SS, et al. Dental morphology and the phylogenetic "place" of *Australopithecus sediba*. *Science*. 2013;340. https://doi.org/10.1126/science.1233062).
60. Churchill SE, Hollyday TW, Charlson KJ, et al. The upper limb of *Australopithecus sediba*. *Science*. 2013;340:1233477.).
61. Kivell TL, Kibii JM, Churchill SE, et al. *Australopithecus sediba* hand demonstrates mosaic evolution of locomotor and manipulative abilities. *Science*. 2013;333:1412–1417.
62. Williams SA, Ostrofsky KR, Frater N, et al. The vertebral column of *Australopithecus sediba*. *Science*. 2013;340:1232996).
63. De Silva JM, Holt KG, Churchill SE, et al. The lower limb and mechanics of walking in *Australopithecus sediba*. *Science*. 2013;340:1232999).
64. Harvati K, Röding C, Bosman AM, et al. Apidima cave fossils provide earliest evidence of Homo sapiens in Eurasia. *Nature*. 2019;**571**:500–504.
65. Posth C, et al. Deeply divergent archaic mitochondrial genome provides lower time boundary for African gene flow into Neanderthals. *Nat Commun*. 2017;8:16046.

66. Smith TM, Tafforeau P, Reid DJ, et al. Earliest evidence of modern human life history in North African early Homo sapiens. *PNAS.* 2007;**104**(15):6128–6133.
67. Parravicini A, Pievani T. Multi-level human evolution: ecological patterns in hominin phylogeny. *J Anthropol Sci.* 2016;94:167–182.
68. Falk D, Zollikofer CPE, Morimoto N, Ponce de León MS. Metopic suture of Taung (*Australopithecus africanus*) and its implications for hominin brain evolution. *PNAS.* 2012;109(22):8467–8470.
69. Cox MP, Mendez FL, Karafet TM, et al. Testing for archaic hominin admixture on the X chromosome: model likelihoods for the modern human RRM2P4 region from summaries of genealogical topology under the structured coalescent. *Genetics.* 2008;178(1):427–437.
70. Hammer MF, Woerner AE, Mendez FL, Watkins JC, Wall JD. Genetic evidence for archaic admixture in Africa. *PNAS.* 2011;108(37):15123–15128.
71. Sept J, Betti L, in Fleagle JG. *Primate Adaptation and Evolution.* 2nd Edition. San Diego: Academic Press; 1999.
72. McHenry HM. Human evolution. In: Ruse M, Travis J, eds. *Evolution: the First Four Billion Years.* Cambridge, MA, USA: The Belknap Press of Harvard University Press; 2009.
73. Zollikofer C, Ponce M, de Leon D, et al. Virtual reconstruction of *Sahelanthropus tchadensis. Nature.* 2005;434:755–759.
74. Klein RG. *The human career human: biological and cultural origins.* 3rd ed. Chicago: University of Chicago Press; 2009. Johnason DC, Edgar B. *From Lucy to Language: Revised, Updated and Expanded.* New York: Simon & Schuster; 2006.
75. Lovejoy CO. The natural history of human gait and posture. Part 1. Spine and Pelvis. *Gait Posture.* 2005;21(1):95–112.
76. Owen LC, McCollum MA. Spinopelvic pathways to bipedality: why no hominids ever relied on a bent-hip–bent-knee gait. *Philos Trans R Soc Lond Ser B Biol Sci.* 2010;365(1556):3289–3299.
77. Lovejoy CO. The natural history of human gait and posture. Part 2. Hip and Thigh. *Gait Posture.* 2005;21(1):113–124.
78. Corriccin RS, Mchenry HM. Hominid femoral neck length. *Am J Phys Anthropol.* 1980;52:397–398.
79. Lovejoy CO. The natural history of human gait and posture. Part 3. The Knee. *Gait Posture.* 2007;25 (3):325–341.
80. DeSilva JM. Functional morphology of the ankle and the likelihood of climbing in early hominins. *PNAS.* 2009;106(16):6567–6572.
81. Fleagle JG. *Primate Adaptation and Evolution.* 2nd ed. San Diego: Academic Press; 1999.
82. Duncan AS, Kappelman J, Shapiro LJ. Metatarsophalaneal join function and positional behavior in *Australopithecus afarensis. Am J Phys Anthropol.* 1994;93(1):67–81.
83. Bramble D, Lieberman D. Endurance running and the evolution of Homo. *Nature.* 2004;432:345–352.
84. Bennett KMR, Harris JWK, Richmond BG, et al. Early hominin foot morphology based on 1.5-million-year-old footprints from ileret. *Science.* 2009;323(5918):1197–1201.
85. Roach NT, Richmond BG. Clavicle length, throwing performance and the reconstruction of the Homo erectus shoulder. *J Hum Evol.* 2015;80(C):107–113.
86. Sylvester AD. Locomotor coupling and the origin of hominin bipedalism. *J Theor Biol.* 2006;242:581–590.
87. de Sousa A, Wood B. The hominin fossil record and the emergence of the modern human central nervous system. In: Kaas JA, ed. *Evolution of Nervous Systems: A Comprehensive Reference. Vol. 4: The Evolution of Primate Nervous Systems.* Oxford: Elsevier; 2007:291–336.
88. Shultz S, Dunbar R. Encephalization is not a universal macroevolutionary phenomenon in mammals but is associated with sociality. *PNAS.* 2010;107(50):21582–21586.
89. Edelman GM. *Second Nature: Brain Science and Human Knowledge.* New Haven, CT, USA: Yale University Press; 2006.
90. Robson SL, Wood B. Hominin life history: reconstruction and evolution. *J Anat.* 2008;212:394–425. https://doi.org/10.1111/j.1469-7580.2008.00867.x.
91. Aiello LC, Wells JCK. Energetics and the evolution of the genus Homo. *Annu Rev Anthropol.* 2002;31:323–338.
92. Shultz S, Nelson E, Dunbar RIM. Hominin cognitive evolution: identifying patterns and processes in the fossil and archaeological record. *Philos Trans R Soc B.* 2012;**367**:2130–2140.

93. Zhang S, Li C-SR. Functional connectivity mapping of the human precuneus by resting state MRI. *NeuroImage*. 2012;59:3548–3562. Bruner E, Preuss TM, Chen X, Rilling JK. Evidence for expansion of the precuneus in human evolution. *Brain Struct Funct*. 2017;222:1053–1060.
94. Prüfer K, Racimo F, Patterson N, et al. The complete genome sequence of a Neanderthal from the Altai Mountains. *Nature*. 2014;505:43–49.
95. Meyer M, Kircher M, Gansauge M-T, et al. A high-coverage genome sequence from an archaic Denisovan individual. *Science*. 2012;338:222–226.
96. Maricic T, Günther V, Georgiev O, et al. A recent evolutionary change affects a regulatory element in the human FOXP2 gene. *Mol Biol Evol*. 2013;30:844–852.
97. McDermott F, et al. New Late-Pleistocene uranium-thorium and ESR dates for the Singa hominid (Sudan). *J Hum Evol*. 1996;3 1:507.
98. Isamah N, Faison W, Payne ME, et al. Variability in frontotemporal brain structure: the importance of recruitment of African Americans in neuroscience research. *PLoS One*. 2010;5(10):e13642).
99. Tang Y, Hojatkashani C, Dinov ID, et al. The construction of a Chinese MRI brain atlas: a morphometric comparison study between Chinese and Caucasian cohorts. *NeuroImage*. 2010;51:33–41.
100. Henneberg M. Evolution of the human brain: is bigger better? *Clin Exp Pharmacol Physiol*. 1998; 25:745–749.
101. Holloway RL. The human brain evolving: a personal retrospective. *Annu Rev Anthropol*. 2008;37 (1):1–19.
102. Balzeau A, Grimaud-Hervé D, Détroit F, et al. First description of the Cro-Magnon 1 endocast and study of brain variation and evolution in anatomically modern Homo sapiens. *Bull Mém Soc Anthropol*. 2013;25:1.
103. Holloway RL, Broadfield DC, Yuan MS. *The Human Fossil Record: Brain Endocasts, Paleoneurological Evidence*. Hoboken, New Jersey: Wiley; 2004.
104. Higham T, Douka K, Wood R, et al. The timing and spatiotemporal patterning of Neanderthal disappearance. *Nature*. 2014;512(7514):306–309.
105. Bastir M, Rosas A, Gunz P, et al. Evolution of the base of the brain in highly encephalized human species. *Nat Commun*. 2011;**2**. Article number: 588.
106. Richards MP, Trinkaus E. Isotopic evidence for the diets of European Neanderthals and early modern humans. *Proc Natl Acad Sci USA*. 2009;106:16034–16039.
107. Green RE, et al. A draft sequence of the neanderthal genome. *Science*. 2010;3(28):710–722.
108. Bailey DH, Geary DH. Hominid brain evolution: testing climatic, ecological, and social competition models. *Hum Nat*. 2009;20:67–79.
109. Donges JF, Donner RV, Trauth MH, Marwan N, Schellnhuber H-J, Kurths J. Nonlinear detection of paleoclimate-variability transitions possibly related to human evolution. *Proc Natl Acad Sci USA*. 2011;108:20422–20427.
110. Hill RA, Lee PC. Predation risk as an influence on group size in cercopithecoid primates: implications for social structure. *J Zool*. 1998;245:447–456.
111. Shultz S, Finlayson LV. Large body and small brain and group sizes are associated with predator preferences for mammalian prey. *Behav Ecol*. 2010;21:1073–1079.
112. Atkinson QD. Phonemic diversity supports a serial founder effect model of language expansion from Africa. *Science*. 2011;332:346–349.

CHAPTER 6

Molecular biology tools used to study hominin evolution

It has often and confidently been asserted, that man's origin can never be known: but ignorance more frequently begets confidence than does knowledge: it is those who know little, and not those who know much, who so positively assert that this or that problem will never be solved by science.

Charles Darwin, The Descent of Man[1]

From genetics to molecular biology

In this chapter, prior to embarking on a description of the molecular biology tools employed in the study of hominins, we discuss the salient events that led to the genesis of molecular biology from its roots in classical genetics. This information is presented in a historical context.

Gregor Mendel (1822–84) reached his conclusions on inheritance clueless on the nature of the "factors" that govern the transmission of the characters that he studied. His work with pea plant strains conducted from 1856 to 1863 established several of the principles that today are the cornerstones of genetics. He conducted his experiments in a 4.9 acre garden lot adjacent to the Abbey of St. Thomas in what is now Brno, Czech Republic, where he was a friar. Mendel delivered his findings in a paper entitled "*Versuche über Pflanzenhybriden*" ("Experiments on Plant Hybridization") to the Natural History Society of Brno in Moravia (Czech Republic) on February 8 and March 8, 1865, and later published in 1866 in German in the obscure journal *Verhandlungen des naturforschenden Vereines* in Brünn.[2]

Due to the limited distribution of the journal, the title of the manuscript referring to "plant hybridization experiments," the lack of an understanding of the physical basis for the postulated genetic factors, and the revolutionary as well as novel nature of the results, the paper had limited exposure and impact, only being cited three times in the following 35 years. In addition, the prevailing ideas of inheritance at the time of Mendel were deeply ingrained in the Classical Greek concept of "blending inheritance," which attempts to explain the transmission of traits through a homogenization or blending process of parental components in offspring. In this blending process, the impact of parental

https://doi.org/10.1016/B978-0-12-819086-9.00012-9

traits on the offspring was thought to average and evenly homogenize. Although today this idea is not scientifically valid, it provided an explanation for a number of observations, such as the intermediate stature of individuals resulting from the marriage of a tall and small person. In fact, considering that most traits are controlled by a multitude of genes (multifactorial inheritance) that tend to average their influence on the way humans look (e.g., skin color), it is not unreasonable to think of a mechanism that would uniformly mix the influence of traits. Even Darwin (1809–82) had difficulties in explaining how heredity works. Borrowing ideas from Hippocrates (circa 460–370 BC) and Classical Greece, he proposed that individual tissues contributed their imprint in the form of "gemmules" to the gonads.[3] Darwin realized that his theory of pangenesis (pan meaning whole and genesis for birth) was a weak link in his theory of evolution. Mendel, on the other hand, was influenced by his mathematical background and experience as a mathematics high school teacher, which contributed to his quantitative approach to the analysis of his data. Expressing his results mathematically, Mendel provided a fresh and revolutionary explanation of inheritance. His quantitative methodology was unique and set him apart from other scientists of the time.

It has been alleged that if Charles Darwin had been aware of Mendel's work, advances in genetic and molecular biology would have occurred earlier, given that Mendel's findings complemented the unifying theory of natural selection, which explains the diversity of life. It is also possible that some of Darwin's apprehension to publish his ideas on evolution and the original scientific rejection experienced by Darwin would have been alleviated if Mendel's work had been widely disseminated to provide support. Thus, Mendel's work is a good example of how new concepts may be buried in the literature due to lack of understanding affecting the path of science.

For his contributions, Mendel is known today as the "founder of genetics," although some believe that "father of scientific misconduct" best describes him, owing to suspicions that he faked some of his data. Mendel's second characterization stems from British statistician and geneticist Ronald Fisher's (1890–1962) statistical analysis of Mendel's original data in a seminal article published in 1936, which concluded that "the data of most, if not all, of the experiments have been falsified so as to agree closely with Mendel's expectations."[4] Fisher's report initiated an intense controversy that still persists today. After decades of considerable inquiry and analysis of Mendel's and Fisher's data by numerous authors, the general consensus is that the data as reported by Mendel is "too good to be true"[4] and that reasonable grounds exist to assume a certain unconscious bias was systematically introduced in Mendel's experimentation.

Mendel's work came to light in 1900 when Hugo de Vries (1848–1935) and Carl Correns (1864–1933), independent of each other and of Mendel, conceived the "laws" of inheritance.[5] Hugo de Vries, a Dutch botanist, introduced the ideas of genes as discrete units of inheritance, mutations, and the mutation theory of evolution. Carl Correns, a botanist from Germany, performed interbreeding experiments similar to Mendel's using the hawkweed plant, unaware of his predecessor's work. He rediscovered the laws of segregation and independent assortment. In his publication of 1900,[6] Carl Correns acknowledged and cited Mendel and Darwin's results, but failed to establish a clear connection between inheritance, natural section, and the theory of evolution.

Another piece of the puzzle regarding how inheritance operates was provided by both Walter Sutton (1877–1916), an American geneticist and physician, studying reproductive cells in grasshoppers, and Theodor Boveri (1862–1915), a German biologist examining sex cells in sea urchins. Independently, these two scientists proposed that the "particles of heredity" or "factors" (as they were referred by Mendel and Darwin due to their unknown nature) that determined inheritance resided inside the nucleus of cells, in specific structures known as chromosomes. This concept is known today as the chromosome theory of inheritance or the Boveri-Sutton chromosome theory. Both of these investigators realized that the laws of segregation and independent assortment first described by Mendel parallel the behavior of chromosomes as they segregate inside sex cells during gamete formation. Sutton published his conclusions in two articles in 1902 and 1903.[7,8] Boveri's insight into the correlation between heredity and chromosomes came about as he realized that all chromosomes were required for normal embryonic development of the sea urchin.[9] Boveri was a remarkable scientist. In addition to his contribution to the field of genetics and specifically inheritance, he postulated that malignancy has its genesis within single cells that develop chromosomal abnormalities triggering them to divide uncontrollably. He further proposed that radiation, chemicals, and pathogens caused metastasis. Thus, with his ideas about carcinogenesis, he also established a connection between inheritance and chromosomes.

During the first quarter of the 20th century, the scientific community increasingly accepted the notion that the so-called factors of heredity resided within chromosomes within the nucleus of cells, but since chromosomes were made up of proteins and nucleic acid, scientists intensely debated over which of these biomolecules carried the genetic information. What was known of the process of heredity at the time placed certain demands on the candidate molecule. It must have sufficient complexity and information to encode for huge numbers of proteins, each thousands of amino acids long, be capable of replicating precisely and distribute identical copies to daughter cells, be very stable to survive a lifetime—centuries for some organisms—and be able to change by mutating.

Since proteins are made of 20 amino acids and nucleic acids are composed of only 4 different subunits (nucleotides), proteins were considered the prime candidates for being the carriers of heredity. To most geneticists, molecules made up of only four unique subunits could not provide for the complexity of heredity and the variability to encode all known traits. As a result, most geneticists during the first quarter of the 20th century were in favor of the notion that heredity resided in proteins.

Yet certain scientific results started to change the opinion of the scientific community. The first of three significant experiments that pointed to DNA and not protein as the material that determines inheritance was performed in 1928 by Fred Griffith (1879–1941).[10]

Griffith was a bacteriologist primarily interested in the epidemiology and pathology of the bacteria that causes pneumonia in humans. In mice, this bacterium is lethal. Griffith worked with a number of strains of this bacterium with different degrees of virulence. The virulent types, when grown in dishes, generated colonies (localized large groups of bacteria) that looked smooth, while strains that proved to be nonvirulent formed rough looking colonies with irregular borders. Subsequent investigations demonstrated that all the smooth bacteria strains possessed an outer cover of polysaccharide, a large sugar molecule, which made them look smooth, while the rough types lacked this capsule. It seems that the polysaccharide cover is capable of camouflaging the microbe, which prevents it from being identified by the immune system of mice. This camouflage allows the bacteria to propagate, eventually killing the rodent, while the rough *Streptococcus pneumoniae*, devoid of the covering, are detected by the surveillance immune system and destroyed. In the course of his experiments, Griffith incidentally noticed that when he killed the smooth virulent bacteria by boiling and coinjected them with live rough nonvirulent bacteria, not only did the mice subsequently die of pneumonia, but he also recovered live smooth microbes of the same specific dead strain that was used to infuse the rodents. To this day, it is not clear what prompted Griffith to conduct such an experiment. Was he testing if some virulent bacteria survived the boiling/killing of the bacteria or was the coinjection simply a negative control in his experiment? His conclusion was that something survived boiling, defiantly surviving death, something that he referred to as the "transforming factor."

Follow-up experiments were conducted in 1944 by Oswald Avery (1877–1955), Colin MacLeod (1909–72), and Maclyn McCarty (1911–2005) (Fig. 1) at the Rockefeller Institute in New York City.[11] Oswald Avery arrived at The Rockefeller Institute Hospital in 1913. In addition to his discoveries on the nature of genetic material, Dr. Oswald Avery was responsible for the development of the first vaccine for pneumococcal pneumonia. Avery, MacLeod, and McCarty built on the experiments by Griffith 16 years earlier. They were interested in accessing the chemical nature of the "transforming principle" detected by Griffith. To accomplish this, after killing the bacteria by boiling, they went on to purify the various types of molecules in the debris from the dead virulent smooth cells and individually tested their capacity to transform the nonvirulent rough cells to virulent smooth cells subsequent to coinjection. First they tested the purified polysaccharides. Yet although the polysaccharides were linked to virulence and pathogenicity, it failed to induce the transformation into the smooth virulent type. Then they tested the other cellular components including proteins and DNA. Only DNA was able to bring about the conversion. From these experiments, they deduced that DNA was essential for determining virulence and that DNA was the transforming factor that eluded Griffith years ago.

Although the experimental results obtained by Avery and collaborators were reliable and conclusive, many investigators were not totally convinced that DNA was the

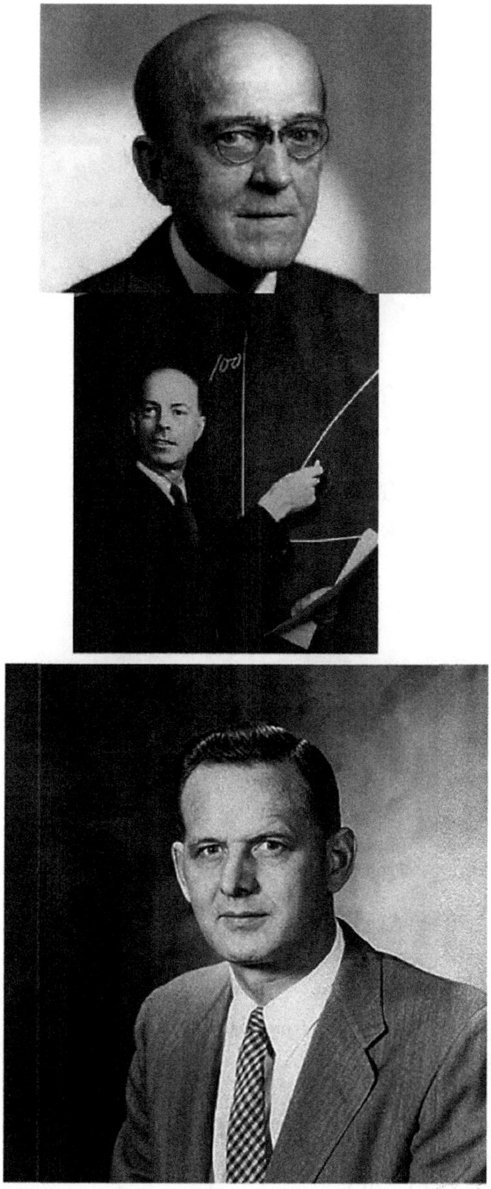

Fig. 1 From left: Oswald Avery (1877–1955), Colin MacLeod (1909–72), and Maclyn McCarty (1911–2005) (https://beckchris.wordpress.com/science-nature-technology/most-important-scientific-discoveries-of-all-time/).

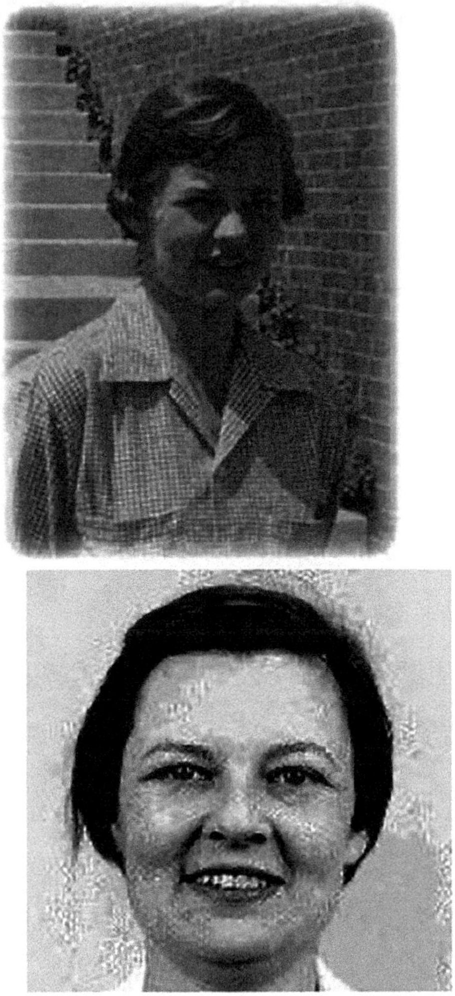

Fig. 2 Martha Chase (https://en.wikipedia.org/wiki/Martha_Chase).

molecule that dictated inheritance. These skeptics were doubtful about the purity of the extracts that the Rockefeller team used in their experiments and pointed out that proteins and/or polysaccharides could still be causing the transformation as contaminants. It was not until 1952 that Alfred Hershey (1908–97) and Martha Chase (1927–2003) (Fig. 2) finally established nucleic acid as the genetic material. This third set of experiments is considered the *coup de grâce* to the notion of proteins being the genetic material.[12]

In their elegant and classical series of experiments, Hershey and Chase used very simple organisms as model systems. One of these model systems, T2, is a linear double-stranded DNA virus with a protective protein coat covering its head. T2 is capable of infecting the bacterium species *Escherichia coli* (*E. coli*) and, after taking over their metabolism, destroy the bacteria by blasting them open, releasing the next generation of viruses. This model system offers the unique opportunity to differentially and individually label the viral protein coat and the internal DNA, and then use the viruses to infect *E. coli*. And in fact, this was exactly what they did. In one culture, they grew the T2 bacteriophages in the presence of radioactive phosphorus (^{32}P) to specifically label the DNA located inside the head of the viruses. In a second culture, they added radioactive sulfur (^{35}S) to label the viral proteins of the head coat. This specificity is afforded by the fact that proteins contain sulfur, but lack phosphorus, while DNA possesses phosphorus, but is devoid of sulfur. They knew that the transforming factor, whatever its chemical nature, needed to be internalized into the bacteria to take over their metabolism and kill them. After allowing some time for the viruses to attach to the bacterial cell membranes and inject the inheritance factor into the bacteria, they subjected the cultures to vigorous shaking by mixing the cultures with a regular kitchen blender. This step was performed to detach the viruses from the bacterial cell membranes. The question was: "What radioisotope, ^{32}P or ^{35}S, was injected into the bacteria to cause their death and destruction?" The presence of the ^{32}P radioisotope inside *E. coli* would implicate DNA as the genetic material. On the other hand, however, if ^{35}S was detected inside the bacteria, then proteins must be the genetic material. The results were conclusive: most of the ^{32}P was found inside the bacteria cells while most of the ^{35}S was found outside the bacterial cells in the phage ghosts (empty viral shell), indicating that the phage proteins never entered the bacterial cells. These results conclusively indicated that DNA is the genetic material and forever changed the path of biological research. For his contribution, Hershey won the Nobel Prize in Physiology or Medicine in 1969. It is also worth noting that Martha Chase was not included as an awardee of the Nobel Prize.

Although the discovery of Hershey and Chase established DNA as the inheritance factor, it was not known how such a simple chemical could accomplish the complex function of dictating all the traits of an organism. Furthermore, it remained to be determined just how the large and abstruse information of inheritance could be stored and passed from generation to generation faithfully. What structure would allow the execution of such complicated functions? The answers to these mysterious questions, it turned out, would have to wait until 1953.

With the functional molecule responsible for inheritance identified as DNA, the structural makeup of the molecule was still unknown. Although most of the credit for the structural model of DNA is given to only two investigators, the American biologist James Watson and the British physicist Francis Crick, a number of scientists contributed substantially to the formulation of the double helix model of DNA, among them and most notably the British chemist Rosalind Elsie Franklin (1920–58) and her colleague, the New Zealand-born British physicist and molecular biologist, Maurice Hugh Frederick Wilkins (1916–2004). From the beginning, the proposed model of DNA structure was surrounded by controversy and allegations regarding scientific misconduct and stealing of data from Franklin's group; issues that still prevail today. Clearly, a substantial amount of information used by Watson and Crick or Crick and Watson (depending on which side of the Atlantic you reside) was derived from the X-ray diffraction images generated by Rosalind Franklin. As the story goes, in 1952 Franklin provided Wilkins Photo # 51, a diffraction X-ray of DNA that she had put aside, with no clear reason why. Wilkins, in turn, showed the image to Watson, who in turn discussed it with Crick and with additional data from Wilkins and Franklin (via a third party, Max Ferdinand Perutz, an Austrian-born British molecular biologist, who won the 1962 Nobel Prize for Chemistry for his work in assessing the structures of hemoglobin and myoglobin) this triggered them to formulate the famous double-helix model in 1953. An accurate accounting of this sequence of events reads more like a spy mystery novel than an account of hard and tedious scientific research. It is notable that in his book "The Double Helix,"[13] James Watson describes Franklin as not being able to interpret her own X-ray diffraction images.[14] Watson's opinions partially stem from the fact that although the X-ray diffraction images generated by Franklin provided the best evidence for the structure of DNA, it seems that she did not realize it at the time. Nevertheless, the first report describing the structure of DNA appeared in 1953 in the prestigious journal *Nature* (Fig. 3), and Franklin, the investigator that generated the critical X-ray, was not an author.[15] In a hand-written letter, Crick communicated to his son at the time in boarding school "In other words we think we have found the basic copying mechanism by which life comes from life."

In 1962, Watson, Crick, and Wilkins collectively were honored with the Nobel Prize in Physiology or Medicine for the formulation of the structural model of DNA. Rosalind Franklin was not acknowledged with the prize in spite of the fact that her images were the single most important data that led to the model. Although there is a Nobel Prize clause that states that no more than three persons can jointly win the price, it is not clear why Maurice Wilkins and not Rosalind Franklin was part of the trio of awardees, considering that she was the one that generated the crucial data. After everything was said and done,

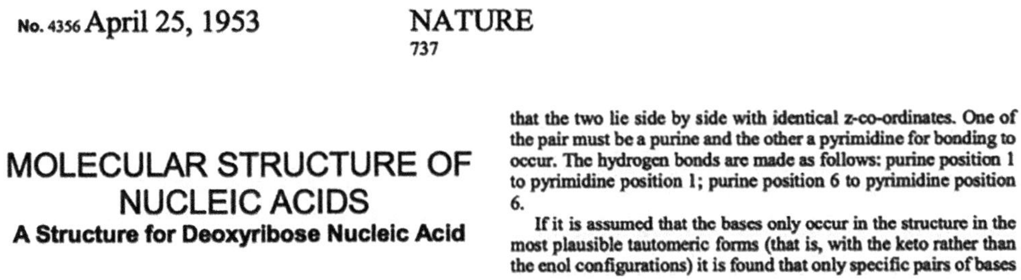
No. 4356 April 25, 1953 NATURE 737

MOLECULAR STRUCTURE OF NUCLEIC ACIDS

A Structure for Deoxyribose Nucleic Acid

that the two lie side by side with identical z-co-ordinates. One of the pair must be a purine and the other a pyrimidine for bonding to occur. The hydrogen bonds are made as follows: purine position 1 to pyrimidine position 1; purine position 6 to pyrimidine position 6.

If it is assumed that the bases only occur in the structure in the most plausible tautomeric forms (that is, with the keto rather than the enol configurations) it is found that only specific pairs of bases

Fig. 3 Article describing the structure of DNA (https://www.nature.com/articles/171737a0).

Franklin died from ovarian cancer in 1958 at the age of 37, without the recognition that she deserved.

It turns out that the model of DNA that we know today as Watson and Crick's double helix is capable of explaining the expected functions of the elusive transforming factor discovered earlier by Hershey and Chase. The double helical structure of DNA allowed for faithful replication. The order of the bases, in addition to the number of bases, makes one gene different from others. Today we know that changes or mutations, spontaneous or induced, in the order of the DNA bases are capable of altering the original sequence and the phenotype (the way organisms look and function).

Genesis of molecular biology

The experiments described in the previous section set the stage for the genesis of molecular biology. In the 1960s, it was clear that the genetic material was nucleic acid, but scientists were not sure how the information stored in DNA was expressed; how the message was transmitted into protein. For example, scholars were uncertain which of the two families of nucleic acids, RNA or DNA, was the molecule that stored the information of traits. Scholars were also not aware of the language of genetics. In other words, they did not understand how 4 nucleotides specified for 20 amino acids and how the genetic transmission system signaled the beginning and end of "sentences," the first and last amino acids of proteins, respectively. Clearly, knowledge of the genetic code was paramount for future developments in genetic engineering, the manipulation of heredity, and molecular biology. Knowledge of the letters, words, grammar, and punctuation was needed to read and use the language of genetics.

The realization that DNA is the genetic material as well as knowledge of its putative structure and function provided the framework for geneticists to envision ways to manipulate it. The experiments by Marshall Nirenberg (1927–2010) and Heinrich Matthaei (born in 1929), a postdoctoral fellow from Germany, at the National Institute of Health (NIH), Bethesda, Maryland, United States, in the 1960s that led to the cracking of the genetic code represent a line of work that focused on manipulating nucleic acid to synthesize proteins.[16] By 1966, Nirenberg's research group had deciphered the 64 RNA 3-letter code words (codons) for all 20 amino acids. In retrospect, their work was the genetics analog to deciphering the Rosetta Stone.

Although these investigations were primarily designed to ascertain the genetic code, Nirenberg and Matthaei were, in fact, synthesizing nucleic acids and manufacturing proteins at will in vitro, outside the cell. These two achievements are two of the cornerstones of molecular biology and genetic engineering.

Nirenberg and Matthaei's experiments paved the way for decoding the meaning of the sequences of the entire human genome a decade later. Nirenberg and Matthaei's work, in essence, confirmed Watson and Crick's model of DNA as well as its function.

They demonstrated that the language of heredity consists of four letters that make up three-letter words or codons and that, in linear order, they constitute a reading frame that specifies proteins. For their achievements, Marshall Nirenberg won the Nobel Prize for Medicine and Physiology in 1968. Heinrich Matthaei was not honored.

Nirenberg remained at the NIH in spite of numerous offers at prestigious research institutions. When asked why he stayed at the NIH, he replied, "because the thing I had least of was time. I figured that if I went to a university I would use a third of my time to write grants. I thought I could use that time more productively by doing experiments," a sentiment shared by many researchers.

Key discoveries in molecular biology

Several pivotal discoveries energized the field of molecular biology and ushered in remarkable advances in a number of fields within biology. Among these advances were the discoveries of restriction enzymes, the capacity to clone (mass produce) pieces of DNA inside cells, the ability to amplify specific sequences of DNA in vitro (in a test tube), sequencing of nucleic acid, and genotyping by micro-array chip technology. Each of these innovations played a crucial role in realizing the power of molecular biology and genetic engineering. In addition, the technologies that stemmed from these breakthroughs and innovations provided the tools to study the DNA of both contemporary and ancient hominins.

Restriction enzymes

Enzymes, by definition, are biological catalysts that function by lowering the energy of activation (energy needed to start a reaction) of specific chemical reactions. Chemical reactions require a certain amount of energy to get started. An analogy from everyday life can be employed to illustrate this concept. Daily, we start the day struggling to get out of bed. Activation energy can be envisioned as the energy people need to invest in getting up, making coffee, drinking it, and, finally, fully waking up. Even for chemical reactions that give out energy (exergonic reactions) a certain initial energy is needed to start the reaction. In nonliving systems, increasing the temperature of the reactants facilitates reactions. Increasing the temperature makes reactants move faster within the medium (e.g., solution), increasing the possibility of collision among them, which starts the reactions. Yet in living systems, that is not an option since high temperatures are not compatible with life. Living things would burn alive! As a result, evolution created enzymes that literally bring together the reactants in specific juxtaposition, thus facilitating the interaction between them in the absence of high temperatures. The mechanism of the reaction is as follows. Enzymes bind to the reactants and bring them close to each other in the correct orientation for their functional groups (the atoms involved in the reaction) to react. In the process of acting as catalysts, enzymes are not changed or degraded; they are recycled and keep doing their job repeatedly, all over again, reaction after reaction. In other words, the "middle-man" role played by biological catalysts allows molecules to interact without increasing their kinetic energy and rate of collision. Enzymes are one of the most remarkable achievements of evolution.

Restriction enzymes are a subclass of enzymes that specialize in cleaving DNA at specific nucleotide sequences. These recognition sites are usually short, about four or six nucleotides long. The restriction enzymes cut the large DNA molecules into fragments. Restriction enzymes were discovered by Swiss microbiologist and geneticist Werner Arber (born in 1929), and American microbiologists Hamilton Othanel Smith (born in 1931) and Daniel Nathans (1928–99) in the late 1960s. Since then, increasing numbers of restriction enzymes have been characterized and put to use. For their accomplishments, the trio was honored with the 1978 Nobel Prize in Physiology or Medicine.

Like many other discoveries, restriction enzymes were found by chance. It turns out that Arber, Smith, and Nathans were not responsible for the original observations. It was renowned Italian American microbiologist Salvador Edward Luria (1912–91) and his graduate student Mary L. Human that in 1952 confronted a number of unexplained phenomena while studying DNA breakdown resulting from viral infection.[17] Salvador Luria, a noted scientist in his own right, won the 1969 Nobel Prize in Physiology or Medicine for his work on the replication mechanism and the genetic structure of viruses. Luria and Human noticed that when they were forced to use the bacteria *Shigella* instead of *Escherichia coli* because their test tube containing phage-sensitive *Escherichia coli* culture broke, they were not able to grow the virus as in *Escherichia coli*. Consequently, further observations indicated that growing the phages in *Shigella* somehow prevented viral growth in *Escherichia coli*, but not in other bacteria species. They went on to notice that the bacterial phages mysteriously reappeared and were able to grow one generation later. Not knowing the scientific basis of their observations, Luria and Human named the phenomenon "host-induced genetic variation," acknowledging that it could not be a mutation since it was not a permanent change. They concluded that somehow the immunity to the virus was caused by the host bacteria. A mechanistic explanation for the observations was provided a decade later in 1962 by Arber (born in 1929) and his doctoral student, Daisy Roulland-Dussoix (1936–2014) (Fig. 4), two molecular microbiologists at the University of Geneva, Switzerland. These researchers provided a brilliant and elegant explanation for Luria and Human's perplexing results. They postulated that bacterial enzymes modify the DNA of the bacterium to protect themselves against the viruses, a type of primitive immune system against the phages. Furthermore, they stated that bacteria produced enzymes that cut viral DNA at specific nucleotide sequences to prevent being killed, and they enzymatically modify their DNA by methylating it (adding methyl groups, CH_4), thus preventing cleavage of their own nucleic acid. The idea is that bacteria were able to destroy viral DNA once it penetrated *E. coli*, before the viral DNA could take over

Fig. 4 Daisy Roulland-Dussoix (https://en.wikipedia.org/wiki/Daisy_Roulland-Dussoix).

the bacterial metabolism, but avoided disintegrating their own DNA by producing restriction enzymes that were incapable of cleaving methylated host nucleic acid.

Even though Daisy Roulland-Dussoix worked with Werner Arber as a doctoral student, probably doing most of the bench work, she was not honored with the Nobel Prize. In fact, the two articles that constitute the basis of the Prize were coauthored by both scientists, as sole authors, with each alternating as first author.[18,19] So, just like in previous cases involving the Nobel Prize, a young female scientist was ignored. Controversy over whether Roulland-Dussoix deserves greater recognition still remains. In a letter by Roulland-Dussoix to her brother she expressed her discontent when she wrote "I am very furious, because apparently he has not even mentioned my name, and I have done half of the work for which he received the Nobel Prize."

Subsequently, in 1968, Werner Arber and his postdoctoral student Stuart Linn isolated and characterized the specificity of the restriction enzyme *Eco*B. Other restriction enzymes followed, such as *Hind* II, isolated from the bacterium *Haemophilus influenzae*. Since then, more than 19,000 restriction enzymes have been purified and characterized. Today these biological catalysts are routinely employed tools used in the mining of genetic information and manipulation of DNA by cloning. Cloning DNA facilitates further analysis by sequencing and is one of the cornerstones of biotechnology.

Complementing DNA restriction enzyme analysis is DNA ligation, or joining. This technique is usually employed subsequent to DNA cleavage by restriction enzymes to assemble the desired combinations of recombinant DNA. Typically, DNA fragments from different sources are put together in a specific fashion to generate chimeric recombinant constructs. The capacity to join foreign pieces of DNA together by ligation, creating novel DNAs from different species, has allowed molecular biologists to mass produce novel combinations of nucleic acids capable of producing unique proteins of scientific or commercial value. These new and often-times unnatural combinations of DNA, or chimeras, are inserted into bacteria as part of plasmids, segments of circular, self-replicating DNA independent of the chromosomes of bacteria and yeast, or bacteriophages for their mass production, production of useful pharmaceutical proteins, and studies of gene expression and evolution. This set of techniques is referred as DNA cloning.

DNA cloning

DNA or molecular cloning is a set of technical procedures that are used to mass-produce nucleic acid within host cells. The essence of DNA cloning, as a tool, is its capacity to mass-produce any DNA. The universal nature of DNA allows for joining DNA fragments from any species. As a tool in studies of hominin evolution, molecular cloning enables investigators to generate enough DNA from extinct or extant populations to perform sequencing of their genomes. This information is paramount for assessing population relationships and migration patterns.

Historically, the groundwork that ushered the field of DNA cloning can be traced to a short period of time in the early 1970s. In 1970, American molecular biologist Morton Mandel (born in 1924) and colleague Akiko Higa demonstrated that foreign DNA (bacteriophage lambda) could be incorporated into *E. coli* by just pretreating the bacteria with a solution of calcium chloride.[20] Subsequently, in 1972, American geneticist Stanley Norman Cohen (born in 1935) and collaborators Annie Chang and Leslie Hsu used this simple method to introduce a plasmid carrying antibiotic resistance into bacteria.[21] The defining experiment was conducted in 1973 when the same group succeeded in inserting the gene for tetracycline (an antibiotic) resistance into bacteria utilizing the pSC 101 plasmid vector.[22] Then, it was in 1977 that the Mexican biochemist Francisco Gonzalo Bolívar Zapata (born in 1948) created pBR322 ("B" after Bolivar and "R" after Rodriguez, a coauthor in the cited manuscript), an artificially created plasmid designed with a man-made piece of DNA or cloning site for incorporating foreign DNA.[23] The artificial cloning site in pBR322 facilitates the incorporation of foreign DNA cut by various restriction enzymes. Most plasmid vectors since then have had components derived from the original pBR322. From then on, the business of DNA cloning took off dramatically.

DNA cloning usually commences after recombinant combinations of DNA from different sources, frequently from different species of plant or animals, are assembled into plasmids or viral DNA, the vectors. These man-made recombinant DNA constructs are then inserted into the host cells, primarily bacteria or yeast. The process takes advantage of the host cell's natural capacity to replicate its plasmid or chromosomal DNA. When the host cell replicates the inserted DNA (recombinant constructs), it generates large numbers of the chimeric DNA. The only thing molecular biologists need to do after allowing the bacteria or fungi to grow is break the host cells open and purify the recombinant DNA.

The resultant cloned DNA is then used for study. Usually, the cloned DNA is first sequenced to assess and verify its identity. Generating the sequences could be the end point of the investigation, since often-time scientists are only interested in phylogenetic/evolutionary relationships. For example, scholars involved in hominin evolution, their migrations and interactions, are particularly focused on comparing the DNA from the different populations to assess relationships. Since sequencing requires relative large quantities of DNA, cloning is a practical tool to achieve these large quantities. In other words, by inserting the constructs into hosts, evolutionary biologists can make the cells work for them as they replicate the foreign DNA of interest. Other studies aim at studying protein structure and function. In those instances, investigators clone the recombinant constructs into expression vectors, plasmids, or bacteriophages that not only replicate the inserted DNA, but also are capable of using the foreign DNA to make mRNA by transcription and proteins by translation. These cloned DNAs can then be purified, examined, and used as pharmaceuticals in the form of gene therapy vehicles as well as for recombinant protein production processes, new vaccines, and basic research.

DNA sequencing

Starting in the 1960s, genetic diversity was assessed by looking at variability at the protein level. In other words, investigators determined diversity among proteins based on their rate of migration during electrophoresis. Electrophoresis is a technique using electrical separation of molecules within a matrix, such as agar, according to charge, size, and shape. Proteins with large net electrostatic charges migrate faster within a gel matrix in a closed electric field, and proteins with different net electrostatic charges, due to their amino acid composition, will appear on the gel at various distances from the origin of the gel where the protein extracts where initially deposited. In these types of experiments, investigators measure the distance protein variants have migrated from the origin. In this way, variant proteins of specific genes can be scored as diversity.

Although this type of methodology has its uses, it suffers from a number of limitations. First, since proteins are being scored, genetic diversity is being detected several steps (transcription, RNA editing, translation, protein modifications) away from the actual source, the DNA. As such, some variability goes undetected due, for example, to silent mutations (mutations that do not change the amino acids of proteins because several nucleotide triples code for the same amino acid). In addition, many of the genes that code for proteins are under selection pressure, since the proteins they produce perform required functions within the organism. Generally, markers under selection pressure are not ideal to study evolutionary relationships because the abundance of these markers (e.g., proteins) are influenced by selection pressure and their abundance in populations are influenced, at least partially, by specific environmental conditions and not by genetic relationships dictated by evolutionary forces, such as gene flow or mutations.

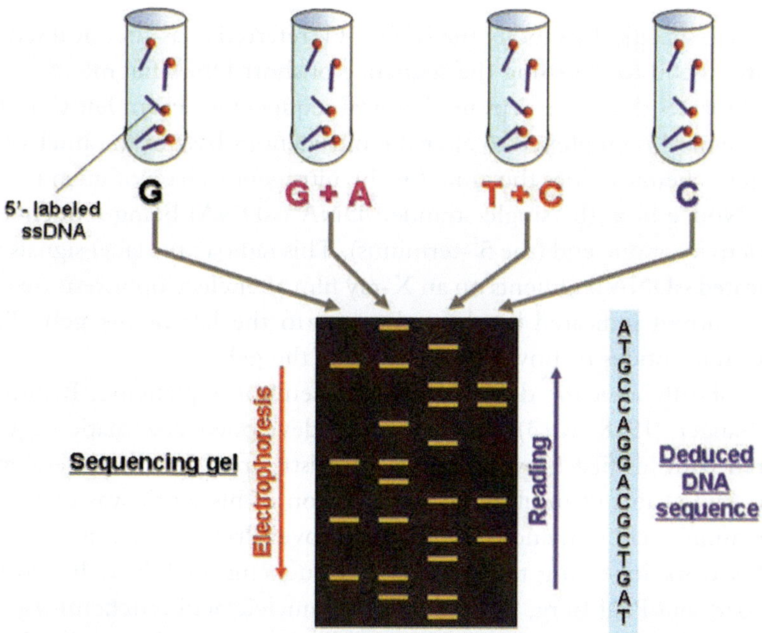

Fig. 5 Chemical sequencing (http://technologyinscience.blogspot.com/2014/07/dna-sequencing-traditional-and-next-gen.html#.XNyFBiMrKRs).

In the 1970s, investigators began to mine genetic diversity by looking *directly* at the DNA by sequencing their nucleotide building blocks. The initial efforts resulted in the Maxam and Gilbert chemical sequencing method (Fig. 5). The American biochemist, physicist, and molecular biologist Walter Gilbert (born in 1932) and Allan Maxam (born in 1944), his student at Harvard University, pioneered the work.[24] For their innovative discoveries that jump-started the field of direct DNA analysis by sequencing, Walter Gilbert was honored with the Nobel Prize in chemistry in 1980. A remarkable scientist, Walter Gilbert is also credited for first proposing the existence of exons and introns (protein-coding and noncoding DNA, respectively), by itself a quantum leap in molecular biology and evolutionary thought. Furthermore, he theorized on the evolution and function of introns and advanced the RNA World Hypothesis for the origin of life.

In chemical sequencing, the approach is to specifically modify each of the four nucleotides of DNA in separate chemical reactions (see the upper portion of Fig. 5 indicating the four reactions, each showing the nucleotides they modify). These nucleotide-specific modifications make the phosphodiester backbone of the DNA, at the site of the alteration, susceptible to breakage. This selective truncation of sequences produces a ladder of DNA fragments of different sizes that are directly related to the order of specific nucleotides along the DNA molecule. The sequence is read from one end of a gel to the other subsequent to electrophoretic separation of the fragments. Since this methodology relies on chemicals to specifically modify the DNA, it is referred to as chemical sequencing and is particularly useful for assessing the sequence of short DNA fragments.

Fig. 5 illustrates the main steps in chemical sequencing. From left to right on top, a particular chemical is employed to alter the nitrogenous base of the nucleotide guanine and a different chemical does the same for the nitrogenous base of guanine and adenine and so on. Notice how the single-stranded DNA (ssDNA) being sequenced is labeled with radioactivity at one end (the 5′-terminus). This radioactive label signals the position of the truncated ssDNA fragments on an X-ray film after electrophoresis (see direction of the electric current indicated by the red arrow to the left of the gel). The deduced sequence of nucleotides is shown to the right of the gel.

In 1977, shortly after the development of chemical sequencing, British biochemist Frederick Sanger (1918–2013) and his group developed enzymatic sequencing. He was honored with his first Novel Prize in chemistry in 1958 for his sequencing of the two amino acid chains of insulin. The implication of his work was that every protein has a finite amino acid sequence. His second Novel Prize in chemistry was granted in 1980 for his work inventing the enzymatic sequencing of DNA. It was shared with Walter Gilbert and Paul Berg, both pioneers in nucleic acid biochemistry.

This approach employs analogs of nucleotides (synthetic, chemically altered nucleotides similar in structure but different in elemental composition), one for each of the four nucleotides, that, when incorporated by the enzyme DNA polymerase, stop additional additions of nucleotides and in doing so truncates the extension of the nascent DNA strand. In this methodology, each nucleotide analog is labeled with a different fluorochrome (fluorescent dye) that allows independent detection of each of the four nucleotides at the end of the truncated strand. Effectively, this procedure, although different in methodology to chemical sequencing, achieves the same outcome, a continuous ladder of halted DNA fragments that is read as a nucleotide sequence (Fig. 6). This technique has been automated and was responsible for the sequencing of the human genome and the genome of sample individuals from various hominin populations.

Currently, a number of novel procedures are being developed as new generation DNA sequencing systems in which DNA fragments are sequenced at random and subsequently are arranged by computers using overlapping sequences. These next generation sequencing procedures have facilitated the output of the entire DNA not only of single individuals, but of populations. Furthermore, since next generation sequencing is less dependent on large pieces of intact DNA, it can tolerate the partially degraded shorter fragments characteristic of ancient DNA typically derived from hominin fossils.

Polymerase chain reaction (PCR)

The in vitro (in a test tube) synthesis of DNA is the essence of PCR. The process is very similar to the mechanism employed by living cells when they replicate DNA.[25] Living

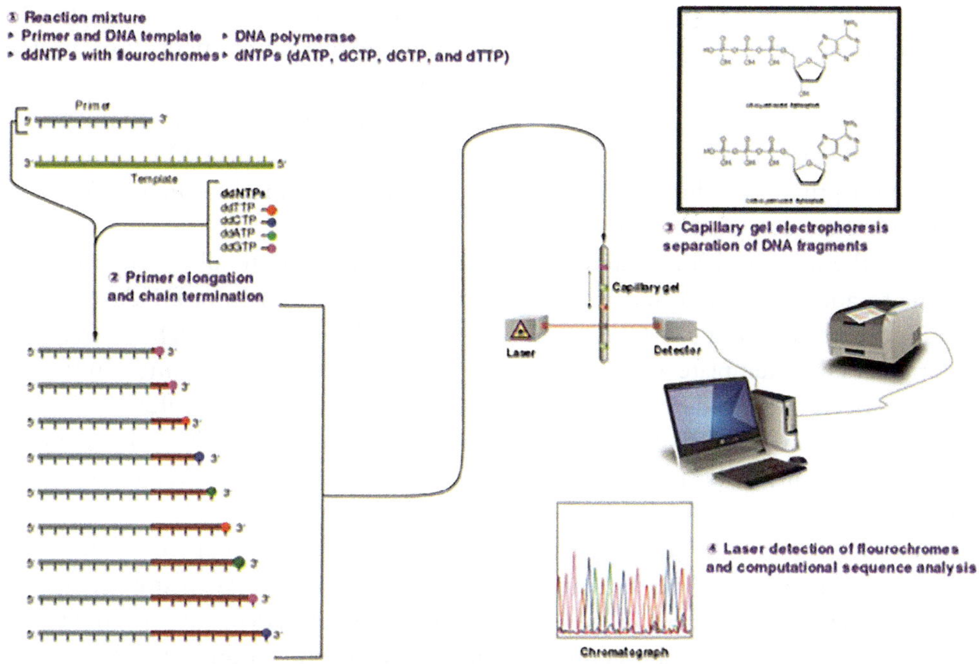

Fig. 6 Enzymatic sequencing (https://upload.wikimedia.org/wikipedia/commons/thumb/b/b2/Sanger-sequencing.svg/1280px-Sanger-sequencing.svg.png).

cells use a variation of PCR to manufacture their DNA, except that instead of using heat to separate the strands of the double helix in order to copy them, living systems employ enzymes. As previously discussed, life is not compatible with the close to boiling temperature used in PCR. Inside cells, separating the two strands of the double helix to expose the nitrogenous bases of nucleotides is the first step in DNA duplication. Next, cells have available short pieces of single-stranded RNA or primers that bind to the single-stranded DNA (the template). Subsequently, the enzyme DNA polymerase attaches one DNA nucleotide at a time to the end of the RNA primer to generate the complementary strand of DNA. The enzyme that makes the complementary strand of DNA can only add new nucleotides to an existing strand of nucleotides, the primer. After the new DNA strand is completed, the short RNA fragment is digested out and substituted with DNA. At that point, the synthesis is finished, resulting in a DNA double helix made up of an old DNA strand (the original template) and a freshly made strand. As a prelude to cell division (cellular reproduction), the cell undergoes the above outlined process of DNA replication.

In PCR, the process of DNA replication is essentially the same as in vivo DNA replication, except that the process is performed in a test tube, heat is used to separate the two strands of DNA, and instead of RNA primers, DNA primers are used.[26] In practice, DNA from an organism is purified from the other cellular components (i.e., proteins, carbohydrates, etc.) and placed in a small test tube containing a mixture of the four require nucleotides and the polymerase enzyme that adds the nucleotides to the primers. Heat is applied to separate the two strands of DNA (denaturation step) and make them available for primer attachment. The primers are allowed to bind to the DNA at a permissive temperature (e.g., 50–60°C).

Then, the mixture is incubated at 37°C for a few minutes to allow the polymerase enzyme to add, one nucleotide at the time, to the growing chain of DNA (elongation step). Subsequently, heat (95°C) is applied to separate the newly made DNA strand from the original DNA template and the process is repeated usually 30–40 times. In this fashion, it has been calculated that after 10 cycles, 1 DNA double helix would generate 1024 copies of itself and after 20 cycles, 1,048,576 would be produced. The potential of this technology is that DNA can be continuously produced as long as the reactants and conditions are provided.

A cursory comparison of DNA duplication inside cells and PCR reveals fundamental mechanistic parallelisms. And the simplicity of the process begs the question, why it took so long for scientists to invent PCR. PCR, like many other discoveries, was the product of teamwork. Yet it took the genius of one person to put a number of known facts together, which were available to all, but somehow missed. That individual was Kary Banks Mullis (born in 1944) (Fig. 7).

Fig. 7 Kary Mullis (https://en.wikipedia.org/wiki/Kary_Mullis).

Although a number of investigators contributed to the proof of concept of PCR, Kary Mullis alone is credited with the idea. Mullis is also credited with the application of the heat-resistant polymerase enzyme of *Thermophilus aquaticus* (*Taq* polymerase) to PCR. *Thermus aquaticus* is a bacterium that lives in hot springs, and thus, it tolerates high temperatures. Furthermore, the enzymes it produces, including DNA polymerase, remain functional at near boiling temperatures, making it ideal for the high temperatures required during PCR. Prior to using *Taq*, PCR was not only tedious to do but also painful, since investigators needed to add heat-sensitive polymerase that degraded during every cycle due to the near boiling temperature required for DNA denaturation. People used to burn their fingers retrieving the small test tubes immersed in hot water to add fresh enzyme at the end of every PCR cycle.

The facts surrounding the genesis of PCR are unusual. At the time, in 1983, Kary Mullis was an employee of Cetus Corporation. He worked as a chemist making short nucleotide fragments for a number of researchers at Cetus. The general public usually associates discoveries with settings such as laboratories, observatories, or archeological digs, etc., yet, the discovery of PCR was different. It occurred while Mullis was driving his little silver Honda Civic one spring night from Mendocino County (just north of the San Francisco Bay area) to his cabin in Anderson Valley, about 100 miles north. Mullis notes the exact moment when the idea came to him as he was driving late at night along Highway 128 halfway into the stretch between Cloverdale and Booneville at mileage marker 46.58, and pulled off the road to record it. The question still remains as to the role the hallucinogenic drug LSD played in the discovery of PCR. Referring to LSD, Mullis himself stated "I found it to be a mind-opening experience. It was certainly much more important than any courses I ever took."[27] During a conversation with a friend, Mullis communicated that LSD had "helped him develop the polymerase chain reaction that helps amplify specific DNA sequences".[28] Yet, it is not clear if Mullis was drugged while driving the little Honda to his cabin.

What is clear is that PCR has revolutionized a number of fields within the realm of biological sciences, including hominin evolution. With its impressive versatility, PCR is applicable to hundreds of uses. Especially in dealing with ancient DNA, the miniscule quantities of nondegraded nucleic acid in fossils are now amenable to analysis by in vitro DNA synthesis using PCR. It can be said that there are two molecular biologies, one before and the other after PCR. For his contribution to the development of PCR, Mullis received a $10,000 bonus from Cetus and Cetus got $300,000,000 from Hoffmann La-Roche when they sold the patent to them.

Subsequent to Mullis's eureka moment, the process of bringing the PCR method to fruition rested mainly on a number of scientists, including Randall K. Saiki (born in 1955) and Henry A. Erlich (born in 1943). They were shifted from other projects to work on the PCR idea and demonstrate that the PCR was feasible and practical. Saiki and Erlich published their work[29] prior to Mullis's PCR paper, which was subsequently rejected by the journals *Science* and *Nature* on the grounds of "information already published" and "it would not be suitable to meet the needs of their readers." Encouraged by James Watson, at the time director of Cold Spring Harbor Laboratory, Mullis submitted and got his papers published.[30,31]

Microchip technology/DNA microarrays

Microchip technology was invented by Patrick O'Reilly Brown (born 1954) in 1981.[32] The basic principle is relatively simple. It is based on adhering single-stranded oligonucleotides, or probes, of known sequence onto a solid surface and allowing for complementary single-stranded DNA, the target, to hybridize. Under stringent hybridization (e.g., high temperature) conditions, only exact matches between the DNA on the slide and the DNA in solution can occur, signaling the presence of specific point mutations (e.g., SNPs). Currently, hundreds of thousands of SNPs across the genome can be routinely genotyped in a single reaction. This technology has revolutionized how individuals are genotyped. In the field of population genetics, prior to the invention of DNA microarrays, years of work by several investigators were required to genotype the same number of SNPs in a single microchip. This methodology is allowing, for the first time, the ascertainment of large number polymorphisms genome-wide in a manner of hours.

Prior to the actual procedure, the microchips need to be designed and made. For this purpose, small amounts (about 6.0×10^{11} molecules) of single-stranded DNA of known sequence are deposited and adhered onto a solid surface or glass slide. Nowadays, these sequences are synthesized directly onto the solid surface of the chip. These dots of DNA are referred to as probes. They also go by the name of reporters or oligos and are usually 25–60 nucleotides long. These oligos are made to correspond to DNA sequences containing specific SNPs or point mutations. For example, if a SNP exists in two forms, one carrying a guanine nucleotide and the other an adenine nucleotide as variants at a certain position, oligos with a "C" and "T," respectively, are dotted in two different spots onto the slide. If the individual being genotyped possesses a "G" in both maternal and paternal chromosomes, his/her DNA will bind to the dot with the "C" nucleotide in the single-stranded oligo.

Commercially made arrays contain up to several millions probes in a single slide. Chips such as the ones in Fig. 8A are made with a glass "window" to allow detection of hybridization between the stationary probe and the complementary single-stranded DNA target. Arrays also can be custom-made to reflect the specific needs of the investigator. Target DNA in population genetic experiments usually reflects the entire genome of individuals. Single-stranded DNA fragments derived from total DNA from individuals of the population in question are labeled with fluorophores (chemical compound capable of emitting fluorescent light when excited by light). The mixture of fluorophore-labeled, single-stranded target DNA is introduced into the micro-array assembly (Fig. 8A) through a pinhole and allowed to hybridized (to pair to form

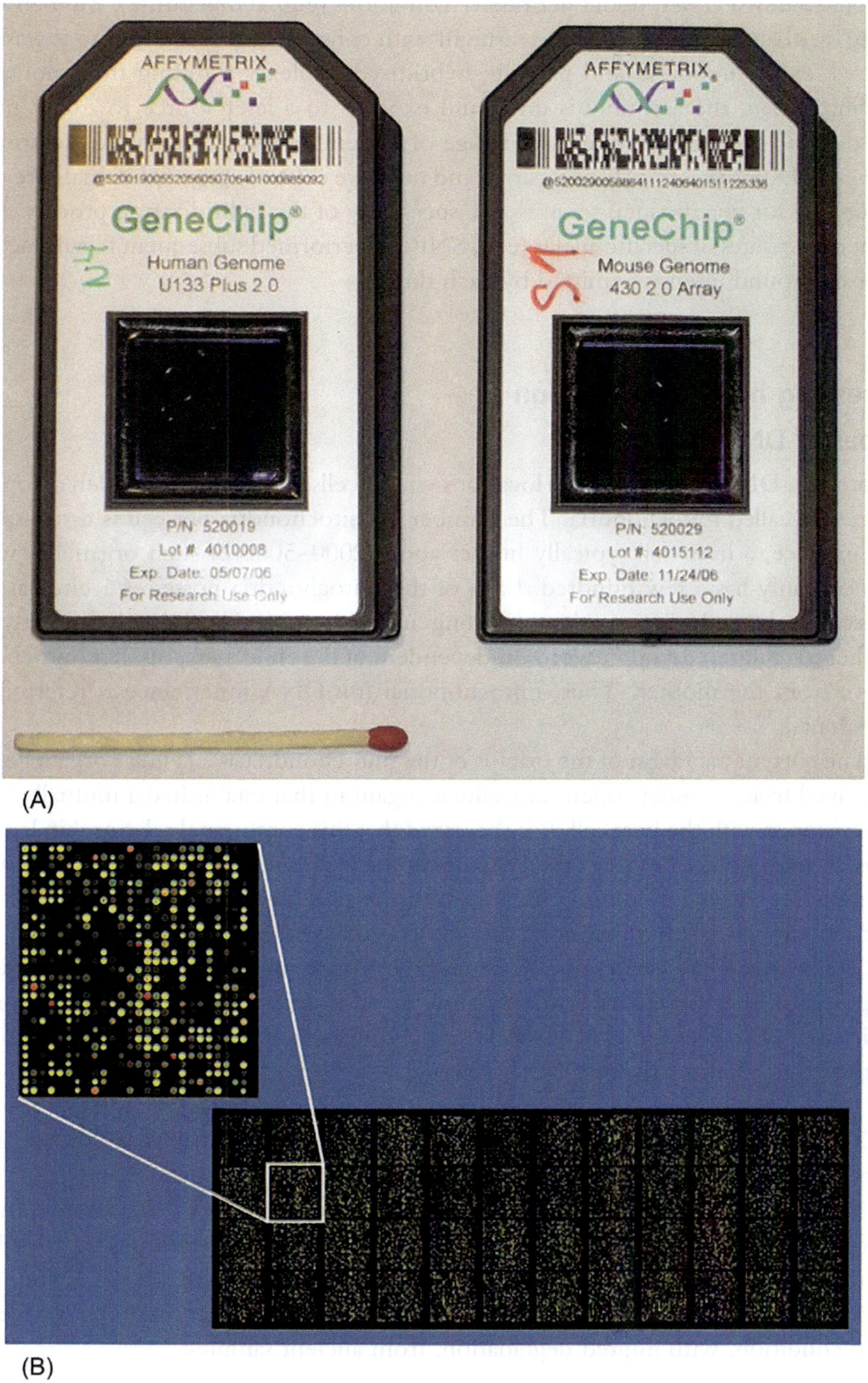

Fig. 8 (A) Two microarray chips (https://en.wikipedia.org/wiki/DNA_microarray#/media/File: Affymetrix-microarray.jpg). (B) Visualization of microarray analysis (http://g2im.u-clermont1.fr/HuGChip/img/biopuce4.bmp).

double-stranded DNA) under permissive conditions (e.g., temperature). After an appropriate incubation period, usually overnight with constant rotation, the chip is washed to remove excess unbound and partially hybridized duplexes resulting from nonspecific binding. Then, the assembly is dried and exposed to a laser source to excite the dye and the grid of dots is scanned. An image of the emitted signals is stored electronically (Fig. 8B). Experiments include positive and negative controls (i.e., individuals previously genotyped for specific loci) to assess the specificity of the hybridization process. Assessment of the states of specific alleles (e.g., SNPs) is performed subsequent to subtraction of the background intensity emitted by each dot.

Assessing hominin evolution

Types of DNA

In humans, DNA is found in two locations within cells, in the nucleus and in cytoplasmic organelles called mitochondria. The number of mitochondria per cell is tissue-specific. For instance, a liver cell typically houses about 2000–5000 of such organelles while a skin cell only has a few hundred. Each of the mitochondria possesses 1 circular DNA molecule, about 16,569 nucleotides long in humans, and is inherited strictly from mother to children. In other words, independent of the child's sex, his/her mitochondria derive from the mother. Thus, mitochondrial (mt)DNA inheritance is referred to as uniparental.[33]

The current paradigm of the origins of the mitochondria as a cellular organelle is that it evolved from an independent unicellular organism that established a mutually beneficial existence with the host cell. It is theorized that this commensal relationship led to the transfer, with time, of most of the genes from the invading cell to the host genome, creating the present condition in which most of the cells' genes are inside the nucleus and neither host cell nor mitochondria can survive independently.

The mitogenome (total mitochondrial DNA) contains a total of 37 genes coding for 13 proteins, 22 transfer RNAs (tRNAs) involved in making proteins, and 2 ribosomal RNAs (rRNAs) that are part of ribosomes (cytoplasmic organelles involved in protein synthesis). In addition, the mtDNA has a regulatory region that controls DNA replication and RNA transcription. This control region is the most genetically variable portion of the mtDNA, and thus is the subject of investigations concerning maternally driven inheritance and matrilineal phylogenetic relationships.

MtDNA is particularly useful in studies of hominin evolution. The large number of mitochondria per cell as well as their double unit membrane (two plasma membranes surround them) contribute to the recovery of sequence-amenable mtDNA from fossils. Compared to nuclear DNA, the small circular mtDNA can more often be recovered in good condition, with limited degradation, from ancient samples.

Another type of DNA is Y chromosome- and male-specific. Although the Y chromosome is located within the nucleus of cells, most of its DNA is inherited uniparentally from father to son. In other words, except for the recombining portions of the Y chromosomes (at the tips or telomeres of the chromosome), which have homology with X chromosomes, the Y chromosome does not recombine and is referred to as the NRY (nonrecombining Y) region. Females do not possess a Y chromosome and therefore Y-specific DNA markers cannot be found. Like mtDNA, the Y chromosome has been extensively examined in studies of hominin evolution. In humans, the Y chromosome encompasses about 58 million nucleotides, approximately 1% of the genome of a male cell, and it contains only about 200 genes, of which just 72 code for proteins. In part, the usefulness of the Y chromosome in studies of evolution derives from the vast amount of its DNA that is thought to be selectively neutral. This is because selectively neutral DNA is not subject to natural selection. As previously discussed in the DNA Sequencing section, natural selection may obscure the true phylogenetic relationship among populations, since it is subject to environmental conditions, which are independent of ancestry.

The third category of DNA is represented by the bulk of the nuclear DNA. Except for the nonrecombining portions of the Y chromosome (NRY) present in the nucleus, most of the nuclear DNA is inherited biparentally and is subject to recombination. This nuclear DNA differs from mtDNA and the NRY in that it is present in duplicate. In other words, it is made up of a maternal and a paternal set of linear chromosomes that separate randomly and exchange DNA or recombine during the creation of sex cells, the eggs and sperm.

The consequences of this reshuffling of DNA are chimeric linear DNAs in subsequent generations made up of intermittent maternal and paternal chromosomes (Fig. 9). In other words, different from the behavior of mtDNA and NRY DNA that are inherited as a whole, most of the nuclear genes separate during egg and sperm formation. If genes are located on different chromosomes or distant from each other (about 50 million bases or more), they assort independently from each other during sexual reproduction. Yet if genes are closer to each other (less than 50 million bases) within the same linear DNA, they do not segregate independently from each other as cells undergo meiosis due to limited recombination. The frequency of separation or recombination between genes is directly proportional to the distance between them. In other words, the greater the distance separating genes, the greater the frequency of recombination events between the two. One of the outcomes of DNA recombination is that it generates large amounts of genetic variability useful in phylogenetic analyses but it does not provide the convenience of mtDNA and NRY markers that allow examination of uninterrupted maternal and paternal lineages generation after generation.

DNA recombination has fundamental consequences impacting evolutionary change. Furthermore, the methodologies employed to study DNA capable of recombination

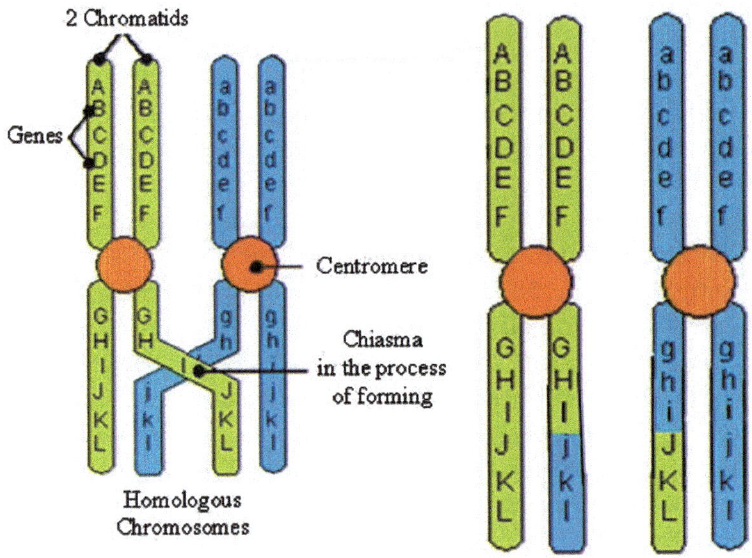

Fig. 9 Consequence of DNA recombination (https://www.wiley.com/en-us/Genomes%2C+Evolution%2C+and+Culture%3A+Past%2C+Present%2C+and+Future+of+Humankind-p-9781118876404).

differ from the algorithms for examining nonrecombining DNA. Algorithms dependent on the principles of independent genetic assortment of genes and recombination would not be appropriate for the analyses of NRY-specific and mitochondrial inheritance since these gene complexes tend to move together as blocks from generation to generation. In the NRY and mitochondrial DNA, de novo mutations are the source of diversity.

Another variable to consider when studying genetic diversity is that the rate of mutation (μ) is not uniform for all types of DNA. For example, the mutation rate of human mtDNA has been estimated to be about 3×10^{-5} per base per 20-year generation, at least 10 times higher than nuclear DNA. This relatively high rate of mutation compared to nuclear DNA is due to a combination of factors, including the low fidelity of the mitochondrial DNA polymerase in replicating DNA, the high concentration of mutagenic intermediary metabolic products of the respiratory transport system (e.g., free radicals) inside the mitochondria, and the less efficient protective proteins and mtDNA repair systems. The mutation rate that affects the number of repeats in short tandem repeat (STR) DNA sequences (nuclear genome) range between 10^{-3} and 10^{-4}. Further, some hyper-mutable, medically important tetranucleotide repeats are capable of expanding their number of repeat units within a few generations.[34] STR mutations affecting the number of repeat units generally lead to increments in numbers of the reiterated sequences, at a rate depending on the actual locus in question. These frequencies are much higher than the point mutations occurring in the nuclear genome that are responsible for substitutions, deletions, and addition of one or a few nucleotides per site per generation, which are in the order of 10^{-6} to 10^{-8}. Mutations in the form of insertion of repetitive elements, such as the members of the *Alu* family, occur at the rate of 1 de novo insertion approximately every 20 births.[35]

Nuclear, mtDNA, and Y-specific inheritance are not only subject to different evolutionary forces resulting from their unique modes of inheritances and mutation rates, but also from the cultural conditions and patterns of behavior that affect males and females of our species differentially. For instance, uniparental genetic markers only allow tracing of migrations of members of one sex or the other. Thus, investigating, for example, only mtDNA would not provide a complete picture of the histories of populations, since the male component of the population is not being considered. Among culturally related factors that may affect maternal- and paternal-specific inheritance is, for example, the marriage of males into female households or vice versa, which would impact patterns of gene flow. Similarly, one should consider the unique roles that male (e.g., hunting, welfare) and female (e.g., child care) hominins have played during evolution, with the corresponding and differential levels of accompanying risks, such as mortality.

Allelic frequencies

Currently most analyses used to assess phylogenetic relationships among hominins are based on comparisons of DNA sequences. Although the number of approaches in the form of computer programs and algorithms continue to grow, most of the data utilized stems from similarities and differences in the sequence of nucleotides. The comparisons performed by these analyses basically compare nucleotide differences among hominins and/or the presence of duplicated or deleted DNA sequences. These qualitative or quantitative differences in DNA may involve single nucleotides or point mutations, or large chunks of DNA or chromosomal aberrations, and every size in between.

In its most basic form, these comparisons ask the question: at a specific position on the DNA molecule, does hominin "A" exhibit the same nucleotide(s) as hominin "B"? Congruency is scored as a character that both hominin groups have in common. Alternatively, the presence of different nucleotides represents a discrepancy and is noted as a character that sets them apart. Disparate forms of DNA are referred to as variants or alleles.

Types of genetic variation

Single nucleotide polymorphisms (SNPs)

A number of DNA-based marker systems are routinely employed to score genetic diversity. DNA diversity can be grouped into single nucleotide polymorphism (SNPs), short tandem repeats (STRs), insertion/deletion polymorphisms (indels), and ancient retroviral insertions. SNPs are the most abundant type of human genetic variation, within and among populations. By definition, in order for a DNA marker to be considered polymorphic, the variant needs to be present in at least 1% of the genome of a population. Based on this criterion, these point mutations are found at a frequency of about 1 in every 300 nucleotide base pairs, totaling approximately 200 million SNPs in the diploid (both parental sets of chromosomes) human genome, which is comprised of 6 billion nucleotides. A given individual has about 5 million total SNPs, which is approximately one for every 1000 nucleotides. In addition to SNPs, some point mutations are rare (<1%). Some are so infrequent that it is difficult to identify and confirm them. In general, polymorphic markers like SNPs are particularly useful in population genetics studies since their frequencies and ubiquitous locations allow for phylogenetic assessment based on their genome-wide distribution. These polymorphisms often take the form of frequency and diversity clines (gradual frequency changes) within geographical regions suggesting past migrational waves. Yet at times, SNPs exist in one or a few populations as private, population-specific variants. Currently, investigators interested in hominin evolution assess hundreds of thousands of SNPs in order to provide a representative picture of the diversity and relationship among individuals and populations. Yet the large amount of SNPs in hominin genomes is still a largely untapped source of genetic diversity. Even among the already-available genome-wide sequenced DNA produced by high-throughput sequencing projects, SNPs still await identification and characterization. The raw data sets generated from these massive genotyping efforts are not interpretable by direct visual observation and require sophisticated computer programs and expertise to summarize and interpret the phylogenetic relationships among individuals and across populations.

It is interesting that although SNPs are found in protein-coding sequences, noncoding regions of genes, and in the intergenic regions between genes, most SNPs are located in unique, nonrepetitive sequences. SNPs are typically visualized with a number of techniques, including direct DNA or RNA sequencing, restriction enzyme digestions (RFLPs), and high-resolution high-performance liquid chromatography (HPLC); a technique that separates and identifies components in a mixture based on their relative affinities to a solid adsorbent material in a column. SNPs can be characterized based on a number of criteria. For example, some SNPs are located within unique DNA sequences of known function, at times affecting protein/enzyme activity (e.g., enzyme activity and/or specificity). SNPs within protein coding sequences could lead to alteration in protein structure and/or function and provide for evolutionary change under selection pressure. It is possible this high concentration of SNPs in coding sequences provides for diversity in a constantly changing environment. Alternatively, SNPs may be located within regulatory elements where proteins bind on the DNA and are involved in RNA production, such as promoters, enhancers, and silencers, and may impact transcription initiation, intron splicing, post transcription modification, and mRNA stability.[36] Yet, it is thought that the majority of SNPs are functionally silent, and therefore, tend to be excellent as markers for population genetic studies since they are not under selection pressure.

In terms of evolutionary studies, SNPs are also used as tags to identify regions within the genome exhibiting linkage disequilibrium (LD) or regions experiencing low levels of DNA recombination. DNA segments affected by LD may represent genes responsible for specific functions; thus, natural selection may favor individuals that do not experience recombination (alterations) in such regions. In addition, since the frequency of DNA recombination is directly proportional to distance between DNA markers (e.g., SNPs), the size of nonrecombined DNA is utilized to assess evolutionary affinity and time since the separation of populations. Medically, tag SNPs are also useful in genome-wide association studies (GWAS) in which hundreds of thousands of SNPs across the entire genome are genotyped, and computer comparisons are performed to identify intact, nonrecombined DNA regions or LDs in individuals affected with certain genetic diseases (e.g., schizophrenia).[37] This approach helps geneticists to zero in into the DNA region(s) responsible for the malady.

Short tandem repeats (STRs)

STR sequences represent one type of repetitive DNA in our genome. The human genome is made up of about 50% reiterated DNA. Fig. 10 lists the various classes of reiterated DNA sequences present in humans. Fig. 10 also provides information on the size

a

Repeat class	Repeat type	Number (hg19)	Cvg	Length (bp)
Minisatellite, microsatellite or satellite	Tandem	426,918	3%	2–100
SINE	Interspersed	1,797,575	15%	100–300
DNA transposon	Interspersed	463,776	3%	200–2,000
LTR retrotransposon	Interspersed	718,125	9%	200–5,000
LINE	Interspersed	1,506,845	21%	500–8,000
rDNA (16S, 18S, 5.8S and 28S)	Tandem	698	0.01%	2,000–43,000
Segmental duplications and other classes	Tandem or interspersed	2,270	0.20%	1,000–100,000

b

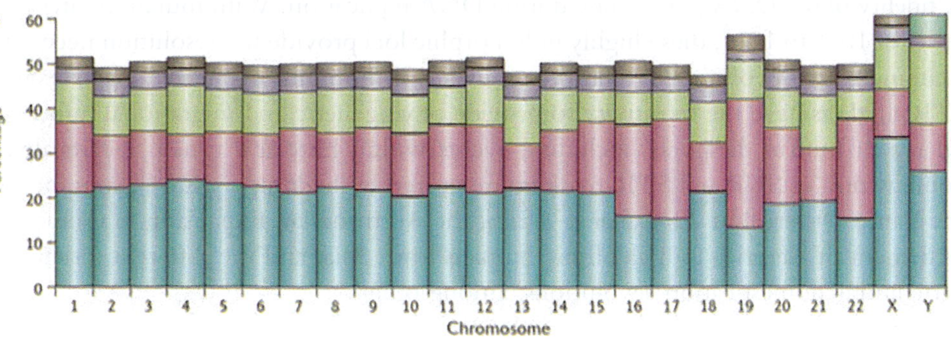

Fig. 10 Families of repetitive elements (https://www.wiley.com/en-us/Genomes%2C+Evolution%2C+and+Culture%3A+Past%2C+Present%2C+and+Future+of+Humankind-p-9781118876404). *(Data adapted from Todd J. Treangen, Steven L. Salzberg,* Nat Rev Genet *2012;13:36–46.)*

of the repeated units, the copy number for each family, and the percentage of the genome occupied by the redundant DNA types. The bar graph in panel B of Fig. 10 illustrates the color-coded proportions of each repetitive class found in each human chromosome. Although STRs represent a pervasive class of DNA, it is not clear how they originated, how they are maintained, or if they perform any function. Nevertheless, STR polymorphisms are routinely employed in population genetics studies as well as in forensic and medical cases.[38] These relative simple sequences are made up of two to six nucleotide units repeated in tandem several to hundreds of times, with each copy arranged head to tail. The nonrecombining region of the Y chromosome in humans is particularly rich in STRs. A similar class of tandem repeats is the VNTRs (variable number tandem repeats). These reiterated sequences are a bit larger than STRs, each repeat being 15 to a few dozen nucleotides, also in tandem and arranged head to tail. They share the same mechanism of mutation that generates the various numbers of tandem repeats in STRs: DNA replication slippage. DNA replication slippage occurs when the replication machinery backtracks and recopies the same repeated unit again as the DNA is being replicated. This mistake results in a mutation that leads to an increase in the number of repeat units within STRs. Due to the simple nature of these reiterated sequences, it is likely that they do not perform functions except possibly as spacer DNA, keeping proper distance among regulatory elements and structural genes (coding DNA). Therefore, most STR and VNTR loci are thought to be selectively neutral. These tandem repeated sequences, in fact, provide two marker systems in one: in addition to the genetic diversity in the form of various numbers of repeats units, internal point mutations within the repeats can also be scored. This dual and superimposed genetic variability provides for greater resolution in population genetic studies and identity assessments.[39]

STR loci have also found their niche in basic population genetics research. STR markers are hypervariable with the different alleles varying in the number of repeating elements. This phenomenon, as in the case of SNPs, is also likely the result of the lack of fidelity of the DNA polymerase during DNA replication. With mutation rates on the order of 10^{-3} to 10^{-4}, these highly polymorphic loci provide the resolution necessary to examine micro-evolutionary and recent events in hominin evolution.[40] For example, STR markers have the sensitivity to provide phylogenetic and migrational information on recent human diasporas, including events within historical times. Specifically, the combination of SNP and STR markers are currently shedding light on issues including the Bantu Dispersal in sub-Saharan Africa, the Austronesian Expansion in Oceania, the settlement of America by Paleolithic humans from Asia, and the different independent agricultural/domestication revolutions, among others.

Indels (insertion/deletions)

Indels represent a broad class of biallelic DNA polymorphisms characterized by the presence (one allele) or absence (the alternative allele) of a sequence of DNA. In its most basic form, segments of DNA in specific locations in the genome are missing in some individuals of a population and present in others. It is likely that these DNA sequences are not vital for the survival of the individual or are marginally under selection pressure. Individual members of specific, reiterated DNA classes (see Fig. 10 for listing) have been useful in phylogenetic studies. For example, recently inserted repetitive members of the short interspersed nuclear elements (SINEs),[41] long interspersed nuclear elements (LINEs),[42] and retroviral[43] families exist in polymorphic states in human populations. SINE and LINE sequences propagate themselves by RNA-mediated transposition using reverse transcriptase.[44] In this process, these repetitive elements are first copied into RNA by transcription and then a replica is made by an RNA-directed DNA polymerase, giving this process the name reverse transcription. SINEs and LINEs constitute approximately 42% of the human genome.[45] Although some areas of the genome are more populated with SINES and LINES than others, these sequences are ubiquitously distributed and generally interspersed throughout all human chromosomes. The most abundant family within the SINE category in humans, *Alus*, occupies about 10% of the human genome and approximately 2000 members exist in a polymorphic state.[46] In other words, not all individuals in a population possess the *Alu* insertion at a given location in the genome. It is likely that many of these polymorphic members represent recent insertion events. As a result of their recent history, these elements are not fixed in the population and remain in a polymorphic state.

As markers, indels such as *Alus* provide for unequivocally identification of the ancestral and derived state since the provability of an independent insertion into exactly the same site on the DNA sequence is highly unlikely and there is no known mechanism for the precise removal of an insertion without leaving residual sequences behind. In other words, SINE/LINE indel polymorphisms are essentially free of homoplasy (trait shared by a set of individuals but not present in their common ancestor) markers. In addition, the ancestral (lack of the insertion) and derived (presence of the insertion) states are always certain to provide directionality in phylogenetic analyses. These characteristics are useful in the study of hominin species. This characteristic of indels allows for polarity of the lack of insertion as the ancestral state. This knowledge is highly valuable in phylogenetic studies because it provides for the directionality of the mutation.

Medically, de novo SINES, LINES, and retroviral insertions are responsible for mutations that lead to a number of human maladies. They do this in two ways. One is the insertion event itself. By inserting into protein-coding, transcriptional-controlling, or spaces in between genes, these elements have the potential to render a protein-coding open reading frame useless or even detrimental/lethal, promoting frame-shift (creating aberrant messages) and mRNA splicing aberrations, as well as generating stop codons or additional coding sequences incompatible with the function of the protein. Secondly, they facilitate illegitimate crossing-over events with the resulting deletions and insertions of DNA during cell division. The abundance, high sequence homology, and ubiquitous locations of these interspersed sequences all lead to recombination between noncorresponding repetitive elements in near and distant regions of the genome. These injurious characteristics of SINES and LINES are responsible for a large number of genetic diseases, some of which are so deadly that they lead to spontaneous early abortions that go undetected.

Independent of their value as markers, these insertion elements represent an enigma in terms of evolutionary biology; the *Alu* SINE family, for example, originated in the primate lineage about 65 million years ago. At this time dinosaurs were becoming extinct. These ancient primates were small and as a group they were beginning to radiate into a multitude of species. It turns out that the *Alu* family started as a duplication of the gene that codes for the small 7 SL RNA moiety of the signal recognition particle of the rough endoplasmic reticulum.[47] It seems that subsequent to this initial gene duplication of the 7 SL RNA gene, the DNA was duplicated again in tandem creating a bipartite sequence with a stretch of poly As in between the 2 units. Since their origin, *Alus* have been amplifying and dispersing rapidly throughout the genome of primates like a plague. *Alus* spread in our genome by LINE-mediated reverse transcription of an RNA polymerase III transcript driven by split internal promoters within the *Alu* sequence. Today, *Alu* elements represent the most abundant family of middle repetitive sequences in humans with approximately 1 million copies per human diploid genome.[41] Yet, no specific singular function has been assigned to *Alus*.[48] While they exhibit high levels of sequence conservation among repetitive elements within primate species as well as among homologous elements in different primate species, they seem to exist and disseminate independent of the rest of the genome, earning them the name "selfish DNA."[49] Thus, the question is: "What mechanism allows for the retention and conservation of 10% of the genome when the DNA has no specific function?" It is likely that these elements provide for evolutionary raw material in the form of DNA that allows for the creation of novel genes as well as for the modification of existing genes and duplicated sequences. This possibility is suggested by the eclectic functions of *Alus*. Specific *Alu* sequences are known to perform a number of different and unrelated functions, such as controlling elements and coding sequences. It has also been suggested that *Alus* and other SINES/LINES are capable of coevolving with preexisting genes near the insertion site, thereby modifying functions or generating novel functions.[50]

Ancient retroviral insertions

Another class of interspersed repetitive elements is the ancient retroviral insertions, also known as endogenous retroviruses (ERVs). Evolutionarily speaking, these reiterated sequences are of particular interest to students of virology and hominin genome evolution. About 100,000 of these sequences, or about 8% of the human genome, is made up of this category of repeated DNA. It is thought that these sequences are capable of mutating and moving to other locations within the genome as well as becoming exogenous pathogenic retroviruses.[51] Although retroviruses infect and insert their DNA mostly into somatic cells, occasionally insertions into the germline DNA occur, enabling their transmission to subsequent generations. Most of our ERVs date back to ancient infections and insertions that occurred millions of years ago and persist in the human genome as truncated defective and infertile proviruses incapable of packaging their genome into virions and infecting other individuals. These attenuated ERVs exist as proviruses about 1 kb long that persist in our genome as single long terminal repeats (LTRs). Yet a minimal number of complete ERVs still exist in our genome, possessing approximately 10,000 bp in size.

Retroviral sequences are typically made up of two LTRs, one located at one end of the ERV and the other at the other end, each containing transcriptional controlling sequences. These LTRs, with their transcriptional controlling sequences, are responsible for altering gene expression of nearby host's structural genes. Although most of these LTR sequences function as tissue-specific alternative promoters and enhancers of host genes, in some instances, such as the tissue-specific expression of salivary amylase and the bile acid-CoA:amino acid N-acyltransferase (integral enzyme in bile metabolism) genes, they work as the major promoters. Thus, as regulators of gene expression, ERV sequences may have played an important role in hominin evolution. Indeed, it has been theorized that some of the evolutionary changes that are attributed to punctuated evolution in humans have been the result of retroviral insertions affecting the host's gene expression dramatically. Along these lines, ERVs are abundantly found in the vicinity of rapidly evolving gene families whose function are largely related to the responses to stress and external stimuli.[52] Therefore, ERVs and especially solo LTRs may have acted as portable cassettes of promoter sequences capable of jump starting major evolutionary events such as the dramatic brain development experienced by hominins in the last 5 million years. This explicit function of ERVs may have driven the retention of some of these sequences in the human and vertebrate genomes. In addition, it is also possible that coevolutionary processes involving ERV sequences with nearby host genes may have been involved in the adoption of novel functions subsequent to insertion. A particularly fascinating issue about ERVs is that is not clear whether these elements originated from infections of exogenous retroviruses, or conversely, whether they were the source of exogenous retroviruses, or both.

In addition to these regulatory effects and their impact as insertional mutations, ERVs have the potential to promote chromosomal rearrangement due to illegitimate recombination between viral sequences at intra- and interchromosomal loci. As with SINES and LINES, these rearrangements involving ERVs generate deleterious/lethal deletions and duplications. These changes are also a source of diversity and raw material for evolution.

Since the majority of ERVs represent ancient insertion events, many are fixed in our genome. Yet one subfamily, HERV-K (HML2), which constitutes less than 1% of the human ERVs, is thought to have integrated recently—within the past few hundred thousand years—and some members are still polymorphic (PRVIs), one allele being the lack of insertion state and the other the insertion state. In other words, these polymorphic ERVs exist in individuals in three possible genotypic states: individuals homozygous for the insertion (both maternal and paternal chromosomes possess the ERV), homozygous for the lack of insertion, or heterozygous. Since individuals and populations differ in the presence and absence, and frequency, respectively, of these polymorphic ERV insertions, PRVI loci have been used for identity and population genetics studies. Specifically, two members of the human-specific subfamily HERV-K (HML2), HERV-K106 and HERV-K116, were actively integrating in the last 800,000 years and HERV-K106 may have infected modern humans as recent as 150,000 years ago, just prior to the Out of Africa migration. It is not clear whether these recent integrations were endogenous retroposition events (secondary insertions within individuals) or external viruses infecting individuals.

In light of this discussion, we cannot help wonder if retroviral epidemics in the past have initiated pandemic infections leading to striking and sudden reduction in hominin population size. It is interesting to contemplate the possibility that some ERVs may have originated as geographically localized or pandemic retroviral infections that drove humanity to the brink of extinction. Could past retroviral infections be responsible for some of the bottleneck episodes that have taken place in recent hominin evolution? And, further, could HIV represent the present-day case of a retroviral epidemic that may evolve to a new ERV?

Analytical methods

Beyond just scoring nucleotide differences, deletions, and duplications, variant frequencies are regularly analyzed to visualize the genetic characteristics of populations. Oftentimes, researches tabulate frequencies of variant forms of DNA of extinct and extant human populations. Yet this type of basic data analysis offers a limited appreciation of the differences and similarities among groups of organisms. Today anthropologists and geneticists are endowed with a large arsenal of methodologies designed to mine and extract information pertaining to population dynamics. The present-day field of population genetics is in possession of massive amounts of data resulting from genome-wide, high-resolution genotyping and whole-genome high-coverage sequencing efforts. Thus, by necessity, the exploration of extinct and extant hominin gene pools requires exceedingly sophisticated computerized systems and software programs.

Tools for analyzing population structure and phylogenetic relationships

To analyze the voluminous amount of raw data generated, population geneticists rely on bioinformatics tools that annotate, store, examine, and compare DNA sequences.[53] Students of hominin evolution benefit from these technological advances by feeding massive amounts of DNA variability data into computer programs based on well-established population genetics principles. Numerous programs with algorithms designed to ascertain phylogenetic relationships among hominin populations are now available. Some of these analytical programs are based on genetic paradigms with origins traceable to Mendel's laws. For example, principles such as the Hardy-Weinberg equilibrium formulation, dating back to the beginning of the 20th century, are the cornerstones of many of these analytical programs. Some software programs are designed to mine the internal genetic structure of populations (intrapopulation parameters). For example, intraspecific bioinformatics tools are routinely used to investigate the internal organization of populations, such as subpopulation structure, that could signals admixture of different groups and/or nonrandom mating within a group. Bioinformatics tools have been employed to study population relationships among extant and extinct hominin groups, such as anatomically modern humans (AMHs), Neanderthals, and Denisovans.[54,55]

Bioinformatics tools are designed to assess a number of parameters. Independent of whether the trait in question is lingual, morphological, physiological, or molecular, variants are identified and scored usually utilizing the counting method.[56] Simply expressed, in this procedure the number of variants of a trait are counted and the frequency of each is determined manually. Variants in genetic studies take the form of alleles, the alternative forms or versions of genes. At the DNA level, alleles or variants are the nucleotide differences that constitute SNPs (point mutations), STRs (number of tandem repeats), and indels (pieces of DNA present or absence). Allelic measurements are the basic raw data that feeds into a number of phylogenetic algorithms employed in population genetics. Allelic frequencies can be utilized individually when genes assort independently (separately) during meiosis. For example, when genes are located on different chromosome pairs (e.g., chromosomes 1 and 7), they segregate independently from each other, generation after generation, and their allelic frequencies should be utilized one by one in the various computational analyses. Conversely, allelic frequencies of genes located on nonrecombinant DNA, including most of the Y chromosome (NRY), mtDNA, and chloroplast DNA (ctDNA or cpDNA) in plants or genes in close proximity to each other on the same chromosome (within 50 Morgan units or about 50 mega bases of DNA), cannot be analyzed as independently segregating units. In such instances, genetic comparisons may be performed at the level of haplotypes. A haplotype is a group of genes that tend to remain together (linked) from generation to generation. In other words, since closely linked genes within haplotypes do not assort independently during meiosis and tend to be inherited together as part of a cluster or cassette, their frequencies may be analyzed as a group instead of individual allelic proportions. Some statistical methods, such as multidimensional scaling (MDS), are capable of taking into consideration the nonrecombining nature of haplotypes by utilizing algorithms in which the input genetic data is in the form of haplotypes or groups of genes. In humans, DNA on the NRY region and mtDNA are routinely utilized as uniparental markers as part of large haplotypes that are transmitted uninterruptedly without recombination, generation after generation, thereby defining lineages.

Internal population genetics parameters

Table 1 illustrates some of the routinely employed bioinformatics programs in population genetics. These programs can be partitioned into two categories: tests addressing intra- or interspecific characteristics. Some of the software are basic and designed to ascertain mutation rates (e.g., μ value) as well as frequency and diversity of variants or haplotypes.

Table 1 Some major analytical methods employed in population genetics studies.

Methodology	Purpose/function	Type of data
Neighbor joining (NJ)	Phylogenetic comparisons based on a bottom-up (agglomerative) clustering method	Phylogenetic tree
Principal component analysis (PCA)	Phylogenetic comparisons based on orthogonal transformation	Plot
Multidimensional scaling (MDS)	Phylogenetic comparisons based on dissimilarities matrix and Euclidean distances	Plot
Hardy-Weinberg equilibrium (HWE)	Impact of evolutionary forces, allele, and genotype frequencies	Allele and genotype frequency changes
Genetic diversity (GD)	Genetic heterogeneity/Inbreeding	Proportion of polymorphic or heterozygous loci or alleles per locus
Heterozygosity	Genetic heterogeneity/inbreeding	Proportion of heterozygotes at a locus
Analysis of molecular variance (AMOVA)	Correlations	Estimation of population differentiation directly from molecular data
Time estimates coalescent time	Age of mutations	Traces alleles to a single ancestral copy
Percent unique haplotypes	Haplotype diversity	Frequency of unique haplotypes
Network analysis	Phylogenetic analysis of haplotypes	Haplogroups' relationships
Contour analysis	Diversity/frequency distribution in space	Geographical representation of alleles
Gst	Genetic distance	For multiple alleles
Fst	Genetic distance	For biallelic markers
Bootstrap	Random sampling with replacement	Statistical significance
Gene counting	Allelic frequency	Allelic abundance
Genotypic frequency	Predicting genotypes	Frequency of combinations of alleles
Haplotype diversity	Uniqueness of particular haplotype	Genetic heterogeneity/ inbreeding

Table 1 Some major analytical methods employed in population genetics studies—cont'd

Methodology	Purpose/function	Type of data
Haplotype discrimination Capacity	Frequency of haplotypes	Genetic uniqueness
R^2 test	Linkage disequilibrium	Nonrandom association of alleles at two or more loci
μ value	Probability of a mutation/generation	Mutation rate
Maximum-likelihood	Estimation of parameters of a statistical model	Phylogenetic comparisons
Demographic expansion	Finite-sites model with heterogeneity of mutation rates	Population expansion
Ewens-Watterson test	Based on the comparison of homozygosity	Assess selection pressure
Shared haplotypes	Percentage of haplotypes in common	Phylogenetic comparisons
F-statistics	Statistically expected level of heterozygosity	Inbreeding test
Exact test	Statistical significance in which all assumptions are met	Assess if data fits the results expected from theory
Assignment test	Likelihood-based ancestry	Assigning genotypes to populations
Mantel test	Correlation between two matrices	Genetic distance correlations
Chi-square test	Statistical significance	Assess if data fits the results expected from theory
Bonferroni correction	Family-wise error rate	Correct errors resulting from multiple comparisons
Watterson estimator	Based on coalescent theory	Population mutation rate
ChromoPainter	Identify identity by descent (IBD) blocks	Chromosome histograms indicating IBD tracks
Globetrotter	Identify, date, and describe admixture events	Genetic maps and bar graphs

Modified from R.J. Herrera, R. Garcia-Bertrand, F.M. Salzano (2016) Genomes, Evolution, and Culture: Past, Present, and Future of Humankind, first ed. John Wiley & Sons, Inc., Hoboken, NJ.

Programs designed to examine intraspecific parameters examine the internal characteristics of taxonomic groups such as species, races, and populations. Some of the indexes examined within populations are based and stem from the Hardy-Weinberg equilibrium expectations: no mutation, random mating, no gene flow, infinite population size, and no selection. Of course, these assumptions are unrealistic and unattainable and are never realized. Therefore, why do experts bother considering such expectations? The reason is that they are useful in identifying the specific evolutionary forces (e.g., selection, migration, etc.) acting on populations that violate the premises of the Hardy-Weinberg formulation.

In the Hardy-Weinberg formulation, expressed mathematically as $p^2 + 2pq + q^2 = 1$, p^2 represents the genotype frequency of the homozygous (the same variant "*A*" was inherited from both parents) individuals carrying the genotype *AA*, q^2 represents the frequency of the homozygous genotype *aa* (the same allele "*a*" was inherited from both parents), and $2pq$ represents the frequency of the heterozygous genotype *Aa* (the variant "*a*" was inherited from one parent while the variant "*A*" was inherited from the other parent). Notice that the sum of all three frequencies adds up to 1% or 100%. To illustrate its function, let us take, for example, a scenario in which selection is acting on a particular allele of a gene designated as "*A*." In such a situation, it would be expected that the frequency of variant "*A*" under positive selective pressure would increase with time and relative higher numbers of individuals carrying the "*A*" allele, compared to those that do not, would be observed. In the Hardy-Weinberg formulation, this would be reflected in relative larger number organisms than expected with the *AA* and *Aa* genotypes. The exact test (Table 1) is routinely employed to accesses the statistical significance of violations of the Hardy-Weinberg equilibrium expectations and the Ewens-Watterson program (Table 1) is used to test for alleles under selection pressure.

A number of analyses designed to ascertain internal genetic characteristics of populations have been derived to test for factors (evolutionary forces) affecting the Hardy-Weinberg equilibrium. Essentially, these tests search for the specific premises of the equilibrium that are being violated and disturbing the equilibrium. With this in mind, population geneticists assess parameters, such as linkage disequilibrium, inbreeding, and selection pressure.

When assessing linkage disequilibrium (LD), for example, investigators are interested in detecting a paucity of recombination within a given stretch of DNA. LDs can be defined as the nonrandom association of alleles of different loci within a fragment of DNA in a given population. This phenomenon is the result of a lack of (or less than expected) recombination. Conversely, "hot spots" of recombination are seen interspersed among LD blocks. The relationship between Hardy-Weinberg equilibrium and linkage equilibrium is as follows. Random mixing of variants among loci provides for Hardy-Weinberg equilibrium, since recombination and mixing across loci brings about linkage equilibrium. When a population experiences LD in a portion of its DNA, the individuals sampled do not represent a single well-mixed population, possibly due to substructure within the group of organisms resulting from nonrandom mating among individuals. Other causes of LDs include positive selection pressure for a specific combination of alleles or simply chance (i.e., random processes). These conditions violate the expectations of the Hardy-Weinberg equilibrium. The R^2 Test (Table 1) is used to examine potential instances of LD.

Inbreeding impinges on the Hardy-Weinberg equilibrium expectations by increasing the frequencies of homozygous individuals (i.e., p^2 and q^2). In instances in which homozygosity involves deleterious or lethal alleles, inbreeding tends to expose rare variants to negative natural selection. Inbreeding is particularly prevalent in small populations. It often results when limited numbers of individuals are isolated, for example, on an island or in a relatively inaccessible region. In humans, certain cultural and/or religious practices bring about inbreeding by promoting marriages between family members. Inbreeding is usually computed using F_{is} (Table 1).

Other factors that impact the Hardy-Weinberg equilibrium are: (1) migration of individuals from genetically different populations (gene flow) and (2) genetic drift (random changes in the frequency of alleles). Migration of organisms into a population could introduce novel variants or alter the frequency of alleles already present in the breeding group. In both scenarios, the Hardy-Weinberg equilibrium would be violated. The impact of genetically distinct migrants moving into a population is accentuated when the receiving group is small and the invaders are numerous. In addition, dispersals out of a population involving a genetically nonrepresentative group of individuals would distort the frequency of alleles in the group left behind as well as in the migrants, violating the Hardy-Weinberg equilibrium expectations in both populations. Information on the internal genetic parameters described above provides clues on the evolutionary histories of populations.

Interspecific population genetics parameters

In addition to intraspecific characteristics of populations, which shed light on the internal evolutionary forces that shape populations, population geneticists are interested in the genetic relationships and interactions among defined groups of organisms. A number of statistical methods aim to assess phylogenetic relationships (e.g., component analysis and multidimensional scaling) as well as genetic distances (e.g., Mantel test) among populations. Other, more specialized analyses provide for correlations between genetic diversity and a number of parameters, such as geographical space (e.g., AMOVA, contour analysis). In addition, statistical methods that identify identical by descent blocks (IBDs) search for identical or similar segments of DNA that have been inherited from a common ancestor population without, or with limited recombination. Since the length of uninterrupted DNA is indirectly proportional to the time of a common ancestor due to repeated recombination events generation after generation, this test is routinely employed to determine the phylogenetic relationship among individuals and populations. IBD routines are also employed to estimate the time elapsed since specific populations shared a common ancestor. Other programs, such as Structure, use multilocus genotype data to examine population structure among populations. Structure is capable of detecting unique populations, assigning individuals to populations, inferring admixture, visualizing hybrid zones, identifying migrants and admixed individuals, and estimating population allele frequencies in situations where many individuals are involved as migrants.

Model-based versus distance-based phylogenies

The algorithms available for the analysis of phylogenetic relationships among populations can be partitioned under various categories. For example, many methods are based on allelic frequency differences and similarities of individual gene variants (e.g., the frequencies of variants "*A*" and "*a*") that are tabulated into matrixes and imputed into the programs. This is a linear approach. Other approaches take advantage of whole blocks of DNA sequence homologies among populations (haplotype-based methods such as identity by descent tracks or IBDs); the larger the segments of DNA in common, the greater the genetic affinity among populations. Phylogenetic programs can also be categorized on the basis of whether they are model-based or distance-based.[59]

Model-based methods assess the likelihood that the data are compatible with a predefined model of population structure. For example, the Structure program requires the investigator to assign specific *K* values a priori as hypothetical numbers of source or ancestral populations. This type of methodology is often criticized for its inflexibility to the range of explanations provided to explain the observed data. These programs tend to limit the interpretations of the results by being restricted to the specific parameters and suppositions of the model, such as specific number of ancestral populations.

On the other hand, distance-based approaches do not require predescribed assumptions or conditions to analyze the data. Distance-based methods are model-free and are designed to identify groups of populations with certain characteristic in common by analyzing matrices of genetic distances or genetic similarities. Therefore, these programs are not limited by specific sets of criteria and provide more flexibility in the interpretation of the data. A number of analysis approaches described in this chapter, such as PCA, MDS, and HCA, rely on distance-based algorithms. Yet some investigators emphasize that distance-based methods possess certain disadvantages. Namely, they rely heavily on subjective visual representations (e.g., trees or plots) and are not rigorous in ascertaining the statistical significance of the phylogenetic relationships or clustering of populations.

PCA

One of the most widely used methods in phylogenetic studies is principal component analysis (PCA). In comparisons among hominin groups, this method is employed to compare hundreds of thousands of polymorphic sites and ascertain interspecies or interpopulation relationships. PCA is capable of utilizing DNA sequencing as well as SNPs (i.e., nucleotide substitutions, deletions, and insertions) as raw data.

In studies involving contemporary and extinct groups of hominins, for example, high-density (large number of markers), high-coverage (depth or multiple numbers of unique reads that include a given nucleotide or SNP) data provides researchers with a reliable picture of genetic relationships. PCA uses Euclidean (straight-line) distances between points to create a similarity matrix. PCA is a dimensionality reduction technique. In other words, in PCA a large number of variables (e.g., SNPs, linguistic, or morphological traits) are reduced to a small set of variables or principal components (PCs). PCA takes an amalgamation approach to examining variables. The intent is to summarize as much of the original variation as possible using the minimum number of principal components or dimensions. In an actual plot, these dimensions are represented by axes (e.g., *X*, *Y*, *Z*, etc.) The first principal component (PC1) generated by the software accounts for the largest amount of the variability of the original data while succeeding components represent decreasing percentages of the remaining diversity. PCA expresses the strength of the analysis in terms of the amount of the total variability represented in its various components (i.e., PC1, PC2, etc.). The higher the percentage of the variability expressed in the graph, from all of the axes, the greater the level of confidence on the relationships illustrated in the depiction. Percentages of 50% or higher for all (total) of the components in a plot is desirable, yet PCA exhibiting lower values are still useful.

Fig. 11 illustrates a PCA graph based on 827 worldwide contemporary humans, Neanderthals, Denisova, and chimpanzees, using a set of 37,758 SNPs. PC1 and PC2 comprise 58.2% and 26.17% of the variance, respectively. PC1 appears on the *X*-axis

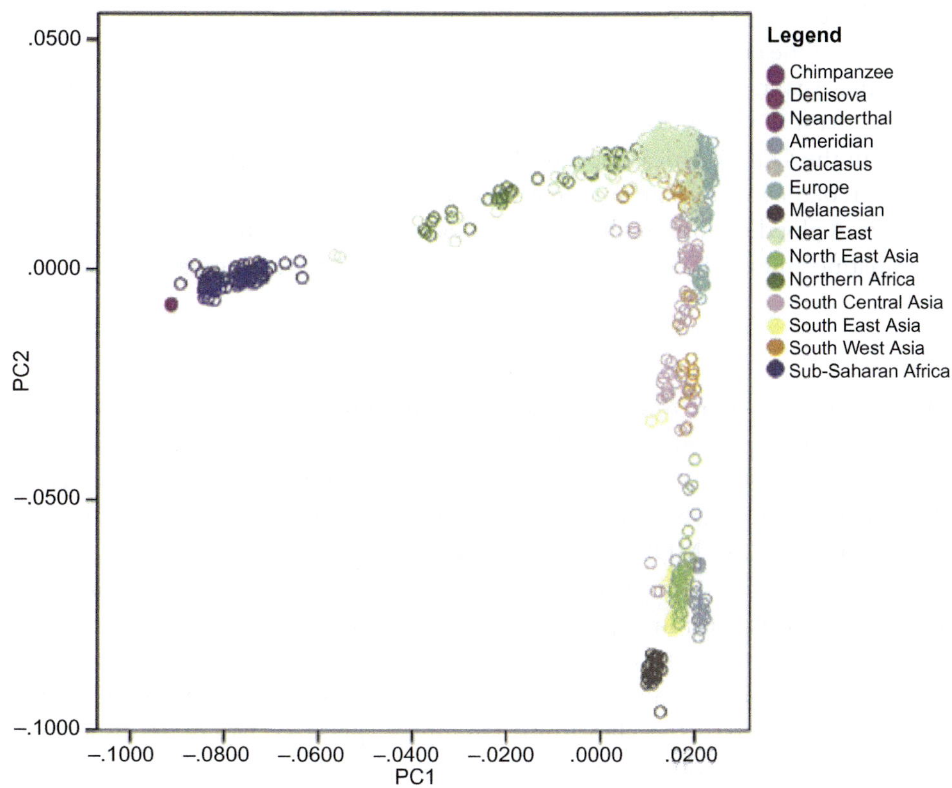

Fig. 11 PCA plot of 827 worldwide contemporary humans, Neanderthals, Denisova, and chimpanzees using the set of all 37,758 SNPs. PC1 and PC2 comprise 58.2% and 26.17% of variance, respectively (https://www.ncbi.nlm.nih.gov/pubmed/23872234).

while the PC2 is on the *Y*-axis. Notice how the chimpanzee (magenta color), representing a related species sharing a common ancestor with *sapiens*, segregates at the extreme upper left portion of the plot. The Denisova (maroon color) and the Neanderthals (dark green color) partition near the chimpanzee. Of the contemporary AMH individuals, the sub-African individuals (dark blue color) cluster closer to the Neanderthals, Denisova, and chimpanzees. The rest of the contemporary AMH populations partition in a clinal (gradual) cascade forming a geographical arch. The geographical order of populations follows a progression from sub-Saharan Africa to North Africa, Near East, South West Asia, South Central Asia, South East Asia, Native Americans, and Melanesia. The order in which the contemporary populations plot follows the migration sequence of AMH migrants out of Africa and toward Asia and Oceania. European groups reside close to and just after the Near East cluster, as expected of a separate and independent dispersal from the Near East to Europe. Of additional note is the bipartite partitioning of the South West Asian Group of populations, one cosegregating with the Near East and the other with South Central Asia. In addition, notice how some European populations are wedged within the South Central Asian and South West Asia populations. This fluidity and intermingling of geographical locations of populations in this part of the world may be indicative of gene flow at the cross roads of East-West migration waves where the continents (Africa, Europe, and Asia) meet. This metaanalysis based on SNPs illustrates the power of PCA in utilizing a massive amount of data to generate a visualization of the sequence of events involved in the Out of Africa dispersal that populated the entire planet.

In terms of the relationship among Neanderthals, Denisovans, and contemporary *sapiens*, this PCA plot demonstrates how two dimensions (components 1 and 2) are capable of discriminating among these three hominin species, one contemporary and two extinct.

MDS

MDS analysis is another commonly used methodology designed to study phylogenetic relationships among populations. MDS analyzes any kind of similarity or dissimilarity among populations as well as correlation values. MDS, like PCA, is a reduction technique in which variables are grouped together into dimensions (e.g., *X* and *Y* axes). In addition, as PCA, MDS employs proximity measures, such as the correlation coefficient or genetic distances (Euclidean distances), to generate a spatial configuration of points in multidimensional space where distances between points reflect the similarity among populations. Thus, both MDS and PCA are also useful tools is visualizing similar groups of populations exhibiting characteristics or traits in common. Unlike PCAs, in MDS analysis the algorithms attempt to preserve pair-wise distances. The reliability of a given MDS analysis in uncovering phylogenetic relationships among populations is reflected in the stress (distortion) value required to generate a given distribution of populations.

In simple terms, MDS works by partitioning the populations and finding the distribution that best approximates the observed variables and genetic distances. In fact, MDS works by moving the populations around in the space defined by the requested number of dimensions in the plot. As this is done, the program checks on how well the distances between populations are reproduced by the various configurations generated. The MDS algorithm systematically evaluates the different configurations with the goal of maximizing the goodness-of-fit expressed as stress values. The lower the stress value (e.g., Kruskal stress) required to generate a given distribution, the more reliable a partition is. In practice, a stress value of zero is not realistic, since a certain degree of stress is expected and tolerable. Stress values under 0.1 are excellent and anything over 0.15 is poor.

Considering the similarities between these two methods, investigators often ask: when is it appropriate to use PCA as opposed to MDS? A fundamental difference between the programs that run PCA and MDS is that PCA presupposes a linear relationship between variables (i.e., variants) and the groups (e.g., populations or species) being analyzed. If this relationship is nonlinear but unimodal, as is the case, for example, when studying available resources within a geographical area, MDS analysis should be employed, since it does not assume a linear relationship. MDS algorithms only attempt to optimize the fit of the dissimilarities among the variables and their Euclidean distances in the MDS plot. As a consequence, although in cases of linear relationships both PCA and MDS may generate similar partitioning of populations, in other instances, depending on the nature of the input data, the results may be very different and in nonlinear scenarios erroneous results should be expected if PCA is used.

Fig. 12 represents an MDS plot that illustrates the partitioning of Taiwanese aboriginal tribes and references worldwide populations. Notice the Kruskal's stress value of 0.152, which represents a goodness-of-fit value somewhere in between fair and poor.

Hierarchical clusters analysis (HCA)

Unlike PCA and MDS, hierarchical cluster analysis (HCA) employs a dissimilarity matrix to generate ranked stratified clusters. This stratified approach usually takes the form of trees or dendrograms. HCA take two possible approaches: agglomerative and divisive. The agglomerative route uses a bottom-up strategy in which populations are organized from the lowest level of a hierarchy to the top. The divisive approach, on the other hand, employs a top-down method in which all populations start in one cluster at the top and then splits successively as one moves down the hierarchy.

In HCA dendrograms, the populations are grouped according to similarities. In other words, HCA assembles together taxa that are more similar to each other than to those that are not. Populations that exhibit mutations that make them similar are separated by small distances in the tree partitioning within the same cluster or branch. Conversely, samples that possess very different traits are very distant from each other in far-off groupings or

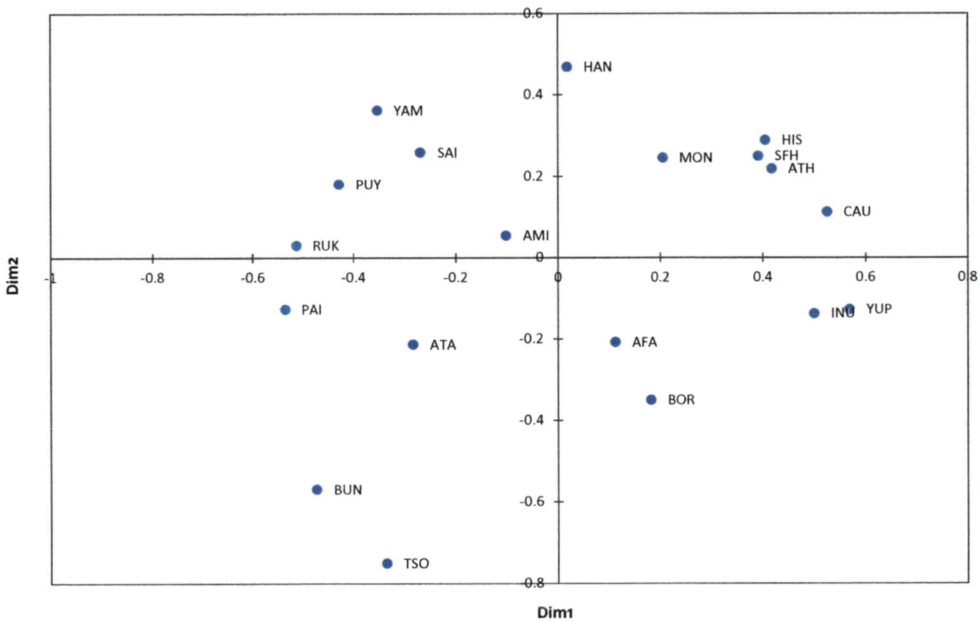

Fig. 12 MDS plot of Taiwanese aboriginal tribes and reference worldwide populations.

assemblages (Fig. 13). HCA converts degree or extent of differences among artifacts to linear distance in a dendrogram. The program is capable of calculating distances (likeness) between all pairs of samples or artifacts using one of a number of measures.

The dendrogram in Fig. 13 represents the partition of eight worldwide extant human populations using STR variants. Notice how the bifurcations in cascading hierarchical format illustrate the different degrees of similarities among the sample populations. Populations that split from a single stem are considered sister groups and reflect close affinities. In addition, notice the degree of statistical support provided by the bootstrap values at the base of each of the bifurcations. Bootstrap values provide a measure of statistical support for each split in the tree. Bootstrapping is based on repeated sampling and resampling of a large body of data, asking the question: "How congruent are the results from each sampling?" This sampling and resampling methodology is performed by randomly obtaining a small portion of the original data set to conduct the analysis, returning the analyzed subset of samples back to the entire data set, obtaining another sample at random, and repeating the sequence of replacement and sampling hundreds or thousands of times. In statistical terms, this is known as random sampling with replacement. This is how it works. The methodology, in essence, tests if the same bifurcation between groups is observed in repeated assessments. If the same bifurcation is seen repeatedly in the results from each independent run, it is interpreted as a sign of robustness and statistical

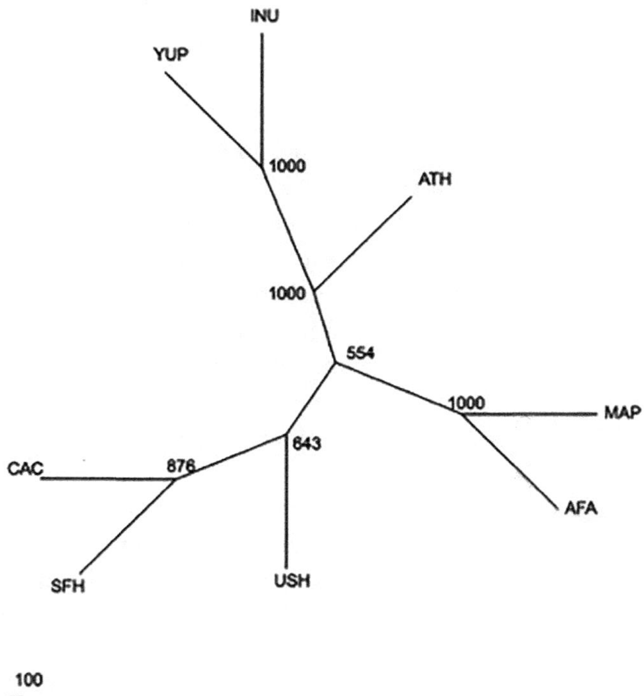

Fig. 13 Phylogenetic tree with bootstrap values. The three-letter acronyms indicate different worldwide populations.

significance. As in Fig. 13, if 1000 bootstrap replications are performed and in all of them a certain split is obtained, then the bootstrap value is 100%. Furthermore, if 1000 replacement and sampling reiterations are performed and bootstrap values of 1000, 554, 643, and 876 (100%, 55.4%, 64.3%, and 87.6%, respectively) are obtained for specific bifurcations, as seen in the tree in Fig. 13, is this statistically significant? Although investigators examine all the bootstrap values in the tree, they also focus on specific bifurcations to determine the statistical significance of a given split. If the same node is recovered in 95% to 100% of the iterations, the bifurcation is highly supported. Bootstraps of about 70% are usually considered as moderate support.

In HCA, a number of different genetic distance measures are regularly employed to construct phylogenetic trees. Neighbor joining (NJ) is one such procedure (Table 1). NJ is a distance-based method that uses a bottom-up, agglomerative clustering approach to generate dendrograms. In this method, genetic distance matrices may be derived from discrete morphological traits or genetic characters, such as DNA or protein sequences. Another frequently employed program is the character-based maximum likelihood (ML) (Table 1) method, which attempts to find a tree in which the maximum number of characters is compatible. Since there can be many equally possible dendrograms, the majority consensus tree of the set of optimal trees is returned. ML incorporates the positions and lengths of branches into the tree that has the highest likelihood of representing the correct phylogenetic relationship among the populations.

Phylogenetic methods that use distance measures to create trees utilize a number of algorithms to calculate genetic distances. This requires the data to be imputed in the form of a distance matrix generated using one of a number of statistical measures or programs. Statistical measures, such as Nei's genetic distance, Cavalli-Sforza, and Edwards, as well as Reynolds, Weir, and Cockerham's genetic distances, are commonly employed (Table 2). The various statistical measures that have been proposed over the years are based on different evolutionary models that incorporate different assumptions. For instance, Cavalli-Sforza genetic distances assume genetic drift only as the evolutionary driving force while Nei's genetic distances are driven by mutation as well as genetic drift (Table 2).

One criticism often heard about this type of analysis is that populations are forced to connect with each other with lines to make the branches of the trees. This shortcoming has motivated a number of investigators to report their data in the form of two or three-dimensional plots (e.g., PCA or MDS), which do not assume explicit linear connections between populations.

Table 2 Some commonly employed genetic distances in population genetics studies.

Name	Parameters	Program
Cavalli-Sforza	Genetic drift only	PHYLIP/ GENDIST
Nei's standard	Driven by mutation and genetic drift	PHYLIP/ GENDIST
Edwards' distance	Angular distance	POPPR
Reynolds, Weir, and Cockerham's	Coancestrality coefficient/genetic drift only	PHYLIP/ GENDIST
Rogers'	Classical Euclidean/fixation index/no mutation	TFPGA
Provesti's	Absolute genetics distance	POPPR
Nei's minimum genetic distance	Driven by mutation and genetic drift	TFPGA
Nei's D_A distance	Driven by mutation and genetic drift/ microsatellite DNA data	DISPAN
Goldstein's	Stepwise mutation for microsatellites	
Bruvo's	Stepwise mutation for microsatellites	POPPR

Modified from R.J. Herrera, R. Garcia-Bertrand, F.M. Salzano (2016) Genomes, Evolution, and Culture: Past, Present, and Future of Humankind, first ed. John Wiley & Sons, Inc., Hoboken, NJ.

Network analysis

Networks are constructed from pair-wise genetic-similarity matrices of all sampled individuals. At times they only involve members of a single population, but often they include samples from different geographical regions. Network analysis is a distance-based approach for assessing subpopulation structure among populations. A subpopulation is defined as a distinct group of individuals within a population who are more genetically related to each other than to individuals outside the subpopulation. Subpopulations are created by a number of evolutionary forces, such as geographic isolation, genetic drift, migration, mutation, and selection. Subpopulation structure is apparent when individuals in a certain portion or lineage of the network partition away from the rest of the graph into distinct subgraphs made up of genetically closely related individuals. Partitioning is interpreted as evidence of subpopulation structure, since the individuals in the subgraphs or branches within the network are more genetically related to each other than to samples in the other lineages of the network. The locations, connections, and proximities of these genetically related clusters or subpopulations within networks might reveal gene flow among subpopulations as well as migration patterns. In the reticulation of a network (Fig. 14), nodes that are more

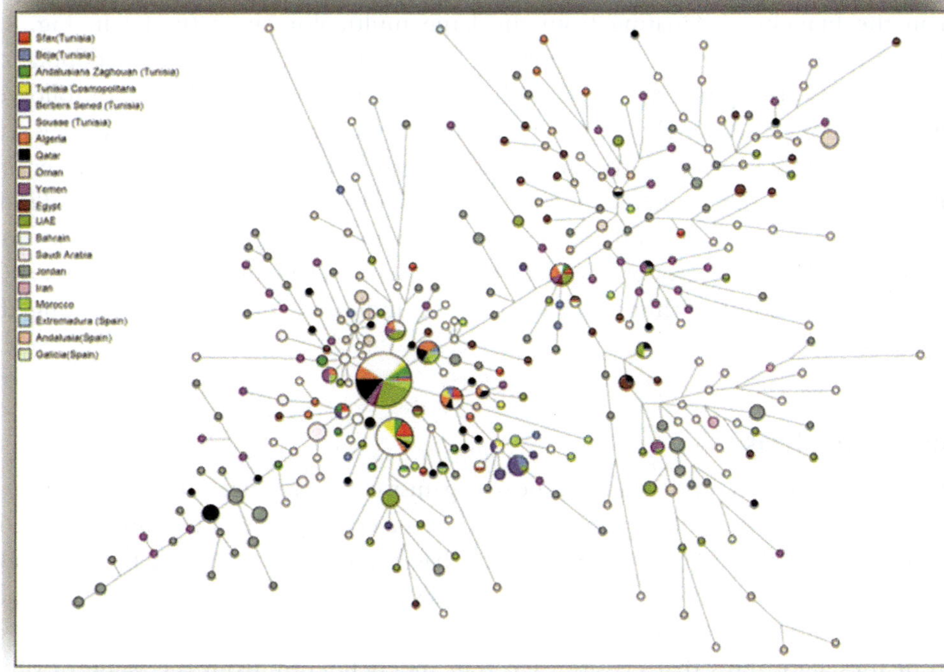

Fig. 14 Network analysis of individuals under haplogroup M267 (https://www.wiley.com/en-us/Genomes%2C+Evolution%2C+and+Culture%3A+Past%2C+Present%2C+and+Future+of+Humankind-p-9781118876404).

densely and closely connected to each other than to outside nodes are referred to as a community. Often, networks take the form of spider-like reticulations.

In practice, networks are well suited for uniparental marker systems, such as Y chromosomal and mtDNA haplotypes.[60] In Fig. 14, representing the phylogenetic partitioning of Y-chromosome haplotypes, the samples are represented as circles or part of circles that are color-coded according to the population they belong to (each population takes a different color or shade), and the size of the circles reflect the number of persons exhibiting a given haplotype. Since the samples are color-coded according to populations, the distribution of samples within the network also provides information on the phylogenetic relationship among the populations. The smallest circles indicate singletons or unique haplotypes represented by one person. Circles with two or more colors indicate sharing of haplotypes among individuals from different populations. The program generates lines connecting haplotypes creating a sort of network, the characteristic from which the name derives. These lines connect individuals that are most closely related and the lengths of the connecting lines are directly proportional to the number of mutational steps. Therefore, the degree of genetic differences between individuals can be ascertained by counting the number of mutational steps and the distance separating the samples.

Networks are particularly informative when they exhibit "spider-type" topology as seen in the branches emanating from the large multicolor circle (node) in Fig. 14. A "spider-type" arrangement may signal the geographical location and/or population where a mutation originated as well as the site of a regional founder effect followed by diversification. The branches represent lineages of sequentially related individuals separated by mutation events with the samples at the terminals of each offshoot representing the most genetically differentiated individuals. Networks that exhibit segregation of individuals of a specific population into particular branches suggest minimal gene flow involving populations or groups in other branches while a lack of compartmentalization of colors (populations) within branches may be indicative of gene flow among the populations.

Structure analysis

In population genetics, understanding how organisms are organized into populations is paramount. The concept of population structure is a cornerstone of evolution. Since its conception in 2000, the program Structure[59] has become a standard method to make different modeling assumptions. Structure analysis is a model-based clustering approach that assumes a model with a certain number of source populations. It is a Bayesian program (it uses algorithms that classify each trait independently from the value of other traits) that assigns individuals to source populations and updates or modifies the probability for a hypothesis as more evidence or data becomes available. It employs multilocus genotype information in the form of SNPs, STRs, and RFLPs (restriction fragment length polymorphisms) to assess population structure. For example, it assesses if some individual members of a population represent a discrete or unique group (i.e., subpopulation), or alternatively, if they are a genetically homogeneous group (no subpopulation structure exists). Thus, the results allow investigators to ascertain the presence of distinct subgroups or subpopulations within a larger group of organisms. It performs this function by detecting allele frequency differences within the data and assigning individuals to specific subpopulations. Structure also allocates individuals to populations, identifies hybrid zones (i.e., geographical points of contacts between or among populations), distinguishes migrants and admixed individuals, and approximates population allele frequencies in situations where many individuals are migrants or admixed. Particularly useful is its capacity to infer and characterize source populations if they are unknown. This feature is capable of alerting investigators to the existence of putative populations in the absence of physical evidence, such as fossils or artifacts.

By examining the genetic similarities and differences among populations and individuals (Fig. 15), Structure is particularly useful as a first-step test to examine population structure prior to further, more sophisticated genetic analyses. At a glance, Fig. 15 provides information on the genetic contribution to populations from source or ancestral groups, and addresses the common ancestry of individuals and populations. This is done

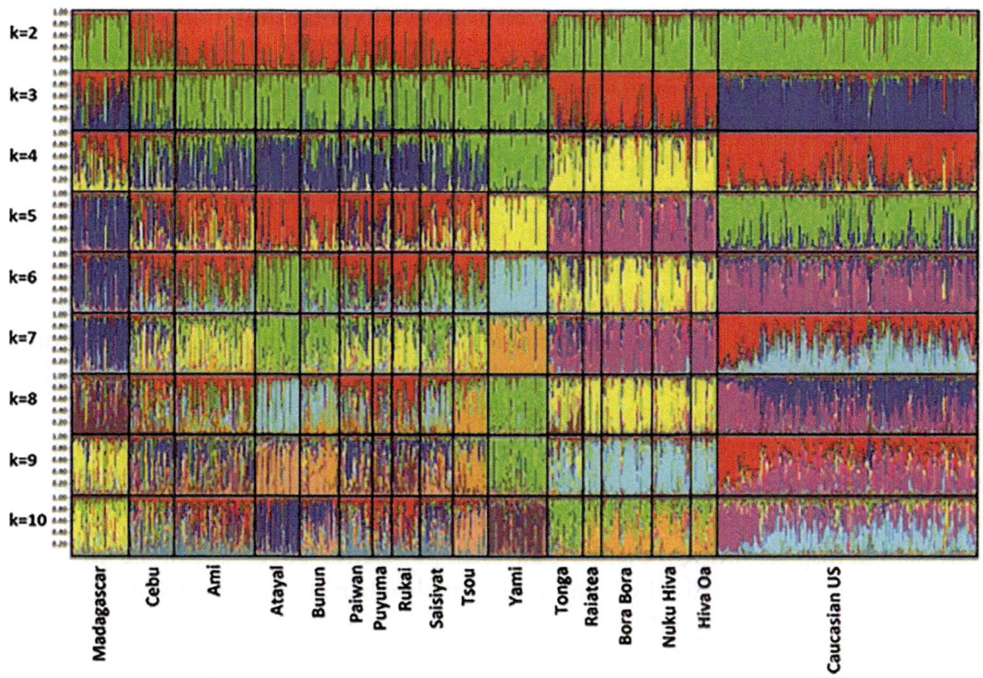

Fig. 15 Structure analysis of Pacific Ocean island populations. Madagascar and US Caucasian are reference groups (https://www.nature.com/articles/s41431-019-0336-2).

utilizing reference populations for comparison. The samples are represented in a bar graph. Each person appears as a bar with colors representing his or her various genetic components and their relative proportions. In this type of analysis, a greater number of genetic markers provides a more comprehensive and thorough analysis, since a greater proportion of the genome contributes to the estimations.

In the graph illustrated in Fig. 15, the individuals are grouped into 17 predetermined populations designated by their names on the *X*-axis. All of the populations in this graph, except US Caucasian (at the far right), were impacted by the Austronesian Expansion thought to have originated in Taiwan about 6000 years ago. Austronesians migrated and populated islands, such as Madagascar in the Indian Ocean, and the entire Pacific Ocean, including Polynesia. This geographical expanse represents two-thirds of the circumference of the world. In this analysis, the US Caucasian group is used as a reference population for comparing the Austronesian populations. The *K* values at the left on the *Y*-axis represent the number of clusters or groups assigned by the investigator. In other words, Structure allows the investigator to designate the number of groups among which the populations are going to be partitioned during the run.

While conducting a Structure analysis, different numbers of clusters are tried in independent runs to ascertain the maximum genetic definition among populations afforded by each *K* value. At $K=2$, for example, the colors indicate the proportions of loci from each person in a population that are derived from two hypothetical source populations. In the case of the graph in Fig. 15, at $K=2$ two types of general profiles based on the amounts of red and green (the two clusters) are displayed by each population. One type of profile is represented by populations from Madagascar, Polynesia (Tonga, Raiatea, Bora Bora, Hiva Oa and Nuku Hiva) and Caucasians from US in which the color green component predominates. The other type of profile, in which the red component prevails, characterizes the populations from Cebu (Philippines) and Taiwan. Although at $K=2$, Structure can discriminate between some populations, it fails to differentiate the populations of Madagascar, Polynesia, and US Caucasian as well as among the Taiwanese populations. Results like this would justify employing higher *K* values to discern better among the populations. Higher resolution and discrimination among populations is in fact seen when the Structure analysis is performed at $K=3$ through $K=10$. At $K=3$, for example, the number of different types of profiles increases, allowing discrimination between Polynesian populations, US Caucasians, and Madagascar. In general, as the *K* values are increased, greater differentiation among population is observed. Fig. 16 illustrates a high magnification of the US Caucasian population at $K=8$. Notice the capacity of Structure to discern individuals within the graph as bars. In this expansion, it is evident that heterogeneity exists within the US Caucasian population with certain individuals exhibiting different color components (e.g., yellow, green, light blue) not observed in other persons. In addition, a portion of the population (people at the far left) possesses a greater abundance of the purple component. This dichotomy within the population

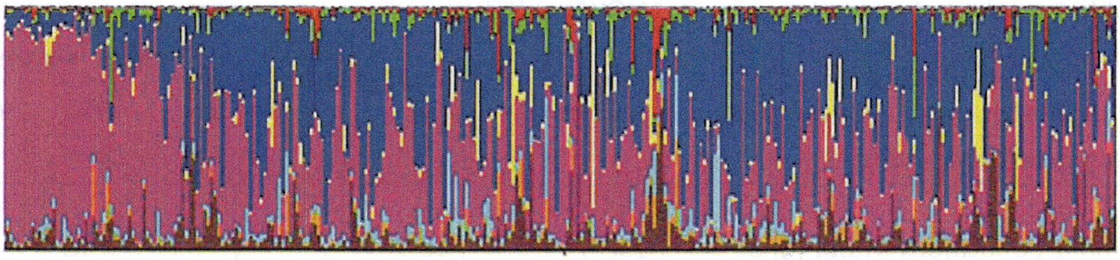

Fig. 16 Detail of Fig. 15 showing an expansion of the US Caucasian population at $K=8$ (https://www.nature.com/articles/s41431-019-0336-2).

may represent some degree of subpopulation structure, which is expected of a genetically heterogeneous group, such as the US Caucasians.

AMOVA (analysis of molecular variance)

AMOVA is a method designed to detect population differentiation, usually based on molecular markers.[57] Executing the program requires two components: a distance matrix derived from the genetic frequency data (e.g., SNPs or STRs), and a separate table with the data partitioned into categories. The program addresses the question of whether genetic differentiation exists among groups and whether it parallels with other characteristics, such as geography or linguistics.

AMOVA calculates variance at three levels: within populations, among populations within groups, and among groups of populations. Investigators using AMOVA are interested in the distribution of the total variation among these three levels. In populations that are engaged in random mating, most of the variance is expected to reside within populations, followed by among groups of populations. The least amount of diversity is usually detected among populations within groups. Deviations from this pattern of variance distribution may be indicative of subpopulation structure (not a random and homogeneous interbreeding population). Using Y-chromosome STR data, Table 3 illustrates how much variance is seen in the three classifications mentioned above. To the right of the table, degrees of freedom are provided as well as the percentage of the total variance in the three categories. The last three columns in Table 3 provide the ϕ statistic values (population differentiation statistics), which should correspond to the percentage of variation among the three levels.

AMOVA can be used to ascertain the statistical significance of geographical (e.g., Europe, Asia, etc.) or linguistic (e.g., Tibetan, Indo-European, etc.) differences as a function of genetic diversity. For example, consider a scenario in which there is an interest in finding out if the geographical locations of certain populations in relation to each other are congruent with the observed genetic frequency distributions. AMOVA is designed to test if the genetic differences among populations correlate with the geographical distances separating them and if the parallelism is statistically significant. This same test can also be employed for any other type of data in which the strength of the correlations needs to be

Table 3 Analysis of molecular variance (AMOVA) of YSTRs.

Classification	DF[a]	Sum of squares	Total variation (%)	Fixation indices		
				*Φ*ST	*Φ*SC	*Φ*CT
Language (four groups)[b]						
Among groups	3	136.1	8.0			0.08
Among populations within groups	5	138.4	16.9		0.18	
Within populations	264	956.6	75.1	0.25		
Geography (seven groups)[c]						
Among groups	6	1948.0	2.9			0.03
Among populations within groups	44	2584.4	7.4		0.08	
Within populations	9864	52,660.5	89.7	0.1		

[a] *DF*, degrees of freedom.
[b] Linguistic partitioning (four groups): 1—Austronesian-Formosan-Atayalic: Atayal; 2—Austronesian-Formosan-Paiwanic: Ami, Bunun, Paiwan, Puyuma, Rukai, Saisiyat; 3—Austronesian-Formosan-Tsouic: Tsou; 4—Austronesian-Malayo/Polynesian-Batanic: Yami.
[c] Geographical partitioning (seven groups): 1—Taiwan aborigines (AMI, ATH, ATW, BUN, PAH, PAW, PUY, RUK, SAI, TSO, YAH, YAL); 2—Northeast Asians (HAN, HSD, HSX, HZJ, JAP, KOR, MON); 3—Southwestern Chinese (GIN, HYN, SHE, YAO, YIG, ZHU); 4—Southeast Asians (IND, MAL, PHI, PHL, THA, VIE); 5—Solomon/Papuan Islanders (BEL, CHO, GEL, GUA, ISA, KOL, MAK, MLT, ONT, RAN, RUS, SAN, SAV, SHO, SIM, REN, TIK, VEL); 6—Australian (AUS); 7—Society Islands, Eastern Polynesian (SIP). All *P*-values are $<.00001$.

assessed. Although often the expectation is that populations that belong to the same geographic or linguistic group would be more genetically similar than populations that are not, it is often important to establish the statistical significance of any relationship.

Linear versus haplotype-based approaches

Most of the methods designed to assess phylogenetic relationships among organisms relay on algorithms that utilize allelic frequencies from individual loci independently. This is referred to by statisticians as a linear type of approach to phylogenetic analysis. With these programs, investigators genotype numerous uniparental and biparental markers to derive allele frequencies. The frequencies are then employed to generate, for example, genetic distances among populations, which are then used to infer phylogenetic relationships.

Haplotype-based methods, on the other hand, involve comparing segments of DNA that are known to share long stretches of sequences. These tracks of conserved pieces of DNA vary in size. The longer the region of DNA that populations share in common, the greater the genetic affinity between the populations. This type of approach has been used to identify the demographic history of populations including bottlenecks, admixture, population histories, common ancestries, dispersals, and interbreeding among hominins including Neanderthals and Denisovans.[61–67]

Identity-by-descent (IBD) tracks

One type of haplotype-based analysis approach involves comparisons of identity-by-descent (IBD) blocks of DNA. Examination of IBD tracks represents a powerful methodology to study hominin genomes, and these methods are increasingly becoming indispensable tools to mine mega quantities of DNA in a large number of populations. Comparisons of DNA blocks exhibiting sequence homologies representing unrecombined segments of chromosomes are finding their way into phylogenetic procedures to study population structure and phylogenetic relationships. IBD methodologies complement the more traditional PCA and HCA algorithms discussed above that seek linear patterns and trends that encompass the entire genome. Furthermore, the haplotype-based approach provided by IBD track mining is capable of capturing more subtle subpopulation structure signals compared to PCA and HCA methods.

To illustrate the biological significance of IBD tracks and how they are used in population studies, a brief explanation on how they form follows. As germ cells undergo meiosis to produce gametes (sperms and ova), as part of sexual reproduction, there is a propensity for homologous chromosomes to pair and engage in a phenomenon known as crossing-over or recombination. This process involves the pairing of homologous chromosomes, one maternally derived and the other paternally derived. This pairing is very precise and occurs at the nucleotide level. This highly efficient coupling of chromosomes is mediated by an intranuclear complex known as the synaptonemal complex. This intimate alignment of chromosomes sets the stage for exchange of genetic material in which pieces of DNA are shared between members of pairs of homologous chromosomes. This reciprocal swap of DNA is generally exact to the nucleotide level. Inexact pairing of nucleotides and exchanges between chromosomes leads to mutations (e.g., frame-shift mutation that usually obliterates the correct sequence of amino acids in proteins). The origin of recombination is a unresolved mystery in biology. Yet it has been hypothesized that meiotic recombination arose from the sexual process of DNA transfer in bacteria and DNA repair. Evolutionarily speaking, recombination serves the dual functions of bringing beneficial alleles together on the same chromosomes and increasing the genetic diversity of organisms by shuffling DNA among chromosomes. Fig. 17 illustrates the consequences of recombination at the chromosome level as homologous chromosomes pair, exchange pieces of DNA, and create novel reshuffled chromosomes with tracks or blocks of DNA derived from the original maternal or paternal chromosomes. In Fig. 17, these regions of DNA, which move and recombine to make the chromosomes of the next generation, are painted in different colors to allow one to follow their trajectories. These chunks of DNA are traced by investigators, generation after generation, as IBD tracks to assess ancestry.

Independent of its origins and original functions, in population genetics studies, recombination is utilized as a tool to measure distances between genes and determine

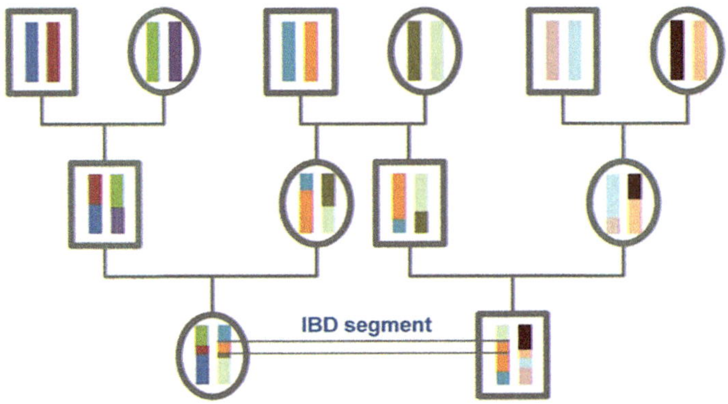

Fig. 17 Creation of identical by descent tracks (IBDs) (https://en.wikipedia.org/wiki/Identity_by_descent).

the time from a common ancestor. These applications are possible because the frequency of recombination between two genes is directly proportional to the distance (in nucleotides) that separates them on the DNA molecule. This makes sense, since recombination occurs in between nucleotides and the greater the number of nucleotides separating genes, the higher the potential for breakage and reunion of DNA fragments.

From the original gene-mapping studies in *Drosophila melanogaster* by Thomas Hunt Morgan (1866–1945) at the beginning of the 20th century, it is known that the further apart two genes are, the greater the recombination frequency between them.[68] Fig. 18 represents a drawing by Morgan and collaborators depicting crossing-over. It shows how they envisioned the mechanism of recombination in 1916. Since Morgan, geneticists have constructed gene maps in different organisms based on recombination frequencies. Although Morgan did not explicitly state this, another implication of his discoveries is that since recombination occurs as a function of time, the greater the amount of time elapsed, the greater the likelihood that DNA recombination occurs between genes. Thus, the number of recombination events between two genes is also directly proportional to time. These principles allow investigators to relate size of undisturbed DNA (no recombination between two genes) to time. In other words, the presence of large unrecombined segments of DNA within chromosomes indicate close phylogenetic relationships between individuals (e.g., cousins) and between populations (e.g., Navajo and Apache Native American populations—two closely related tribes).

In practice, investigators assess the sizes of unrecombined DNA fragments by sequencing or genotyping markers such as SNPs, STRs, indels, etc. from the genome of organisms. Markers (alleles) in common delineate the limits where unrecombined DNA persists. Thus, if through SNP genotyping it is ascertained that the average length of undisturbed DNA fragments shared by population "A" and "B" is 3 cM (centimorgans or about 3 million base pairs) and between population "A" and "C" is 1 cM, the data supports a

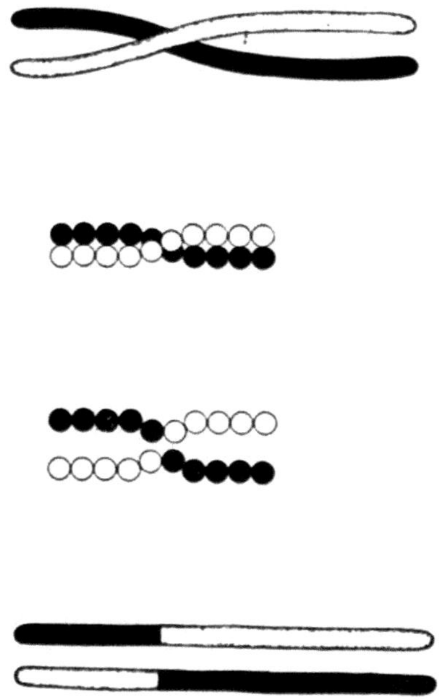

Fig. 18 Crossing over and recombination according to Morgan et al. (https://upload.wikimedia.org/wikipedia/commons/0/0e/Morgan_crossover_1.jpg).

greater phylogenetic affinity between the "A" and "B" populations than between "A" and "C". In other words, the larger the size of the undisturbed DNA segments, the closer the relationship between individuals or populations. Studies have employed this type of approach to shed light on the phylogenetic relationships of contemporary human[69] and extinct hominin populations such as Neanderthals and Denisovans.[61]

The length of unrecombined DNA segments used in this type of assessment is routinely expressed in terms of IBD track sizes. To illustrate how this type of analysis works, we are going to examine the data obtained by Botiguéa and colleagues,[69] which reports on the genetic impact that North African populations had on European populations as seen in the IBD track lengths of North African origin. Fig. 19 illustrates the population-specific IBD track size sharing (in centimorgans) between North Africa and Europe populations. The IBD track sizes (*Y*-axis) in the figure are scaled by 100 for better visualization. The European populations are represented on the *X*-axis and each North African population and the Qatari are represented by colors and symbols in the map on the upper right corner of the figure. Notice the larger IBD tracks shared between North African populations and South West European populations as compared to other European populations. This type of data suggests greater gene flow and genetic affinity between North Africa and South West Europe than between North Africa and the remainder of Europe.

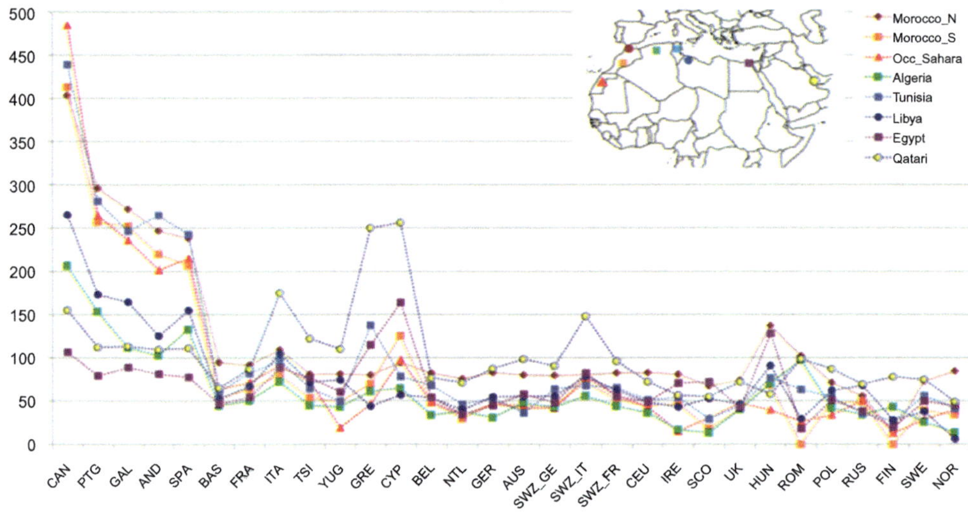

Fig. 19 IBD tracks sharing (in centimorgans, cM) between North Africa and Europe (https://www.pnas.org/content/110/29/11791).

IBD track sharing has also been employed to examine phylogenetic relationships among ancient hominin populations such as Neanderthal and Denisova and contemporary humans. Using DNA sequences from the 1000 Genomes Project,[70] 160,000 different short IBD tracks with a median length of 23 kb (1000 base pairs) were identified.[61] These IBD tracks were detected by searching for rare nucleotide variants, which provide more high-resolution information on IBD than common variants. Since these short IBD tracks are thought to represent old fragments of DNA that have persisted and remained intact since ancient times, they were used to investigate the remote evolutionary history of hominins. The results of these experiments revealed that the IBD tracks mined from the 1000 Genomes Project that match the Denisovan or the Neanderthal genomes are found significantly more frequently in Asian and European populations than in African populations, yet some are also shared with Africans. In addition, while some Denisova-matching IBD segments are exclusively shared with Asian populations, many are shared exclusively with African individuals. Often the IBD that match Neanderthals also match Denisovans, indicating that they predate the origin of the two. Furthermore, some IBD tracks present in Neanderthals are exclusively shared with African populations. These results suggest that some of these short IBD segments predate the Out of Africa human dispersal, to the time of the ancestors of humans, Neanderthals, and Denisovans. Illustrative of the type of data obtained when archaic species are examined is the relationship between IBD tracks from the 1000 Genomes Project that match the Denisova genome and are private (only seen) in African populations compared to IBD tracks not detected in

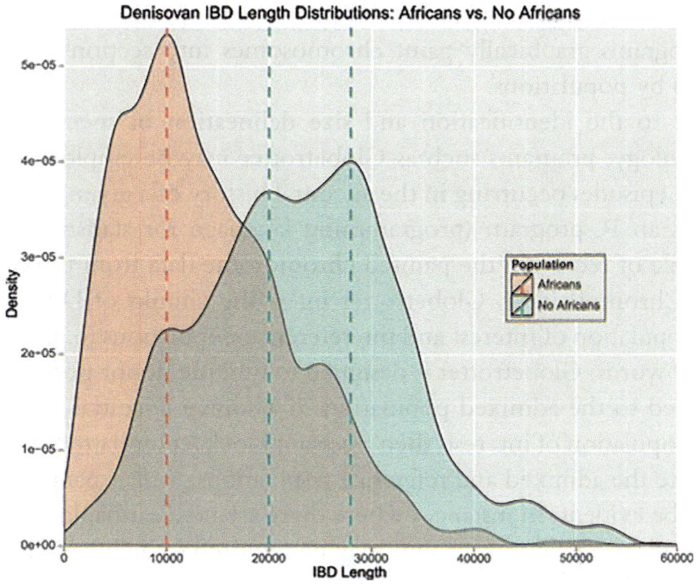

Fig. 20 Densities of lengths of IBD segments that match the Denisova genome and are private to African populations versus IBD segments that are absent in African populations (https://www.ncbi.nlm.nih.gov/pmc/articles/PMC3905877/).

African populations (Fig. 20). Notice that IBD tracks that are not shared with African populations are larger than IBD tracks that are private to African populations, indicating that the later represent older IBD tracks.

The globetrotter algorithms to detect admixture

As described above, genome-wide IBD tracks, as a whole, are routinely employed to establish phylogenetic affinities among populations. This application does not require selection and pairing of *specific* homologous tracks among individuals and populations to perform the comparisons. This approach just compares the total number and sizes of IBD blocks in populations. In addition to assessing these genome-wide phylogenetic relationships among populations, IBD tracks are utilized to ascertain admixture events by examining *specific* segments of DNA for homology.[62] For this purpose, investigators employ a number of algorithms such as Germline or ChromoPainter (Table 1) to mine genomes and identify IBD shared segments among populations.[71] Briefly, using sequencing data or SNPs as well as other polymorphic markers, these programs seek out short, exact pair-wise matches between individuals from different populations and then search on both sides of these small segments for additional sequence homologies in an effort to extend the area of comparison and include longer tracks that are indicative of IBDs. The outputs of these programs are contiguous chunks of DNA or haplotypes that are visualized using different-color bars of various sizes proportional to the length of the IBDs. The programs graphically paint chromosomes into sections that are indicative of IBDs shared by populations.

Subsequent to the identification and size delineation of specific tracks of DNA sequence homology, programs such as Globetrotter may be employed to identify and date admixture episodes occurring in the ancestral history of a given target population.[58] Globetrotter is an R program (programming language for statistical computing and graphics) capable of receiving the painted chromosome data from multiple populations generated by ChromoPainter. Globetrotter infers the chunks of DNA that are shared between the population of interest and the reference populations (e.g., ancestral populations). In other words, Globetrotter is designed to find the donor population(s) that may have contributed to the admixed population. If a source ancestral group is genetically similar to the population of interest, then a certain block-color pattern in common likely would dominate the admixed and reference populations, a clear parallelism in the color pattern would be evident. In instances where there are no clear haplotype matches to the target population, several donor groups may be required to delineate haplotype-sharing patterns. These more complex haplotype patterns are more difficult to discern since they may result from ancient admixture events. In cases of admixture, a linear array of colored chunks indicates the source of the haplotypes in the reference populations. A mixture of different-source haplotypes indicates several donor populations. The graph can be cleaned and simplified by reducing the number of noncontributing reference populations that do not contribute haplotypes (IBDs) to the population of interest. An examination of the size of the haplotypes in common between the population of interest and donor groups may allow an estimation of the admixture event date.

Data visualization tools

Heat maps

The phylogenetics analysis tools described above attempt to describe how organisms have evolved. The output data generated from these programs provide visual representations in the form of trees, networks, and tables. Yet this type of representation is, at times, difficult to interpret and understand conceptually. With this in mind, a number of visualization tools have been developed; among them are the heat maps. Heat maps are color-based intensity representations of ranges of values that allow for the visual comparison of multiple characteristics such as large numbers of genetic markers simultaneously. The differences in color and intensity also work to draw the human eye to the part of the graph that deserves more attention.

Fig. 21 represents a typical heat map in which the genotypes of 450 organisms are summarized. The *X*- and *Y*-axes provide the same phylogenetic tree, showing the partitioning of the 450 individuals based on 22,377 SNPs. The tree was generated using a matrix of Roger's genetic distances (Table 2). Degree of relatedness based on the SNP

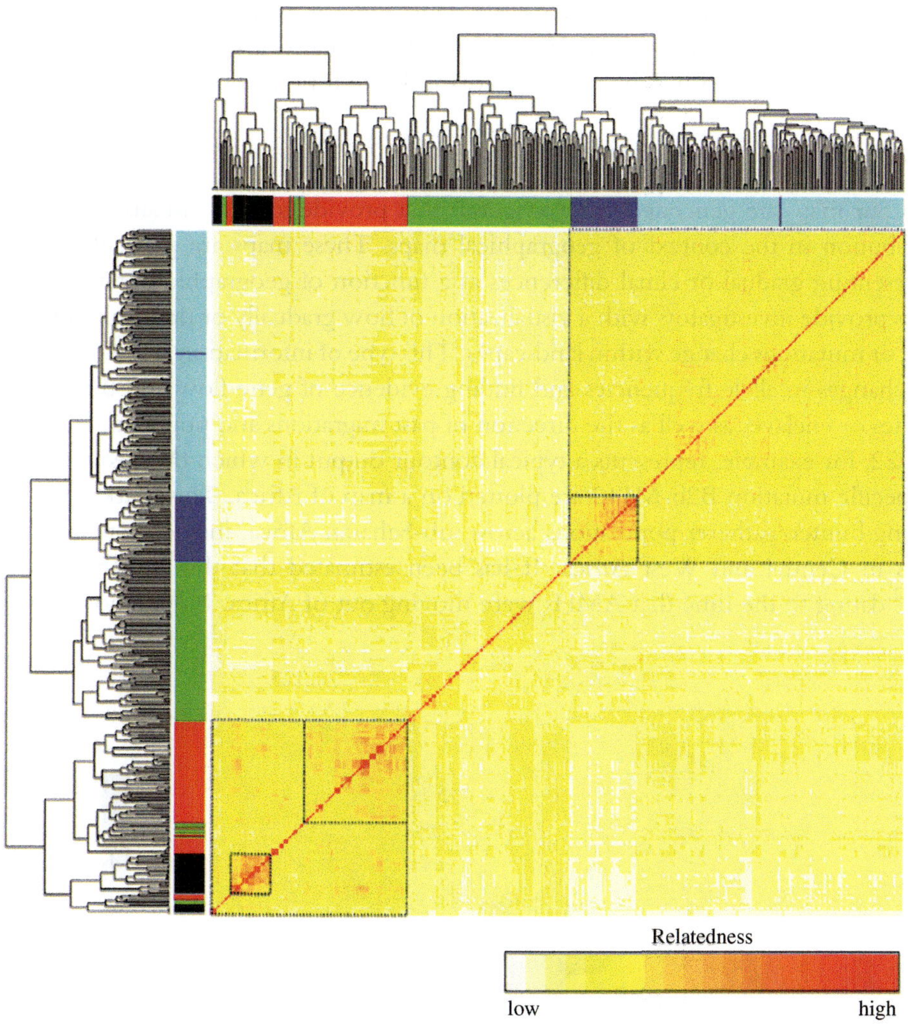

Fig. 21 Genetic relatedness heat map (https://dl.sciencesocieties.org/publications/tpg/articles/8/2/plantgenome2015.03.0013).

genotypes is summarized in the heat map in the form of colors from light yellow (no relatedness) to red (strong relatedness). The diagonal line within the graph provides the color signal of each genotype compared to itself. Colors next to trees indicate the *k*-means clustering assignment. This methodology partitions genotypes into clusters with each genotype assigned to the cluster with the nearest mean. This relatedness heat map reveals several distinct areas exhibiting intense red color indicating higher levels of relatedness possibly reflecting population substructure within the group of 450 individuals. To illustrate how the information in the heat map is interpreted, notice the color and strength of the signals as a genotype from the *X*-axis is cross-matched with a given genotype on the *Y*-axis. The signals differ depending on which genotype on the *X*-axis and on the *Y*-axis are compared. Robust genetic similarities are seen as dark red colors while comparisons with light yellow colors indicate little or no relatedness.

Contour maps

Contour maps are generated with programs that provide for the visualization of genetic distribution in the context of geographical maps. These maps are particularly useful in representing gradual or clinal differences as a function of geographic distances. Contour maps provide investigators with a visualization of how gradually or drastically specific variants or mutations change within landscapes. This type of information is useful in interpreting changes in allele frequencies and finding evidence of gene flow and the presence of isolates or enclaves as well as the direction of past migrations and population interactions. Fig. 22, for example, represents a typical contour output in which the frequencies of the Y-specific mutation B2b-M112 are plotted on a map of Africa. This marker is abundant among hunter/gatherer populations across sub-Saharan Africa, including pygmy populations in Central and West Africa. It has been estimated that this mutation is >55 ky old[72] dating to the time that AMHs were moving out of Africa. Notice how this variant

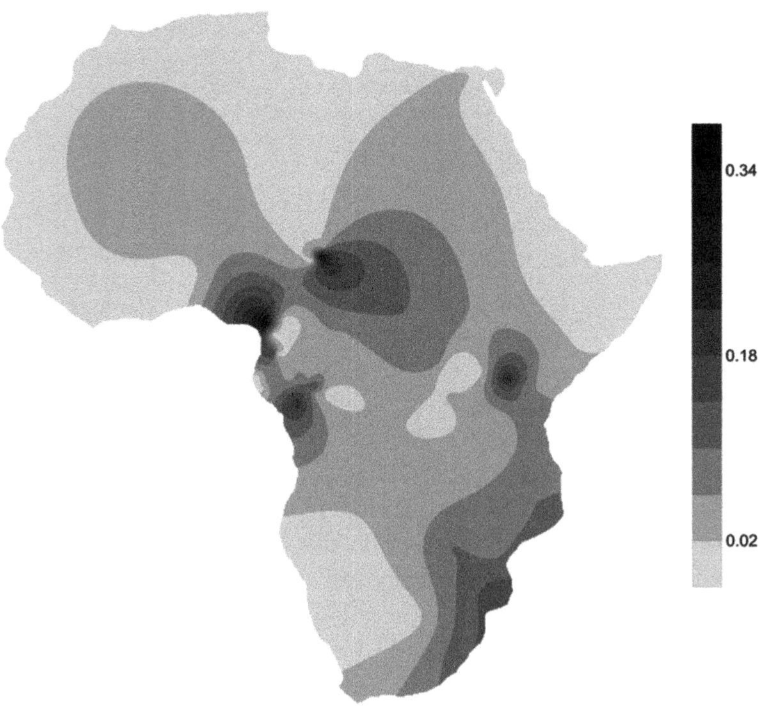

Fig. 22 Frequency contour map of Y haplogroup B2b in Africa. Scale at the right of map represents percentages of DNA variants (https://www.elsevier.com/books/ancestral-dna-human-origins-and-migrations/herrera/978-0-12-804124-6).

exhibits a number of high-frequency foci in sub-Saharan Africa. This contour map also clearly shows demic gradients of decreasing frequencies emanating in various directions from each foci. These clinal waves likely represent signals of past dispersals.

References

1. Darwin C. *The Descent of Man, and Selection in Relation to Sex*. London, UK: John Murray Press; 1871.
2. Mendel JG. *Versuche über pflanzenhybriden, verhandlungen des naturforschenden Vereines in Brünn, Bd. IV für das Jahr. Abhandlungen*; 1866:3–47. For the English translation, see: Druery CT. Bateson, W. 1901. Experiments in plant hybridization. Journal of the Royal Horticultural Society. 26: 1–32.
3. Holterhoff K. The history and reception of Charles Darwin's hypothesis of pangenesis. *J Hist Biol*. 2014;47:661–695.
4. Fisher RA. Has Mendel's work been rediscovered? *Ann Sci*. 1936;1:115–137.
5. Bowler PJ. *Evolution: The History of an Idea*. Berkeley, USA: University of California Press; 2003.
6. Correns C. G. Mendel's Regel iiber das Verhalten der Nachkommenschaft der Rassenbastarde. *Ber Deut Bot Ges*. 1900;18:158–168.
7. Sutton WS. On the morphology of the chromosome group in Brachystola magna. *Biol Bull*. 1902;4:24–39.
8. Sutton WS. The chromosomes in heredity. *Biol Bull*. 1903;4:231–251.
9. Baltzer F. Theodor Boveri. *Science*. 1964;144:809–815.
10. Griffith F. The significance of pneumococcal types. *J Hyg*. 1918;27:113–115.
11. Avery OT, MacLeod CM, McCarty M. Studies on the chemical nature of the substance inducing transformation of pneumococcal types. Induction of transformation by a desoxyribonucleic acid fraction isolated from Pneumococcus type III. *J Exp Med*. 1944;79:137–158.
12. Hershey AD, Chase M. Independent functions of viral protein and nucleic acid in growth of bacteriophage. *J Gen Physiol*. 1952;36:39–56.
13. Watson JD. *The Double helix: A Personal Account of the Discovery of the Structure of DNA*. London, UK: Weidenfeld & Nicolson Press; 1968.
14. Elkin LO. Rosalind Franklin and the double Helix. *Phys Today*. 2003;56:42–48.
15. Watson JD, Crick FH. Molecular structure of nucleic acids: a structure for deoxyribose nucleic acid. *Nature*. 1953;171:737–738.
16. Matthaei H, Nirenberg M. Characteristics and stabilization of DNAase-sensitive protein synthesis in *E. coli* extracts. *Proc Natl Acad Sci U S A*. 1962;47:1580–1588.
17. Luria SE, Human ML. A nonhereditary, host-induced variation of bacterial viruses. *J Bacteriol*. 1952;64:557–569.
18. Dussoix D, Arber W. Host specificity of DNA produced by *Escherichia coli*: II. Control over acceptance of DNA from infecting phage λ. *J Mol Biol*. 1962;5:37–40.
19. Arber W, Dussoix D. Host specificity of DNA produced by Escherichia coli: I. host controlled modification of bacteriophage λ. *J Mol Biol*. 1962;5:18–36.
20. Mandel M, Higa A. Calcium-dependent bacteriophage DNA infection. *J Mol Biol*. 1970;53:159–162.
21. Stanley NC, Chang AC, Hsu L. Nonchromosomal antibiotic resistance in bacteria: genetic transformation of *Escherichia coli* by R-factor DNA. *Proc Natl Acad Sci U S A*. 1972;69:2110–2114.
22. Cohen SN, Chang AC, Boyer HW, Helling RB. *Proc Natl Acad Sci U S A*. 1973;70:3240–3244.
23. Bolivar F, Rodriguez RL, Greene PJ, et al. Construction and characterization of new cloning vehicles. II A multipurpose cloning system. *Gene*. 1977;2:95–113.
24. Maxam AM, Gilbert W. A new method for sequencing DNA. *Proc Natl Acad Sci U S A*. 1977;74:560–564.
25. Huberman JA, Riggs AD. On the mechanism of DNA replication in mammalian chromosomes. *J Mol Biol*. 1968;32:327–341.
26. Saiki R, Scharf S, Faloona F, et al. Enzymatic amplification of beta-globin genomic sequences and restriction site analysis for diagnosis of sickle cell anemia. *Science*. 1985;230:1350–1354.

27. Schoch R. *Q&A—A Conversation with Kerry Mullis*. *California Monthly*. vol. 105. Berkeley, CA: California Alumni Association; 1994:20.
28. Harrison A. *LSD: The Geek's Wonder Drug?* ; 2006. https://www.wired.com/2006/01/lsd-the-geeks-wonder-drug/.
29. Saiki KS, Arnheim N, Erlich HA. A novel method for the detection of polymorphic restriction sites by Cleavage of Oligonucleotide Probes: application to sickle-cell anemia. *Bio/Technology*. 1985;3:1008–1012.
30. Mullis K. Specific enzymatic amplification of DNA in vitro: the polymerase chain reaction. *Cold Spring Harb Symp Quant Biol*. 1986;51:263–273.
31. Mullis K, Faloona F. Specific synthesis of DNA in vitro via a polymerase-catalyzed chain reaction. *Methods Enzymol*. 1987;155:335–350.
32. Schena M, Shalon D, Davis RW, Brown PO. Quantitative monitoring of gene expression patterns with a complementary DNA microarray. *Science*. 1995;270:467–470.
33. Anderson S, Bankier AT, Barrell BG, et al. Sequence and organization of the human mitochondrial genome. *Nature*. 1981;290:457–465.
34. Fresko I, Soy M, Hamuryudan V, et al. Genetic anticipation in Behçet's syndrome. *Ann Rheum Dis*. 1998;57:45–48.
35. Cordaux R, Hedges DJ, Scott WH, Batzer MA. Estimating the retrotransposition rate of human Alu elements. *Gene*. 2006;373:134–137.
36. Duan Z, Shen H, Lee JE, et al. DNA repair gene XRCC3 241Met variant is not associated with risk of cutaneous malignant melanoma. *Cancer Epidemiol Biomark Prev*. 2002;11:1142–1143.
37. Manolio TA. Genomewide association studies and assessment of the risk of disease. *N Engl J Med*. 2010;363:166–176.
38. Tautz D. Hypervariability of simple sequences as a general source for polymorphic DNA markers. *Nucleic Acids Res*. 1989;17:6463–6471.
39. Martinez L, Reategui EP, Fonseca LR, et al. Superimposing polymorphism: the case of a point mutation within a polymorphic *Alu* insertion. *Hum Hered*. 2005;59:109–117.
40. Rowold DJ, Herrera RJ. Inferring recent human phylogenies using forensic STR technology. *Forensic Sci Int*. 2003;133:260–265.
41. Batzer MA, Deininger PL. Alu repeats and human genomic diversity. *Nat Rev Genet*. 2002;3:370–379.
42. Socorro A, Fernandez I, Masleh M, et al. Phylogenetic information in polymorphic L1 and Alu insertions from East Asians and Native American populations. *Am J Phys Anthropol*. 2005;128:171–184.
43. Herrera RJ, Lowery RK, Alfonso A, McDonald JF, Luis JR. Ancient retroviral insertions among human populations. *J Hum Genet*. 2006;51:353–362.
44. Dombroski BA, Feng Q, Mathias SL. An in vivo assay for the reverse transcriptase of human retrotransposon L1 in Saccharomyces cerevisiae. *Mol Cell Biol*. 1994;14:4485–4492.
45. Lander ES, Linton LM, Birren B. Initial sequencing and analysis of the human genome. *Nature*. 2001;409:860–921.
46. Rowold DJ, Herrera RJ. Alu elements and the human genome. *Genetica*. 2000;108:57–72.
47. Mighell AJ, Markham AF, Robinson PA. Alu sequences. *FEBS Lett*. 1997;417:1–5.
48. Szmulewicz MN, Novick GE, Herrera RJ. Effects of Alu insertions on gene function. *Electrophoresis*. 1998;19:1260–1264.
49. Labuda D, Striker G. Sequence conservation in Alu evolution. *Nucleic Acids Res*. 1989;17:2477–2491.
50. von Sternberg RM, Novick GE, Gao GP, Herrera RJ. Genome canalization: the coevolution of transposable and interspersed repetitive elements with single copy DNA. *Genetica*. 1992;86:215–246.
51. Belshaw R, Pereira V, Katzourakis A, et al. Long-term reinfection of the human genome by endogenous retroviruses. *Proc Natl Acad Sci U S A*. 2004;101:4894–4899.
52. van de Lagemaat LN, Landry JR, Mager DL, Medstrand P. Transposable elements in mammals promote regulatory variation and diversification of genes with specialized functions. *Trends Genet*. 2003;19:530–536.
53. Levitt M. The birth of computational structural biology. *Nat Struct Biol*. 2001;8:392–393.
54. Green RE, Krause J, Briggs AW, et al. A draft sequence of the Neanderthal genome. *Science*. 2010;328:710–722.
55. Lowery RK, Uribe G, Jimenez EB, et al. Neanderthal and Denisova genetic affinities with contemporary humans: introgression versus common ancestral polymorphisms. *Gene*. 2013;530:83–94.

56. Ceppellini R, Siniscalo M, Smith CAB. The estimation of gene frequencies in a random-mating population. *Ann Hum Genet.* 1955;20:97–115.
57. Excoffier L, Smouse PE, Quattro JM. Analysis of molecular variance inferred from metric distances among DNA haplotypes: application to human mitochondrial DNA restriction data. *Genetics.* 1992;131:479–491.
58. Hellenthal G, Busby GBJ, Band G, et al. A genetic atlas of human admixture history. *Science.* 2014;343:747–751.
59. Pritchard JK, Stephens M, Donnelly P. Inference of population structure using multilocus genotype data. *Genetics.* 2000;155:945–959.
60. Wei W, Ayub Q, Xue Y, Tyler-Smith C. A comparison of Y-chromosomal lineage dating using either resequencing or Y-SNP plus Y-STR genotyping. *Forensic Sci Int Genet.* 2013;7:568–572.
61. Hochreiter S. HapFABIA: identification of very short segments of identity by descent characterized by rare variants in large sequencing data. *Nucleic Acids Res.* 2013;41(22), e202.
62. Gusev A, Palamara PF, Aponte G, et al. The architecture of long-range haplotypes shared within and across populations. *Mol Biol Evol.* 2011;29:473–486.
63. Palamara PF, Lencz T, Darvasi A, Pe'Er I. Length distributions of identity by descent reveal fine-scale demographic history. *Am J Hum Genet.* 2012;91:809–822.
64. Botigue LR, Henn BM, Gravel S, et al. Gene flow from North Africa contributes to differential human genetic diversity in southern Europe. *Proc Natl Acad Sci.* 2013;110:11791–11796.
65. Ralph P, Coop G. The geography of recent genetic ancestry across Europe. *PLoS Biol.* 2013;11(5), e1001555. Tyler-Smith, C. (ed.).
66. Gravel S, Zakharia F, Moreno-Estrada A, et al. Reconstructing Native American migrations from whole-genome and whole-exome data. *PLoS Genet.* 2013;9(12), e1004023. Williams, Scott M (ed.).
67. Ringbauer H, Coop G, Barton NH. Inferring recent demography from isolation by distance of long shared sequence blocks. *Genetics.* 2017;205:1335–1351.
68. Morgan TH, Sturtevant AH, Muller HJ, Bridges CB. *The Mechanism of Mendelian Heredity.* New York, USA: Henry Holt Press; 1915.
69. Botiguéa LR, Hennb BM, Gravelb S, et al. Gene flow from North Africa contributes to differential human genetic diversity in southern Europe. *PNAS.* 2013;110:11791–11796.
70. The 1000 Genomes Project Consortium. An integrated map of genetic variation from 1,092 human genomes. *Nature.* 2012;491:56–65.
71. Gusev A, Lowe JK, Stoffel M, et al. Whole population, genome-wide mapping of hidden relatedness. *Genome Res.* 2009;19:318–326.
72. Tishkoff SA, Gonder MK, Henn BM, et al. History of click-speaking populations of Africa inferred from mtDNA and Y chromosome genetic variation. *Mol Biol Evol.* 2007;24:2180–2195.

CHAPTER 7

Continuity of traits among hominins

The little religion that I have clung to—that what matters most is the continuity of life, and its improvement from one generation to another.

David O. Selznick[1]

Continuity is fundamental to life and evolution

Continuity is an integral and essential part of biological evolution. The process of biological evolution requires the inheritance or transmission of traits to be passed from generation to generation, and only on rare occasions, changes or mutations provide some degree of deviation from the norm. These relatively stable traits are transmitted to the next generation in a code: the genetic code. In all known organisms, this code is chemical in nature and relies on the sequence of units or nucleotides, each made up of a nitrogenous base, a pentose (five-carbon) sugar, and a phosphate group. It is the impressible stability of the genetic code that provides the continuity that is observed among living things. Independent of whether unicellular or multicellular organisms are examined, all survive using their genetic information to direct their metabolisms and chemical reactions that would allow them to survive and reproduce. It is likely that life originated only once and that all living things are linked to the initial primordial cell. Thus, life, as we know it, has depended on constancy and continuity since its conception.

Yet this continuity is periodically interrupted. At the individual level, it is clear that diversity exists. Even identical twins are not genetically identical. Subsequent to conception and separation of identical embryos, mutations transform them into distinct individuals as they grow by cell division. At the level of populations, whether they are part of the same or different species, diversity exists. Generally speaking, when this diversity reaches a certain level, it interferes with sexual reproduction, bringing about speciation. It is envisioned that the genesis of new species occurs when a population with a certain amount of internal diversity splits into subgroups. This separation of members of a population often occurs as a result of geographical partition. For example, a group of organisms could separate when a river changes its course, splitting the population into two. Alternatively, a population in the process of expansion may become disconnected and isolated in a region. Such speciation events may lead to lineages that remain independent and, with time, evolved into higher-level taxa such as genera, family, etc. Alternatively, the

Sex and Cohabitation Among Early Humans
https://doi.org/10.1016/B978-0-12-819086-9.00005-1

separated groups could reunite again into a single population if the geographical barrier is eliminated, or partially assimilate or fuse via introgression.

The process of speciation facilitates diversity among living things as the physical and biological separation of groups of organisms brings about increasingly independent differentiation of their gene pools, which in turn, with time, brings about reproductive isolation. After reproductive isolation, gene flow between groups stops.

Continuity as a function of time

It is understood that speciation generally brings about the truncation of genetic flow between organisms (and groups of organisms). This cessation of gene flow eventually leads to the genetic, physiological, and anatomical differentiation of populations into species. When species are created due to geographic isolation by physical barriers such as a mountain range, river, or desert, the process is referred to as allopatric speciation. Speciation does not require the physical separation of groups of organisms. Mutations such as gross chromosomal aberrations (e.g., chromosomal inversions, polyploidy, or multiple copies of chromosome sets) could lead to species formation within a single population without geographic isolation. This type of species formation is known as sympatric speciation.[2]

All species experience changes in their gene pool or genetic makeup as a function of time. DNA is known to change or mutate spontaneously as a function of time. This innate tendency of the genetic material to change not only transforms the DNA but also alters the anatomy and physiology of organisms. For example, the human genome-wide mutation rate is approximately 0.5×10^{-9} per base pair per year.[3] Chemical and radiation mutagens increase the mutation rate. Mutation rates also vary in relation to DNA types. For these reasons, species are always changing. Taking *Homo sapiens* as an example, the species 200,000 years ago (ya) at its inception was not the same as contemporary humans are today. Our species has constantly changed. In addition, to this unavoidable and constant drift experienced by the DNA sequence that disrupts continuity, the environment has been continuously selecting for mutations that favor individuals' and populations' survival. Differences in intensity and type of selection pressure imposed by the environment impact the course of evolutionary change within populations disrupting continuity. For instance, in pastoral and agricultural societies, a mutation that allows milk (lactose) digestion in adults would be favored, selected for, and incorporated into the gene pool rapidly. That was exactly what happed in Northern/Central Europe. A mutation that allowed lactose consumption in adults took place, and in less than 20,000 years, the mutation spread across European populations. In evolutionary terms, such a dramatic event is referred to as a selective sweep. In great apes, this phenomenon is known to increase with population size.[4]

Even in the absence of physical or reproductive separation, or isolation, time will transform groups of organisms and create novel species if sufficient time is allowed to pass. Therefore, it is possible that if somehow two groups of organisms, separated only by time, are brought together, they would not be able to reproduce. Thus, continuity is relative.

Yet despite mutations, selection pressure, and physical and reproductive isolation, continuity of traits is fundamental to life. This continuity is apparent not only among populations within species but also among species. Thus, continuity of traits persists among fully independent species. This should not be a surprise since, at one point in the past, sister species such as great apes showing continuity of traits had the same gene pool, and part of their DNA, now partition in different species, is still shared. Contemporary humans and chimpanzees are almost 99% identical in their DNA.

Furthermore, since all living things are evolutionary linked, and this relationship extends to the beginning of life, continuity of traits extends deep in time. The more recent two or more species share a common ancestor, the greater the number of traits in common is expected between them and the greater continuity between them.

Ancestral and derived traits

An ancestral trait is inherited from the common ancestor of a group of organisms and has experienced little change since then. A trait that originates within a group of organisms for the first time is referred to as derived. When ancestral traits are passed from generation to generation, the feature is said to experience continuity. Often, continuity of traits is observed in the form of mosaicisms in which a mixture of ancestral and derived traits is present in an organ or part of an organism's anatomy (see Chapter 5 for discussion on mosaicism). To account for this phenomenon, certain parts of the anatomy must be evolving faster than others, resulting in structures that retain the ancestral condition while others are experiencing change.

Interestingly, different hominin species exhibit different patterns of mosaicism. In other words, various combinations of ancestral and derived characteristics are seen in different species and populations. It seems as if certain traits that are part of structures or organs have changed at different rates responding to unique selection pressures. It is as if specific selection pressures dictate the appropriate combination of traits suited to survive in a given environment and, in the process, combine specific ancestral and derived traits. Since the combination of ancestral and derived traits are observed in specific body parts (e.g., dentition, hand anatomy, hip morphology, skull, etc.) and the mosaicisms differ among species and anatomical units (see "Mosaicism" section in Chapter 5), it seems that specific structures are semiautonomously controlled by inheritance reflecting subsets or modules. These modules of different developmental origins respond differentially to selection pressure and possibly reflect an underlying hierarchical organization based on differential genetic and developmental origins.

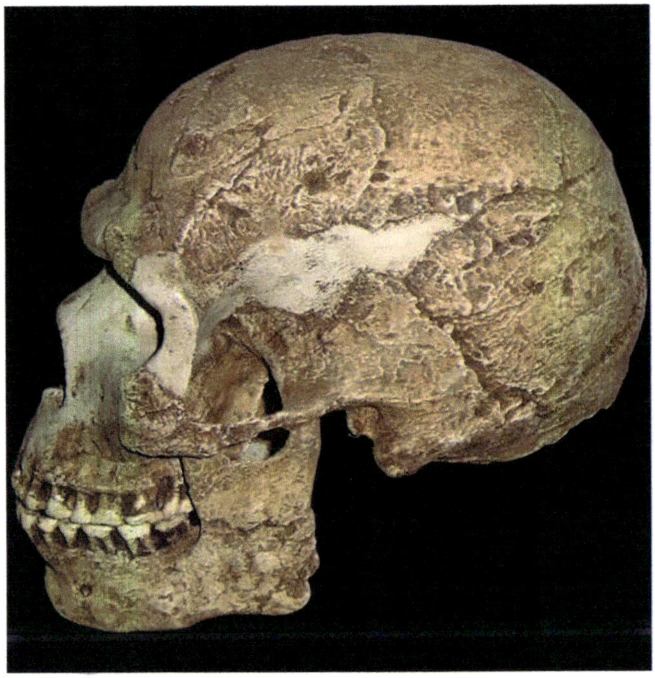

Fig. 1 Mixture of traits. Skhul V dated at about 80,000–120,000 years old, exhibiting a mix of archaic and modern traits (https://en.wikipedia.org/wiki/Homo_sapiens#/media/File:Skhul.JPG).

Mixtures of ancestral and derived traits are evident throughout hominin evolution. Middle and Lower Paleolithic hominins with transitional features intermediate between *Homo erectus*, Neanderthals, and early AMH are generally referred to as archaic *Homo sapiens* and often exhibit various combinations of mosaicism.[5] Good examples of mosaicism are seen in the Skhul/Qafzeh hominin fossils.[6] For instance, Skhul 5, one of the hominin fossils discovered in the Skhul Cave in the slopes of Mount Carmel in Israel is today classified as an early *Homo sapiens* exhibiting a mixture of ancestral features including prominent supraorbital ridges (superior margin of the bony orbit) and jutting jaw as well as derived traits such as a rounded braincase as in modern humans (Fig. 1).

The retention of ancestral traits across species and, at times, across genera represents examples of continuity of traits (Chapter 5). If the commonality of traits among different species is not the result of convergent evolution, continuity of traits brought about by genetic links is responsible for the transmission of ancestral traits.

Continuity and the multiregional model of hominin evolution

Although the concept of continuity of traits is fundamentally linked to the process of evolution in general, the term is frequently associated with the multiregional hypothesis of

Fig. 2 Milford Howell Wolpoff (https://www.sigmaxi.org/programs/lectureships/past-lecturers/2003-2004).

recent hominin evolution. In its original form, the hypothesis was formulated in 1984 by Milford Howell Wolpoff (1942–present).[7–9] Since its conception, Wolpoff (Fig. 2) has been an active figure in the evolution of the ideas associated with this theory. The multiregional hypothesis, also known as the polycentric hypothesis, postulates that *Homo sapiens* started evolving from *Homo erectus* stock about 2 million years ago (mya) in different parts of the world quasiindependently. Subsequently, the various hominin lineages have evolved semiautonomously in the different regions with some periodic gene flow episodes among them.[10] It is theorized that these occasional interregional gene flow events have prevented speciation and kept *sapiens* as one species. The multiregional theory stipulates that every hominin species since *Homo erectus*, including Neanderthals and Denisovans, as well as all modern human forms leading to AMH, evolved regionally (Fig. 3). According to the multiregional hypothesis, transitional (mosaic) specimens should be common. In this model of recent hominin evolution since *Homo erectus* the progression from archaic to contemporary humans took place in parallel in various regions of the world, and the integrity of the species was maintained by sporadic gene flow among the regions. In fact, in order for the overall species similarities to be maintained, considerable amounts of gene flow among the geographical regions were needed as the independent regional lineages evolved.

The out of Africa (OOA) hypothesis, single-origin hypothesis, replacement hypothesis, or recent African origin model provides an alternative explanation for the origin and distribution of AMHs. It posits that modern humans had a unique origin in sub-Sahara Africa and subsequently migrated out of the continent into the Levant and/or Arabia and then into Asia and the rest of the world.[11] If this theory is correct, it means that all contemporary humans are of African descent. In its current form, the OOA model proposes that *sapiens* undertook several out of Africa dispersals, possibly beginning as early as 270,000 ya. Recent evidence from Greece suggests that AMHs reached southeastern

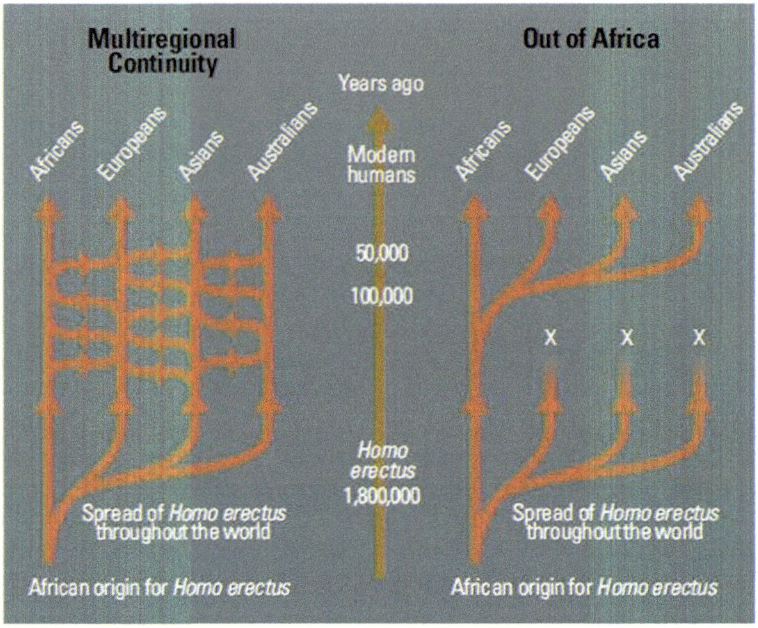

Fig. 3 Multiregional origin of modern humans (https://www.google.com/search?q=multiregional+hypothesis&oq=multiregional+&aqs=chrome.0.0l2j69i57j0l3.17475j0j7&sourceid=chrome&ie=UTF-8).

Europe approximately 215,000 ya[12] and then migrated out of Africa again about 130,000 to 115,000 ya.[13] Currently, it is not clear if these initial migrants contributed to the gene pool of contemporary *sapiens*, became extinct, or returned to Africa, since by 80,000 ya AMHs are not seen in Southwest Asia until about 55,000 ya.[14,15] Yet one migration wave that clearly led to the settlement of Asia and Europe took place about 70,000–50,000 ya via the Horn of Africa and/or the Levant.[16] This exodus rapidly moved along the coast of Asia and reached Australia by around 65,000–50,000 ya.[17] The settlement of Europe from Southwest Asia was somewhat delayed, taking place around 30,000–35,000 ya.[18]

Several important details set the OOA model apart from the multiregional hypothesis. One is the very recent origin of *sapiens* postulated by the OOA model. Although a number of recent discoveries have pushed the genesis and Eurasian dispersal of AMHs back in time,[19] the time period according to OOA is still one order of magnitude younger than the time proposed for the migration of *Homo erectus* out of Africa (2 mya) that supposedly eventually lead to the regional origins of *sapiens* in various parts of the world. Another premise of the OOA model is that it relies on the complete replacement of previously existing species in the area of occupation. In other words, in the OOA model, no regional continuity of traits from preexisting populations occurs. Furthermore, the OOA model minimizes the impact of independent parallel evolution in generating the AMH traits and posits that transitional fossils exhibiting mosaicisms of traits would not occur outside of Africa. This last premise of the OOA is clearly not met.

A number of lines of evidence support the out of Africa theory. One is the fossil record. The oldest fossils of *sapiens* have been found in Africa or nearby areas. In chronological order, the oldest identifiable *sapiens* fossils are the remains from Jebel Irhoud, Morocco, which date back to around 300,000 ya. The Jebel remains include skulls, a humerus, a hip bone, and a lower jaw from up to five different individuals, including three young adults, an adolescent, and a child about 8 years of age.[20] Next in line is the Florisbad skull from South Africa, which has been dated to about 260,000 ya.[21] More recent are the maxilla and associated teeth recovered from the Levantine Misliya Cave in Israel, dated at approximately 177,000 to 194,000 ya.[22] Altogether, these recent discoveries strongly suggest that the cradle of AMHs encompasses Africa and neighboring areas, and the lack of AMH fossils of comparable ancient age from other world regions argue against the parallel genesis of *sapiens*.

Genetics also provides data in favor of the OOA hypothesis. In the 1980s, a group of investigators including Allan Charles Wilson (1934–91), Rebecca L. Cann (1951–present), and Mark Stoneking (1956–present) (Fig. 4) reported in a groundbreaking article (Fig. 5) that contemporary human populations had diverged recently from a single population in Africa. Working out of the University of California at Berkeley, this research group utilized mitochondrial DNA (mtDNA) markers to date the most recent common ancestor of modern human populations, which was dubbed "Mitochondrial Eve." As part of their report, they concluded that all the mitochondria in contemporary humans worldwide derive from one woman who lived about 200,000 ya.[23] Based on these markers, investigators have estimated that AMHs left Africa 150,000 to 60,000 ya.[24] Furthermore, it has been determined that out of all the mtDNA types (known as haplogroups) that were present in Africa at the time of the OOA event, only one, L3, is found outside Africa. It is possible that as a consequence of the difficult exodus across the Red Sea across the Bab-el-Mandeb Strait (Gate of Tears), a bottleneck event restricted the number of individuals and mtDNA types migrating into Southwest Asia. Based on mitochondrial and nuclear DNA markers, it has been estimated that about 50,000–70,000 ya, a small group of individuals (approximately 150–1000 people) carrying the L3 mtDNA crossed the Bab-el-Mandeb Strait and migrated from East Africa across the Red Sea into the Near East.[25] This group and its descendants traveled along the coast of Arabia through what is today Yemen and Oman, transversed the Strait of Hormuz, and subsequently reached India and then East Asia. During this migration, the Strait of Hormuz was a dry valley.[26] Both mtDNA and Y-chromosome markers suggest that most of the gene pool that characterizes *sapiens* today originated in Africa.

In addition to the mtDNA and Y-chromosome data, which explore the transmission of maternal and paternal DNA lineages, respectively, studies based on autosomal DNA (DNA in nonsex chromosomes) markers indicate greater genetic diversity among

Fig. 4 Alan Charles Wilson, Rebecca L. Cann, and Mark Stoneking (https://www.google.com/search?sxsrf=ALeKk00OCXQ5v4Rw1Pgzri2JnjZFwPxwgw:1593966943202&source=univ&tbm=isch&q=Allan+Wilson,+Rebecca+L.+Cann+and+Mark+Stoneking+images&sa=X&ved=2ahUKEwi8k7vLxbbqAhUGneAKHcdaB7AQsAR6BAgFEAE&biw=1440&bih=743#imgrc=CNK_MepY7z2zhM).

Africans than among Europeans and Asians.[27] This type of result is expected if Africans represent the ancestral population that migrated out to populate the rest of the world. The reduction in genetic diversity as *sapiens* migrated away from Africa and sequentially settled uninhabited regions is a logical outcome of the OOA dispersal since bottleneck events in tandem must have occurred as humans migrated eastward. The African origin theory is today the most accepted model for the origin and widespread distribution of AMHs.

The multiregional model rests heavily on the concept of continuity of traits since, according to the model, hominins have evolved regionally and semiindependently to

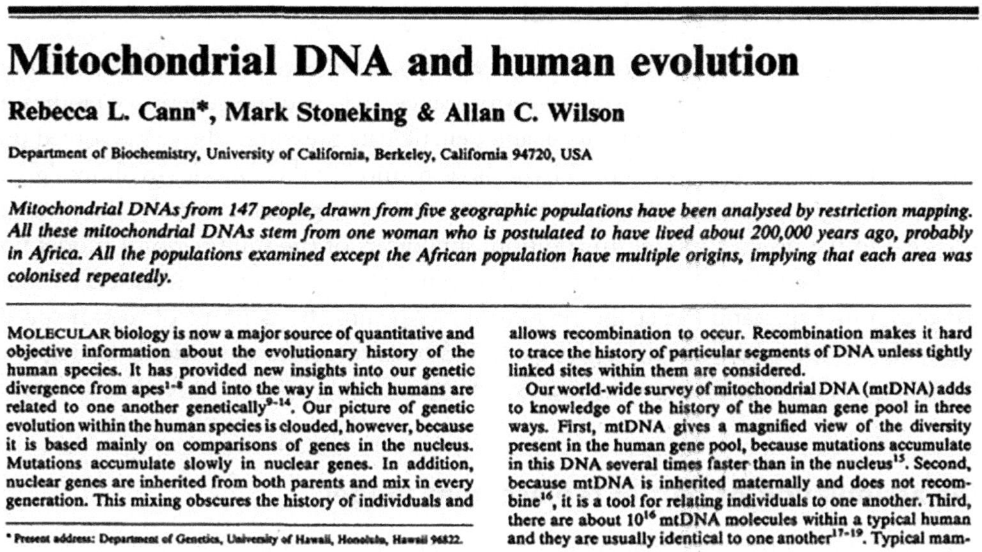

Mitochondrial DNA and human evolution

Rebecca L. Cann*, Mark Stoneking & Allan C. Wilson

Department of Biochemistry, University of California, Berkeley, California 94720, USA

Mitochondrial DNAs from 147 people, drawn from five geographic populations have been analysed by restriction mapping. All these mitochondrial DNAs stem from one woman who is postulated to have lived about 200,000 years ago, probably in Africa. All the populations examined except the African population have multiple origins, implying that each area was colonised repeatedly.

MOLECULAR biology is now a major source of quantitative and objective information about the evolutionary history of the human species. It has provided new insights into our genetic divergence from apes[1-8] and into the way in which humans are related to one another genetically[9-14]. Our picture of genetic evolution within the human species is clouded, however, because it is based mainly on comparisons of genes in the nucleus. Mutations accumulate slowly in nuclear genes. In addition, nuclear genes are inherited from both parents and mix in every generation. This mixing obscures the history of individuals and allows recombination to occur. Recombination makes it hard to trace the history of particular segments of DNA unless tightly linked sites within them are considered.

Our world-wide survey of mitochondrial DNA (mtDNA) adds to knowledge of the history of the human gene pool in three ways. First, mtDNA gives a magnified view of the diversity present in the human gene pool, because mutations accumulate in this DNA several times faster than in the nucleus[15]. Second, because mtDNA is inherited maternally and does not recombine[16], it is a tool for relating individuals to one another. Third, there are about 10^{16} mtDNA molecules within a typical human and they are usually identical to one another[17-19]. Typical mam-

* Present address: Department of Genetics, University of Hawaii, Honolulu, Hawaii 96822.

Fig. 5 Mitochondrial Eve article (http://www.kushima.org/is/?p=54503).

account for region-specific traits. In such a scenario, regional evolutionary change connecting populations through time in a given area is required to generate and maintain the continuity of traits. Region-specific continuity would not exist if all *sapiens* derived from a single recent stock, as the "out of Africa" hypothesis contests. If a single recent worldwide dispersal gave rise to all *sapiens* on the planet, as the OOA model Africa posits, there would not be region-specific continuity since entire the species derived from the same recent African lineage. Instead, contemporary humans would be very similar to each other independent of their geographical location. Indeed, the proportion of human genetic diversity due to differences among regional populations is modest, and in fact, people from different populations can be genetically more similar than individuals from the same population.[28]

The association between the multiregional hypothesis and the concept of continuity of traits is somewhat misleading since the continuity of traits could occur whether *Homo sapiens* originated independently in different regions of the world or only in Africa. This connection between the two stems from the fact that continuity of traits is expected during evolution within geographical areas (e.g., within South East Asia) but not as much among regions (e.g., South East Asia, Africa, and Europe). The rationale is that since hominins within geographical regions are semiautonomously breeding and evolving separately from hominins in different areas, it is expected that they would share more of their gene pool among themselves than between members from different regions and thus exhibit more continuity of traits.

The revised multiregional model

The multiregional theory has been revised over the years to account for the overwhelming fossil record of early *sapiens* discovered in Africa since 1984 and the genetic data indicating an African origin of AMHs.[29] The fossil and genetic data presented above has forced proponents of the original multiregional hypothesis to modify their model to allow Africa's much more significant role in hominin evolution and less regional continuity than was first proposed in the original model. In addition, supporters of the multiregional model now acknowledge the role of gene flow in maintaining species-specific similarities, integrity, and preventing speciation. Yet the revised model retains the concept of region-specific traits. The OOA model, on the other hand, has been forced to concede that integration of region-specific traits by way of introgression and continuity of certain features is a reality.

The assimilation or partial replacement model

The OOA model proposes that AMHs originated *only* in Africa and dispersed as a new species all over Eurasia and in the process *totally* replaced archaic Eurasians, including *Homo erectus*. The overwhelming amount of evidence indicates that *sapiens* originated in Africa. However, clearly, the second premise of the OOA model, which states that *all* traces of previous hominins were eliminated during the dispersal leading to total replacement of the locals, has failed. To accommodate this reality, a number of scholars have proposed the OOA and hybridization model, which no longer adheres to the idea of complete replacement of the original indigenous premodern populations of the occupied regions and allows *some degree* of assimilation. In this model, the exact extent of admixture remains undetermined. Still, other scholars think that although *sapiens* originated in Africa, the level of assimilation was quite substantial, and replacement, in the strict sense of the word, of one population by another did not occur. This model emphasizes the role of gene flow, admixture, and region-specific selection pressures in bringing about directional morphological change. It could be argued that the difference between the OOA and hybridization and the assimilation models is a matter of degrees of assimilation, with the latter being more obtuse.

Even with the currently limited body of anatomical and genetic evidence, it is untenable to dismiss some degree of regional input from archaic *Homo* to the migrant *Homo sapiens*. In the strict interpretation of the OOA model, total replacement of all archaic indigenous populations in all Eurasian regions is required. Demonstration of continuity in just one region (e.g., East Asia) would invalidate the notion of total recent population replacement. The questions that remain unanswered are: how much of an input took place and who were the indigenous ancestors in the various regions of contact (i.e., Europe, Asia, and Oceania)? In an attempt to accommodate the growing amount of conflicting evidence relating to the strict multiregional and OOA theories,

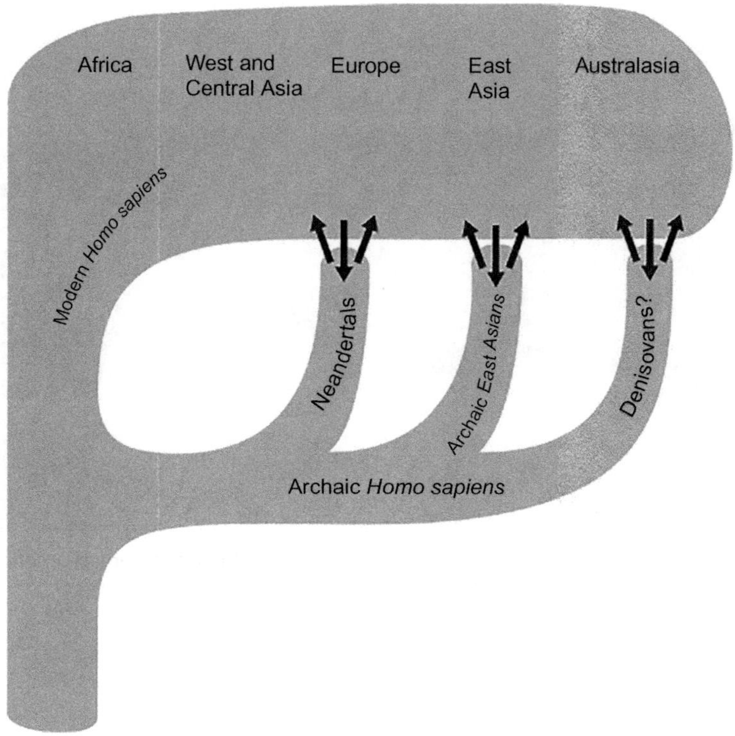

Fig. 6 The assimilation model (https://www.sciencedirect.com/science/article/abs/pii/S104061821630386X).

a variation of the Assimilation Model (AM) (Fig. 6) mentioned above was developed in the late 1980s by American anthropologist Fred H. Smith (Fig. 7). The model was formally published in 1989.[30] The AM was initially formulated to explain the observed regional continuity of anatomical traits, and states that the archaic contributions, though relatively small, were not insignificant.[31] Today, the model is equally applied to genetic evidence. Smith's original inspiration for formulating his assimilation ideas originated when he noticed several anatomical features that contemporary Europeans share with Neanderthals. He proposed that these traits were derived from Neanderthals. Among these traits, he noticed a bump in the back of the skull known as the occipital bun (Fig. 8). This characteristic of Neanderthals was also found in high frequencies in modern Europeans but it was not seen among African populations. At the time, Smith concluded that the most likely explanation is that this was evidence for some input from Neanderthals into that modern human gene pool. Since those initial anatomical observations, the number of traits proposed to link contemporary *sapiens* to archaic humans has increased. As the amount of evidence accumulates from various geographical regions of Eurasia and Oceania, indicating continuity of traits among

Fig. 7 Fred H. Smith (https://news.illinoisstate.edu/2018/11/fred-h-smith-named-american-association-for-the-advancement-of-science-fellow/).

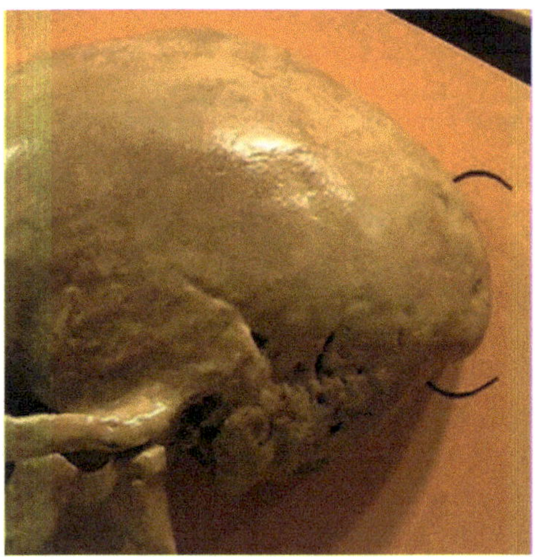

Fig. 8 Occipital bun (https://commons.wikimedia.org/wiki/File:OccipitalBun.jpg).

different populations of archaic and modern humans, the AM is emerging as an attempt to accommodate the data.[32]

What does the evidence say?

The features that are evaluated to ascertain the continuity of traits fall into two categories: they are either anatomical or genetic. Since the early work of Weidenreich (Fig. 9) in the

Fig. 9 Franz Weidenreich (https://alchetron.com/Franz-Weidenreich).

mid-20th century,[33,34] anthropologists have been noticing certain morphological traits in common between regional specimens of *Homo erectus* and contemporary populations. Franz Weidenreich (1873–1948) was a German anatomist and physical anthropologist who pioneered this work recognizing and recording these anatomical similarities. He developed an interest in the numerous morphological features that Peking Man had in common with modern Chinese individuals. These observations led him to formulate his polycentric evolution model of human origins that later developed into today's multiregional model. In those early years, Weidenreich had the insight to envision gene flow as an integral part of the theory.[35] Many of the traits later described by Milford Wolpoff to argue for the multiregional theory and Marta Mirazon Lahr (1965–present) for the OOA model were first described by Weidenreich in his original publications.

Table 1 illustrates a battery of 30 anatomical traits that were reevaluated for regional continuity by Marta Mirazon Lahr (Fig. 10) in 1994.[36] Some of the traits listed in this table are metric or quantitative (e.g., human height), which are scored as part of a continuous range of variability. The others are nonmetric or discrete (e.g., presence or absence of an additional root in a molar), and are simply based on the presence or absence of a feature. Over the years since the original publication of these traits, they have been examined and reexamined repeatedly by a number of experts, each providing their personal opinion on the validity of the features to access continuity as well as whether they indicate continuity. Unsurprisingly, there has not been a consensus regarding the validity

Table 1 List of continuity traits, description, means of scoring, and region studied (EA, East Asia features; AUS, Southeast Asian/Australian features).

Trait	Description
EA1	Sagittal keeling (grade scores: SK1, SK2, SK3)
EA2	Mandibular exostoscs/torus (+/−)
EA3	Course of the nasofrontal and frontomaxillary sutures (measured as the vertical distance between nasion and the orbital edge of the frontomaxillary suture)
EA4	Profile of the nasal saddle and nasal roof (grade scores: NS1, NS2, NS3, NS4)
EA5	Rounded frontals (measured as frontal angle, Howells[38])
EA6	Orbital shape (scored as horizontal or oblique inclination of the superior, lateral, and inferior margins, combined to form orbital shapes)
EA7	M3 agenesis (+/−)
EA8	Reduced posterior dentition (as a comparison of M1, M2, and M3 buccolingual, mesiodistal, and area size among regions)
EA9	Facial flatness or lack of facial prognathism (measured as Gnathic Index, and differences between prosthion, subspinale, and nasion radii)
EA10	Lateral facial flatness (measured as the bimaxillary and naso-frontal subtenses, and the relative values of the zygoorbital, frontomalare, ectochonchion, and zygomaxillare radii. The four radial measurements were highly intercorrelated, and transformed into a single factor through PCA, and the factor scores of each case used as representing overall lateral facial orientation)
AUS1	Development of a supraorbital torus (grade scores: ST1, ST2, ST3, ST4, ST5)
AUS2	Lambdoid depression (+/−)
AUS3	Size of pterion articulation
AUS4	Profile of the infraglabellar notch (grade scores: IN1, IN2, IN3, IN4)
AUS5	Frontal flatness (measured as frontal angle, Howells[38])
AUS6	Position of minimum frontal breadth (taken as the distance between minimum frontal breadth along the temporal crest and both the superior margin of the orbital rim and the frontomalar sutures)
AUS7	Horizontal superior orbital margin (as for orbital shape)
AUS8	Presence of a prebregmatic eminence (+/−)
ASU9	Position of maximum parietal breadth (distance curyon-porion, transformed into a proportion of vertex radius)
AUS10	Facial prognathism (measured as for EA9)
AUS11	Development of a zygomaxillary tuberosity (grade scores: ZT1, ZT2, ZT3, ZT4)
AUS12	Eversion of the lower border of the malars (modified from Larnach and Macintosh,[39] and measured as the difference between (a) the distance between the orbital edge of one zygomatico-maxillary suture to the same point on the other side, and (b) the distance from a point approximately halfway along the lower free edge of the malar bone on its outer surface to the corresponding point on the opposite side)
AUS13	Type of narial margins (scored as 10 different categories)
AUS14	Degree of curvature of the maxilla (measured as the vertical difference observed laterally between (a) the distance prosthion-Frankfort Plane, and (b) the distance from the posterior end of the maxillary margin and the Frankfort Plane)
AUS15	Development of the zygomatic trigone (grade score: TR1, TR2, TR3, TR4)

Table 1 List of continuity traits, description, means of scoring, and region studied (EA, East Asia features; AUS, Southeast Asian/Australian features)—cont'd

Trait	Description
AUS16	Presence of a suprameatal tegmen (measured as the difference between biauricular breadth and intermeatal breadth, divided by two. The latter measurement taken from the most lateral margin of the tympanic plate on each side)
AUS17	Angle between the petrous and the tympanic (measured as the angle formed by the tympanic plate and petrous bone with and intermeatal line, the petro-lympanic angle derived from these)
AUS18	Development of an occipital crest (grade score: OCRI, OCR2, OCR3)
AUS19	Development of an occipital torus (grade score: OT1, OT2, OT3, OT4, OT5, OT6, OT7)
EA/AUS20	Rounding of the inferolateral margin of the orbit (grade score: RO1, RO2, RO3)

Modified from Marta Mirazon Lahr. The multiregional model of modern human origins: a reassessment of its morphological basis, *J Hum Evol*. 1994;26:23–56.

Fig. 10 Marta Mirazon Lahr (https://www.arch.cam.ac.uk/directory/mbml1).

and the signal provided by the traits. And to the lay observer, the debate appears as an exercise in futility, which leads to no meaningful conclusion. For example, although experts such as Milford Wolpoff may consider the criteria in Table 1 valid and indicative of regional continuity, others such as Marta Mirazon Lahr regard them otherwise.[37] For instance, of the 30 metric/nonmetric features in Table 1 (modified from Ref.[36]), Mirazon Lahr recognizes 7 (23.3%) as providing no significant regional distribution, 12 (40.0%) as showing trends not proposed by the multiregional model, and 11 (36.7%) indicating the trends proposed by the multiregional model. Mirazon Lahr also concludes that "the results of this study indicate that the morphological basis for the Multiregional Model is incorrect, and that alternative interpretations of the observed regional differences in modern human cranial morphology should be sought".[37] Some authors share her opinion while others are in total disagreement, indicating that most of the traits listed show regional continuity of features.

In the 1980s and 1990s, proponents and critics of the multiregional model have relied on anatomic traits such as those listed in Table 1 to investigate whether gene flow occurred between modern humans and archaic hominins to argue the validity of the theories. Unfortunately, this approach has not been very fruitful. One reason for this is that the metric traits are inherently subject to ambiguity due to their subjective nature. As a result, scholars' opinion is divided as to what constitutes, for example, a "large" or "small" structure. Nonmetric traits are also subject to disagreement because investigators differ on whether the traits represent instances of regional gene flow between archaic and modern forms or retention of common ancestral polymorphisms (features shared because they were mutually inherited from a common ancestor).[40] Many nonmetric traits are also disputed on the grounds of whether they are homologous or not. In other words, experts argue on whether or not the character in the archaic fossil represents the *same* feature as in the modern *sapiens*, or whether they result from homoplasy or parallel evolution. Often, it turns out to be a matter of opinion whether or not a certain character represents regional continuity or not. Investigators that support the multiregional model or the assimilation model usually indicate the character as evidence of genetic continuity. In contrast, scholars who favor the replacement model see the trait as being different or merely noninformative. To the outsider, it seems that the experts on both sides of the aisle are highly dogmatic as they debate the issues.

Although the controversy persists over the signals provided by specific anatomical markers, mainly due to their subjective nature and/or the limited number of specimens examined, some studies have adopted more objective scientific approaches to the study of metric data, including rigorous statistical analyses. A case study is provided in a 2001 article published in the journal *Science*.[41] In this investigation, M.H. Wolpoff and colleagues examined the relationship between putative ancestral populations as stipulated by the OOA theory to two regional early AMH populations in peripheral areas far from Africa. The early human specimens studied were the early modern Australian cranium from Willandra Lakes named WLH-50 (dated to approximately 15,000 ya) and an adult male cranium from Mladeč (dated to about 31,000 ya) in Moravia, Czech Republic. The early *sapiens* from Australia was compared to earlier modern populations from the Levant and/or Africa. This group of fossils represents the putative ancestor population that migrated as part of the OOA dispersal and replaced archaic local groups in East Asia. The WLH-50 was also compared to the Ngandong fossils of Java (considered by many to be *Homo erectus*). Ngandong represents the local archaic group that may have contributed to the modern WLH-50 specimen's regional ancestry. The Mladeč crania were compared to European Neanderthals as the local archaic group that may have contributed to the regional ancestry of Mladeč and the Qafzeh and Skhul specimens as the closest potential source of a replacing population as part of the OOA migration.

In this investigation, the complete replacement model anticipates a closer anatomical affinity between the early modern humans (i.e., the Australian WLH-50 and Mladeč samples) and the older Levantines and/or Africans fossils. If, on the other hand, the

Australian WLH-50 and Mladeč samples exhibit significant similarities to the regional populations (i.e., Ngandong and Neanderthals, respectively), then there is reason to question the validity of the complete replacement theory (OOA). Expectations of significant affinities of the early modern groups to their respective older Ngandong or Neanderthals regional populations would be suggestive of dual ancestry from the original OOA migration and the regional populations.

A total of 16 unambiguous nonmetric traits were scored for the Australian WLH-50 cranium, including angular torus; coronal keel; sagittal keel on frontal; lateral frontal trigone; *linea obliquus* strongly developed; mastoid crest; sagittal keel on parietal; postlambdoidal eminence; prebregmatic eminence; projecting inion; sulcus dividing the medial and lateral elements of the supraorbital torus or superciliary arches; superiormargin of the orbit blunt (as opposed to sharp); suprainiac fossa; supramastoid crest; temporal line forms a ridge; and transversely extensive nuchal torus. The Australian WLH-50 cranium exhibits a number of features, which over the years investigators have associated with older Indonesian hominins, including the Ngandong fossils of Java.[42] The early *sapiens* from Mladeč, which resemble earlier Neanderthals were compared using 30 non-metric traits.[43] These nonmetric traits were scored using pairwise differences: 1 for the presence and 0 for the absence of specific traits. The results indicated no evidence that the Australian WLH-50 fossil could be grouped exclusively with the African or Levantine hominins (the OOA ancestors) to the exclusion of the regional Ngandong fossil or that the Mladeč samples could be linked exclusively to the Qafzeh and Skhul fossils (the OOA ancestors) and not the regional European Neanderthals. The authors concluded that the data refute the complete replacement model.

Indonesia-Australia

Three geographical regions have attracted attention when considering the continuity of traits: Indonesia/Australia, China, and Europe. The region encompassing Indonesia and Australia involves hominins such as late *Homo erectus* that inhabited the area and may have introgressed their DNA into archaic AMHs. An important figure who promoted the multiregional theory in the Indonesia/Australia region was Australian anthropologist Alan Gordon Thorne (1939–2012). Thorne's involvement in the evolution of ideas in the area of human evolution began when, after a career as a journalist, he started an association with his future Ph.D. mentor, anthropologist Neil Macintosh (1906–77). In 1968, Thorne started excavating the Kow Swamp site located 10 km southeast of Cohuna in the central Murray Valley of Australia as part of his doctoral project. He and his team unearthed partial remains from 22 individuals. The site has provided optically stimulated luminescence (OSL) (see Chapter 4 for methodology) dates of about 22,000–19,000 ya.[44] In the initial description of the crania found at Kow Swamp, Thorne found a number of archaic ancestral features not seen in recent or contemporary Aboriginal crania. These include a receding frontal squama (frontal bone), massive supraorbital

Fig. 11 Kow Swamp 1 (https://ccschmitt.github.io/Kow%20Swamp%201.html).

regions (thickening and raised ridges), and a supraglabella fossae (slightly depressed bone that joins the two brow ridges) (Fig. 11). At the time, he concluded that these traits "preserved an almost unmodified eastern *erectus* form".[45] In the Kow Swamp crania, Thorne was able to trace an evolutionary sequence starting with the earliest *erectus* fossils from Sangiran (1.6–1.0 mya) (Fig. 12) through Ngandong (109,000–106,000 ya) (Fig. 13), both from Java, ending in prehistoric and recent Aboriginal Australians. Moreover, the sharing of these traits was seen as evidence supporting a continuity of specific anatomical features through time linking Indonesian *Homo erectus* of the terminal Pleistocene and the more recent Australian *Homo sapiens*. In other words, the commonalities between the Javanese *erectus* and the Kow Swamp crania provided Thorne with evidence in support of the multiregional hypothesis in the Indonesia/Australia region.

Subsequent, anatomical analyses by Pietrusewsky in 1979 indicated that morphologically the samples from Kow Swamp are distinct from prehistoric or regional contemporary samples from Australia.[46] To Thorne, Pietrusewsky's data suggested the survival of *Homo erectus* features in Australia until as recently as 10,000 ya. On the other hand, to other investigators, including the British archeologist/anthropologist Donald Brothwell (1933–2016), the Kow Swamp samples indicated that the vault size and shape were similar to various Indonesian *erectus* because of artificial cranial deformation. He contested that the specific similarities in cranial anatomy between *erectus* and the Kow Swamp crania resulted from a cultural practice. That is, parents gradually altered the shape of the heads

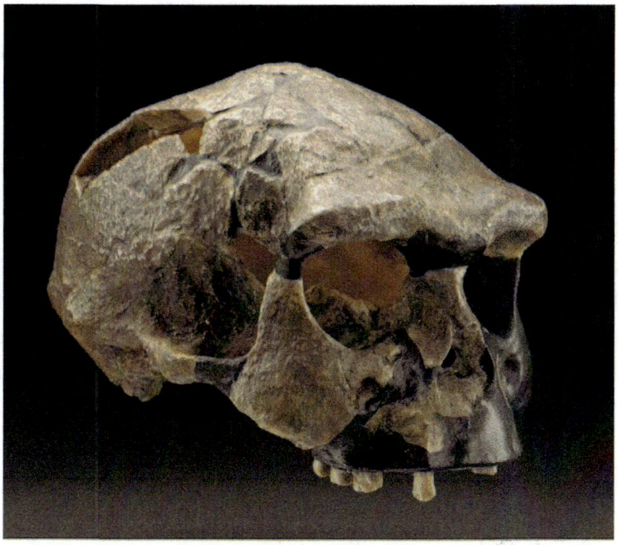

Fig. 12 Sangiran-17 (https://humanorigins.si.edu/evidence/human-fossils/fossils/sangiran-17).

Fig. 13 Solo man (Ngandong) (https://upload.wikimedia.org/wikipedia/commons/f/f8/Ngandong_7-Homo_erectus.jpg).

of their newborn infants by modifying the growth pattern of the skull.[47] Brothwell's contentions have been criticized because artificial growth modifications are not expected to mimic the specific shape and structures of an archaic hominin skull as opposed to just producing aberrant anatomy. In other words, how could arbitrarily altering the growth pattern of the skull generate specific structures such as massive brow ridges?

More recently, 24 years after the discovery of the Kow Swamp remains, American anthropologist Andrew Kramer tested 17 morphological features that have been proposed to support regionalism in the Indonesia/Australia region. Kramer contrasted Sangiran-17 (Fig. 12) from Sangiran, Central Java, Indonesia to aboriginal *sapiens* from Africa and Australia. The OOA model would be supported if the African and Australian aboriginal *sapiens* group together, while the multiregional model would be supported if the Australian aboriginal *sapiens* partition with the Sangiran skull. He found that 8 of the 17 nonmetric features associate Sangiran-17 to modern Australians, while only 3 exclusively group the humans from Africa and Australia. Kramer concluded that the results support morphological continuity within the region, which implies the presence of a genetic continuum in Australasia dating back at least 1 million years and suggests multiregional evolution.[48] Other authors examining the same Indonesian *erectus* and Kow Swamp skulls have concluded some degree of regional continuity. Phillip J. Habgood, for example, identified four ancestral traits unique and continuous in the Indonesia-Australia region. These traits—a sagittally flat frontal bone with a posterior position of minimum frontal breadth, prominent facial prognathism, and zygomaxillary tuberosities—appear uniquely in fossils of the Indonesia/Australia area, but not in any other region.[49] After restoring the face of Sangiran 17,[50] Baba and colleagues concluded that its degree of similarity with Kow Swamp is not as extensive as originally proposed by Thorne and Wolpoff in 1981.[51] Yet in their study, Baba et al. only examined 7 of the 17 traits investigated by Thorne and Wolpoff. In 2007, archeologist Darren Curnoe also compared the Kow Swamp samples to the Ngandong (Solo) *erectus calvarii* (Fig. 13) from Indonesian.

To evaluate the Kow Swamp specimens' possible relationships to Late Pleistocene African and Levantine skulls (about 120,000–324,000 ya), Curnoe included the Herto, Kabwe, and Laetoli, as well as the Skhul IV and Skhul V, respectively. To prevent sexual dimorphism bias, all fossils examined were adult males. Calvarial length metrics were compared, including maximum cranial length, frontal chord, frontal subtense, maximum frontal breadth, parietal chord, occipital chord, and biasterionic breadth. These measurements have been used widely in paleoanthropology studies. Although this study failed to support the multiregional and the complete replacement models, the authors suggest that the data supports the assimilation model for the origins of the Kow Swamp population.[52] Curnoe and collaborators further state that their data indicate that admixture reached a high level in Australasia and that this region likely provides the most robust case for assimilation. Other investigations, on a more global scale, have echoed their conclusions regarding the Kow Swamp skulls. For example, in a review of the fossil evidence involving modern human origins from around the world, Trinkaus concluded in favor of the assimilation model, indicating that overall the evidence among early modern human populations suggests some degree of admixture with groups outside of Eastern Africa.[53] He further states that assimilation was neither rare nor trivial during hominin evolution and that it represented a geographically dominant force.

In the meantime, many investigators have criticized the conclusions linking the Kow Swamp sample and the Indonesian *erectus* skulls morphologically and invoking a regional continuity. Colin P. Groves (1942–2017), for example, has indicated that Kramer's study was inconclusive since he did not establish the polarity of the traits examined. The issue of polarity in phylogenetic analyses and studies is not trivial. When two states (types) exist, for example brown eyes and blue eyes, the question in terms of polarity is which is the ancestral and which is the derived state. Another way of expressing it, in the example given, do blue eyes evolve from brown eyes or vice versa? If a trait's polarity is unknown, sometimes it is difficult to assess if a given population could be the descendant or the ancestor. It may be difficult, as well, to determine if two traits are homologous (related by ancestry) or analogous (the result of convergent evolution). Polarity is usually ascertained by assessing the state of the character in a population outside the group of organisms in question (e.g., outside the hominin group). For example, if the feature in question involves hominins, the state in chimpanzees representing the ancestral condition may help establish the much-needed polarity. And Phillip J. Habgood, although he described four ancestral traits unique and continuous in the Indonesia-Australia region, also considers that some of the traits originally described by Thorne as evidence of regional continuity are retained primitive features present on *Homo erectus* and archaic *Homo sapiens* skulls in general (not the result of regional gene flow).[54]

East Asia

In 1934, the Cenozoic Laboratory of the Geological Survey of China invited Franz Weidenreich to study a number of hominin fossils from the district of Zhoukoudian, which later became known, as a group, as Peking Man.[55] During the period extending from 1927 to 1937, a limestone cave system in Zhoukoudian, southwest of central Beijing, yielded approximately 200 hominin fossils from about 40 individuals and more than 10,000 lithic artifacts. These fossils were later identified as *Homo erectus* and were later dated to a range of 680,000–780,000 ya using the $^{26}Al/^{10}Be$ method (see Chapter 4 for a description of the methodology).[56] Today it is thought that the Zhoukoudian site was occupied from 780,000 to 230,000 ya. When Weidenreich examined *Sinanthropus pekinensis*, as the fossils were originally known, he noticed certain parallelisms in the fossils' anatomy and those of modern populations from China.[57] Even in the absence of intermediate fossils connecting Peking Man with contemporary Chinese, Weidenreich claimed to identify four cranial traits in common in both groups. These initial four traits were the *torus mandibularis* (bony growth in the lower mandible along the surface nearest to the tongue), shovel-shaped incisors, Inca bones (accessory bones found in human skulls due to ossification failure), and sagittal keeling (thickening of the midline of the frontal bone). Later Weidenreich increased the number of East Asian regional features to twelve.[58] At the time, Weidenreich wrote:

> *Not less than twelve peculiarities, characteristics of the cranial and limb bones of Sinanthropus and of the teeth also—some having a high percentage of occurrence in the available specimens—occur with a relatively high degree of frequency in Mongolian groups.*

Weidenreich's early work set the stage for a school of thought that advocated for regional continuity of traits from *erectus* to more recent hominins such as Dali, Liujiang, and other recent Chinese fossils.

More recently, an additional trait has been implicated with multiregionalism. An unusual dental feature known as a 3-rooted lower second molar (3RM) is rare (less than 3.5% occurrence) in non-Asian contemporary humans, however, exceeds 40% in China and Native Americans.[59] The presence of 3RM in two Mid-Late Pleistocene hominins, one from Xiahe (northeast edge of the Tibetan Plateau), China, and the other Penghu 1 from Taiwan, dated at 160,000 and 190,000, respectively, suggests an ancient Asian origin for this trait, and one that precedes *Homo sapiens* in the region. Xiahe has been identified as a Denisovan, and as such, provides evidence that modern East Asian populations acquired the 3-rooted lower molar via introgression. It also offers morphological evidence of a strong link between archaic and recent East Asian *Homo sapiens* populations.

Initially, scholars, including Chinese anthropologists and archeologists, did not accept Weidenreich's original ideas of regional continuity. Since its discovery and for the rest of the first half of the 20th century, Chinese scientists did not recognize Peking Man as a direct ancestor of contemporary Chinese populations or modern humans as a whole.[60] Yet as the time gap in the Chinese fossil record between *Sinanthropus pekinensis* and modern Chinese populations became increasingly populated, Peking Man started being universally accepted in China as the people's ancestor. During the second half of the 20th century and up to the present, new discoveries with well-established, reliable dates have helped establish a more evenly represented fossil sequence and chronology. Table 2 provides a listing (as of July 2013) of hominin fossils from East Asia, while Fig. 14 indicates the location of the main

Table 2 Listing of hominin fossils from East Asia.

Site	Fossils	Age in thousands of years
Yuanmou	Upper incisors	1700 (paleomag)
Guojiabao	Lett tibia	1000 minimum
Gongwangling, Lantian	Partial skull	750–1150 (paleomag)
Chenjiawo, Lantian	Mandible	650–500 (paleomag)
Yunxian (Quyuan River Mouth)	Two crania	581 ± 93 (ESR) 870–830 (paleomag)
Zhoukoudian (Locality 1)	Six skullcaps, other dental, gnathic, and postcranial material	230–578 (ESR amino acid racemization, fission track, thermoluminescence, and uranium series)

Table 2 Listing of hominin fossils from East Asia—cont'd

Site	Fossils	Age in thousands of years
Hexian	Skullcap, cranial fragments, mandible, isolated teeth	150–270 (uranium series)
Jinniushan	Complete skull and much of skeleton	260 (ESR, uranium series)
Chaoxian	Occipital	160–200 (uranium series)
Dali	Complete skull	281 (ESR/uranium series)
Changyang	Partial maxilla	170–220 (uranium series)
Dingcun	Parietal, teeth	160–210 (uranium series)
Tongzi	Teeth	113–181 (uranium series)
Zhoukoudian (locality 4)	Premolar	171–124 (uranium series)
Maba	Skullcap	135–129 (uranium series)
Dadong	Four teeth	130–300 (uranium series, fauna, ESR)
Xujiayao	Cranial bones of 10 individuals	125–204 (uranium series)
Zhirendong	Mandible, two molars	>100 (uranium series)
Huanglong Cave	Seven teeth	103–44 (uranium series and ESR)
Liujiang	Complete skull and associated postcrania	Late Pleistocene
Tianyuan Cave	Mandible, associated postcrania	42–39 (AMS radiocarbon)
Chuandong	Two skulls	Late Pleistocene
Jianping	One humerus	Late Pleistocene
Lijiang	One cranium	Late Pleistocene
Miaohoushan	Two parietal fragments and one radius	28 (radiocarbon)
Jingchuan	Cranium without facial or basilar portion	48–15 (OSL)
Shiyu	Occipital fragment	28.9 (radiocarbon)
Salawusu	Cranial bones, limb bones	50–37 (uranium series)
Wushan (Yuanyuang)	Skullcap, other cranial bones	38.4 (radiocarbon)
Laibin	Cranial base	Late Pleistocene
Minatogawa	5–9 skeletons	18 (radiocarbon)
Ziyang	Incomplete skull	39–37 (radiocarbon)
Salkhit	Skullcap	Probably late Pleistocene
Maludong	Range of cranial and postcranial remains	13.6–14.3 (cal. yr. radiocarbon)
Longlin Cave	Partial cranium	11.5 K ± 255 (cal. yr. radiocarbon)
Upper Cave Zhoukoudian	Three complete skulls and associated material	34–29 (radiocarbon)

Modified from Karen R. Rosenberg, Xinzhi Wu, *The Origins of Modern Humans: Biology Reconsidered*, 2nd ed. In: Fred H. Smith, James C.M. Ahern. John Wiley & Sons, Inc.; 2013. Published 2013 by John Wiley & Sons, Inc.

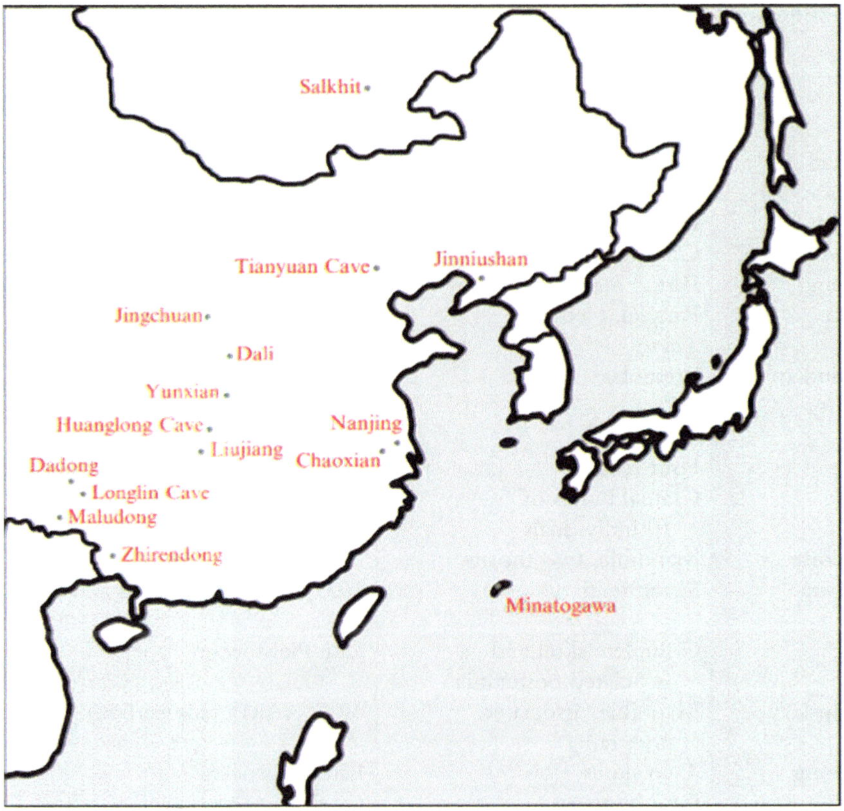

Fig. 14 Location of important fossil sites in East Asia as reported by Karen R. Rosenberg and Xinzhi Wu in 2013 (https://www.researchgate.net/publication/278318820). *(From Karen R. Rosenberg, Xinzhi Wu.* The Origins of Modern Humans: Biology Reconsidered, *2nd ed. In: Fred H. Smith, James C. M. Ahern (eds.). John Wiley & Sons, Inc.; 2013. Published 2013 by John Wiley & Sons, Inc.)*

archeological sites as reported by Rosenberg and Wu in 2013.[61] Today, most Chinese archeologists and anthropologists are convinced that there is continuity in the Chinese fossil record that is traceable all the way back to Peking Man.[62] Yet the majority of their colleagues outside China are not persuaded, and to many observers of the argument, it seems that an ideological divide exists, separating Chinese archeology from the West.

Outside of China, a good number of investigators indicate that the traits suggested as evidence of regional continuity are, in fact, indicative of affinities with hominins from the West and not from East Asia[29] and simply indicate that the features represent common ancestral traits. In general, Western scholars are unconvinced of the traits' transitional nature and point that additional well-characterized and precisely dated fossils are required before claiming regional continuity. Lahr, utilizing the previously implicated traits, has criticized the contention of continuity on anatomical grounds. Specifically, referring to the

East Asian material, Lahr also has pointed out that the so-called regional features are present in other geographical areas as well. In some instances, they are more frequent outside East Asia,[63] but have been negated as diagnostic of continuity based on the argument that they are not seen in many modern Chinese crania.[64] On more general grounds, Bae has indicated that the Chinese hominin fossil record is plagued with problems affecting reliable dating and classification of remains to specific species or populations.[65] The general attitude among many scholars outside China is that morphologically the multiregional evolution model is based on weak evidence and thus is dubious and should be further reviewed.[66]

On the other hand, proponents of the multiregional model indicate that several tendencies characterize East Asian fossils dating from the time of the Peking Man to the present. These distinctions include: the existence of regionally predominant traits in East Asia that are not observable as frequent in other parts of the world. These include an abundance of mosaicism involving ancestral and derived traits, as would be expected from a gradual transition between chronospecies (species derived from a sequential development pattern) rather than a replacement of one species by another; the ancestral features occur earlier than expected based on the chronology of a complete replacement of indigenous populations by migrant humans from Africa as stipulated by the OOA model; traits considered *sapiens*-specific appear in East Asia long before AMH are expected to arrive and replace archaic human populations; the presence in East Asia of some anatomical features characteristic of other regions (e.g., Europe); and absence of abrupt discontinuity of technology/art in the archeological record that would indicate the impact of a novel migrant population introducing a new culture.

Regarding technological transition in a newly occupied territory, the expectations are that the migration of a new culture would introduce novel cultures and traditions, especially if the new tools are superior and provide an advantage to the residents that previously practiced an antiquated technology; yet non-Acheulean lithics persisted in East Asia through the time period of the OOA dispersal and beyond. No intrusive cultures or technologies have been detected in the many decades of archeological research in Asia. The same Early Paleolithic style continues being made way into the Late Pleistocene, which ended 11,700 ya. A complete replacement scenario by modern AMH Africans displacing the archaic indigenous people should have impacted East Asia's traditions and technologies, and such a change should be apparent in the archeological record. This has not been the case.

On the issue of mosaicism, it is important to emphasize that in many of the East Asian fossils since Peking Man, the degree of admixture of ancestral and derived traits is so acute that it has been difficult to assign remains to specific taxons with any degree of certainty, forcing scholars to state, for example, as is the case of the Dali cranium, "the skull belongs to a late *Homo erectus*, *Homo heidelbergensis* or archaic *Homo sapiens*."

Most of the hominin fossils recovered from the time of Peking Man are claimed to exhibit region-specific traits as well as mosaic combinations of ancestral and derived states, and some even provide morphological influences from the West. Three important fossils (in chronological order: Hualongdong ~300,000 kya, Dali ~268,000 kya, and Xuchang ~125,000 kya) are discussed below as case studies of Middle to Late Pleistocene human evolution in East Asia. The sites were selected due to their completeness, state of preservation, and relative lack of distortion. Together and in addition to other remains from East Asia, they highlight a pattern of anatomical variation within the region and alleged cases of morphological continuity.

Hualongdong

During excavations in 2006 and then from 2014 to 2017 in the Hualong Cave near the village of Pangwang in South Eastern China, 16 hominin fossils were discovered.[67] This hominin assemblage included eight cranial fragments, seven isolated teeth, three femoral pieces, and sizeable portions of an adolescent skull, named Hualongdong 6 (HLD 6). A virtually reconstructed image of the skull is provided in Fig. 15 with mirror-imaged portions shown in gray. This skull was classified as an archaic *Homo sapiens*. From the dentition, it was determined that the individual was a 13 to 15-year-old adolescent individual. Its brain size was 1150 cc. The remains were found in the context of a bountiful collection of megafauna and an assemblage of small lithic cores and flakes. Ailuropoda (e.g., giant panda) and Stegodon (an extinct, elephant-like mammal) were the main animal fossils in the cave. Radiometric dating of fossil teeth using the U-series method (see Chapter 4 for methodology) and the associated fauna provided dates between 275 and 331 ky before the present.[68]

HLD 6 is considered by proponents of regional continuity in East Asia as the most recent challenge to the OOA theory of human evolution. The almost complete skull

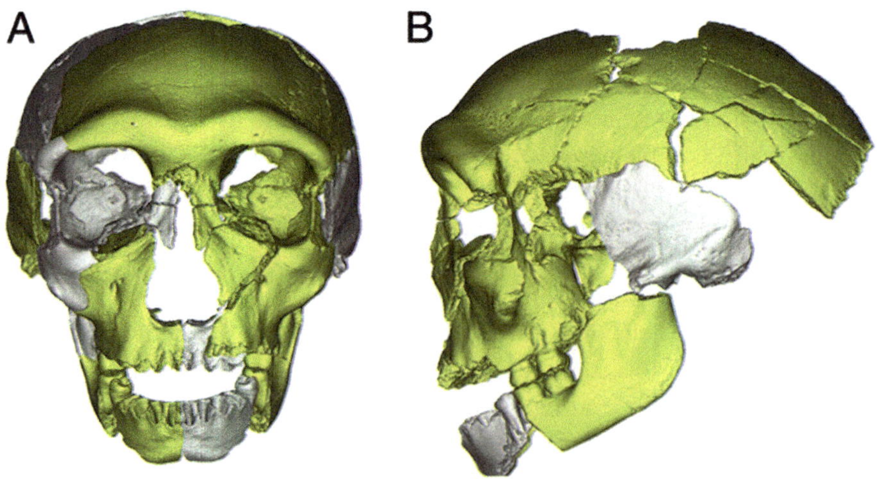

Fig. 15 Hualongdong 6 (www.pnas.org/cgi/doi/10.1073/pnas.1902396116).

presents a unique combination of features that allows investigators to draw definite conclusions regarding hominin evolution in East Asia. HLD 6 exhibits the general trends of mosaicism of ancestral and derived traits. Among the derived traits, HLD 6 exhibits encephalization (increase in brain size) and craniofacial gracilization (decrease in robustness). Specifically, HLD 6 provides many features repeatedly seen in other archaic human fossils from East Asia. These include a low and wide cranial vault, a low and wide nasal aperture (opening in the skull to the nose), and reduced or absence of third molars. The fact that these traits are common in other hominins in East Asia, evolutionarily older and younger, suggests to many a regional continuity of traits. In addition, HLD 6 has a number of traits that are transitional into the path leading to *sapiens*. Some of the cranial and dental features of HLD 6 foreshadow the transition to AMHs. These include a chin. A clear diagnosis of AMHs is the presence of a chin. In addition, HLD 6 lacks the ancestral projecting face and instead exhibits a flat face. These evident *sapiens* traits in East Asia about 300,000 ya are significant because the OOA model posits that AMHs migrated out of Africa not earlier than about 180,000 ya. So, if *sapiens* solely originated in Africa, how can AMH features be seen in East China 120,000 years earlier? This is clearly a conundrum. Advocates of the regional continuity model point to HLD 6 as one of many smoking guns.

Dali

The Dali skull was excavated in 1978 near the village of Jiefang in the Dali district, Shaanxi (Shensi) province of northwestern China.[69] Dali is a complete, well-preserved Late Middle Pleistocene cranium (Fig. 16). Recent dating of the site's fluvial deposits using OSL (see Chapter 4 for methodology) indicates that the age of Dali is between 267.8 and 258.4 kya.[70] Early studies describe the skull as more similar to *Homo erectus* than Neanderthals, likely a later stage of *erectus* that was more derived than Peking Man.[71] The Dali skull possesses a collage of features and is usually described as an intermediate or transitional hominin. Some of its features resemble *Homo erectus*, such as an overall robust vault, a cranial length that is much longer than the cranial breadth, relatively flat frontal bone, prominent brow ridges, receding forehead, a ridge along the rear of the skull, small bregmatic angle (angle form by the intersection of the coronal and sagittal sutures on the top of the skull), and thick cranial walls. On the other hand, the facial skeleton is more like *sapiens*, with low and thin cheekbones, absence of the supraorbital tubercle (small rounded projection above the eye socket), and large brain volume, among other things. The curvature of the brow ridges is similar to the condition in archaic humans from Europe and Africa. Other features are reminiscent of Neanderthals. Yet certain features such as the shape of the brow ridge, the cheekbone and upper jaw regions, the upper and lower facial skeleton, and details of the frontal vault distinguish Dali from the Neanderthals of Europe and Western Asia. Overall, Dali resembles a *sapiens* skull in possessing a

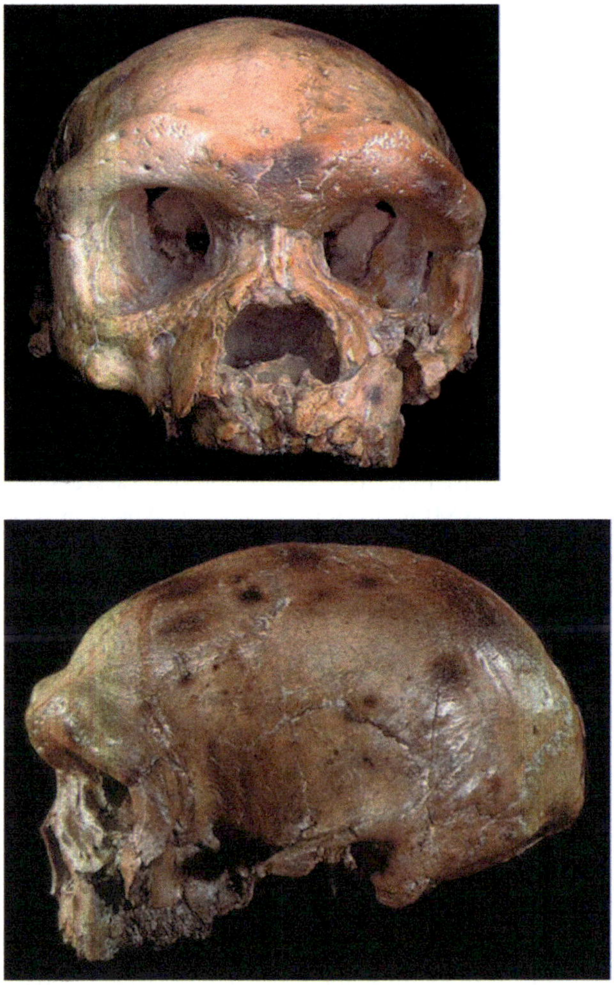

Fig. 16 Dali Man (https://www.researchgate.net/publication/278318820; http://www.peterbrown-palaeoanthropology.net/dali.html).

globular, rounded cranium as opposed to a lower, elongated skull typical of *erectus*. Its cranial capacity is 1120 cc, which is intermediate between *erectus* and *sapiens*.

Due to its unique combination of ancestral and derived traits reflecting different stages of East Asian hominin evolution since Peking Man and its good state of preservation, Dali has been the focus of intense discussions (and arguments) relating to the migration and settlement of humans in the region. Proponents of the multiregional theory claim that Dali encompasses the signals of continuity of hominins in East Asia; it shows the presence of local predominant traits in East Asia that are claimed not to be observable as frequently in other parts of the world. It clearly represents a population exhibiting high levels of mosaicism. In addition, it exhibits ancestral features earlier than expected based on the OOA chronology of invading *sapiens* from Africa and shows *sapiens*-specific traits in East Asia long before AMH were expected to arrive and replace archaic native groups. Furthermore, it demonstrates the presence in East Asia of some anatomical features characteristic of other regions (e.g., from Europe), suggesting interregional gene flow. The presence of associated small stone artifacts with the remains—mainly scrapers characteristic of *erectus*, a well-established technology in the region—and the absence of abrupt discontinuity of the old lithic tradition suggest the absence of an invading novel migrant population introducing a new culture.

According to many proponents of the multiregional theory in East Asia, the Dali population is the result of admixture in which *Homo erectus* traits were retained while incorporating features from migrating western Eurasian groups such as Neanderthals and *sapiens*. As proposed by Athreya and Wu,[72] Dali's anatomy suggests that its population was significant in the origin of early AMHs in China, not excluding the contribution of *Homo erectus*. In addition to issues relating to the multiregional hypothesis, Dali is being discussed concerning the settlement and fate of *erectus* in East Asia,[73] the features that define *Homo heidelbergensis,*[74] and its phylogenetic connections to Denisovans.[75]

Xuchang

Also crucial in the context of the issue of continuity of traits in East Asia are two fossilized crania from Lingjing, Xuchang, Henan Province, just 366 km east of Shaanxi Province from where Dali was found. The remains are two crania named Xuchang 1 and Xuchang 2, indirectly dated by OSL between about 105,000 and 125,000 ya.[76] It is important to point out that the crania from Xuchang and Dali are temporarily separated by about 135,000 years. Just as with Dali and many of the other East Asian fossils since Peking Man, Xuchang 1 and Xuchang 2 represent a collage of ancestral and derived traits in addition to a number of Neanderthal traits as well as features that are specifically associated with Eastern Eurasia hominins.

Among the derived traits, the Xuchang crania exhibit marked encephalization. Xuchang 1, the more complete of the two, has an endocranial volume of approximately 1800 cm^3, which corresponds to the high end of Neanderthals and early modern human ranges. The large cranial size of Xuchang 1 may suggest that it was a male. In general, the skull exhibits considerable gracilization (less robust overall morphology). Both crania have a lightly built cranial vault (where the brain sits), and their brow ridges are less pronounced than in other archaic humans, including Neanderthals. In terms of ancestral traits, as seen in more archaic hominins from Eastern Eurasia, the two crania exhibit low and broad braincases that round onto the inferior skull as well as low sagittally flat and inferiorly broad neurocrania. The Neanderthal features include the shape of the back

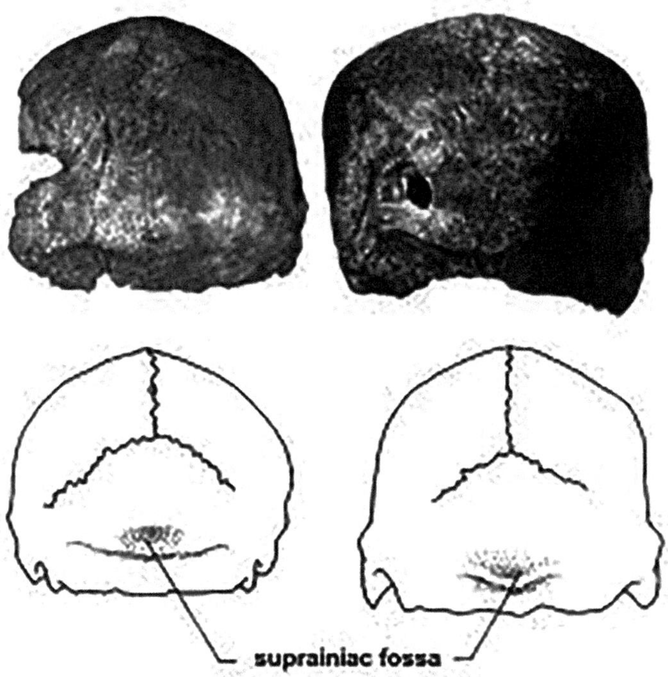

Fig. 17 Suprainiac fossa (https://www.jstor.org/stable/26292737?seq=1; https://www.google.com/search?q=suprainiac+fossa+neanderthals&tbm=isch&hl=en&sa=X&ved=2ahUKEwjmz4SY3rvrAhXOhIMKHUnPBUUQrNwCKAB6BQgBEMcB&biw=1440&bih=753#imgrc=WiHoTyA-58t0mM).

of the skulls, the presence of a suprainiac fossa (dent in the back of the head rarely seen in AMHs) (Fig. 17), and the morphology of the bony labyrinth of the inner ear, specifically, the configuration of the semicircular canals.

This extensive and complex collage of ancestral and derived traits as well as Neanderthal-like traits has not been seen before in a specimen and underscores the fact that recent human evolution was not a simple process. It was not just a genesis in Africa and a dispersal of fully evolved *sapiens* worldwide. The amalgamation of these complex combinations of traits from different times and sources suggests a complex evolutionary process. In addition, the use of the term transitional to describe the observed mosaic of ancestral and derived traits in a fossil is too simplistic. It does not satisfy the need for a more complete explanation to describe hominin evolution satisfactorily. Xuchang's crania indicate that an OOA dispersal by itself is not enough to explain the above described observations. Furthermore, it suggests there may be an element of truth in the regional continuity model. According to many proponents of the multiregional model, the overall combination of traits argues for regional continuity and interregional population dynamics. Erik Trinkaus (1948–present), one of the authors of the original study that describes the crania, stated:

The features of these fossils reinforce a pattern of regional population continuity in eastern Eurasia, combined with shared long-terms trends in human biology and populational connections across Eurasia. They reinforce the unity and dynamic nature of human evolution leading up to modern human emergence.

(100,000-year-old human skulls from East Asia reveal complex mix of trends in time, space March 2, 2017, The Source. Washington University of St. Louis*).*

In addition, the Xuchang remains are perplexing considering that Neanderthal traits ended up in Central China when Neanderthals have not been found east of Siberia. Undoubtedly, it is more likely that gene flow facilitated the morphological affinities observed as opposed to convergent evolution (evolution of similar-looking features independently, not derived from a common ancestor), and this suggests some degree of east-west population gene flow across Eurasia.

In terms of technological tradition, the lithic industry associated with the Xuchang fossils is characterized by a scarcity of prepared cores. It presents a modest and unique technological progress at the site compared to locations outside of China.[77] Like other sites in East Asia, this indicates limited technological input from outside the region. In East Asia and Xuchang, there is a lack of typical Levallois (tradition characteristic of AMHs to the west) prepared core technology. These findings corroborate previous data that characterize the progress in lithic technology in China between about 300 and 40 ka as slow and conservative, with little difference from simple cores, flakes, and tools that lack standardization. Although at the Xuchang site the technological influences from the west are evident and gene flow from Neanderthal is evident, the stone artifacts recovered illustrate that lithic technology followed a unique developmental trajectory independent of the west, at least in this part of Central China. Yet the lithic profile at Xuchang still exhibits finite impact from outside technology, suggesting limited influence by novel migrant groups.

As with most East Asian hominins since *erectus*, Xuchang exhibits the alleged multiregional evolution indicators. The two Xuchang crania exhibit the East Asia regional traits, mosaicism of ancestral and derived traits, *sapiens*-specific features in China before AMH were expected to arrive in East Asia, a mixture of East Asian and traits from other regions such as Europe indicating interregional gene flow, and the association with lithics with a unique developmental trajectory independent of the west suggesting some degree of regional isolation and no evidence of a massive replacement event by a foreign population.

Europe

In Europe, the continuity between archaic and modern humans involves two groups of hominins, Neanderthals, and AMHs. The discussions involving the regionality of traits in Europe include several specific morphological features, fossils that exhibit a mosaic of traits resembling a composite of both species, and DNA that was introgressed in both directions. Proponents of the multiregional hypothesis claim that the available data demonstrates that total replacement of archaic by modern humans did not happen. This group of scholars advocates for some degree of gene flow between the two species, although the extent of interbreeding varies from expert to expert. Some scholars indicate that assimilation was minimal (a few percentages of the current total human genome). In contrast, others suggest that Neanderthals were totally incorporated into the *sapiens* gene pool due to Neanderthals' small numbers and regional isolation as well as the overwhelming numbers of OOA invaders.

Proponents of the OOA theory, on the other hand, interpret the observed morphological similarities between Neanderthals and AMHs as different types of artifacts or simply mistaken identity. Hard-core enthusiasts of the OOA model site a number of reasons for what they consider false continuity of traits. Among the most frequently mentioned is homoplasy (traits shared by different species but not present in their common ancestor and thus resulting from convergent evolution from independent origins), features inherited from a common ancestor, widespread distribution of the traits in question (nonregional distribution), and common misinterpretation and/or bias.[24,78] That said, many of the fossils in question are poorly preserved, and some authors have suggested that these anatomical traits are in fact shared by other *Homo* species worldwide and thus were not unique to any specific regions. In other words, they are not regional.[29,79] The anatomical parallelisms suggested by proponents of the multiregional hypothesis have also been criticized based on confusing polarity (not establishing clearly the ancestral and the derived state of a character). As previously stated, this is a valid criticism considering that the polarity of traits allows for assessing who the ancestor is and who is the descendant in two related species or states of a trait. The knowledge of the polarity of features also helps determine if similar features are related by descendants (inherited from a common ancestor) or convergent evolution (similar structure but different evolutionary origins).

The lineages of Neanderthals and sapiens

The lineages of Neanderthals and AMHs separated between 500,000 and 300,000 ya.[80,81] Neanderthals evolved in Europe from a branch of *Homo heidelbergensis* about 400,000 ya while *Homo heidelbergensis* evolved from *Homo erectus* in Africa around 600,000 ya, prior to migrating to Europe.[82] It is thought that the *sapiens* lineage separated in Africa from a *Homo heidelbergensis* ancestor between 350,000 and 400,000 ya. If this chronology is correct, Neanderthals and *sapiens* have been evolving separately for about 16,000 generations (assuming a 25 years generation time) in two separate continents under distinct habitats: Neanderthals in a temperate zone, AMHs in a tropical region.

Most of the available anthropological and genetic evidence suggests that the bulk of human DNA is of recent African descent, and AMHs migrated out of Africa relatively recently. Thus, *Homo sapiens* is, for the most part, the evolutionary product of an African habitat with the corresponding selection pressures. AMHs evolved in a tropical environment under tropical selection pressure, and its general body plan reflects tropical adaptations. Moreover, the evolution of australopithecines and hominins has been restricted to the continent of Africa under tropical conditions for about 7 million years. Thus, it is not surprising that AMHs represent a predominantly African mode of evolution. Whatever physical type has been recently introduced from species such as Neanderthals, which evolved in temperate Eurasia, was superimposed on an African body plan when the two came into contact in Eurasia. This is reflected in the anatomy of fossils that are claimed to possess Neanderthal anatomical features. It is also very apparent that after about 30,000 years after contact, the Neanderthal general body plan has disappeared from Eurasia, and only limited examples of Neanderthal traits are claimed to survive in contemporary humans.[83]

One of the most compelling pieces of evidence that suggests that *Homo sapiens* originated in Africa is the abundance and the old dates reported for the hominin fossils discovered there. And although the majority and greatest diversity of australopithecines and hominin fossils have been unearthed in East Africa, more recent discoveries suggest a more heterogeneous and pan-African origin for humans. Recent findings indicate that early African *sapiens* with various combinations of ancestral and derived traits exhibiting mosaicism were living all over Africa at different times.[84] The picture that is beginning to emerge is one of the high levels of hominin diversity throughout the continent. And although AMHs are today relatively genetically homogeneous throughout the world, the species had diverse origins—multiple African origins.

Presently, the oldest recorded *Homo sapiens* fossil was discovered in western North Africa about 50 km southeast of the city of Safi in East Morocco.[85] The skull was dated at 315,000 ya by thermoluminescence (see Chapter 4 for a description of the methodology) and initially was thought to represent the skull of a Neanderthal. Currently, it is considered a *Homo sapiens*. In addition to the almost complete skull, a jawbone, teeth, and limb bones from three adults, a juvenile, and a child approximately 7 years old have been discovered. These findings push back the genesis of *sapiens* in Africa by 100,000 years. Together with many other early *sapiens* fossils, such as the Florisbad Skull from South Africa and Omo Kibish from Ethiopia, dated at 260,000 ya and 200,000 ya, respectively,[20] they not only support the species' origin in Africa but also suggest a complex pan-African process. Not unlike other early African archaic humans, the Jebel Irhoud skull exhibits a mosaic of features including facial, mandibular, and dental morphology that signals affinities with early anatomically modern humans while possessing ancestral neurocranial and endocranial morphology. Specifically, the relationship of the face to the vault is modern, lacking prognathism (bulging out or protrusion of the lower jaw). In terms of ancestral features, it has a relatively large lower jaw, a very thick and continuous supraorbital torus (brow ridge), and an elongated braincase.[21] When the Jebel Irhoud skull was initially described, the combination of ancestral and derived features was so striking that it was characterized as a possible Neanderthal/*sapiens* hybrid.[86]

Early enclaves of *sapiens* in Southwest Asia and Europe

Apidima

Similarly, the *Homo sapiens* that migrated out of Africa present a complex combination of traits. The fossils found in southwest Asia and Europe represent a mosaic of features with various mixtures of ancestral and derived traits. Let's consider, for example, Apidima 1, the oldest *sapiens* fossil found outside of Africa. The fossil was discovered in a small cave located on the western coast of the Mani Peninsula in Southern Greece. The site, excavated in the late 1970s, yielded two partial skulls minus the lower jaw.[12] They were named Apidima 1 and Apidima 2 and dated at 210,000 and 170,000, respectively, using the uranium-thorium method (see Chapter 4 for methodology). Apidima 1 presents a mosaic of ancestral and modern human features. The cranium is rounded like a soccer ball, a typical trait of *sapiens*, and it lacks the characteristic Neanderthal features, such as the distinctive occipital bun (Fig. 8) in the back of the skull. The second skull, Apidima 2, exhibits typical Neanderthal-like morphological features. Subsequent to the initial discoveries, the Apidima caves have provided additional Neanderthal and AMH fossils and artifacts. This data is significant in several ways. First, it pushes back the presence of *sapiens* in Europe by 100,000 years. Second, it shows that the site was populated by early *Homo sapiens* prior to being occupied by Neanderthals, who arrived 40,000 years later. It also suggests co-habitation and possible interaction of the two species for a protracted period of time and an eventual replacement of humans by Neanderthals, not the other way around. Furthermore, it supports a scenario in which AMHs dispersed out of Africa multiple times during various periods and points out that the demographics of *sapiens* was not a simple single dispersal out of Africa, but it occurred in a complex series of migrational waves.

The Apidima findings also raise questions regarding the fate of this early *sapiens* enclave as well as all the others that follow before the well-known Africa dispersal of 70,000–50,000 ya (OOA). In other words, what happened to these populations? Given that these *sapiens* existed when there was a substantial Neanderthal presence in Southwest Asia and Europe, it is possible that AMHs were unable to compete successfully with Neanderthals, especially in the unstable climate of the time. It is possible that these AMHs failed to leave their genetic imprint in contemporary human populations. Also possible is that some of these *sapiens* may have interbred with the local indigenous groups, such as Neanderthals, being absorbed into their gene pool, contributing to the proposed continuity of traits.

Zuttiyeh

The Apidima 1 fossil is not unique in indicating that *sapiens* migrated out of Africa in several waves long before the out of Africa dispersal of 70,000–50,000 ya that is typically associated with the peopling of the world. Other examples include the Zuttiyeh, Misliya, Skhul, and Qafzeh AMH remains. All of these enclaves suggest intricate scenarios as *sapiens* attempted to migrate out of Africa—an evolutionary history far more complicated than proposed by the OOA model. The Zuttiyeh specimens consist of a craniofacial fossil discovered in Israel in 1925 and dated radiometrically from the archeological context between 200,000 and 500,000 ya.[87] And although the window of time is substantial, it may represent one of the earliest *sapiens* outside the African continent. Zuttiyeh has been described over the years as a Neanderthal, early anatomically modern *Homo sapiens*, and a generalized specimen ancestral to both. Some scholars believe that the Zuttiyeh fossil is a *Homo heidelbergensis*.[88]

Misliya

Misliya is represented by the jaw of a modern human from Israel first described in 2018. It was dated directly to 177,000–194,000 ya using the uranium-thorium method.[22] The Misliya maxilla was excavated in the context of associated Levallois lithics suggesting that the emergence of this technology is linked to the presence of *Homo sapiens* in the region and previously in Africa. Just as Apidima 1, Misliya indicates that *sapiens* departed Africa much early than previously envisioned and, just as Apidima 1, suggests that AMHs dispersed at least as 220,000 ya.

Skhul and Qafzeh

Other early modern human fossils have been found at Skhul and Qafzeh in Southwest Asia in Israel dating to about 130,000–90,000 ya. Recent age estimates indicate that the history of this wave of occupation was more prolonged and complicated than initially envisioned. This recent data suggest that Skhul was occupied about 130,000–100,000 ya, while Qafzeh was active from around 110,000–90,000 ya.[89] The Skhul remains were excavated between 1929 and 1935 on Mount Carmel. Seven adults and three children were recovered. Anatomically, the Skhul individuals exhibit more ancestral features compared to the Qafzeh remains. The Skhul crania are a mosaic of ancestral and derived traits. Ancestral traits include an overall robustness with prominent brow ridges and jutting jaws, while their braincases are rounded as in modern humans. When first described, this combination of traits was so confusing that various experts classified the fossils as Neanderthals, AMHs, or hybrids of the two. These assertions were not far-fetched considering that Neanderthal fossils dating to 61,000–48,000 years ago have been discovered in the vicinity at the Kebara Cave.[90]

At Qafzeh, a total of 15 individuals have been found, including 8 children. The remains were recovered in the context of Mousterian artifacts, some suggesting burials. Although Mousterian lithics are associated with Neanderthals, they were also produced by early AMHs. The morphology of the Qafzeh fossils also provided a mixture of ancestral and derived traits. For example, one of the specimens, Qafzeh 9, shows a high forehead and a *sapiens*-like nonprognathic face (jaw does not project forward) and lacked an occipital bun and a distinct chin. Interestingly, in a nearby site of Mount Carmel known as the Tabun Cave, a set of teeth and a skeleton that belonged to Neanderthals, dated at 90,000 and 122,000 ya, respectively, were discovered.[91,92] These dates suggest that these Neanderthals were contemporaneous with the Skhul and Qafzeh *sapiens* and likely interacted technologically and biologically by interbreeding, again explaining the many cases of fossils with mosaic features and regional continuity. The population dynamics of the Skhul and Qafzeh *sapiens* and Neanderthal populations from the Levant indicates that their occupation was neither short nor culturally uniform.

Were Neanderthals and sapiens at the fringe of reproductive isolation?

Suppose the Neanderthal and *sapiens* lineages split approximately 500,000 ya and then came into contact again in Southwest Asia and Southeast Europe about 210,000 ya in places such as Apidima. In that case, their gene pools may have experienced gene flow approximately 290,000 years after separation. Potentially bringing these two lineages into physical proximity may have provided for interbreeding. Clearly, there is evidence that some technological and DNA transfer took place.[93] Yet a lingering question that remains is how reproductively compatible these two species were after such a protracted period of separation.

It is expected that Southwest Asia/Europe and Africa's unique environmental conditions likely provided unique sets selection pressures resulting in unique evolutionary paths for the two lineages during their isolation. Considering their extensive partition and the very different habitats in which they evolved, it is likely that when AMHs and Neanderthals met in South West Asia and then in Europe, they may have been at the edge of biological incompatibility. In other words, at the time of their contact, their capacity to produce fertile hybrids may have been small or minimal. Therefore, it is no surprise that reduced male fertility and rapid natural selection to remove the Neanderthal-derived DNA may have acted to create a partial reproductive barrier between the two species.[94] Additional evidence of partial biological/reproductive incompatibility between these two species is supported by data indicating that an early AMH dated 40,000 ya from East China (the Tianyuan individual) possessed about the same proportion of Neanderthal or Denisovan DNA in its genome as present-day people from the same area. This suggests that the loss of Neanderthal and/or Denisovan introgressed genetic material from early AMHs occurred soon after interbreeding.[95] And if indeed Neanderthals and *sapiens* were at the verge of reproductive isolation at the time of contact, the interbreeding evidence that is seen today in the fossil record and ancient and contemporary DNA represents only a fraction, an underestimation of the actual level of introgression that occurred during contact.

Furthermore, it has been determined that natural selection has acted against Neanderthal genes in Neanderthal-*sapiens* hybrids.[96] In turn, that has resulted in low Neanderthal ancestry distributed throughout the genomes of many contemporary human populations.

It is thought that this "purifying selection" acted against Neanderthal DNA when it was introduced into much larger *sapiens* gene pools. It is theorized that this condition likely imposed a heavy genetic load on the early-generation human/Neanderthal hybrids, eliminating them rapidly. Considering the smaller population size of Neanderthals, this situation diminished not only the hybrids but also eliminated an already decreasing Neanderthal population by absorbing them into the *sapiens* gene pool. All of these factors together likely contributed to the rapid diminution of Neanderthal genetic contributions and the elimination of Neanderthal's features a few thousand years after contact. This type of assimilation process and rapid negative selection against hybrids would act to obscure the actual regional gene flow between species making it difficult, thousands of years later, to detect regional continuity. Based on these results, it is clear that assessments of continuity of traits may be impacted by the time that passed subsequent to the contact between the various populations. Therefore, estimates of interbreeding between groups are best done by examining specimens close to the period of contact.

Regional continuity in Europe

A number of specific morphological traits in Neanderthals and *sapiens* have been reported in the literature, suggesting some regional continuity in Europe. Initially, in a series of articles in the early 1990s, a small group of scholars that included David W. Frayer (1947–present), reported on a total of four traits that in combination were identified as signs of regional continuity in Europe.[97,98] These included a horizontal-oval shaped mandibular foramen (opening on the interior medial surface of the lower mandible that allows the entrance of a nerve and vessels) (Fig. 18), anterior mastoid tubercle (a small rounded projection on the mastoid bone behind the ear), suprainiac fossa (dent in the back of the head) (Fig. 17), and reduction of the nasal breadth (distance between the two most lateral points on the rim of the nasal opening) associated with tooth-size reduction.

In 2005, Fred H. Smith and colleages[32] published further observations on the topic of regional continuity in Europe and pointed out two additional features linking Western Neanderthal with early modern European populations. In addition to the features previously described by Frayer and collaborators, Smith et al. discuss at length upper midfacial prognathism (the forward projection of the upper-middle facial region, including the nose) and the occipital bun (Fig. 8). Regarding prognathism, it was noticed that while the earliest modern Europeans show significant reduction of this trait relative to Neanderthals, they exhibit more of the feature than Skhul and Qafzeh *sapiens* from the Near East. In this context, it is important to emphasize that the Skhul and Qafzeh specimens are not only older than the earliest modern Europeans, but are also thought to represent their ancestors. Thus, the early modern Europeans exhibit a distinct characteristic that is not often seen in their presumed Near Eastern forebears. If prognathism is a condition inherited from the hominin African ancestors by way of the Southwest Asian OOA migrants,

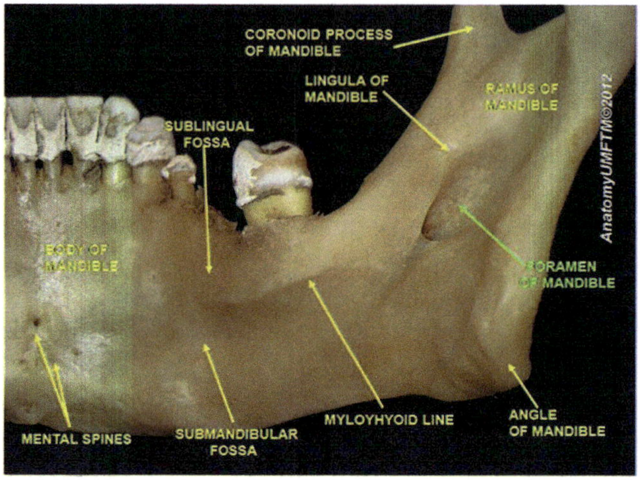

Fig. 18 Mandibular foramen (https://www.dentalcare.com/en-us/professional-education/ce-courses/ce325/inferior-alveolar-nerve-block).

the expectations are that the feature would be prominent in the Near East. Thus, based on these observations, Smith and colleagues suggested that some degree of Neanderthal ancestry might explain the observed prognathism. Similarly, early modern Europeans also exhibit a high incidence of the occipital bun (Fig. 8), while neither the Skhul or Qafzeh nor early modern African specimens show this trait at all. Since the Skhul, Qafzeh, and African early moderns are regarded as the ancestors of the early modern Europeans, the absence of the bun in the former three groups suggests that the trait in early modern Europeans cannot be derived from them and raises the possibility that "bunning" was inherited from Neanderthals.

Further comparative morphological work by Erik Trinkaus (1948–present) and collaborators in 2007 has added to the list of anatomical features that link AMHs to Neanderthals in regional continuity in Europe.[99] In their work, Trinkaus and colleagues selected only European fossils classified as early modern humans (e.g., Brassempouy, Cioclovina, Mladeč, Muierii, Oase, Les Rois, and La Quina Aval), all older than 33,000 before the present, and Gravettian human remains, and compared them with Western Neanderthals and African Middle Paleolithic modern fossils. The early modern humans represent the first *sapiens* to populate Europe, and the Gravettian, the group that succeeded them. The African Middle Paleolithic modern humans are regarded as the ancestors of the early modern Europeans. In selecting these groups, Trinkaus et al. wanted to assess traits present in early modern humans, Gravettians, and Neanderthals but absent in African Middle Paleolithic modern humans. The rationale was to identify features that were not inherited from the African forefathers to the European *sapiens* but may have been transmitted from Neanderthals. They uncovered a number distinctive Neanderthal and/or plesiomorphic (ancestral) features that had been lost among the African Middle Paleolithic modern humans. These traits included aspects of cranial anatomy, external morphology of the base of the skull, shape of the mandible, size and shape of teeth, morphology and size of clavicles, scapulae, shape and size of metacarpals (intermediate part of the skeletal hand), and limb proportions. Trinkaus et al. concluded, "the ubiquitous and variable presence of these morphological features in the European earlier modern human samples *can only be parsimoniously explained* as a product of modest levels of assimilation of Neanderthals into early modern human populations as the latter dispersed across Europe."

Does DNA show regional continuity?

mtDNA

Ever since the seminal report of Cann, Stoneking, and Wilson in 1987 demonstrating that the mtDNA of 147 worldwide individuals was of recent African origin, most of the scientific community, as well as the public at large, have been convinced that the cradle of humanity was Africa.[23] Although the notion that *sapiens* originated in Africa can be traced back to Charles Darwin (1809–82) in his book *The Descent of Man, and Selection in Relation to Sex*[100] more than 150 ya and Thomas Henry Huxley's (1825–95) comparative anatomy studies,[101] demonstrating the similarities between humans and great apes, it was not until modern DNA studies such as Cann et al.'s mtDNA work that the OOA model became a household phrase. Cann et al.'s data fueled the interest in the origin of humans and specifically the OOA model.

Rebecca Cann and collaborators' work was restricted to the DNA of the mitochondria, an organelle that is located in the cytoplasm of cells. mtDNA is inherited strictly from mother to children; thus it is transmitted uni-parentally from generation to generation. It is not inherited like nuclear DNA from both parents. They only examined the DNA of contemporary humans. Since mutations accumulate in the DNA as a function of time, Cann and her group used the number of mutations (DNA diversity) to assess the time since the origin of *sapiens*. They assumed that all contemporary humans shared a common maternal ancestor. In other words, they determined how frequent mutations occurred and created a molecular clock to translate the number of mutations into the number of years since the origin of humans.

In doing the calculations, Cann and colleagues made some assumptions:

(1) All contemporary people shared a common maternal ancestor with one mtDNA sequence.

(2) The mutation rate is constant.

(3) mtDNA is not under selection pressure.

(4) There is no dropout of mitochondria lineages.

These assumptions are not always valid and may compromise the conclusions that all of the mtDNA lineages that are seen today originated in Africa and that all the contemporary lineages of mtDNA shared a common ancestor 280,000–117,000 ya.

The premise of a single common maternal ancestor from a single region of Africa (e.g., East Africa) with identical mtDNA sequence has recently become a point of contention resulting from the work of Eleanor M.L. Scerri of the Max Planck Institute and colleagues. In their recent article, they argue that although *sapiens* evolve in Africa, archeological and genetic data suggest that *Homo sapiens* evolve all over the continent; not in a single location.[102] The idea of a single region of origin is too simplistic based on the available data. They argue for a far more complicated history of humanity in which human groups from all over the continent contributed to that process. They describe a dynamic, interconnected patchwork of populations engaging in gene flow. In this scenario, the likelihood of a single maternal ancestor for all AMHs may not be realistic. As indicated by M.H. Wolpoff, the mitochondrial Eve extrapolated by Cann et al. was "not the most recent common ancestor of all living people" since "mitochondrial history is not population history".[103]

The assertion that mutation rate is constant in time is not correct. Mutation rate is dependent on the type of mutation (e.g., single nucleotides versus chromosomal translocations), location in the genome (e.g., mitochondria versus the nonrecombinant portion of the Y chromosome), and time (2 mya versus the present). For example, mutation rate is thought to have increased since the beginning of the Holocene 11,000 ya.[104]

The assumption that mtDNA is not under selection pressure is also incorrect. Although it was thought that mtDNA was neutral as far as selection pressure is concerned, today, it is known that mitochondrial genes are under strong selective pressures to maintain mitochondrial function via removal of deleterious mutations via purifying selection.[105] Strict selection is required to sustain cellular respiration and metabolism. Thus, the mtDNA mutations that are observed in organisms are not the result of a random process. Detrimental and deleterious mutations are selected against. Some mtDNA mutations are incompatible with life interfering with development and promoting spontaneous abortions. Specifically relating to recent hominin evolution and brain growth, it is known that the brain is a high-maintenance organ, and has been under strong selection pressure for larger size for the last 200,000 years.[106] Based on these observations, it is reasonable to expect that the accumulation of mtDNA mutations has changed during human evolution, especially in sequences involved in energy production.

Mitochondria are also highly susceptible to dropout events, which lead to the loss of mtDNA lineages.[107] The impact of such an event is that archaic human mitochondrial types from various interbreeding events could have been lost due to genetic drift or a selective sweep. In other words, random elimination of mitochondrial lineages may occur during mating, especially if the number of individuals with a given mitochondria type is small. This happens because certain individuals with a given mtDNA type may fail to reproduce. This phenomenon may provide the false impression of mitochondrial homogeneity when, in fact, in the past, a larger number of mtDNA lineages existed. This, in turn, would lead to incorrect assessment of origins, number of ancestors, and time since a common ancestor.

Genetic expectations of the OOA model

As it relates to DNA, one of the expectations of the OOA model is that genetic diversity would be greater in Africa compared to Eurasia. The expectation is that as *sapiens* dispersed out of the continent, random and sequential partitioning of the original African DNA would tend increasingly to reduce the genetic variability in the migrating wave. Thus, the further away from Africa the migrants traveled, the less genetically diverse they became. Well, the genetic diversity in African populations is greater than among Europeans and Asians. In a study sequencing 50 noncoding (not coding proteins or RNAs) DNA segments, each about 500 bp long, in 10 Africans, 10 Europeans, and 10 Asians, it was determined that the average nucleotide diversity for the 50 segments is only 0.061% ± 0.010% among Asians and 0.064% ± 0.011% among Europeans but almost twice as high (0.115% ± 0.016%) among Africans.[27] The data indicated that genetic diversity is greater among Africans than among Eurasians and that genetic diversity among Eurasians is largely a subset of that among Africans, supporting the OOA model. Yet not all lines of evidence support a recent origin of the entire modern human gene pool in Africa. In this regard, one report points out that AMH populations from Africa could have ultimately replaced archaic human populations in Asia and Europe. However, the data does not rule out some degree of assimilation of genetic material and morphological traits from archaic resident populations.[108] This model proposes that AMHs migrated out of Africa in a continuous expansion wave, not in a single migration or several isolated pulses. The admixing with local archaic groups occurred with natural selection or genetic drift, eventually favoring the modern type (the African type).

Inconsistencies involving the OOA model

Since the publication by Cann et al., the human-origin paradigm, for the most part, has been that *Homo sapiens* is entirely an African species. It was not until the genomic DNA (nuclear DNA, not mtDNA) sequences from Neanderthals were assessed, and interbreeding between humans and Neanderthals discovered, that the dogma of "100% African DNA in *sapiens*" was challenged.[109] Today the genetic data indicates that the bulk of DNA in the contemporary human gene pool derives from Africa. Yet the notion of *Homo sapiens* being an entirely recent African product is simply false. The reason for this assertion is simple, not all the data, morphological or genetic, support a recent origin of the entire modern human gene pool. In addition to the morphological evidence presented in this chapter from various geographical regions and the well-publicized Neanderthal and Denisovan interbreeding events, genetic data now argues that non-African DNA, attributed to ancient interbreeding events, is part of the contemporary human genome. Thus, *sapiens* is not entirely of recent African origin.

According to the OOA model and the time of dispersal set by the mtDNA sequences, AMHs migrated out of Africa about 60,000–70,000 ya.[110] Additional estimates based on many different types of DNA sequences such as Y-chromosome, autosomes (nonsex chromosomes), whole genome as well as mtDNA indicate that the exodus occurred sometime between 40,000 and 100,000 ya.[111] These estimations are calculated by assessing the amount of genetic diversity in specific DNA sequences (e.g., the mtDNA sequences of a species), and from those levels of diversities, coalescence time or time to the most recent common ancestor is generated. Coalescence time indicates when the earliest ancestor of the species existed. As previously stated, the OOA model posits that subsequent to the migration out of Africa, archaic hominins in Eurasia were totally substituted by the newcomers.

Yet several inconsistencies have been reported concerning the time of the exodus as stipulated by the OOA model. For example, in an analysis of 15 noncoding DNA locations on the X chromosome, it was found that the most recent common ancestor (coalescence time) of each DNA site dates back to different times in the past, and some are much older than expected by the mtDNA OOA predictions.[112] Some of the DNA sequences provide levels of diversity that indicate an origin about 2 mya, suggesting an ancient population split around the time hominins first emerged from Africa as *Homo erectus*, rather than more recently as indicated by the mitochondrial data. In other words, the data suggest that some of the DNA in contemporary humans is derived from hominins as old as *Homo erectus*. These old pieces of DNA in contemporary AMHs are inconsistent with the predictions of the OOA model and a total replacement of archaic populations. Although most of the 15 DNA locations exhibit greater diversity in Africa, which suggests an African origin, some of the DNA segments originated in Asian populations supporting an Asian source for the sequences. Furthermore, four of the sites that exhibit greater diversity in Africa reveal regional diversity incompatible with a *single* expansion from Africa as stipulated by the mtDNA Eve hypothesis. The regional and time of origin differences for the various African X-chromosome sites contradict a single recent migration.

Similarly, further investigations reported in a series of articles by Alan R. Templeton have reported coalescence times that contradict the dates and frequency of migrations as previously generated from the OOA Eve mtDNA data. For example, in a 2002 paper, Templeton explored DNA regions that included mtDNA, Y-chromosomal DNA, two X-chromosome regions, and six autosomal sites.[113] The data again reinforced the dominant role that Africa played in shaping the *sapiens* gene pool by indicating a total of three major dispersals out of Africa. The results signal the initial OOA expansion event of *Homoerectus* about 1.9 mya that, according to the data, contributed to the contemporary *Homo sapiens* gene pool. A second OOA dispersal took place around 700,000 ya that involved interbreeding with at least some Eurasian populations, and a third OOA migration that occurred about 100,000 ya also involved interbreeding. The data suggest that interbreeding was not the exception but the rule since *erectus* times. Overall, the data demonstrate that some DNA variants present in humans today were already in existence in Asia hundreds of thousands of years ago before the 100,000 ya migration. In other words, even if *sapiens* uni-parental DNA markers (i.e., mtDNA and Y chromosome) migrated recently during the OOA 100,000 ya dispersal, archaic humans already possessed older genes from previous migrations out of Africa.

Templeton presents additional information on inconsistencies involving the OOA model in a second 2005 paper, which describes the origin of *sapiens* DNA sequences long before the recent 100,000 ya migration.[114] In this publication, 25 genetic sequences located on the X chromosome and autosomes were examined for their coalescence times and gene flow between Africa and Eurasia. Of the 25 genetic sites, 19 reflect continuous gene flow for at least 1.46 mya, and only 5 are associated with the recent 100,000 ya OOA migration. Three are linked with the 2 mya expansion of *Homo erectus* out of Africa into Eurasia and seven with an intermediate dispersal involving people practicing the Acheulean tool tradition associated with *Homo erectus* and derived species such as *Homo heidelbergensis*. The coalescence time distributions of these 25 genetic locations are illustrated in Fig. 19. A number of back to Africa migrations were also detected during the period subsequent to the most recent OOA dispersal. Of particular significance in this study was the rejection of the complete replacement hypothesis of archaic by modern hypothesis with the statistical certainty of greater than 99% ($P < 10^{-17}$).

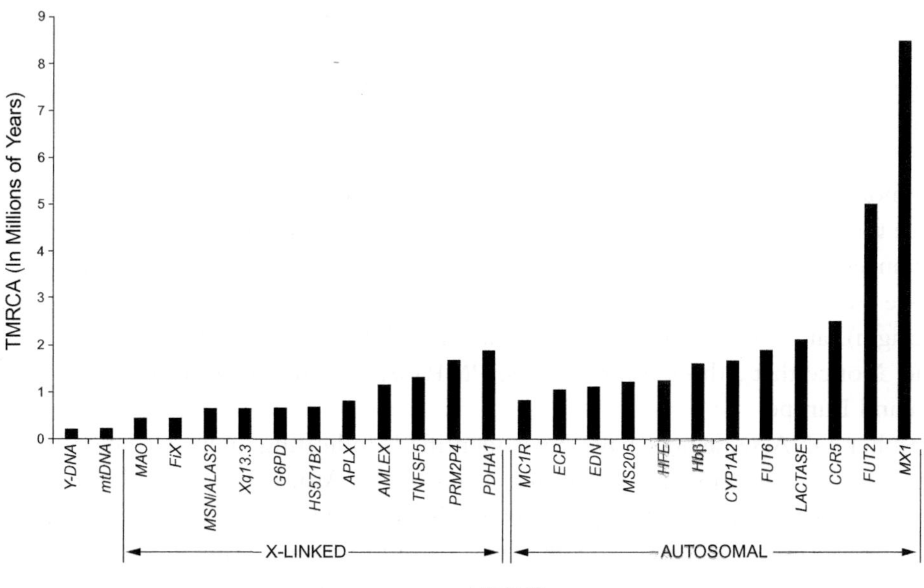

Fig. 19 Estimated coalescence times for 25 human DNA regions (https://www.semanticscholar.org/paper/Haplotype-trees-and-modern-human-origins.-Templeton/b2a1bc95354ae2a2677b0d9a9fec0531af19b305/figure/4). *(Details and references for DNA regions studied are given in Templeton AR. Population biology and population genetics of pleistocene hominins. In: Henke W, Tattersall I, eds.* Handbook of Paleoanthropology. *Berlin: Springer.)*

Table 3 Polymorphism in worldwide DNA sequence data, time back to the most recent common ancestor (TMRCA), and birthplace of the most recent common ancestor (PMRCA).

Gene or site	TMRCA (Myr)	PMRCA
Plpc	1.28	Af/As
Hprtc	0.53	Af/As/Eu
Gkc	0.41	As
Pdha1c	1.05	As/Af/Eu
Pdha1	1.59	Af
dys44c	1.35	Af
Zfx	0.93	Af
Xq13.3	0.56	Af
Ace	1.11	–
Lpl	0.91	Af/Eu
b-globin	1.36	Af/As/Eu
Mc1r	0.71	Af/As
mtDNAf	0.20	Af
YAPg	0.19	Af

Adapted from Naoyuki Takahata, Sang-Hee Lee, Yoko Satta. Testing multiregionality of modern human origins. *Mol Biol Evol.* 2001;18(2):172–183. https://academic.oup.com/mbe/article/18/2/172/1079265.

Over the years, some other investigators have reported ancient origins for genes with DNA sequences reflecting coalescence times dating back all the way to *Homo erectus* and suggesting admixture involving archaic and recent groups outside of Africa. Table 3 illustrates a partial listing of such genes and sites compiled by Naoyuki Takahata (1946–present) and collaborators.[115] The time estimations are provided as a time to the most recent common ancestor (TMRCA) or coalescence time, and they range from 0.19 to 1.59 mya, while the places for the most recent common ancestor (PMRCA) or origin of the DNA sequences are given as Af (Africa), As (Asia), and Eu (Europe). The PMRCAs are based on the level of DNA sequence diversity of the genes (higher diversity suggesting location of origin), and multiple locations indicate that different continents share the ancestral gene. Notice that although most of the PMRCAs are in Africa, some show origins in Asia and Europe.

Other DNA sequences have been identified, indicating ancient coalescence times dating to the expansion time of *Homo erectus* out of Africa. For example, the gene for CMP-*N*-acetylneuraminic acid hydroxylase or CMAH in contemporary human has a coalescence time of 2.9 mya.[116] The gene exists in two forms: one is confined to Africa, suggesting that some African populations remained isolated in the continent, while the second form migrated multiple times to Eurasia, likely with *Homo erectus*. Furthermore, Eurasia's CMAH diversity pattern suggests interbreeding between the 2.9 my-migrants and the resident archaic human groups. The data signals an ancient dispersal without a complete replacement of archaic inhabitants by modern humans.

Similarly, the gene for the X chromosome-linked pyruvate dehydrogenase *E1* or PDHA1 exhibits deep coalescence time dating back to 1.86 ya, and the diversity pattern is inconsistent with a single origin of modern humans.[117] In addition, unlike mtDNA data, PDHA1 does not show evidence of human population expansion from Africa and thus is inconsistent with a recent species origin.

The gene for *N*-acylsphingosine amidohydrolase or ASAH1 is involved in mental processes in humans. It exists in *sapiens* in two forms, V and M, with their coalescence time being 2.4 mya. The distribution and diversity pattern of the two forms suggest that subsequent to a recent population expansion from Africa, a population carrying the V lineage hybridized with archaic humans outside Africa. It is speculated that the V form spread into the entire archaic resident population by positive selection. The data presented by Kim and Satta[118] is incompatible with a recent origin of a modern human species that replaced archaic forms without interbreeding.

Arylamine N-Acetyltransferase or NAT1 also exhibits a deeply rooted tree with a coalescence time of 2.0 mya.[119] NAT1 shows its origins in Eurasia, not Africa. The authors of the study conclude that the simplest explanation for the data is partial hybridization between modern humans expanding from Africa and the preexisting archaic Eurasia populations and a rejection of the total replacement hypothesis.

Some concluding remarks

Based on the evidence presented in this chapter, the total replacement hypothesis fails to explain the available information. In addition, as additional data becomes available in the form of morphological and DNA data, continuity of traits would likely become more apparent. Multiple dispersals of hominins from Africa to Eurasia and back to Africa have occurred since the *Homo erectus* migration of 2 mya, and some have contributed to contemporary humans' gene pool. Thus, *sapiens* is not purely of recent African origin. As it stands now, AMHs are still considered predominantly of African descent but with the incorporation of various amounts of archaic forms. If the percentage of admixture with resident archaic DNA is just about 10%, the assimilation was modest, and for some, that is still OOA, while for others, it shows that the multiregional model is valid. It is likely that in the next few decades, with additional analysis of extant and extinct groups, a much better understanding of the human genetic makeup will be attained. The available data based on frequent instances of coexistence (see Chapter 1) among multiple hominin groups suggest that the details of interactions, admixture, and continuity were likely complex and region-specific. Thus, a more comprehensive and extensive survey of the worldwide genetic landscape should provide a more precise assessment of the continuity of traits among populations.

References

1. Selznick DO. In: Behlmer R, ed. *Memo from David O. Selznick (reprint ed.)*. New York: Modern Library; 2000.
2. Fuller ZL, Leonard CJ, Young RE, Schaeffer SW, Phadnis N. The role of chromosomal inversions in speciation. *bioRxiv*. 2017;21177.
3. Scally A. The mutation rate in human evolution and demographic inference. *Curr Opin Genet Dev*. 2016;41:36–43.
4. Kiwoong N, Munch K, Mailund T, et al. Evidence that the rate of strong selective sweeps increases with population size in the great apes. *PNAS*. 2017;114:1613–1618.
5. Ayala FJ, Conde Camilo JC. *Processes in Human Evolution: The Journey from Early Hominins to Neanderthals and Modern Humans*. UK: Oxford University Press; 2017.
6. Trinkaus E. Femoral neck-shaft angles of the Qafzeh-Skhul early modern humans, and activity levels among immature near eastern Middle Paleolithic hominids. *J Hum Evol*. 1993;25(5):393–416.
7. Wolpoff MH, Wu X, Thorne AG. Modern *Homo sapiens* origins: a general theory of hominid evolution involving the fossil evidence from East Asia. In: Smith FH, Spencer F, eds. *The Origins of Modern Humans: A World Survey of the Fossil Evidence*. New York: Liss; 1984.
8. Wolpoff MH, Hawks J, Caspari R. Multiregional, not multiple origins. *Am J Phys Anthropol*. 2000;112:129–136.
9. Hawks JD, Wolpoff MH. Sixty years of modern human origins in the American Anthropological Association. *Am Anthropol*. 2003;105:89–100.
10. Wolpoff MH, Spuhler JN, Smith FH, et al. Modern human origins. *Science*. 1988;241:772–774.
11. Stringer C. Human evolution: out of Ethiopia. *Nature*. 2003;423(6941):692–693.
12. Harvati K, Röding C, Bosman AM, et al. Apidima Cave fossils provide earliest evidence of *Homo sapiens* in Eurasia. *Nature*. 2019;571:500–504.
13. Bae CJ, Douka K, Petraglia MD. On the origin of modern humans: Asian perspectives. *Science*. 2017;358:eaai9067.
14. Been E, Hovers E, Ekshtain R, et al. The first Neanderthal remains from an open-air Middle Palaeolithic site in the Levant. *Sci Rep*. 2017;7:2958.
15. Liu W, Martinón-Torres M, Cai YJ, et al. The earliest unequivocally modern humans in southern China. *Nature*. 2015;526:696–699.
16. Haber M, Jones AL, Connell BA, et al. A rare deep-rooting D0 African Y-chromosomal haplogroup and its implications for the expansion of modern humans Out of Africa. *Genetics*. 2019;212:1421–1428.
17. Clarkson C, Jacobs Z, Marwick B, et al. Human occupation of northern Australia by 65,000 years ago. *Nature*. 2017;547:306–310.
18. Macaulay V, Hill C, Achilli A, et al. Single, rapid coastal settlement of Asia revealed by analysis of complete mitochondrial genomes. *Science*. 2005;308:1034–1036.
19. Hammond AS, Royer DF, Fleagle JG. The Omo-Kibish I pelvis. *J Hum Evol*. 2017;108:199–219.
20. Richter D, Grün R, Joannes-Boyau R, et al. The age of the hominin fossils from Jebel Irhoud, Morocco, and the origins of the Middle Stone Age. *Nature*. 2017;546:293–296.
21. Hublin J-J, Ben-Ncer A, Bailey SE, et al. New fossils from Jebel Irhoud, Morocco and the pan-African origin of *Homo sapiens*. *Nature*. 2017;546:289–292.
22. Hershkovitz I, Weber GW, Quam R. The earliest modern humans outside Africa. *Science*. 2018;359:456–459.
23. Cann R, Stoneking M, Wilson A. Mitochondrial DNA and human evolution. *Nature*. 1987;325:31–36.
24. Stringer C. Why we are not all multiregionalists now. *Trends Ecol Evol*. 2014;29:248–251.
25. Zhivotovsky LA, Rosenberg NA, Feldman MW. Features of evolution and expansion of modern humans, inferred from genomewide microsatellite markers. *Am J Hum Genet*. 2003;72:1171–1186.
26. Quinn JA, Woodward SA. *Earth's Landscape*. California: ABC-CLIO Press; 2015.
27. Yu N, Chen FC, Ota S, et al. Larger genetic differences within Africans than between Africans and Eurasians. *Genetics*. 2002;161:269–274.
28. Witherspoon DJ, Wooding S, Rogers AR, et al. Genetic similarities within and between human populations. *Genetics*. 2007;176:351–359.

29. Stringer CB, Andrews P. Genetic and fossil evidence for the origin of modern humans. *Science.* 1988;239:1263–1268.
30. Smith FH, Falsetti AB, Donnelly S. Modern human origins. *Yearb Phys Anthropol.* 1989;32:35–68.
31. Ahern JCM, Jankovic I, Voisin J-L, Smith FH. Modern human origins in Central Europe. In: Smith FH, Ahern JCM, eds. *The Origins of Modern Humans. Biology Reconsidered.* Hoboken: Wiley Blackwell; 2013.
32. Smith FH, Janković I, Karavanić I. The assimilation model, modern human origins in Europe, and the extinction of Neandertals. *Quat Int.* 2005;137:7–19.
33. Weidenreich F. Morphology of Solo man. *Anthropol Pap Am Mus Nat Hist.* 1951;43:205–290.
34. Weidenreich F. Six lectures on *Sinanthropus pekinensis* and related problems. *Bull Geol Soc China.* 1939;19:1–110.
35. Washburn SL. The origin of races: Weidenreich's opinion. *Am Anthropol.* 1964;66:1165–1167.
36. Mirazon LM. The multiregional model of modern human origins: a reassessment of its morphological basis. *J Hum Evol.* 1994;26:23–56.
37. Mirazon Lahr M, Foley R. Multiple dispersals and modern human origins. *Evol Anthropol.* 1994;3:48–60.
38. Howells WW. *Cranial Variation in Man. A Study by Multivariate Analysis of Patients of Difference Among Recent Populations.* Peabody Museum of Archaeology & Ethnology, Cambridge, MA: Harvard University Press; 1973.
39. Larnach SL, Macintosh NWG. *The Craniology of the Aborigines of Coastal New South Wales. Oceania Monographs, no. 13*; 1966. Sydney, Australia.
40. Lowery RK, Uribe G, Jimenez EB, et al. Neanderthal and Denisova genetic affinities with contemporary humans: introgression versus common ancestral polymorphisms. *Gene.* 2013;530:83–94.
41. Wolpoff MH, Hawks J, Frayer DW, Hunley K. Modern human ancestry at the peripheries: a test of the replacement theory. *Science.* 2001;291:293–297.
42. Hawks J, Oh S, Hunley K, et al. An Australasian test of the recent African origin theory using the WLH-50 calvarium. *J Hum Evol.* 2000;39:1–22.
43. Jelinek J, Wolpoff M, Frayer DW. Evolutionary significance of the quarry cave specimens from Mladeč. *Anthropologie.* 2005;XLIII:215–228.
44. Stone T, Cupper ML. Last Glacial Maximum ages for robust humans at Kow Swamp, southern Australia. *J Hum Evol.* 2003;45:1–13.
45. Thorne AG, Macumber PG. Discoveries of Late Pleistocene man at Kow Swamp. *Nature.* 1972;238:316–319.
46. Pietrusewsky M. Craniometric variation in Pleistocene Australian and more recent Australian and New Guinean populations studied by multivariate procedures. *Occas Pap Hum Biol.* 1979;2:83–123.
47. Brothwell D. Possible evidence of a cultural practise affecting head growth in some late Pleistocene East Asian and Australasian populations. *J Archaeol Sci.* 1975;2:75–77.
48. Kramer A. Modern human origins in Australasia: replacement or evolution? *Am J Phys Anthropol.* 1991;86:455–473.
49. Habgood PJ. The origin of the Australians: a multivariate approach. *Archaeol Ocean.* 1986;21:130–137.
50. Baba H, Aziz F, Narasaki S. Restoration of the face of Javanese Homo erectus Sangiran 17 and re-evaluation of regional continuity in Australasia. *Acta Anthropol Sin.* 2000;19:34–40.
51. Thorne AG, Wolpoff MH. Regional continuity in Australasian Pleistocene hominid evolution. *Am J Phys Anthropol.* 1981;55:337–349.
52. Curnoe D. Modern human origins in Australasia: testing the predictions of competing models. *Homo.* 2007;58:117–157.
53. Trinkaus E. Early modern humans. *Annu Rev Anthropol.* 2005;34:207–230.
54. Habgood PJ. The origin of anatomically modern humans in Australasia. In: Mellars P, Stringer CB, eds. *The Human Revolution: Behavioural and Biological Perspectives in the Origins of Modern Humans.* Princeton: Princeton University Press; 1989.
55. Weidenreich F. The new discovery of three skulls of *Sinanthropus pekinensis. Nature.* 1937;141:376–379.
56. Shen G, Gao X, Gao B, Granger D. Age of Zhoukoudian Homo erectus determined with (26)Al/(10) Be burial dating. *Nature.* 2009;458:198–200.
57. Weidenreich F. The new discovery of three skulls of *Sinanthropus pekinensis. Science.* 1937;85:316–317.

58. Weidenreich F. The skull of *Sinanthropus pekinensis*: a comparative study on a primitive hominid skull. *Palaeontol Sin New Ser D*. 1943;no. 10. whole series no 127.
59. Baileya SE, Hublin J-J, Antón SC. Rare dental trait provides morphological evidence of archaic introgression in Asian fossil record. *PNAS*. 2019;116:14806–14807.
60. Schmalzer S. *The People's Peking Man: Popular Science and Human Identity in Twentieth-Century China*. Chicago: University of Chicago Press; 2008.
61. Rosenberg KR, Wu X. In: Smith FH, James CM, eds. *The Origins of Modern Humans: Biology Reconsidered*. 2nd ed. Hoboken: John Wiley & Sons, Inc; 2013.
62. Liu L, Chen X. *The Archaeology of China: From the Late Paleolithic to the Early Bronze Age*. Cambridge: Cambridge University Press; 2012.
63. Field JS, Petraglia MD, Lahr MM. The southern dispersal hypothesis and the South Asian archaeological record: examination of dispersal routes through GIS analysis. *J Anthropol Archaeol*. 2007;26:88–108.
64. Lieberman DE. Testing hypotheses about recent human evolution from skulls. *Curr Anthropol*. 1995;36:159–196.
65. Bae CJ. The late middle Pleistocene hominin fossil record of Eastern Asia: synthesis and review. *Yearb Phys Anthropol*. 2010;53:75–93.
66. Koesbardiati T. *On the Relevance of the Regional Continuity Features of the Face in East Asia* [Ph.D. dissertation]. University of Hamburg; 2000.
67. Dong W. Biochronological framework of Homo erectus horizons in China. *Quat Int*. 2016;400:47–57.
68. Wu X-J, Pei S-W, et al. Archaic human remains from Hualongdong, China, and Middle Pleistocene human continuity and variation. *PNAS*. 2019;116:9820–9824.
69. Wu XZ. A well-preserved cranium of an archaic type of early *Homo sapiens* from Dali, China. *Sci Sin*. 1981;24:530–541.
70. Sun X, Yi S, Lu H, Zhang W. TT-OSL and post-IR IRSL dating of the Dali Man site in central China. *Quat Int*. 2017;434:99–106.
71. Wang Y, Xue X, Jue L, Zhao J, Liu S. Discovery of the Dali fossil man and its preliminary study. *Kexue Tongbao*. 1979;24:303–306. In Chinese.
72. Athreya S, Wu X. A multivariate assessment of the Dali hominin cranium from China: morphological affinities and implications for Pleistocene evolution in East Asia. *Am J Phys Anthropol*. 2017;164:1–23.
73. Dennell R. Palaeoanthropology: *Homo sapiens* in China 80,000 years ago. *Nature*. 2015;526:647–648.
74. Rightmire GP. Homo in the Middle Pleistocene: hypodigms, variation, and species recognition. *Evol Anthropol*. 2008;17:8–21.
75. Massilani D, Skov L, Hajdinjak M, et al. Denisovan ancestry and population history of early East Asians. *bioRxiv*. 2020.
76. Li ZY, Wu XJ, Zhou LP, et al. Late Pleistocene archaic human crania from Xuchang, China. *Science*. 2017;355:969–972.
77. Li H, Li ZY, Gao X, et al. Technological behavior of the early Late Pleistocene archaic humans at Lingjing (Xuchang, China). *Archaeol Anthropol Sci*. 2019;11:3477–3490.
78. Stringer C, Brauer G. Models, misreading and bias. *Am Anthropol*. 1994;96:416–424.
79. Stringer CB. Replacement, continuity and the origin of *Homo sapiens*. In: Brauer G, Smith FH, eds. *Continuity or Replacement? Controversies in Homo sapiens Evolution*. Balkema Rotterdam; 1992.
80. Stringer C. The status of Homo heidelbergensis (Schoetensack 1908). *Evol Anthropol*. 2012;21:101–107.
81. Gómez-Robles A. Dental evolutionary rates and its implications for the Neanderthal–modern human divergence. *Sci Adv*. 2019;5:eaaw1268.
82. Ko KW. Hominin interbreeding and the evolution of human variation. *J Biol Res (Thessalon)*. 2016;23:17.
83. Skov L, Coll MM, Sveinbjörnsson G, et al. The nature of Neanderthal introgression revealed by 27,566 Icelandic genomes. *Nature*. 2020;582:78–83.
84. Scerri EML, Thomas MG, Manica A, et al. Did our species evolve in subdivided populations across Africa, and why does it matter? *Trends Ecol Evol*. 2018;33:582–594.
85. Smith TM, Tafforeau P, Reid DJ, et al. Earliest evidence of modern human life history in North African early *Homo sapiens*. *Proc Natl Acad Sci U S A*. 2007;104:6128–6133.

86. Smith FH. In: Bräuer G, Smith FH, eds. *Continuity or Replacement: Controversies in Homo sapiens Evolution*. Rotterdam, Netherlands; Brookfield, VT: A. A. Balkema; 1992.
87. Freidline SE, Gunz P, Jankovic I, Harvati K, Hublin JJ. A comprehensive morphometric analysis of the frontal and zygomatic bone of the Zuttiyeh fossil from Israel. *J Hum Evol*. 2012;62:225–241.
88. Cartmill M, Smith FH. *The Human Lineage*. Hoboken: John Wiley & Sons; 2009.
89. Groucutt HS, EML S, Stringer C, et al. Skhul lithic technology and the dispersal of *Homo sapiens* into Southwest Asia. *Quat Int*. 2019;515:30–52.
90. Bar-Yosef O, Vandermeersch B, Arensburg B, et al. The excavations in Kebara Cave, Mt. Carmel. *Curr Anthropol*. 1999;33:497–550.
91. Grun R, Stringer CB. Tabun revisited: revised ESR chronology and new ESR and U-series analyses of dental material from Tabun C1. *J Hum Evol*. 2000;39:601–612.
92. Harvati K, Lopez EN. A 3-D look at the tabun C2 jaw. In: Marom A, Hovers E, eds. *Human Paleontology and Prehistory. Vertebrate Paleobiology and Paleoanthropology*. Berlin, Germany: Springer; 2017.
93. Villanea FA, Schraiber JG. Multiple episodes of interbreeding between Neanderthals and modern humans. *Nat Ecol Evol*. 2019;3:39–44.
94. Sankararaman S, Mallick S, Dannemann M, et al. The landscape of Neandertal ancestry in present-day humans. *Nature*. 2014;507:354–357.
95. Fu Q, Meyer M, Gao X, et al. DNA analysis of an early modern human from Tianyuan Cave, China. *Proc Natl Acad Sci USA*. 2013;10:2223–2227.
96. Juric I, Aeschbacher S, Coop G. The strength of selection against Neanderthal introgression. *PLoS Genet*. 2016;12, e1006340.
97. Frayer DW, Wolpoff MH, Thorne AG, et al. Theories of modern human origins: the paleontological test. *Am Anthropol*. 1993;95:14–50.
98. Frayer DW. Evolution at the European edge: Neanderthal and the Upper Paleolithic relationships. *Préhistoire Européene*. 1992;2:9–69.
99. Trinkaus E. European early modern humans and the fate of the Neandertals. *PNAS*. 2007;104:7367–7372.
100. Darwin C. *The Descent of Man, and Selection in Relation to Sex*. London: John Murray; 1871.
101. Huxley TH. *Evidence as to Man's Place in Nature*. London: Williams and Norgate, London; 1863.
102. Scerri EML, Chikhi L, Thomas MG. Beyond multiregional and simple out-of-Africa models of human evolution. *Nat Ecol Evol*. 2019;3:1370–1372.
103. Thorne AG, Wolpoff MH. The multiregional evolution of humans. *Sci Am*. 2003;13:46–53.
104. Loogvali EL, Kivisild T, Tõnu M, Villems R. Explaining the imperfection of the molecular clock of hominid mitochondria. *PLoS ONE*. 2009;4, e8260. O'Rourke D. (ed.).
105. Shtolz N, Mishmar D. The mitochondrial genome–on selective constraints and signatures at the organism, cell, and single mitochondrion levels. *Ecol Evol*. 2019;7, 342.
106. Anderson ML, Finlay BL. Allocating structure to function: the strong links between neuroplasticity and natural selection. *Front Hum Neurosci*. 2014;2013(7):918.
107. Relethford JH. Genetic evidence and the modern human origins debate. *Heredity*. 2008;100:555–563.
108. Eswaran V. A diffusion wave out of Africa: the mechanism of the modern human revolution? *Curr Anthropol*. 2002;43:5.
109. Green RE, Krause L, Briggs AW, et al. A draft sequence of the Neandertal genome. *Science*. 2010;328:710–722.
110. Wallace DC, Brown MD, Lott MT. Mitochondrial DNA variation in human evolution and disease. *Gene*. 1999;238:211–230.
111. López S, van Dorp L, Hellenthal G. Human dispersal out of Africa: a lasting debate. *Evol Bioinformatics Online*. 2015;11(Suppl 2):57–68.
112. Hammer MF, Garrigan D, Wood E, et al. Heterogeneous patterns of variation among multiple human x-linked Loci: the possible role of diversity-reducing selection in non-Africans. *Genetics*. 2004;167:1841–1853.
113. Templeton AR. Out of Africa again and again. *Nature*. 2002;416:45–51.
114. Templeton AR. Haplotype trees and modern human origins. *Yearb Phys Anthropol*. 2005;48:33–59.
115. Takahata N, Lee SL, Satta Y. Testing multiregionality of modern human origins. *Mol Biol Evol*. 2001;18:172–183.

116. Hayakawa T, Aki I, Varki A, et al. Fixation of the human- specific CMP-N-acetylneuraminic acid hydroxylase pseudogene and implications of haplotype diversity for human evolution. *Genetics*. 2006;172:1139–1146.
117. Harris EE, Hey J. X chromosome evidence for ancient human histories. *Proc Natl Acad Sci*. 1999;96:3320–3324.
118. Hl K, Satta Y. Population genetic analysis of the N-acylsphingosine amidohydrolase gene associated with mental activity in humans. *Genetics*. 2008;178:1505–1515.
119. Patin E, Barreiro LB, Sabeti PC, et al. Deciphering the ancient and complex evolutionary history of human arylamine N-acetyltransferase genes. *Am J Hum Genet*. 2006;78:423–436.

CHAPTER 8

An African-wide origin of *Homo sapiens*

Checking a box on a form for race—"Caucasian," "Hispanic," "African-American," "Native American," or "Asian-American"—is untenable and ridiculous....how far back does one go in history? Native Americans are really Asians, if you go back more than twenty or thirty thousand years to before they crossed the Bering land bridge between Asia and America. And Asians, several hundred thousand years ago probably came out of Africa, so we should really replace "Native American" with "African-Asian-Native American." Finally, if the Out of Africa (single racial origin) theory holds true, then all modern humans are from Africa. (Cavalli-Sforza now thinks this may have been as recently as seventy thousand years ago.) Even if that theory gives way to the Candelabra (multiple racial origins) theory, ultimately all hominids came from Africa, and therefore everyone in America should simply check the box next to "African-American."

Michael Shermer[1]

Issues to consider when thinking of hominin evolution

Any model of human evolution must provide explanations to a number of issues regarding the origins of *sapiens*. One crucial question that needs to be addressed by models of human evolution is whether present-day modern anatomy evolved rapidly in a matter of thousands of years among earlier populations of hominins or gradually over a protracted period during the last 400,000 years. In addition, with the current body of anatomical, archeological, and genetic evidence available to scholars, it is not clear which hominin species fits best as the ancestor to contemporary humans, or when ancestral populations diverged from each other and how much gene flow occurred among their descendant populations after separation. Did anatomical modern humans (AMHs) completely replace archaic forms without interbreeding? The resolutions to these important issues are complicated by the possibility that some of the features seen in present-day humans may not represent characters from a recent species such as *Homo heidelbergensis* but retained ancestral traits from more ancient hominins such as *Homo erectus*. The presence of such ancient traits dating back to 1 million years ago may result from the retention of characters and/or introgression. Nevertheless, having a better idea of the identity of our ancestors and their characteristics is essential to understanding *sapiens*' origins.

It is also important to have a better idea of the geographical area(s) that contributed to the genesis of *sapiens*. Did *sapiens* evolve from a single ancestral population in East Africa,

Sex and Cohabitation Among Early Humans
https://doi.org/10.1016/B978-0-12-819086-9.00011-7

South Africa, or North Africa? Or was it a Pan-African phenomenon with various populations contributing their DNA and ideas during a protracted period? And if so, what was the nature of their interactions, and how long did each of them last? Did ecological/environmental forces drive this process or just chance? These are issues that will be considered in this chapter.

Furthermore, considering that some archaic species of hominins such as *Homo naledi* and *Homo heidelbergensis* survived until recent times in Africa, is it possible that some of *sapiens*' traits were inherited directly from them? Is it possible that ancestral species will only be detected in the DNA of extant *sapiens* populations—especially if the ancestral groups lived in poorly investigated forested regions where fossils and DNA degrade easily? In addition, it is possible that *sapiens* arose from numerous interbreeding events. If so, were those events instances of admixture of species or reintegration of populations?

In terms of the Out of Africa event(s) of AMHs, were population bottlenecks associated with the migrations? If so, how dramatic was the reduction in population size? What were the sizes of ancestral populations during hominin evolution? Did population sizes fluctuate dramatically in time and space?

This chapter addresses some of the issues mentioned above related to the genesis and interactions of *sapiens* populations in Africa.

Single region of origin models

Ever since the early years of evolutionary thought, scholars have envisioned relationships among different forms of life in a hierarchical format. The idea of a branching scheme to arrange living things can be traced back to Plato and Aristotle's Historia Animalium.[2] This concept was later developed into the *Great Chain of Being* during medieval times as part of the Christian faith.[3] In the Chain of Being, God is depicted at the top as the creator of the Universe, followed by angels, who, like him, are entirely spirits. Then humans, animals, plants, and minerals follow in progression (Fig. 1).

The concept of a Chain of Beings was adopted by Edward Hitchcock (1793–1864) (Fig. 2) in his tree of life (Fig. 3) of 1840.[4] Hitchcock's tree of life is the earliest known illustration of paleontological and geological information in a branched format. In his *Elementary Geology*,[5] the diagram depicts plants and animals against a geological background. More recently, evolutionists have adopted tree schemes to illustrate evolutionary relationships among related species (Fig. 4). In these more contemporary depictions, ancestors of AMHs are diagrammed as a function of time linearly one after the other in an evolutionary "conga line" indicating a direct relationship among populations, one giving rise to the next.

The questions presented in the previous section have been traditionally addressed using evolutionary models, which portray hominin species in linear one-to-one

Fig. 1 Great chain of Being (https://en.wikipedia.org/wiki/Great_chain_of_being#/media/File:Great_Chain_of_Being_2.png). A 1579 drawing of the Great Chain of Being from Didacus Valades, *Rhetorica Christiana*.

relationships in a tree of life similar to the one in Fig. 4. This vision of hominin evolution has dominated the way of scholarly thought since the days of Darwin. These models present individual species giving rise to a more modern form as a function of time. Initially, these chronological phylogenetic relationships were derived from simple observations of quantitative or qualitative morphological traits. In other words, species with the most significant number of characteristics in common were placed adjacent to each other. In contrast, species exhibiting different traits were distant from each other in a separate branch of the tree (Fig. 4). As technology advanced during the 20th century, biochemical/genetic characters or markers such as proteins and DNA became increasingly employed as criteria to assign relationships among species. In conjunction with incorporating these new generations of markers, several sophisticated statistical analyses became

Fig. 2 Edward Hitchcock (http://blog.hmns.org/2014/06/hitchcock-dinosaurs-feathers).

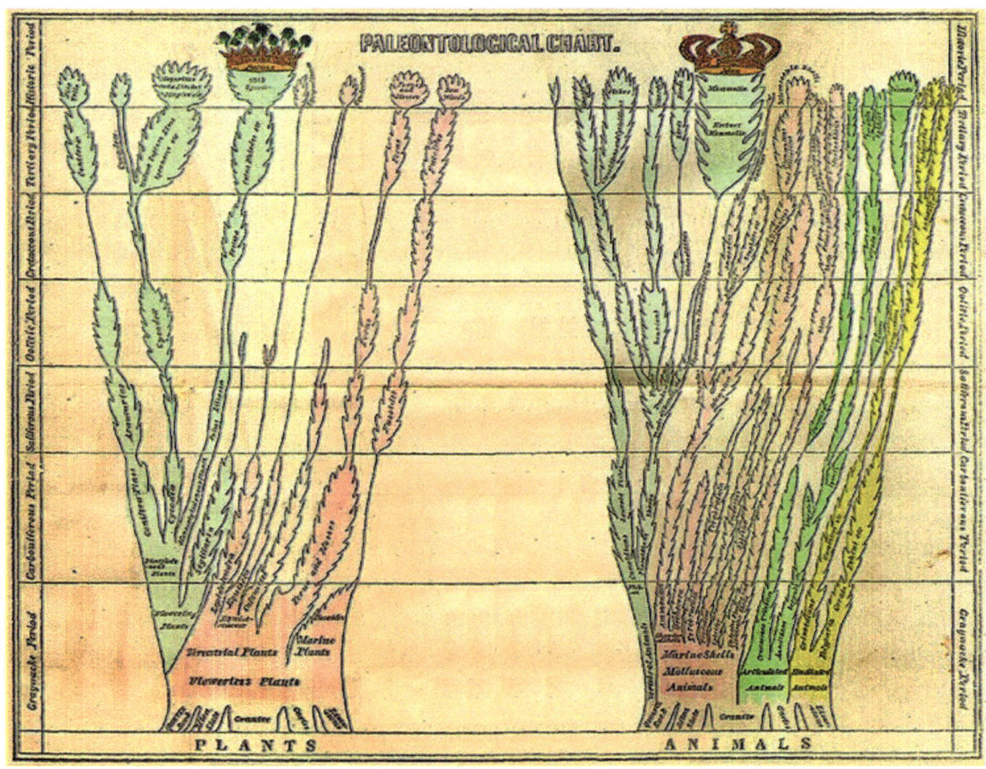

Fig. 3 Edward Hitchcock's tree of life (https://en.wikipedia.org/wiki/Edward_Hitchcock#/media/File: Edward_Hitchcock_Paleontological_Chart.jpg).

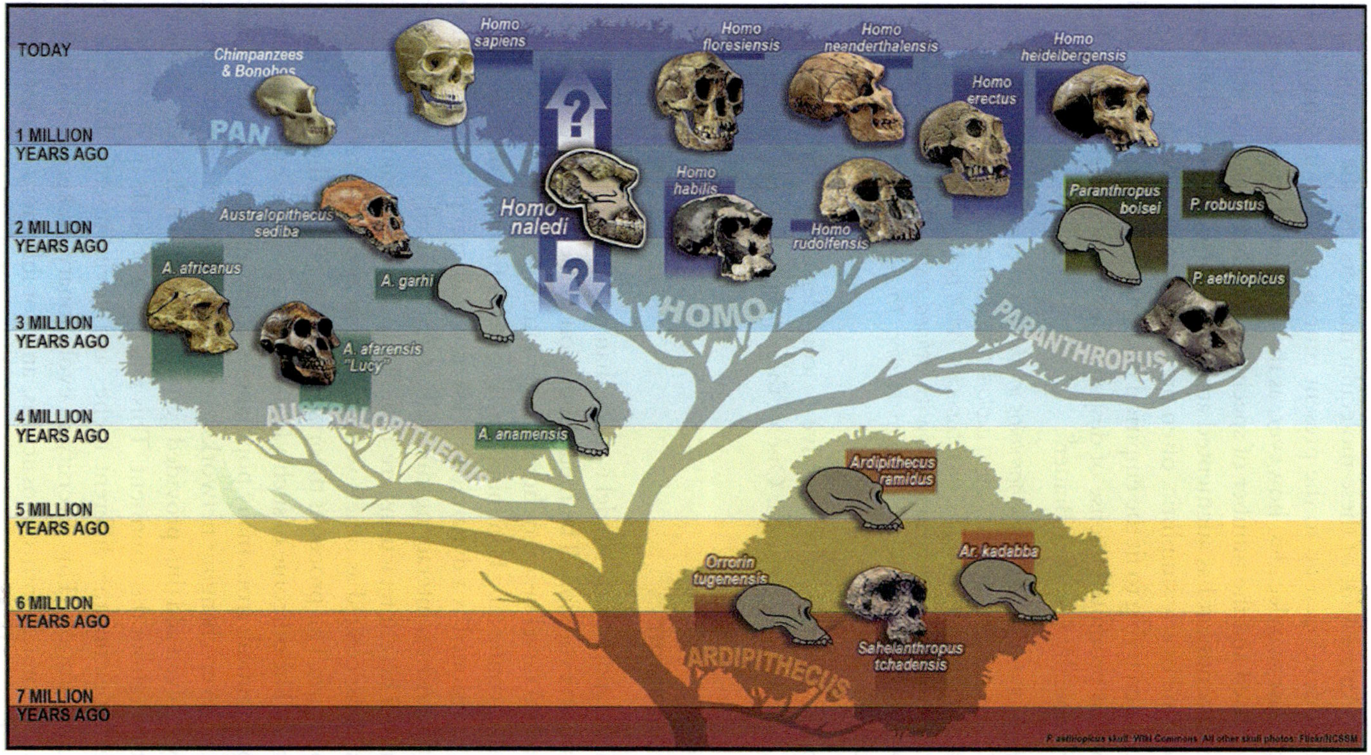

Fig. 4 Hominin phylogenetic tree (http://trapped-by-the-box.blogspot.com/2013/05/human-evolution-asian-or-african.html).

available to assess the evolutionary relationship and the linear order of the species in phylogenetic trees. An account of these molecular markers and the computational and statistical methods developed up to the present time is provided in Chapters 4–6.

During the last two decades, however, as the number of fossils, artifacts, and genetic data have increased and the number of specimens exhibiting similar traits discovered all over the Africa continent has augmented attempts to organize hominin species in linear trees have become overly simplistic, often problematic and at times senseless. This sense of frustration when trying to fit the empirical data with traditional linear models of hominin evolution has prompted a sense of despair, as expressed in the passage below from a review 2008 paper by Gunther Brauer[6]:

> *Clearly, Middle and Late Pleistocene hominins in Africa and Europe show morphological differences in time and space. It is necessary to explore the evolutionary pattern by detecting groups, grades, steps, paleo-demes, morphs, or subspecies. Yet it only causes confusion, and is indeed unsatisfactory, to call any unit, no matter how it is defined, a species and, based on this, to suggest phylogenetic trees that are unrealistic and misleading. This even becomes worse when it enters popular science.*

This situation has provided the initiative to consider alternative models to explain the increasingly complex body of data. One of these proposals is the metapopulation model.

A shift in paradigm

The African metapopulation model is more than one additional hypothesis that attempts to explain human evolution. The metapopulation model is unique in that it provides an alternative radical novel vision of hominin evolution in which there is no strict one-to-one relationship connecting hominin populations as a function of time. Instead, hominin evolution is envisioned as a dynamic process in which groups of populations travel over distances within Africa, interacting culturally/technologically and interbreeding in complex networks. These interactions are thought to have constantly been changing with connections being created as others were truncated at times driven by ecological/environmental pressure. This process is theorized as taking place at the level of the entire continent of Africa. The outcome of such a Pan-African process is a very extensive gene pool receiving contributions from many populations at any given time. The individual populations in these ever-changing networks were the evolutionary products of the specific niches. They evolved and thus provided genetic and cultural diversity to the entire metapopulation throughout the continent. This was made possible by the diverse evolutionary processes reflecting the different niches and environments that gave rise to each contributing population. This diversity in evolutionary backgrounds likely contributed to the many anatomical forms, mosaicisms, and combinations of observed traits in the hominin fossil record. This evolutionary scheme likely existed at various levels of interactions, some regions involving geographically nearby populations while others

Fig. 5 Popular view of human evolution (https://www.radiologybusiness.com/topics/imaging-informatics/advanced-imaging-reveals-human-evolution-neck).

demonstrated long-range conductivity involving gene flow and technological transfer, all levels connected in a hierarchy of interactions forming a continuum.

Implied in this model is the expectation of an extensive gene pool that allowed gene flow over long geographical distances. This, in turn, provided for considerable genetic diversity and adaptive flexibility in a changing environment. It also provided for a technological and cultural highway that facilitated the transfer of information.

Over the years, students of human evolution have been conditioned to think of human evolution as a linear order with species fixed in cubicles as a function of geological time. This model has become widely accepted by the public, and people, in general, have become comfortable with this vision of human evolution to the point that it is now part of our culture (Fig. 5). Although we have become used to such a notion, being conceptionally simple, the accumulating body of paleontological and genetic data is increasingly becoming incompatible. And although the metapopulation model represents a radically new way of thinking—one that is less tangible and may be unsettling to some—it can better explain the current body of evidence. With this novel model, we are required to think of ancestors not as discrete and confine forms of life restricted in space, which lived during finite periods but as dynamic groups of organisms, that cumulatively shared their DNA and traditions over extensive periods of time and space.

In the traditional view, *Homo sapiens* emerged in one region and/or population of Africa. Instead, new data suggest that various transitional human groups, with a mosaic of primitive and derived features, may have lived over an extensive area from Morocco to South Africa between >300 and 12 ka.

Genesis of the Pan-African theory of hominin evolution

As discussed in Chapter 7, Cann, Stoneking and Wilson's seminal article in 1987[7] that reported on mitochondrial Eve and the African origin of AMHs set the stage for the Out of Africa (OOA) dispersal theory. The mitochondrial Eve model postulated all

AMHs descended from a lineage of early humans, which evolved in Africa around 150,000 years ago and then spread worldwide, replacing local archaic hominin populations. Yet estimates on the time of origin, departure, and number of dispersals have changed dramatically in recent years. Currently, the fossil record suggests that AMHs existed in different regions of Africa as early as about 300,000 years ago. Numerous DNA segments and genes indicate that several early human lineages contributed to contemporary humans. The origins of some of these genes are traced to as early as 2 million years ago. Information is now available that suggests that *sapiens* have Pan-African origins with multiple sites of origin within the continent, that *sapiens* migrated out of Africa and back numerous times, and that different levels of assimilation or admixture occurred between OOA AMH migrants and local archaic populations in a region-specific manner.

The first idea that a Pan-African metapopulation existed and gave rise to *sapiens* was formulated based on genetic data, not fossil remains. Early in the 2000s, as molecular/genetic diversity in contemporary humans was realized, biological anthropologists such as Rosalind Maria Harding noticed that the view of a single founding population of early *sapiens* being ancestral to modern humans was incompatible with the available genetic variation observed in contemporary human populations.[8] She observed that the genetic variability data indicated that specific genes dated to very different ages and indicated different genealogies and histories. Genetic data from existing human populations suggested that various DNA sequences originated at other times during the Middle to Late Pleistocene (300,000–50,000 years ago). This information was not compatible with a single recent ancestral population that underwent a bottleneck event as it migrated out of Africa and gave rise to AMHs. Proponents of the OOA dispersal theory postulated such a bottleneck event to explain a hypothetical reduction in size and genetic diversity of a population to a small number of survivors. Such a reduction in the number of individuals in a population should be detectable in the genetic imprint of contemporary human populations, but it is not. Instead, the observed genetic diversity was consistent with an older African metapopulation that existed all over the continent that underwent repeated episodes of regional extinctions, re-colonization, expansions, and admixture. The DNA evidence suggests that some of our genes originated in various archaic populations at different time periods, not just in a single lineage in Africa or at one time.[9]

Two years later, Garrigan and Hammer (Fig. 6) provided genetic data indicating that contemporary AMH populations descended from multiple African archaic subpopulations.[10] They derived this conclusion from the unexpected antiquity and age of different DNA segments located on the X chromosome and autosomes (nonsex chromosomes). They indicated that the observed range in age among DNA segments in present-day *sapiens* resulted from a diverse ancestry of multiple archaic human subpopulations in Africa. In this ancient subpopulation structure model within Africa, the origin of *sapiens* is pushed back in time. They formulated a number of models with various numbers of interacting Pan-African early human populations dating back to 200 kya and concluded

Fig. 6 Michael F. Hammer (https://cancercenter.arizona.edu/person/michael-hammer-phd).

that the data was not compatible with the single origin of *sapiens* and is best explained by an ancestral population structure with low migration (as opposed to high migration) among African-wide populations.

Subsequently, morphological data have corroborated the information provided by the observed ancient age and sequence diversity of genes. When neurocranial morphology was assessed in a large number of early modern human (EMH) samples from 200,000 to 60,000 years ago and compared to older hominin fossils dating back to 1.8 million years ago and recent modern humans, it was found that the early modern group exhibits more shape variation than any of the other groups studied.[11] Based on several hundred neurocranial landmarks, the researchers also found that the early modern humans are more morphologically similar to contemporary humans than to archaic groups. The greater anatomical variation among Pan-African early modern human crania and their similarities to geographically dispersed contemporary populations is compatible with a more complex ancestral population structure than the one suggested by the classic out-of-Africa model. These anatomical findings are more consistent with a model in which African populations are dynamically interconnected by population expansion and contraction events. In such an African-wide model, regional populations are temporarily connected by migration and severed by extinction. This scenario is also agreeable with the genetic evidence presented in previous paragraphs.

More recently, the idea of an interbreeding metapopulation of early *sapiens* that encompassed the entire African continent began to take form into a unifying theory. In their influential 2018 paper in the journal *Trends in Ecology and Evolution*, Professor Scerri[12] and collaborators ask the question, "Did Our Species Evolve in Subdivided Populations across Africa, and Why Does It Matter?" In this publication, they formally challenge the notion that our species, *Homo sapiens*, evolved within a single population

Fig. 7 Eleanor M. L. Scerri (http://www.eleanorscerri.com).

and region of Africa. Eleanor M. L. Scerri (Fig. 7) and colleagues elaborate on a novel model of human evolution that attempts to explain several specific genetic and anatomical diversity patterns not described by previous traditional linear models. They integrate their ideas into a theory that they refer to as the Structured Metapopulation Model. In this model, they see *Homo sapiens* as the evolutionary product of the entire African continent and not of any given geographical area. This model proposes that the entire African continent was home to many ancestral, genetically diverse groups of populations that interacted dynamically to give rise to AMHs eventually. These interactions were not just interbreeding but are thought to include the cultural-technological transfer of information. These investigators explain that the traditional narrative of human evolution has become overcomplicated to present the increasingly complex body of data and stay relevant. This less geographically constrained model posits that the interactions among hominin populations changed with time during early human evolution. These interactions may be envisioned as nonstatic networks or lattices of gene flow and culture that allow early hominins to share DNA and ideas within the entire continent. Scerri et al envision *sapiens* evolution as changes in connectivity among many hominin populations. They provide for mixing of genomes and technology and posit that the model further explains the cultural, anatomical, and genomic diversity observed over wide geographic areas in Africa. The metapopulation model frees students of human evolution from ambiguous, constrained phylogenetic trees, ever-increasing naming of species, speciation events, and hypothetical ancestral populations.

Fossil evidence

A review of African hominin fossils indicates that the anatomical characteristics associated with AMHs did not appear sequentially in any single region of the continent. Instead, *sapiens*-specific traits such as a globular braincase, small and delicate face, and a chin are seen across Africa in mosaic-like fashion, emerged at various times and in different combinations with diverse ancestral features. Wide anatomical variation in early modern

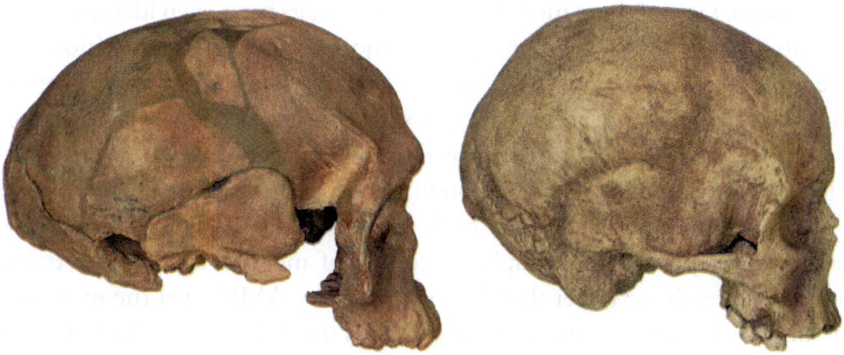

Fig. 8 Skulls of a Jebel Irhoud individual and an approximately 20,000-year-old *H. sapiens* fossil (https://www.researchgate.net/publication/317834005).

humans (300,000–50,000 ya) throughout Africa ranging from skulls such as Florisbad and Jebel Irhoud (Fig. 8) to Ngaloba and Herto are observed (see Herto discussion below). For comparison, the figure also includes crania from Skhul and Qafzeh. This array of fossils shows differing combinations of ancestral and derived traits and shows some of the continent's diversity. In other words, *sapiens* traits first appear separately at various times in different locations across Africa. Then they came together in the past 100,000–40,000 years to generate the modern type recognized as the contemporary anatomy. It seems as if *sapiens*-derived features evolved separately in different hominin lineages and then came together through dynamic gene flow processes among the subpopulations within the continent. This flexible metapopulation model avoids the senseless assignment of species names to different fossils exhibiting a unique combination of traits. This scenario is compatible with a fragmented and structured continental-wide evolutionary process involving various African populations toward the modern human form. The anatomy of several African-wide fossils is discussed below to illustrate this metapopulation model.

Jebel Irhoud

With the discovery of additional fossils at Jebel Irhoud in Morocco, North Africa, and the reliable dating of the remains and tools at the site, the idea of an African-wide metapopulation gained additional popularity among the anthropological community.[13] Jebel Irhoud is made up of the remains of at least five individuals: three adults, one adolescent, and one infant about 7.5 years old. Although all the recovered fossils at the site seem to derive from the same population and deposition layer dated at about 315 ka,[14] their characteristics are not uniform. In other words, the derived and ancestral traits differ among the individuals. It is not clear whether this phenomenon results from sex differences (sexual dimorphism), developmental (age) differences. and/or polymorphism within the population. In the

Irhoud fossils, the face is where modern human-like traits have been identified. Among the derived traits is the small gracile nonprognathic (nonprotruding) faces positioned under the braincase (Fig. 8). Another derived trait is the unremarkable brow ridges observed. Although the expression of the brow ridges varies among the Jebel Irhoud skulls, overall they are less pronounced than more archaic fossils. The recovered jaws are morphologically *sapiens*-like with prominent chins, although overall larger. Their heights decrease from front to the back, and the mandibular body is narrow like in AMHs.

The dental morphology also displays a collage of ancestral and derived traits. Overall the Jebel Irhoud teeth are larger than in contemporary AMHs, yet the morphology parallels the observed derived pattern of other EMHs from Africa and the Levant. The post canine teeth are smaller compared to older hominins. The molars, for example, are even smaller in relation to other early humans, and their crowns are similar to those of *sapiens*. The molars are morphologically complex and reminiscent of other African early humans with features such as accessory cusps. The lower incisor and canine roots retain the ancestral large size, but the shapes are within the range of contemporary AMHs. Analysis of the enamel-dentine junctions indicates the retention of the ancestral pattern. From the infant dentition recovered, the investigators assessed that the developmental pattern of teeth eruption was similar to that of *sapiens*.

Contrary to the face and dentition, which represent a collage of ancestral and derived traits, the braincase and endocast of the Jebel Irhoud skulls possess an overall ancestral morphology. The archaic neurocranium seen in early hominins and in Jebel Irhoud is elongated, resulting from the low configuration of the occipital squama (concave bone at the base of the cranium posterior to the foramen magnum) and elongated temporal bones at the sides and base of the skull. When seen from the side, skulls with this ancestral shape look like a football (oval or spheroid in shape), while the early human (*sapiens*) skulls look like soccer balls (Fig. 8). The Jebel Irhoud foreheads slope is indicating the ancestral condition. The dichotomy between the mosaic morphology of the face and teeth on the one hand and the ancestral braincase on the other indicate that the evolution of the two anatomies toward the modern *sapiens* condition progressed at different rates in Jebel Irhoud and that the endocast evolved rapidly more recently.

According to Hublin and collaborators,[13] the globular braincase evolved in the last 300 ky, driven by a series of genetic changes that affected brain connectivity organization and development. A parallelism has been observed among brain globularity, behavioral modernity, and the appearance and fixation (carried by everyone in the population) of specific genes involved in axonal and dendritic growth or synaptic transmission. It has been demonstrated that specific genes were subject to positive selection contributing to their rapid fixation into the AMH gene pool prior to the Upper Paleolithic (50,000 to 12,000 ya).[15] These genes include sequences such as NOVA1, SLITRK1, KATNA1, LUZP1, ARHGAP32, ADSL, HTR2B, and CNTNAP2 that arose during this time period in the AMH lineage, and are all involved in nervous system functions.[16]

Florisbad

Thomas Frederik Dreyer (1885–1954) discovered the Florisbad fossils 1932 in Florisbad, South Africa.[17] The remains include a partial adult skull (the right side of the face, frontal bone, part of the maxilla, and an upper right, third molar) (Fig. 9). The fossil was found in

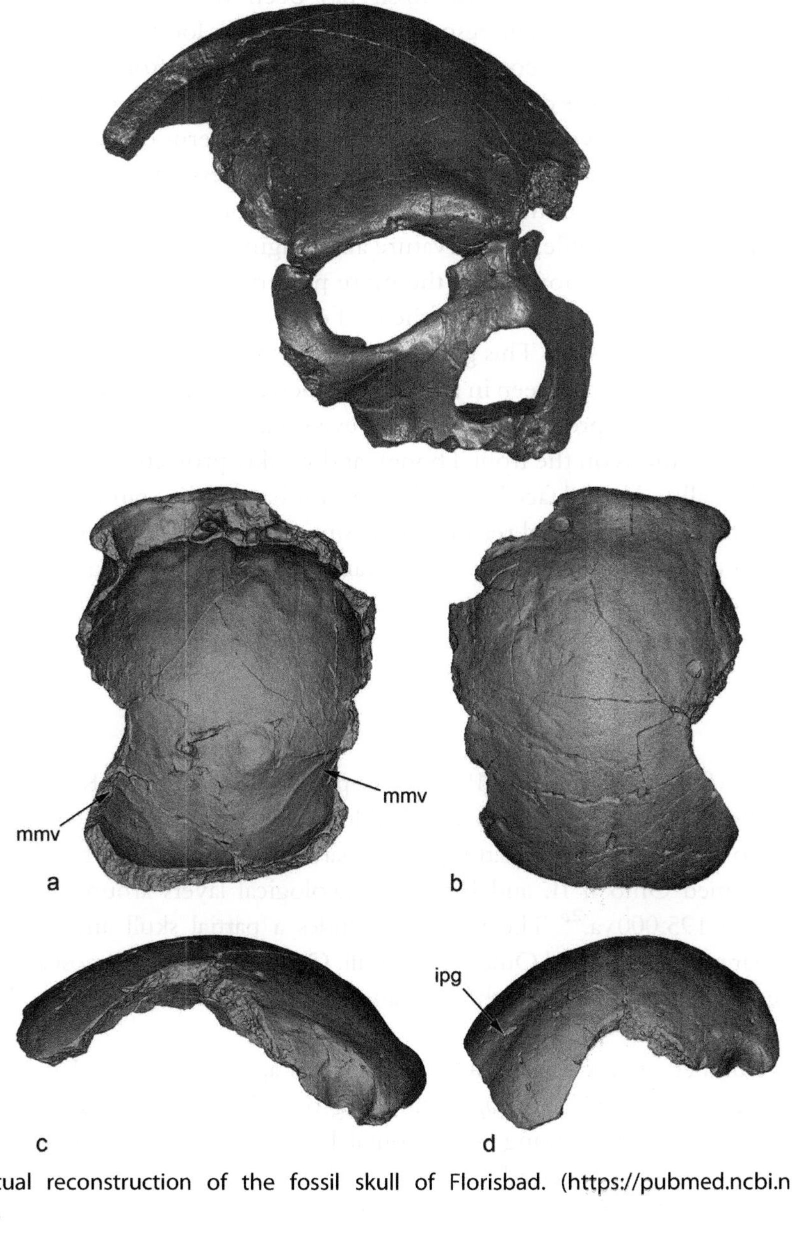

Fig. 9 Virtual reconstruction of the fossil skull of Florisbad. (https://pubmed.ncbi.nlm.nih.gov/33130632).

association with Levallois tools. The numerous and diverse fauna found at the site suggest a river or lake and grassland near the site.[18] Due to the combination of ancestral and derived traits, the skull has been classified as either *Homo heidelbergensis* or *Homo sapiens* over the years. More recently, in 2016, Chris Stringer has argued that the fossil is an early *Homo sapiens*.[19] Similarly, other investigators have recently classified the skull as an early *sapiens*.[12,13,20] The enamel from the fossil has been dated between 294,000 and 224,000 ya directly by electron spin resonance.[21] Florisbad is located about 7600 km from the Jebel Irhoud site in Morocco and approximately 4500 km from Herto in Ethiopia.

Traditionally, some of the difficulties in interpreting the position of the Florisbad skull in hominin evolution derive from the complex mosaic patterns of ancestral and derived traits that it exhibits. In this collage of traits, Florisbad possesses more derived characteristics in the anterior portion of the skull, while the posterior end of the crania shows more ancestral features. For example, the curvature and bulging of the frontal bone are within contemporary human variation, while the more posterior parietal (the flat cranial bone that covers the mid-posterior portion of the roof of the skull) and the vascular pattern of the endocranium are ancestral. This general tendency of derived features seen anteriorly and ancestral posteriorly is also seen in Jebel. Other derived traits seen in Florisbad include a distinct canine fossa (depression beneath the eye socket), a fully formed tuber frontale (two rounded elevations on the frontal bone), and weakly projecting brow ridges as well as a relatively small and broad face below the anterior base of the cranium (cranial fossa), a modern face-to-braincase spatial relationship. Among the ancestral traits, Florisbad possesses a very broad cranial fossa (floor of the cranial cavity), and it lacks the characteristic swelling of the dorsal parietal regions seen in AMHs. An idiosyncratic trait of Florisbad is the interparietal groove (depression between the parietal bones, in their posterior region).

Omo

About 5500 km from Jebel Irhoud in 1967, a group of investigators from Kenya led by Richard Leakey discovered many fossils that looked like *sapiens*. Since the discovery took place near the Omo River in Southern Ethiopia, the partial skulls recovered between 1967 and 1974 were named Omo I, II, and III.[22] The geological layers around the fossils were argon-dated to 195,000 ya.[23] The Omo I includes a partial skull and skeleton, while Omo II is a large braincase, and Omo III is a skull. Omo I exhibits the most advanced traits of the three while Omo II possesses the most ancestral characteristics, and Omo III shows an intermediate state.

After a number of reconstructions, investigators agree that Omo I had a high and voluminous cranial vault with a *sapiens*-looking occipital (back of the head) anatomy but retained an ancestral-looking wide frontal bone and strong but partitioned brow ridges. Omo I lacks the ancestral sagittal keel that is present in Omo II. Specifically, among the advanced traits, Omo I exhibits a short, flat, and broad face, a high forehead, fully formed chin, U-shape palate, modern size/shape teeth, modern-looking cochlea and semicircular canals of the inner ear, a small nuchal plane for neck muscle attachment, no sagittal keel on top of the skull, a limited occipital torus (see Fig. 9, Chapter 7), and globular braincase. In addition, the parietals (bones at the side and roof of the cranium) are long and curved and are the location where the cranium reaches its maximum breadth. The postcranial bones of Omo I include the shoulder, arm, hand, ribs, part of the spine, and legs. The anatomy of the postcranial bones is as in modern humans.

Although the Omo II skull is from the same strata and age as the Omo I crania, its morphology is considerably different. Overall, Omo II has a modern human anatomy, but with more ancestral features compared to Omo I. Specifically, the cranial bones are more robust, the cranial vault is lower than in Omo I, and its forehead is not high as in Omo I but sloped, the braincase is narrow (not globular) with parietals that are not superiorly expanded but parallel, and the occipital torus in the back of the skull is more prominent than in Omo I. Omo II also exhibits a sagittal keel or thickening of the midline of the frontal bone and the attachment site for neck muscle in the back of the skull is extensive. The tympanic bone is massive.

The Omo III skull incorporates some of the traits found in Omo I and II. It resembles Omo I in being large and high with a rounded modern-looking occipital (posterior) region, but looking much more like Omo II when seen from the rear, showing high, vertical walls.

As with the Jebel Irhoud fossils, the Omo skulls exhibit a range and different combinations of ancestral and derived traits. Concerning the Pan-African metapopulation model, it is noteworthy that at least Omo I and II are contemporary and excavated from the same general location, suggesting considerable diversity in time and space. Yet it is unknown whether Omo I, II, and III are members of the same population. If they are, do the differences result from sexual dimorphism and/or polymorphisms within the species?

When comparing the Jebel Irhoud and Omo skulls, two groups separated by about 100 ky and 5500 km, various combinations of ancestral and derived traits are seen. Both the Jebel Irhoud and Omo skulls possess limited brow ridges, prominent chins, and short faces. Yet the skulls from Jebel are elongated (nonglobular), the teeth and roots are larger than in AMHs, while in Omo I, the skull is globular, and the dentition is more like in AMHs. It is noteworthy that Omo II and Jebel share elongated braincases with parietals that are not superiorly expanded. It could be argued that a time differential of 100 ky may be partially responsible for some of the more advanced traits seen in Omo I.

Herto

Herto is an important hominin fossil site in the Middle Awash of the Afar Triangle in Northeast Ethiopia. The site is located about 620 km from the Omo fossil site in Southwest Ethiopia. In 1997, the remains of at least 12 hominins were discovered and subsequently classified as AMHs and dated to a period from 160,000 to 154,000 ya by $^{40}Ar/^{39}Ar$ technology.[24] The fossils include an almost complete skull of an adult (Fig. 10), a nearly complete skull of a 6- or 7-year-old child, and parts of adult cranial vaults, including parietal bone fragments as well as a right upper molar and an upper premolar. Several fossils exhibit man-made cut marks and other alterations made with a sharp tool, possibly evidence of postmortem ritualistic practices. Along with the bones, an assemblage of tools belonging to the Levallois or Acheulean traditions was found. A considerable number of hippos and bovine kills were also collected from the site. Considering the size and aggressiveness of hippos, it is unclear whether the hominins were hunting or scavenging the animals from the lakeside site.

The almost complete skull exhibits several ancestral, derived, and idiosyncratic traits. All the remains belong to robust, large individuals, especially in the back of the skulls. The capacity of the skull is about 1450 cm^3, and the length of the skull is outside the range of more than 5000 contemporary skulls.[25] In addition to being very elongated, the skulls

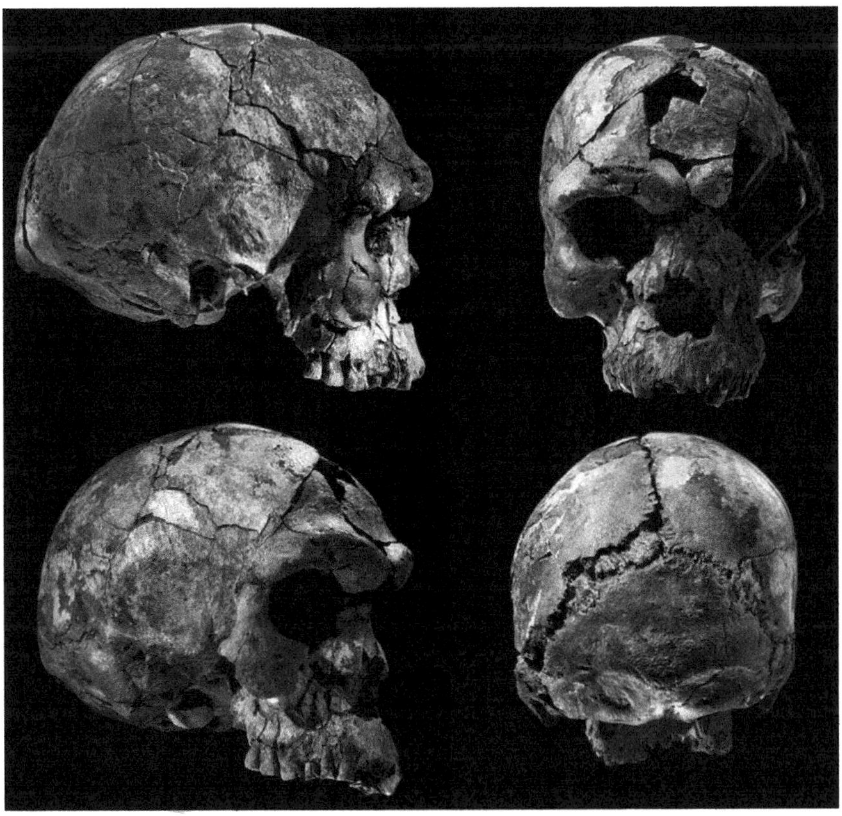

Fig. 10 Herto skull (file:///Users/ReneJHerrera/Documents/Herto%20White-TD-et-al-2003-Nature-Herto.pdf).

possess exceptionally large vaults and deep, tall, and broad faces. Other ancestral characteristics include a slightly projecting lower face, prominent zygomatics (cheekbones), projecting brow ridge (although less pronounced than in pre-*Homo* and split into halves), eye sockets that are more separated than in AMHs, large and projecting mastoid processes (bone behind the ear that serves for the attachment of head), prominent and projecting occipital bun just like in Jebel Irhoud 2, nonglobular occipital bone, and weakly curved parietal bones.

In terms of advanced traits, the Herto skulls exhibit a number of features, including a globular shape (except for the occipital or posterior end), a high cranial vault greatest interparietal breadth high on the vault, flat face with minimal facial projection, a raised forehead, and cheekbones that look more modern than archaic *sapiens* but are still prominent.

Although the Herto cranium possesses a globular shape anteriorly like Omo I, the skull is still elongated at the occipital or posterior end as in the Jebel skulls. It also shares a projecting occipital bun, just like in Jebel Irhoud 2. Similar to Omo I, Herto has large vaults, strong and partitioned brow ridges, as well as a broad and flat face.

Overall the cranial descriptions of the skulls from Jebel Irhoud, Florisbad, Omo, and Herto illustrate great anatomical complexity and diversity in the context of combinations of ancestral and derived traits in the form of mosaicism. Although the combination of ancestral and derived characteristics varies from site to site, during this time period of 150,000 to 300,000 ya, they share, for the most part, many of the features that eventually coalesce into what it is recognized as the AMH type today. This common theme of features is Pan-African exhibited in fossils from Northwest, Northeast, and South Africa. It is not restricted to any given region. For example, the globularity of the skull is first seen anteriorly, and it is not until about 150,000–200,000 ka that it is seen in its fuller form in Omo I (195,000 ka) and Herto 1 (160,000 ka), both in Ethiopia. From the above account, it could be surmised that no single population in a particular location is responsible for the origin of *sapiens*. It is easier to explain the observations by postulating the existence of a number of dynamically interacting African-wide early human lineages. In this scenario, it can be envisioned that many of these early populations contributed unevenly to the contemporary gene pool while some became evolutionary dead ends that failed to contribute to AMHs.

A good candidate for an evolutionary dead end may be represented by *Homo naledi*, a small-brained archaic-looking *Homo* species with an unusual mosaic of ancient and modern characteristics. Dated to 335,000–236,000 ka,[26] this fossil is a good example of the anatomical variation and complexity of *Homo* species at this time in Africa. The presence of such an archaic-looking *Homo* population living in Africa with more derived species such as Jebel Irhoud, Omo, and Herto and side by side in South Africa with Florisbad suggests that *naledi* represented a group of a different *Homo* lineage in Africa, a hominin that may not have contributed to modern *sapiens*, that was not associated with any population ancestral to *Homo sapiens*.

Phylogenetic modeling reconstruction—A virtual human common ancestor

In a recent report based on the phylogenetic modeling reconstruction by Aurélien Mounier and Marta Mirazón Lahr, evidence supporting the Pan-African model was provided.[27] These researchers designed a conceptionally unique approach and developed a virtual representation of the early human common ancestor. In brief, they predicted the cranial anatomy of the last common ancestor of all contemporary *sapiens*, which they name LCA. This virtual prototype was constructed using a total of 780 landmarks or measurements from key diagnostic locations in the skulls (Fig. 11). To build the hypothetical last common ancestor of all contemporary *sapiens* (LCA), they imputed measurements from a total of 263 crania, 245 from extant modern humans, and 18 from fossils. Seven of the fossils were Early Pleistocene skulls representing *Homo habilis*, *Homo ergaster*, and *Homo georgicus*, while eight were Neanderthals from various geographical populations (Near East, South Europe, and West Europe), and three were early *Homo sapiens* (Qafzeh 6 and 9, Skhūl V) from Southwest Asia. The extant 245 modern human skulls were from 21 populations (ten Africans, eight Eurasians, two Inuits, and a North American group).

One of the virtual LCA generated is illustrated in Fig. 12. Overall, it is less robust than the actual early human fossils discussed above and compared to Florisbad, Irhoud 1, and Omo II, it exhibits a more receding frontal and a more protruding occipital bone.[27] In addition, the virtual last common ancestor remarkably showed many of the derived traits typical of *sapiens*. The skull is globular with parietal eminences possessing tuber frontale, discontinuous brow ridge, and the base of the cranium is flexed. The face is retracted, showing a slightly projecting zygomatic and a visible maxillary canine fossa with two strongly marked curves. Yet it also exhibits some ancestral traits, including a slightly receding frontal bone, projecting brow ridges separated from the rest of the frontal bone by a groove, some prognathism, a small occipital bun, and a wide interorbital distance. Thus, the virtual LCA captured the repertoire of ancestral and derived traits seen in the geographically dispersed early *sapiens* such as Jebel, Florisbad, Omo, and Herto.

When the virtual LCA was compared to three of the early human skulls discussed above (Florisbad, Irhoud 1, and Omo II), overall, the image reflects a complex scenario responsible for the evolution of *sapiens* during the late Middle Pleistocene (300,000–50,000 ya). The comparisons demonstrated that Mounier and Lahr's phylogenetic modeling approach was able to identify specific geographically localized populations and lineages all over Africa that contributed to sapiens, although not all equally and some not at all. For example, the skull from Florisbad shares the most phenotypic affinities with the virtual LCA, and some of the other early human fossils examined exhibit different phenotypic profiles not seen in the virtual model, which indicate

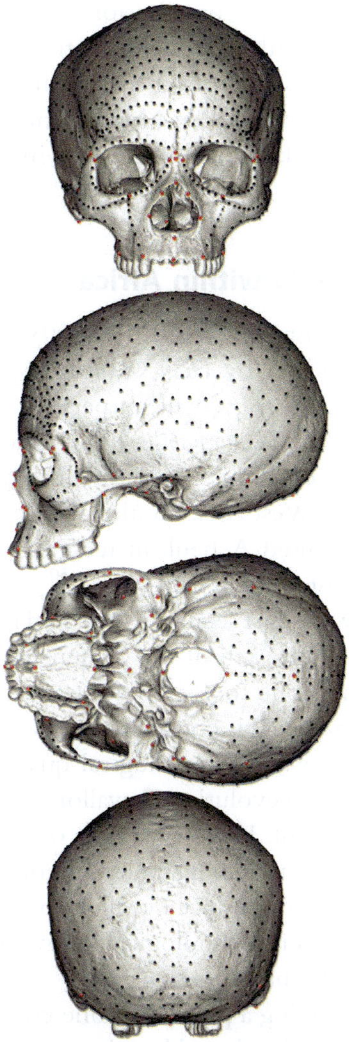

Fig. 11 Virtual representation of the early human common ancestor. The virtual prototype was constructed using a total of 780 landmarks or measurements from key diagnostic locations in the skulls (https://doi.org/10.1038/s41467-019-11213-w).

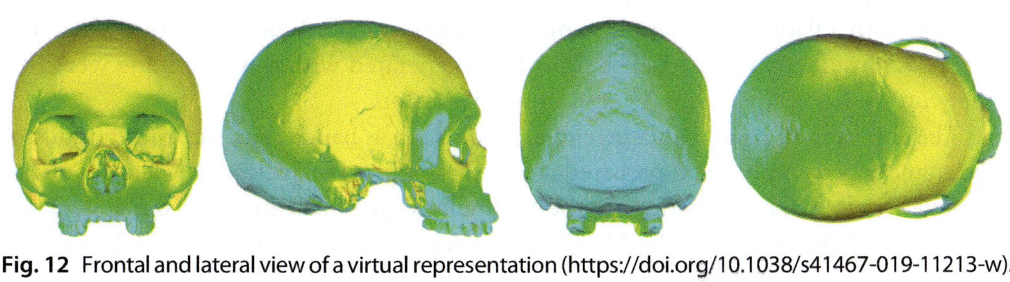

Fig. 12 Frontal and lateral view of a virtual representation (https://doi.org/10.1038/s41467-019-11213-w).

different degrees of similarities and contributions from several African early human populations and lineages. The authors concluded that modern humans are the result of a process that amalgamated populations from South and East Africa. They also concluded that early human populations from North African represented by Irhoud from Morocco might have been absorbed by introgression into Neanderthals during the late Middle Pleistocene.

Technological homogeneity within Africa

Prior to the Levallois tool industry (300,000–100,000 ya), the Acheulean tradition was practiced by *Homo erectus* and later by *Homo heidelbergensis* during the period between 1.76 million and 130,000 ya.[28] The Acheulean period was preceded by the Oldowan technology (2.6–1.7 my) used by *Homo habilis*.[29] Acheulean lithics are characteristically oval and pear-shaped hand-axes. Lithics made in this style are found in archeological sites across Africa and much of West Asia, South Asia, East Asia, and Europe. In Europe and West Asia, Neanderthals used Acheulean technology before transitioning to the more recent Mousterian industry about 160,000 ya.[30]

The Levallois technology was first seen in Africa, and it is thought that it originated in archaic/early humans as an outgrowth of the Acheulean hand-axe technology.[31] The Levallois approach to making stone tools is radically different from what was practiced before during the Acheulean, and it is considered a revolutionary event in the technological progress of humans. It was a technological quantum lead, and it is thought to reflect a higher level in cognitive evolution. Levallois is not just a different way to trim a rock to make it sharp and useful. It is a clear sign of behavioral modernity. The procedure is premeditated by nature, involving stages, forethought, and planning. It has been proposed that teaching the skills to produce Levallois tools requires the use of language.[32] It has been argued that this cultural-technological innovation paralleled the development of modern human anatomy and physiology.

The procedure involves striking a prepared stone core with another hard rock. This generates an edge that is further developed by trimming off pieces around the border of the intended lithic flake. At this point, the core looks like a dome. Then the core is hit internally to separate a central lithic flake. In this process, the edges of the released flake remain sharp, resulting from the previous trimming. The detached flake is plano-convex in shape with a flat or plane facial on one side and a rounded surface on the other. This process is illustrated in Fig. 13. The lithic tools generated with this technique were routinely utilized as scrapers, knives, and projectile points.

The abolition of handheld large cutting tools such as hand-axes (Acheulean industry) and an increased emphasis on prepared core technologies and hafting (Levallois industry) marked a radical transformation of the technology practiced by early humans. These technological changes signal the Middle Stone Age (MSA) transition, which lasted from

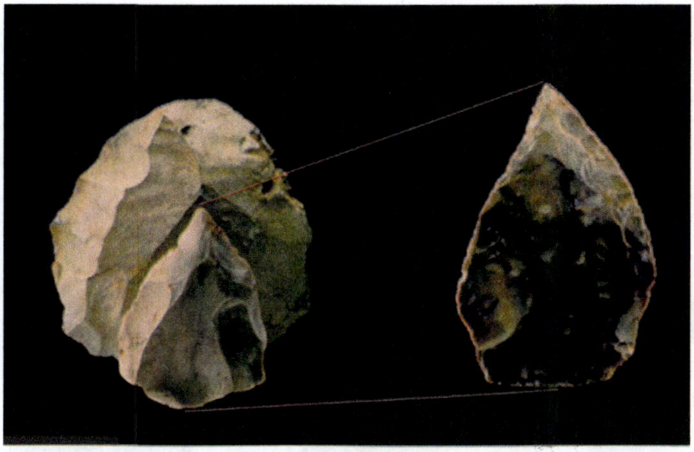

Fig. 13 The Levallois technique (https://www.turkanabasin.org/2020/04/learning-origins-of-stone-tool-technology).

280,000 ya and ended about 50,000 ya. Archeologists have noticed that these technological changes took place in different parts of Africa at about the same time. In addition to the Jebel in North East Africa site dated at 295,000–320,000 ya, Levallois tools have been found in Olorgesailie in East Africa[33] and in Florisbad in South Africa with dates of 280,000 ya.[21]

Although Levallois tools exhibit some diversity in the shape of the raw cores that were used, the general structure and surface of the flake that the technology produces are very uniform.[34] And although there is certain regional-specific variation in the tools, there is an overall continuity throughout much of the Middle Stone Age, and in several African-wide regions, the Levallois lithics do not exhibit specific regional characteristics and therefore can be characterized as generic.[35] Fig. 14 illustrates an assortment of Levallois tools from Jebel Irhoud.

The association of Levallois lithics in several sites during the period of 300,000–50,000 ya with its characteristic distinctive forms have been interpreted to indicate a wide-ranging Levallois culture resulting from the expansion of early *Homo sapiens* within and out of Africa. Furthermore, it has suggested to some investigators that the genesis of this technological tradition signals the emergence of the hominin evolutionary lineage that led to AMHs and implies that there were several early human populations in Africa at this time communicating with each other and exchanging ideas. In other words, the observed congruency between technological innovation and the associated fossil evidence suggests a complex Pan-African process of connectivity among populations starting before or around 300,000 ya, and argues against theorizing the origin of *sapiens* in any given region of the continent. And although it is possible that this uniformity in technology could be the result of independent origins and evolution of similar techniques

Fig. 14 Levallois tools from Jebel (https://www.pbs.org/newshour/science/ancient-fossils-morocco-mess-modern-human-origins).

resulting from parallel cognitive evolution in different lineages of early modern humans, it is easier to envision that similarities came about by occasional contact among populations.[28] With this in mind, a metapopulation model may provide a satisfactory explanation for Africa's technological uniformity during early human evolution.

Genetic diversity and the age of genes

Early studies based on mtDNA and Y-chromosome DNA portrayed a rather simple vision of human evolution in which AMH originated recently about 150,000 years ago in a small population in East Africa. Since these early studies that used uniparental (inherited from one parent, mother, or father) genetic markers, additional investigations using more representative genome-wide DNA are gathering information from the more complex part of the genome, the X chromosome, and the autosomes (the nonsex chromosomes). These studies are providing a very different picture of the evolution of early humans.

Any model that attempts to describe human evolution needs to explain satisfactorily the patterns of genetic variation seen in present-day humans. The traditional single origin model fails to explain the observed patterns of genetic variation and distribution in humans today. When the observed patterns of genetic diversity are traced back in time, investigators see not a single ancestral East African population but contributions from several widely distributed archaic African populations. Thus, the genetic data is congruent with the previously discussed anthropological and archeological data discussed in this chapter.

The main expectation of a single origin model in which a single lineage of hominin provided the totality of the DNA seen in *sapiens* today is that the observed genetic diversity can be traced back to a *single* ancestral population, for example, in Africa. In other words, in this scenario, any other hominin population or lineage that may have existed in the past failed to contribute DNA to the genome of present-day humans. On the other hand, if more than one lineage contributed to present-day *sapiens*, several sources of DNA from various ancestral lineages would be expected. If so, it is likely the various DNA sources were introduced into the human lineage at different times during hominin evolution. Geneticists can test the possibilities of a single or several ancestral populations contributing to contemporary humans by assessing *if* and *when* ancestral population (s) contributed specific genes or sequences to the human lineage. To do so, investigators determine the history or genealogy of particular sequences or genes in the human genome. Part of the genealogy or history of specific human sequences is when a gene was present in a common ancestor in the past. Since DNA diversity is directly proportional to the time the sequences are separated from a common ancestor, the older is the split of the lineages or the time to the most recent common ancestor (TMRCA) of the two lineages. In other words, the TMRCAs are based on the assumption that the number of nucleotide mutations separating two sequences is a linear function of time. Thus, DNA diversity is used as a clock to measure time. In this way, geneticists can determine when a DNA sequence started diverting independently in different lineages.

In a hypothetical scenario, if a gene exhibits an elevated level of diversity among AMHs, it indicates that it has been part of the genome for a longer time than a gene with minimal diversity. Furthermore, since the rate of change or mutation in DNA is known, investigators routinely ascertain when specific sequences became part of the genome by quantitating the amount of DNA diversity among individuals. The more diversity a gene exhibits within a lineage, the older it is. In this way, investigators compare the ages of genes within the genome of species and the time when the human lineage inherited the gene from a common ancestor shared with a different lineage.

The expectation of a species that derives from a single origin or lineage is that all of its DNA sequences or genes exhibit comparable TMRCAs. If, on the other hand, a population derives from more than one ancestral population, the chances are that its genes reflect different TMRCAs and ancestries. This heterogeneity in TMRCAs of genes within a genome (e.g., human) signals a multiorigin for the species.

Using this approach, investigators have ascertained the TMRCAs of specific DNA segments that have become part of the human lineage. To do this, they assess the distribution of ages of genes throughout the genome. If the lineage is of single origin, the TMRCAs of DNA sequences in the AMH genome would exhibit a unimodal distribution. Fig. 15 illustrates the expected unimodal curves for the frequency distribution of ages (TMRCAs) of three types of human DNA (red curve, autosomes or nonsex

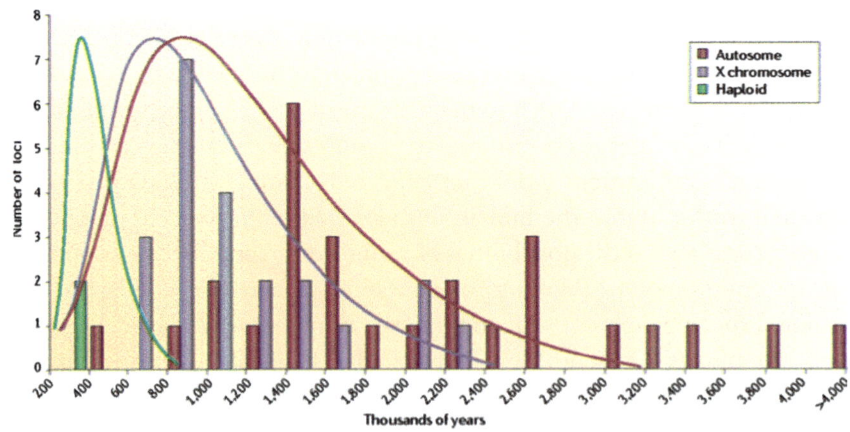

Fig. 15 Frequency distribution of times to the most recent common ancestor (TMRCA). The three color-coded curves are the unimodal expected probability distributions of TMRCAs. Colored bars show the frequencies of TMRCAs of three types of human DNA (*red*, autosomes or nonsex chromosomes; *purple*, X chromosomes; *green*, mtDNA and Y chromosome) (https://www.researchgate.net/publication/6867539).

chromosomes; purple, X chromosomes; green, mtDNA, and Y-chromosome). When the TMRCA dates indicating when specific genes split from a common ancestor are plotted against the theoretical distributions expected under the single origin model, several regions of the genome (e.g., autosomes) exhibit older dates than expected (Fig. 15). In Fig. 15 the red-colored bars (autosomes) representing genes (loci) plot in the upper tail of the curve (ancient TMRCAs). These older dates indicate that many genes have ancestries and genealogies incompatible with a single origin model because they suggest that a number of ancestors contributed to the *sapiens* lineage at various times.[10] The TMRCAs of a number of the DNA segments present in the *sapiens* lineage are provided in Fig. 22 and Table 3 of Chapter 7, and illustrate a wide range of ages from 1.59 my to 190,000 ya. Therefore, the genome seems to be made up of pieces of DNA with widely different evolutionary histories.[36]

In addition, several genes indicate very ancient TMRCAs dating back to *Homo erectus/habilis* times. For example, the gene for CMP-N-acetylneuraminic acid hydroxylase or CMAH in contemporary human has a coalescence time (split time of lineages) of 2.9 mya,[37] the X chromosome-linked pyruvate dehydrogenase *E1* or PDHA1 exhibits a date of 1.86 ya,[38] *N*-acylsphingosine amidohydrolase or ASAH1 was dated at 2.4 mya, and arylamine *N*-acetyltransferase or NAT1 at 2.0 mya.[39] All of these ages are inconsistent with a single origin of modern humans.

Genomic inference from linkage disequilibrium

Linkage disequilibrium occurs when a portion of DNA experiences less recombination than expected. As a result, linkage disequilibrium or LD gives rise to blocks of DNA that remain relatively unaltered (DNA sequences are less susceptible to recombination and change by homologous recombination during DNA pairing during meiosis). In fact, it is the striking preservation or conservation of specific DNA segments that signals the presence of LD in a region of DNA. This limited frequency of recombination within segments of DNA can result from a lack of sequence similarity between pairing sequences. Normally, recombination occurs when homologous or corresponding maternal and paternal sequences in an organism's genome exchange segments during meiosis. LD can occur for different reasons, including natural selection for certain sequences, which will lead to their preservation. Another way to bring about LD is by introducing a piece of DNA into a population or lineage (e.g., of hominins) after the lineages have had time to change during a relatively long protracted period of separation. In such an event, the introduced DNA and the endogenous counterpart may have changed so much that they would not be able to recognize each other, pair, and recombine. Since this admixed DNA remains relatively intact in the introgressed genome, investigators can identify it, sequence it, assess its origin (i.e., assess the donor species or population), and, by determining how much it has changed by mutation from the source gene since the admixture event, determine when the introgression event took place.

By using this approach, many investigators have been able to identify several DNA sequences that have been introgressed into the human lineage at different times in the past and, by doing so, demonstrating gene flow among hominin populations in Africa and that humans descended from more than one African population or lineage. The first of such reports was published in 2005 and describes a 17,500 bp (base pair) sequence at the Xp21.1 location (i.e., position 21.1 on the short arm of the X chromosome) that has been in LD for at least 1 my.[40] Other locations exhibiting LD and rejecting the single origin model have been described in the DNA sequences CD209/CD209L.[41,42]

In a more recent study explicitly aimed at detecting DNA locations under LD and admixture among archaic African populations, three sites were identified.[43] The group of African populations examined included San Bushmen (hunter-gatherer of South Africa), Xhosa (a Bantu group from the Eastern Cape region of South Africa), and Mbuti (forest pygmy population of Eastern Congo). One of the identified segments of DNA is a 31,400 nucleotide region located on the small arm of chromosome 4 known as 4qMB179. This segment of DNA was found in LD in several pygmy populations of Central Africa, and it is thought that it descended from an archaic segment that entered the contemporary pygmy population(s) by admixture. Analysis of the 4qMB179 DNA segment indicates that the lineage of hominins that donated the introgressed DNA and the human lineage diverged about 1.5 mya from each other. It was also determined that the DNA fragment got admixed into the human lineage approximately 20,000 ya. A second introgressed segment, 13qMB107, is located on the long arm of human chromosome 13. It is a 7000-nucleotide-long DNA segment on LD identified in two San individuals. Analysis of the sequence suggests an estimated divergence time of 1 mya and a recent introgression time of 20,000 kya. The third site in LD is 18qMB60, located on the long arm of human chromosome 18. It was found in the San Bushmen, Xhosa of Southern African, as well as in the Mbuti people of Eastern Congo. The Mbuti represent the only population examined that exhibits the three introgressed DNA segments. The geographic location of the Mbuti in the forest region of Central Africa suggested to the researchers that the source of the introgressed DNA might have been a now extinct archaic population that hybridized with the human lineages that have survived to the present.[43] Unfortunately, Central Africa is a region poorly represented in genetic studies, and there is a very poor fossil record making comparative studies difficult. Fig. 16 shows the distribution of the introgressed 4qMB179, 18qMB60, and 13qMB179 DNA fragments in 14 African populations.

Overall, the genomic inference data from linkage disequilibrium analysis suggest that the hominin lineage that gave rise to AMHs did not evolve in genetic isolation and point to specific times and populations contributing to admixture events. And although the sequences under LD discussed indicate more recent events compared to the Middle to Late Pleistocene (33,000–50,000 ya) fossils of Jebel, Florisbad, Herto, and Omo, they

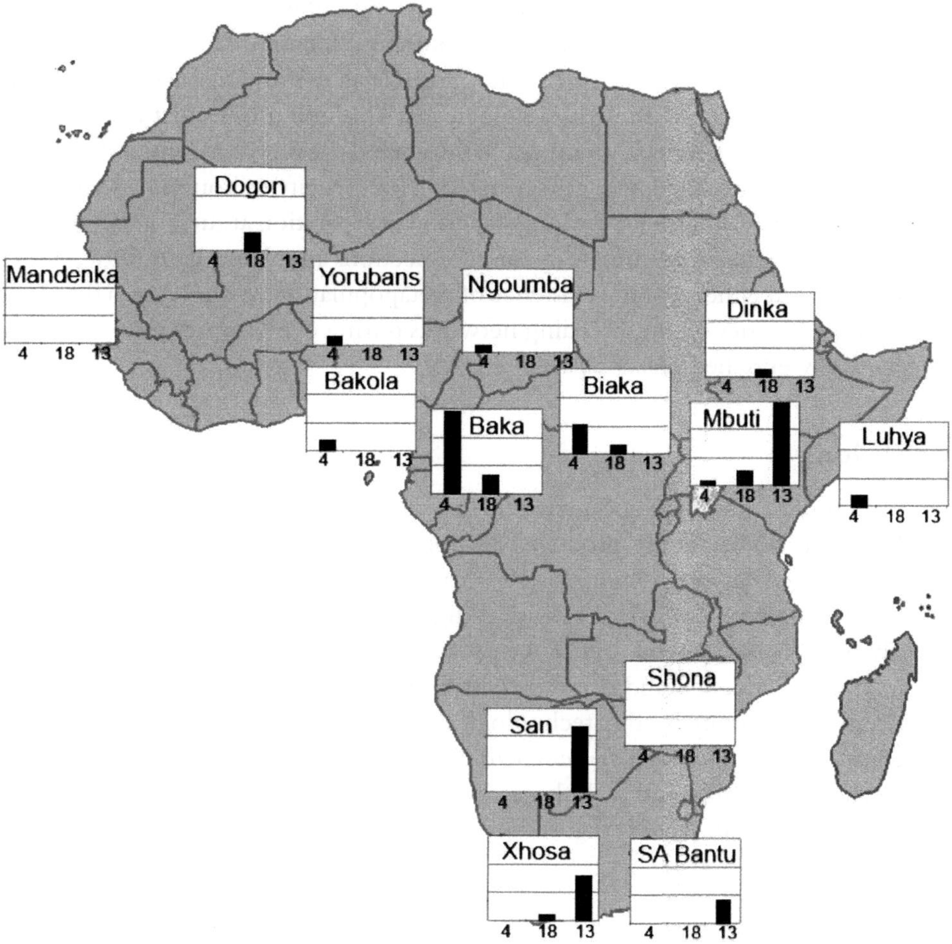

Fig. 16 Frequency of introgressive variants within three sequenced regions in an expanded sample of 500 sub-Saharan Africans. The filled bar represents the frequency of a variant marking the divergent haplotype at 4qMB179 (left), 18qMB60 (center), and 13qMB179 (right) in each of 14 population samples. Each *horizontal line* on the bar charts represents a frequency of 5% (www.pnas.org/cgi/doi/10.1073/pnas.1109300108).

suggest that hominins interbred and produced fertile hybrid offspring. Overall, the genetic data support the contention derived from the fossil and tool tradition that *sapiens* derived from an amalgamation of distinct archaic populations and fails to support the notion of a single-origin/location model.

Furthermore, it is noteworthy that all three sequences under LD signal separation events of hominin lineages dating back 1.0–1.5 mya and that only recently admixture leading to introgression has occurred. In other words, all three pieces of DNA signal estimated divergence between the human lineage and different hominin lineages dating back to when *Paranthropus* species lived in South and Central Africa, *Homo habilis* and *Homo ergaster* existed in Kenya, East Africa, and *Homo erectus* and *Homo antecessor* were in East Africa.[44] Therefore, these data suggest that hominin groups that lived as separate species (no gene flow between them) could interbreed, possibly in Central Africa, 1 million years or more after their lineages separated. This capacity of hominin species to mate and exchange DNA after such a prolonged period of separation argues for the potential of hominins to participate in numerous interconnections involving gene flow. This also underscores the likelihood of the existence of metapopulations within Africa that dynamically, through a series of interbreeding networks during evolutionary time, contributed and genetically enriched *Homo sapiens*.

Conclusions

The fossil, tool technology, and genetic data presented in this chapter corroborate the hypothesis that AMHs is the product of a complex, protracted evolutionary series of admixture events. Although useful for 100 years, the traditional concept of linear human evolution is simplistic, constricting, misleading, and incapable of explaining the increasing body of fossil and genetic data. Taken together, the numerous examples of different combinations of ancestral and derived traits frequently in the form of mosaicisms seen throughout the continent, the technological parallelisms in the tool traditions seen throughout African, and the genetic signals from DNA of various time periods admixing with the human lineage and in linkage disequilibrium provide strong support for the multiregional origin of AMHs that involved the entire continent.

References

1. Shermer Why M. *People Believe Weird Things: Pseudoscience, Superstition, and Other Confusions of Our Time*. New York: Holt Paperbacks; 2002.
2. Mahoney EP. Lovejoy and the hierarchy of being. *J Hist Ideas*. 1987;48:211–230.
3. Lovejoy AO. *The Great Chain of Being A Study of the History of an Idea*. MA: Harvard University Press; 1971.
4. Archibald JD. Edward Hitchcock's pre-Darwinian (1840) tree of life. *J Hist Biol*. 2009;42:561–592.
5. Hitchcock E, C. *Hitchcock Elementary Geology*. New York: Ivison, Phinney & Co; 1860.
6. Brauer G. The origin of modern anatomy: by speciation or intraspecific evolution? *Evol Anthropol*. 2008;17:22–37.
7. Cann R, Stoneking M, Wilson A. Mitochondrial DNA and human evolution. *Nature*. 1987;325:31–36.
8. Harding RM, McVean G. A structured ancestral population for the evolution of modern humans. *Curr Opin Genet Dev*. 2004;14:667–674.
9. Wakeley J. Metapopulation models for historical inference. *Mol Ecol*. 2004;13:865–875.
10. Garrigan D, Hammer MF. Reconstructing human origins in the genomic era. *Nat Rev Genet*. 2006;7:669–680.
11. Gunza P, Booksteina FL, Mitteroecker P. Early modern human diversity suggests subdivided population structure and a complex out-of-Africa scenario. *PNAS*. 2009;106:6094–6098.
12. Scerri EML, Thomas MG, Manica A, et al, A. Did our species evolve in subdivided populations across Africa, and why does it matter? *Trends Ecol Evol*. 2018;33:582–594.

13. Hublin J-J, Abdelouahed B-N, Bailey SE. New fossils from Jebel Irhoud, Morocco and the pan-African origin of Homo sapiens. *Nature*. 2017;546:289.
14. David Richter R, et al. The age of the hominin fossils from Jebel Irhoud, Morocco, and the origins of the Middle Stone Age. *Nature*. 2017;546:293–296.
15. Prüfer K, Racimo F, Patterson N, et al. The complete genome sequence of a Neanderthal from the Altai Mountains. *Nature*. 2014;505:43–49.
16. Meyer M, Kircher M, Gansauge M-T, et al. A high-coverage genome sequence from an archaic Denisovan individual. *Science*. 2012;338:222–226.
17. Dreyer TF. A human skull from Florisbad, Orange Free State, with a note on the endocranial cast, by CU Ariens Kappers. *Proc Koninklijke Akad Wetenschappen te Amsterdam*. 1935;38:3–12.
18. Lewis PJ, Brink JS, Kennedy AM, Campbell TL. Examination of the Florisbad microvertebrates. *S Afr J Sci*. 2011;107:7–8.
19. Stringer C. The origin and evolution of Homo sapiens. *Philos Trans R Soc Lond B Biol Sci*. 2016;371:20150237.
20. Mounier A, Lahr M. Deciphering African late middle Pleistocene hominin diversity and the origin of our species. *Nat Commun*. 2019;10:3406.
21. Grün R, Brink JS, Spooner NA, et al. Direct dating of Florisbad hominid. *Nature*. 1996;382:500–501.
22. Fleagle JG, Assefa Z, Brown FH, Shea JJ. Paleoanthropology of the Kibish Formation, southern Ethiopia: introduction. *J Hum Evol*. 2008;55:360–365.
23. Mcdougall I, Brown FH, Fleagle JG. Stratigraphic placement and age of modern humans from Kibish, Ethiopia. *Nature*. 2005;433:733–736.
24. White TD, Asfaw B, DeGusta D, et al. Pleistocene *Homo sapiens* from Middle Awash, Ethiopia. *Nature*. 2003;423:742–747.
25. Stringer CB. Out of Ethiopia. *Nature*. 2003;423:693–695.
26. Dirks PHGM, Roberts EM, Hilbert H, et al. The age of Homo naledi and associated sediments in the Rising Star Cave, South Africa. *eLife*. 2017;6:e2423.
27. Rightmire GP. Middle and later Pleistocene hominins in Africa and Southwest Asia. *Proc Natl Acad Sci*. 2009;106:16046–16050.
28. Adler DS, Wilkinson KN, Blockley SM, et al. Early Levallois technology and the Lower to Middle Paleolithic transition in the southern Caucasus. *Science*. 2014;345:1609–1616.
29. Wood B. *Human Evolution A Very Short Introduction*. Oxford, UK: Oxford University Press; 2005.
30. Callaway E. Neanderthals: bone technique redrafts prehistory. *Nature*. 2014;512:242.
31. Clark G. *World Prehistory: A New Synthesis*. Cambridge, UK: Cambridge University Press; 1969.
32. Lycett SJ, von Cramon-Taubadel N. A 3D morphometric analysis of surface geometry in Levallois cores: patterns of stability and variability across regions and their implications. *J Archaeol Sci*. 2013;40:1508–1517.
33. Brooks A. Long-distance stone transport and pigment use in the earliest Middle Stone Age. *Science*. 2018;360:90–94.
34. Groucutt HS, Scerri EML, Lewis L, et al. Stone tool assemblages and models for the dispersal of *Homo sapiens* out of Africa. *Quat Int*. 2015;382:8–30.
35. Thompson J, Mackay A, Nightingale S, et al. Ecological risk, demography, and technological complexity in the Late Pleistocene of Northern Malawi: implications for geographical patterning in the Middle Stone Age. *J Quat Sci*. 2017;33:261–284.
36. Pääbo S. The mosaic that is our genome. *Nature*. 2003;421:409–412.
37. Hayakawa T, Aki I, Varki A, et al. Fixation of the human-specific CMP-N-acetylneuraminic acid hydroxylase pseudogene and implications of haplotype diversity for human evolution. *Genetics*. 2006;172:1139–1146.
38. Harris EE, Hey J. X chromosome evidence for ancient human histories. *Proc Natl Acad Sci*. 1999;96:3320–3324.
39. Patin E, Barreiro LB, Sabeti PC, et al. Deciphering the ancient and complex evolutionary history of human arylamine N-acetyltransferase genes. *Am J Hum Genet*. 2006;78:423–436.
40. Garrigan D, Mobasher Z, Kingan SB, et al. Deep haplotype divergence and long-range linkage disequilibrium at Xp21.1 provide evidence that humans descend from a structured ancestral population. *Genetics*. 2005;170:1849–1856.

41. Barreiro LB, Patin E, Neyrolles O, et al. The heritage of pathogen pressures and ancient demography in the human innate-immunity CD209/CD209L region. *Am J Hum Genet*. 2005;77:869–886.
42. Koda Y, Tachida H, Pang H, et al. Contrasting patterns of polymorphisms at the ABO-secretor gene (*FUT2*) and plasma α(1,3)fucosyltransferase gene (*FUT6*) regions in human populations. *Genetics*. 2001;158:747–756.
43. Hammer MF, Woerner AE, Mendez FL, et al. Genetic evidence for archaic admixture in Africa. *Proc Natl Acad Sci USA*. 2011;108:15123–15128.
44. Sawyer GJ, Deak V, Sarmiento E, et al. *The Last Human*. New Haven: Yale University Press; 2007.

CHAPTER 9

The first out of Africa migrants

Migration is an expression of the human aspiration for dignity, safety and a better future. It is part of the social fabric, part of our very make-up as a human family.

Ban Ki-Moon

The first hominin migrants out of Africa

The first nonhuman hominin expansion out of Africa is documented by the fossils from the genera *Graecopithecus* and *Ouranopithecus* found in Greece and Anatolia, respectively, and dated to around 8 mya.

Ouranopithecus is a genus of extinct Eurasian great ape that some researchers consider to be the last common ancestor of apes and humans. A controversial theory that is not widely accepted is that *Ouranopithecus* is a forerunner to australopithecines and humans.[1]

Graecopithecus is most likely a member of the African apes (homininae) and was originally identified by a single mandible found in Greece (Fig. 1). Analysis of tooth specimens, dated to 7.2 mya, has led to suggestions that *Graecopithecus* may have been the oldest direct ancestor of *Homo* excluding the chimpanzee lineage, or alternatively the last common ancestor to both humans and chimpanzees. It was found to be around 200 ky older than *Sahelanthropus tchadensis*, the oldest known hominid found in Africa.[2–4]

The human expansion out of Africa has important implications for understanding the complexity of modern human origins and diversity as well as the genetic and phenotypic structure of extant populations. The dates for the earliest dispersal of human and human-like ancestors from Africa continue to get pushed back in time, as archeologists and paleontologists continue to find ancient artifacts and fossil evidence. Recent findings of stone tools in Shangchen, China, dated 2.12 mya implies that hominins left Africa much earlier than previously thought and that humans first traveled east into Asia, and then much later west into Europe.[5] Prior to this, the oldest evidence of humans out of Africa had been located in Georgia, which was around 1.8 million years old. The Shangchen stone tools predate the Georgia tools by 270 ky.

Later, evidence of northerly human expansion around 120 kya suggests that early hominins were able to adapt to somewhat colder regions. However, when conditions became much colder, their populations likely retreated to warmer areas in the south or their populations died out leaving only traces of their existence.

https://doi.org/10.1016/B978-0-12-819086-9.00015-4

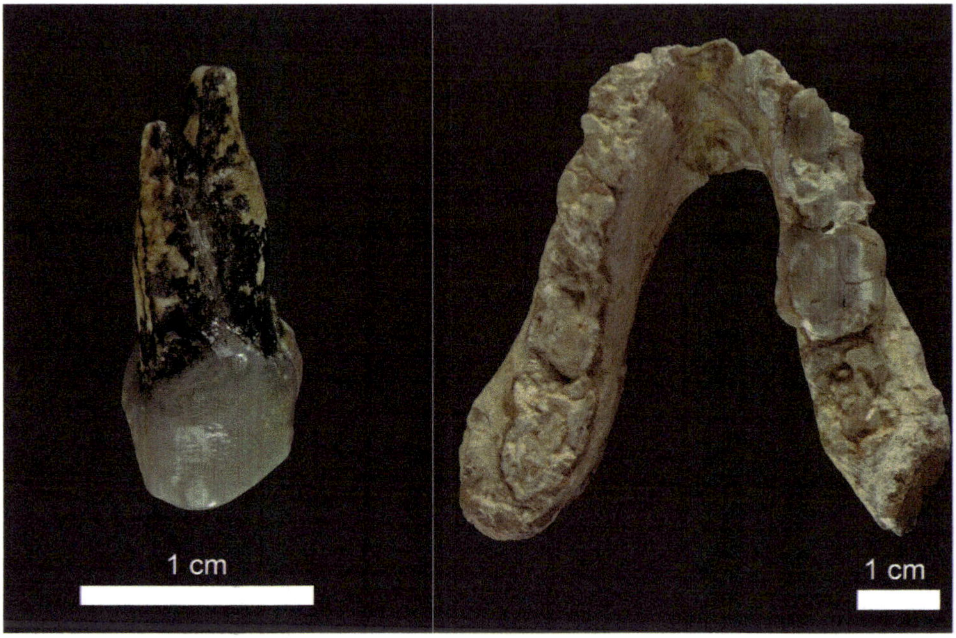

Fig. 1 Left: 7.24-million-year-old upper premolar of *Graecopithecus freybergi* from Azmaka, Bulgaria. Right: 7.175-million-year-old lower jaw of *Graecopithecus freybergi* from Pyrgos Vassilissis, Greece. *(Data from Wolfgang Gerber/University of Tübingen. http://www.sci-news.com/othersciences/anthropology/graecopithecus-freybergi-hominin-04888.html.)*

Before our ancient human-like ancestors could spread out of Africa they needed to develop the physical and mental capabilities that would enable them to survive in Middle East and Asian environments where unique food sources and fresh water were scarce and seasonal. Bipedalism evolved in Africa in some of our ancient ancestors over 6 mya, which allowed for easy movement on land over long distances. Migration would have also required an increased level of intelligence to cope with unfamiliar environments and improved technology to aid subsistence. Travelers would also need to adapt to a diverse diet that included more meat to increase options in seasonally arid environments.

The earliest hominid lived at the end of the Miocene and into the Pliocene epoch 4.5–6 mya. These early hominids had limited capabilities; however, over time, they evolved larger cranium sizes, which in turn improved organization, the development of complex communication, and eventually bipedal movement. For years, bipedal walking was considered the defining quality of the modern human lineage; however, as scientists began to find a variety of fossils of early bipedal species, they concluded that brain development was probably the most substantial change that made early humans unique from their ancient ancestors.[6]

Bipedal movement

There are many theories attempting to explain the evolution of bipedal movement. Increased speed cannot be a logical explanation since many quadrupedal animals are much faster than humans. One explanation that Darwin proposed was freedom of hands which allowed for tool development for defense, and hunting. This theory fails to explain the discrepancy between the evolution of bipedalism 6–7 mya and the use of stone tools about 3.3 mya; however, some have argued that earlier tools were made of wood or other perishable materials that cannot be recovered.

A number of theories have tried to explain the development of bipedalism in hominins. Some involve the climatic changes and habitat alterations taking place at the time of the hominin separation from great apes. The savanna theory, for example, suggests that a number of geological events including the creation of the East African Rift (Fig. 2) and Ridge prompted progressive ecological changes in the regional habitat transforming wet forest to arid grasslands and savannas. It is postulated that the mountains acted as barriers, keeping rain precipitation on the west side of the East African Ridge. The lack of rain east of the range transformed the land into the dry savanna seen today. As a result, it is envisioned that early arboreal hominin populations gradually came under strong selection pressure for traits that would allow them to survive in the new flat habitat. Therefore, hominins needed to respond to the novel challenges in order to survive as species. They needed to evolve physiological and anatomical adaptations that would allow them to survive in the new dry plains of Northeast Africa. Even though the savanna theory is popular, many aspects of the savanna model do not appear to describe the conditions under

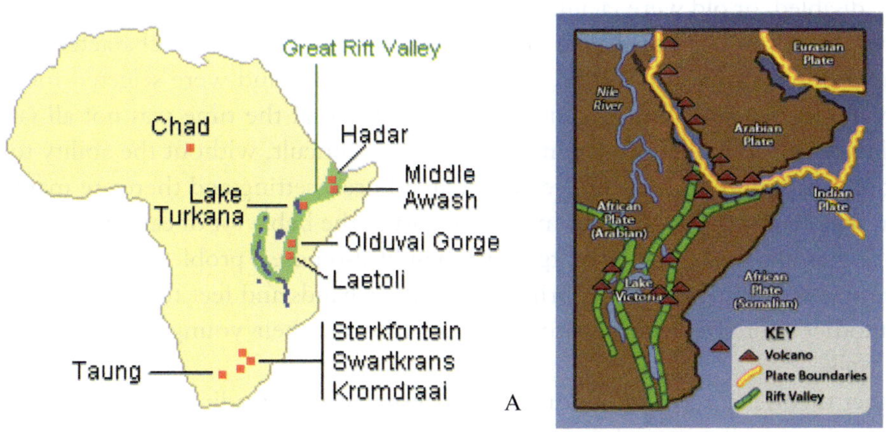

Fig. 2 (A) Shows the Great Rift Valley and Olduvai Gorge. (B) Shows the East African Rift Zone and some of the volcanic activity and plate boundaries (http://www.eclp.com.na/demo/geography/mod1page18.html). *(From (A) Herrera RJ, Garcia-Bertrand R.* Ancestral DNA, Human Origins, and Migrations. *London, United Kingdom: Academic Press.)*

which bipedalism evolved. All of the sites before 3 million years ago seem to have been partially or fully wooded, and no early hominids are known from full savanna environments. In addition, the fossil forelimbs of early hominids demonstrate the continued importance of climbing.

The transition from an arboreal existence to upright posture and bipedalism is observed in the fossilized skeletons of early hominins such as the australopithecines more than 4 mya. Australopithecines display a medley of traits reflecting both arboreal and bipedal locomotion. In particular, members of the different species of the genus *Australopithecus* exhibited a number of lingering ancestral traits such as curved fingers designed for grasping to branches in a forest habitat while at the same time experiencing the foramen magnum toward the base of the skull to allow for an upright posture.[7]

Most of the models that have been put forward in an attempt to explain the evolution of bipedalism are based on the geological and climatic episodes that changed forest to savannas east of the Northeast African Range. One of the explanations for the evolution of bipedalism, for example, is that early hominins found it advantageous to adopt an upright posture to detect and frighten potential predators. This theory suggests that aposematism (advertising by an animal to potential predators), a scheme designed to scare attackers away by appearing larger and as menacing as possible, played a major role. Also, an upright posture may have been advantageous in the savanna as a method of surveillance against intruders as well as attracting mates during sexual advancements. However, upright posture could have also been a disadvantage that made it easier for predators to detect individuals.

Bipedalism allows for the use of the arms and hands for various tasks other than and during locomotion. For instance, activities that are associated with hominin evolution such as tool making, hunting, and gathering/processing food as well as caring for the young, disabled, or old were facilitated due to having free hands. In the rather structureless, flat landscape of grasslands, morphological and behavioral characteristics that increased the probabilities of survival were advantageous and were selected for due to the greater fitness provided. Because the environment at the time was not all savanna, and bipedalism makes climbing considerably more difficult, without the ability to grasp with the feet, hominids were less secure in an arboreal setting and there are many indications that climbing remained an important part of the behavior of early hominids. The lack of a grasping foot in bipedal organisms would also pose a problem for child-rearing. The young of chimps and other primates use their hands and feet to grasp and cling to their mother's fur. Bipedalism forced parents to carry their young possibly until they could walk.

Other theories for the evolution of bipedalism include: wading to forage for aquatic food; standing in tall grass to detect predators; sexual display; following migrant herds; less exposure to direct sunlight; easier release of heat by being above ground; and conserving energy. Regardless of the various theories for the evolution of bipedalism and the disadvantages, no theory completely explains the impetus for bipedal evolution.

Some researchers assume that prebipedal primates were terrestrial quadrupeds or knuckle-walkers like chimpanzees. These may have been facultative bipeds that were adapted for running bipedally through the forest, standing upright to forage overhead, and climbing trees. One example of a facultative bipedal primate is the gibbon, which routinely engages in these activities and when on the ground, runs bipedally. Gibbons have the same number of lumbar vertebrae as humans, a human-type chest, and they stand up straighter than chimpanzees, who are also occasionally bipedal. Constitutive terrestrial bipedal movement probably did not occur until hominins became less dependent on trees for refuge and began to forage by trekking seasonally over long distances.

As primates increased in size and their frequency of bipedal movement, sustained erect standing on lower limbs would favor the evolution of the human hip, knee, and foot. The foot would probably have gone through the most dramatic change, from a prehensile organ to a heel-supported appendage. Increased body size would enhance locomotor efficiency, compared to smaller animals, by more effective use of the elastic energy of tendons and muscles, and fewer strides to cover a given distance. Increased size may have also led to the evolution of the distinctive human calf muscle (gastrocnemius muscle, also called leg triceps). Unlike those of great apes, the leg triceps is heavily tendinous, adding to its function as an energy-conserving spring during walking and running. The muscle provides the propelling force in running and jumping by pulling the heel up and thus extending the foot downward.

Frequent squatting during foraging may have further selected for strong heels and for even weight distribution between the heel, forefoot, and the closely placed feet. Frequent squatting would also add to the development of the hamstring, buttock, and anterior thigh muscles.

Other major morphological features diagnostic of bipedalism include: presence of a bicondylar angle, or valgus knee; a more inferiorly placed foramen magnum; presence of a reduced or nonopposable big toe; a higher arch on the foot; a more posterior orientation of the anterior portion of the iliac blade; a relatively larger femoral head diameter; an increased femoral neck length; and slightly larger and antero-posteriorly elongated condyles of the femur. Bipedal walking uses an inverted pendulum movement, while running uses a spring-mass movement.[8]

The fossil record reveals three Miocene candidates with potential hominin affinity and the first to show bipedal movement: *Ardipithecus kadabba*, *Orrorin tugenensis*, and *Sahelanthropus tchadensis*. The *Sahelanthropus tchadensis* fossils are the oldest hominin remains currently known. Remains were found in West-Central Africa and dated to 6–7 mya. Fossil evidence indicates cranial and dental features that suggest hominin affinity. A small brain, sloping face, prominent brow ridges, and small canine teeth are some of the human-like features. The presence of the anteriorly placed foramen magnum suggests a bipedal gait due to the similar basicranial morphology of more recent hominids. The foramen magnum is a large opening in the cranium from which the spinal cord exits the brain. *S. tchadensis* has one that is located farther forward than that of any other primates except humans. This placement indicates that the head was held on an upright body, which can be associated with walking on two legs. This evidence makes *S. tchadensis* the first facultative biped. Upright walking may have helped this species survive in forest and grassland habitats by being able to see further and avoid predators.[9,10]

Fossils of *Orrorin tugenensis*, another of the oldest human-like ancestors, are dated to 5.8 6.0 mya. These individuals share several apomorphic features with modern humans, as well as some with australopithecines. They were approximately the size of chimpanzees and had small teeth with thick enamel, similar to modern humans. The femur of *Orrorin tugenensis* differs from those of apes and *Homo*, and most strongly resembles that of *Australopithecus* and *Paranthropus*, indicating they were bipedal and showed an upright posture.[11]

Ardipithecus kadabba is dated to between 5.2 and 5.8 and shows hominin affinities such as evidence of bipedalism and canine reduction. She had both opposable big toes and thumbs in order to climb trees. Fossil remains of her legs, feet, pelvis, and hands suggested that she walked upright on the ground but was a quadruped when moving around trees.[12]

The discovery of *Pithecanthropus erectus* provided anthropologists with an intermediate form that had both ape and human qualities.[13] Although the cranial size of *P. erectus* was small, it had a femur that was uniquely similar to modern bipedal humans, helping to question the belief that brain size evolved before bipedalism. This discovery and others that followed forced scientists to question what defined the genus *Homo*.

Australopithecines were also among the first to indicate bipedal movement. These individuals had a face with forward-projecting jaws and a conical rib cage as in great apes. The relatively long extremities and curved fingers and toes suggest the persistence of some degree of arboreal existence in spite of the absence of opposable big toes. Yet it is possible that some of the ancestral arboreal characters were retained due to their selective neutrality or their beneficial secondary functions. The foramen magnum was at the base of the skull and their pelvis was more human-like compared to *Ardipithecus* with a short and broad ilium and a wide sacrum located in the back of the hip joint, all indicative of upright posture. In addition, the femur of australopithecines pointed frontally toward the knee, suggesting bipedal locomotion.

Members of one of the species of australopithecines are responsible for a striking set of footprints on volcanic ash created about 3.7 mya in Laetoli, Tanzania, 45 km south of Olduvai Gorge. The footprints were found by Mary Leakey in 1976. The imprints are seen in a stretch of 24 m and were made by three adult individuals walking in the same direction at a speed of 1 m/s, the regular pace of a modern human. It was observed that the smaller footprints exhibit signs of an uneven burden on one side. It has been suggested that a female carrying a child on her hip as she walked was responsible for the imprints. Although the species that made the tracks has not been established, most investigators attribute the footprints to *A. afarensis*.

The depressions represent the oldest evidence of bipedal locomotion by a hominin. No knuckle impressions indicative of quadruped locomotion are evident. Noteworthy in these imprints are the absence of an opposable big toe and the presence of a very human-like arch. In addition, the topology of the impressions reveal that the heels touched the ash first and then the weight of the body was moved to the front of the foot prior to pushing off the toes. This pattern is very reminiscent of the human gait.

Considering the relative small brain size of australopithecines, in the range of 420 to 500 cm^3, the indentations indicate that upright posture and bipedal locomotion evolved prior to the spur in brain size. In relation to brain size, it is interesting that the SRGAP2 gene, a gene linked to neuronal movement and differentiation as well as nerve connections, started duplicating itself about 3.5 mya during Australopithecine speciation and replicated a second time 2.5 mya coinciding with the appearance of the first *Homo* species from an australopithecines ancestor and the first spur in brain growth in hominins (Fig. 3).

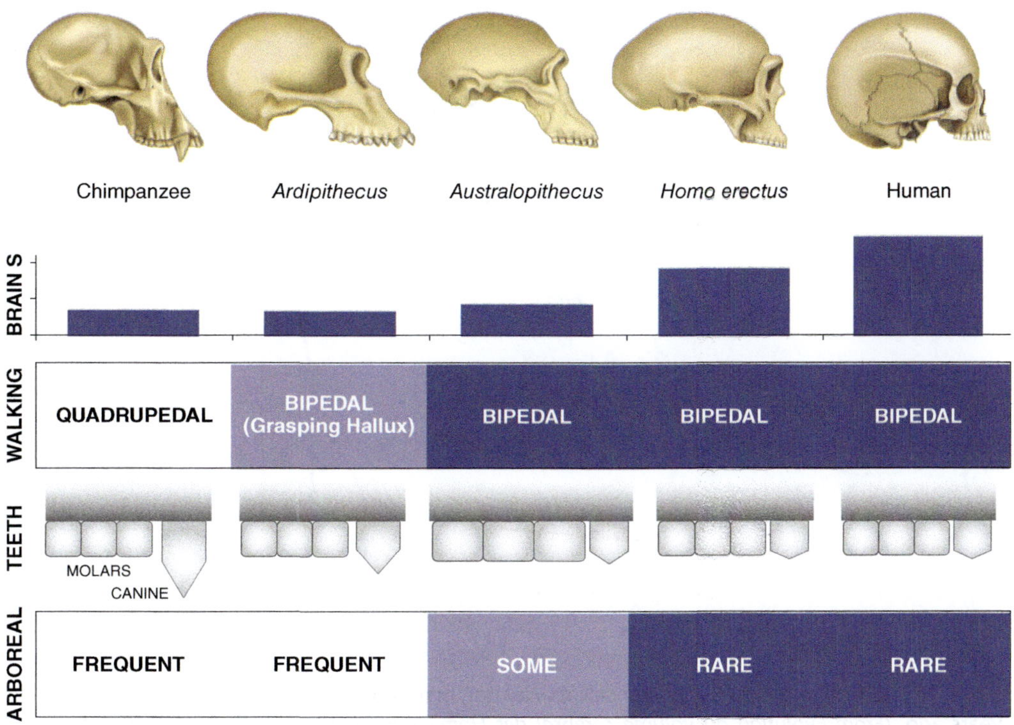

Fig. 3 Timelines of bipedalism and brain size in hominins. *(From https://www.nature.com/scitable/knowledge/library/overview-of-hominin-evolution-89010983/.)*

The evolution of the human brain

In the last 2 million years, the human brain has nearly tripled in size (Fig. 3) Hominid brain size increased slowly until the arrival of *H. habilis*, upon which a trend of tremendous brain growth began (Fig. 4). Hominin brain expansion tracks so strongly with refinements in tool technology that some researchers disregard other factors that may have contributed to increasing brain size. The fossil record shows evidence of widespread stone tool use following *H. habilis* emergence. Appropriately named "handyman," *H. habilis* was the first known hominid to make stone tools, and is believed to have habitually used toothpicks, hand axes, and sharp stone implements to carve meat off carcasses. Thousands of tools such as these were found in the Olduvai Gorge in Tanzania and are attributed to *H. habilis*.[14] While tool use may have contributed to an increase in brain size, investigators also point to social complexity, foraging strategies, communication, and

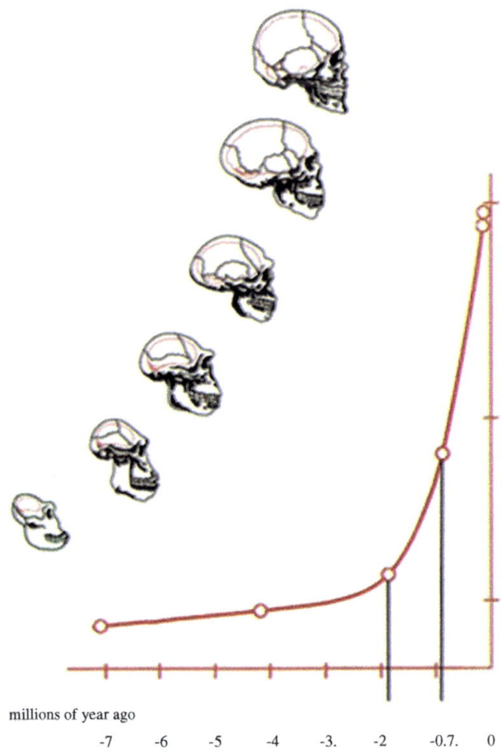

Fig. 4 The growth rate of hominid cranial capacity began to rapidly increase, starting around 2 million years ago. *(From Evolution du volume cerebral des Hominides.* Le Journal du Net. *2010. http://www.linternaute.com/science/biologie/dossiers/06/0608-memoire/8.shtml. Copyright 2011 Benchmark Group—69–71 avenue Pierre Grenier 92517 Boulogne, Billancourt Cedex, France.)*

cultural behaviors that may have led to brain expansion but left no or few archeological remnants.

The next million years following the appearance of *H. habilis* saw continued encephalization, with the human brain growing half again in size by the time of *Homo erectus* (1.9 mya), who had a brain of about 900–1000 cm^3. *H. erectus* is generally considered to have been the first human species to emigrate out of Africa and is possibly the longest-lived hominid, having existed for nearly 2 my.[15]

Between their tall bodies and relatively large brains, *H. erectus* individuals needed larger quantities of energy at more frequent intervals than their ancestors did. Meat consumption evolved as a solution to this problem, with meat being a more energy-dense food than plants. Although *H. habilis* is believed to have scavenged meat off animal carcasses, *H. erectus* was the first hominid to rely on meat as a significant part of its diet. Eating meat was a way to absorb nutrients with a shorter digestive tract, which made energy more readily and more quickly available. Additionally, *H. erectus* is regarded as the first hominid to use fire, which allowed for the cooking of meat and even easier, faster digestion.[15,16]

H. erectus is also associated with the Acheulean stone tool industry, which gave rise to the invention of large cutting tools like hand axes and cleavers. The increased use of a more diverse set of tools may have been responsible for *H. erectus*' long-lived existence despite changing climates and harsh environmental conditions.[17] The use of tools also contributed to the ability of *erectus* to migrate to different locations and eventually out of Africa.

The human brain reached its largest size 400 kya with the appearance of *Homo neanderthalensis*. Neanderthal brain size reached an average of 1400 cm^3. Modern-day humans have an average brain size of 1350 cm^3 but the range of brain size can vary among different populations and depends on the size of an individual. Based on fossil evidence, Neanderthals and modern humans from the same time period had similar brain sizes.

Although Neanderthal brain size at birth was similar to *Homo sapiens*, most likely due to birthing constraints (head size and vaginal expansion), Neanderthal brain growth rates during early infancy were higher, resulting in larger adult brain sizes but not an earlier completion of brain growth. Because large brains growing at high rates require large, late-maturing mothers, it is likely that Neanderthal life history was similarly slow, or even slower-paced, than for *H. sapiens*.

It appears that humans achieve their large adult brains primarily through higher growth rates early during ontogeny, rather than through a markedly extended period of growth. Primate brain growth occurs mainly during the prenatal period, which indicates a high maternal investment in its development; however, brain growth also occurs during large portions of the postnatal period. Overall, these adaptations represent components of distinctive life-history adaptations, with potentially important implications for the evolution of primate cognition. Studies indicate that modern human-like ontogenetic traits such as increased neonate brain size and postnatal brain growth rates were already present in *H. erectus*.[18–20]

Although the human brain has the same overall morphology as the brains of other mammals, its cerebral cortex is by far more developed. A number of other animals possess larger brains, but when body sizes are taken into consideration, humans exhibit a relative brain size two to three times larger than those of the bottlenose dolphin and chimpanzee, respectively. The distinctive feature of the brain in modern humans compared with other animals is a very large brain relative to body mass. Compared with those of other species, the human brain is not only bigger but also has higher energy demands. The brain consumes 22 times the amount of energy as muscle tissue, approximately 25% of the glucose metabolized by the body, and 15% of the oxygen intake.[21]

The human brain is about 1200 cm^3 on average, contains approximately 100 billion neurons, more than 100,000 km of interconnections, and has an estimated storage capacity of 1.25 trillion bytes. The distinctive feature of the human brain is extensive corticalization (wrinkling of the brain's cortex) with a six-layered cerebral cortex. The surface area of the modern brain is between 1500 and 2000 cm^2 (about the size of two full pages of a US newspaper). To fit such a large surface area inside the cranium, the cortex is highly folded into regions called gyri and grooves referred to as sulci.[12] At the most basic level, the brain is an organ of soft tissue encased by the vertebrate skull that functions as the coordinating center of sensory information and of intellectual and nervous activity. The human brain is made up of three major parts: the forebrain, midbrain, and hindbrain (Fig. 5). The forebrain is mainly comprised of the cerebrum, which consists of right and left hemispheres connected by a bridge of nerve fibers. The cerebrum integrates sensory and neural functions, and initiates and coordinates voluntary activity in the body. The outer layer of the cerebrum is made up of gray matter, whereas the inner part is composed of white matter. Gray matter contains the cell bodies of neurons, whereas white matter contains the dendrites and axons that spread out from the cell bodies to connect to other neurons.[12] The two cerebral hemispheres are each divided into four lobes: the frontal lobe, responsible for thinking, making judgments, planning, decision-making, and conscious emotions; the parietal lobe, associated with spatial computation, body orientation, and attention; the temporal lobe, associated with hearing, language, and memory; and the occipital lobe, associated with visual processing.[22]

The midbrain, or the upper part of the brain stem, contains nuclei that link the various parts of the brain responsible for motor functions, eye movement, and auditory control. The hindbrain, or the lower part of the brain stem, is involved in the autonomic nervous system and facilitates unconscious actions and processes such as breathing and regulating heart rate. The medulla is the part of the brain stem that controls breathing and heartbeat. The cerebellum is also part of the hindbrain and coordinates and regulates timing of action with available sensory information. Located at the front of the cerebellum, the pons facilitates the timing coordination established by the cerebellum. The pons is

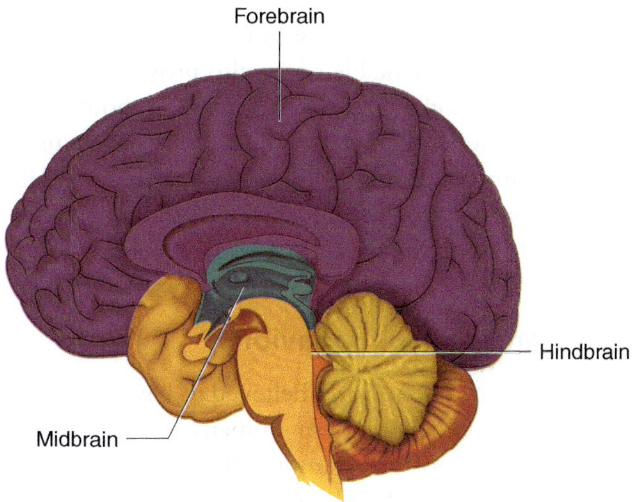

Fig. 5 The diagram shows basic brain structure and its 3 main regions; **Forebrain**, **Midbrain** and **Hindbrain**. Each region consists of different parts that are indicated in the non-bold font.

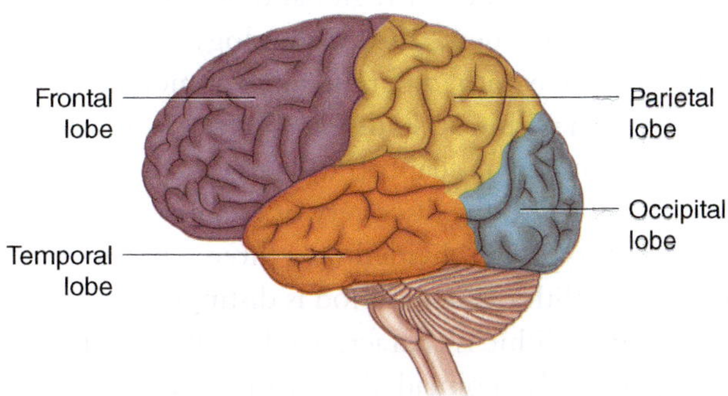

Fig. 6 Diagram shows the hemispheres of the brain which are divided into 4 lobes: the frontal lobe, responsible for thinking, planning, decision-making, and conscious emotions; the parietal lobe, associated with spatial computation, body orientation, and attention; the temporal lobe, associated with hearing, language, and memory; and the occipital lobe, associated with visual processing.

primarily the white matter sending signals between the cerebrum and spinal cord to the cerebellum, which then routes output for action to these two areas[23] (Fig. 6).

Much of the human brain, which encompasses about 86 billion neurons, is part of the cortex and frontal lobes, areas known for a number of higher-level functions such as reasoning, abstract ideas, and consciousness. The cortex is responsible for neuronal activities associated not only with self-awareness but also with mindfulness, creative thoughts, and cosmic awareness. This remarkable organ also provides for emotions, empathy, and altruism. The cortex itself is highly furrowed made up of about 10 billion nerve cells, each interconnected to other neurons by 100 trillion synapses or junctions. These neurons communicate using neurotransmitters firing at a rate of about 10 billion times per second. Although research is continuously making advances in the neurobiology of humans and other species, it is likely that the human brain will remain as an intellectual enigma.[24]

Tool technology and human behavior

The origin of modern human behavior is believed to have coincided with the beginning of the Upper Paleolithic, when modern humans replaced Neanderthals throughout Europe. The need for stone tools arose among Neanderthal and modern human populations as local food sources depleted with increased competition. Early humans needed to develop technology selected for long-distance transport and foraging in order to pursue large, dangerous terrestrial animals. Increased stone tool production served as the primary way through which early humans adapted to their ecological surroundings. The transition from Oldowan assemblages to Acheulean tools is representative of a significant stage for early human behavior development. The transition coincided with the origin of *Homo sapiens* and is the first evidence of regional differentiation.

In addition to bipedal movement, tool technology was an important attribute for hominins being able to move into different environments. Movements throughout the Stone Age occurred both within and outside of Africa with varying demands on individuals and populations in order to adapt to the new environments.[6]

The Stone Age coincides with the discovery of the oldest known stone tools, dated to 3.3 mya. The Stone Age is typically divided into three separate periods: the Paleolithic, the Mesolithic, and the Neolithic. Each period is distinguished by the use and degree of tool sophistication. The Paleolithic is characterized by the use of stone tools, wood tools, and bone tools. This period also coincided with the genus *Homo* developing fire, cave paintings, and burials of the dead (Table 1).

Australopithecus and *Paranthropus* were two hominins that used stone tools. *Australopithecus* is a genus of early hominins that existed in Africa during the Late Pliocene and Early Pleistocene. *Australopithecus* emerged in the Afar region of East Africa, around 4 million years ago. Fossil remains in this region reveal the early events in human evolution including the theoretical transition from *Australopithecus* to *Homo*, the emergence of *Paranthropus*, and the advent of manufactured stone tools. *Australopithecus afarensis* is the most abundantly preserved hominin from the Afar region between 3.8 and 2.9 mya. Some of the earliest known stone chopper tools were found in Lomekwi, Kenya, and date back to 3.3 mya. The tools may be the product of *Australopithecus garhi* or

Table 1 Cultural period, geological age, stone tools, and mode of subsistence in the Indian subcontinent.

Terminology	Geological age	Typical Indian stone tool types	Main subsistence base
Lower Paleolithic	Lower Pleistocene	**Pebble** and **core tools** like hand axes, **cleavers**, and **chopping tools**	Hunting and gathering
Middle Paleolithic	Middle Pleistocene	Flake tools, including those made by prepared core techniques such as the **Levallois technique**	Hunting and gathering
Upper Paleolithic	Upper Pleistocene	Blade tools made on flakes, e.g., parallel-sided blades and burins	Hunting and gathering
Mesolithic	Holocene	**Microliths**	Hunting, gathering, fishing, with instances of animal domestication in a few places
Neolithic	Holocene	**Celts** (ground and polished hand axes)	Food production based on animal and plant domestication

From Herrera RJ, Garcia-Bertrand R. *Ancestral DNA, Human Origins, and Migrations*. London, United Kingdom: Academic Press.

Paranthropus aethiopicus, two known hominins that used stone tools to cut and break open animal bones. *Paranthropus* is classified as an early robust australopithecine; however, it is debated whether or not *Paranthropus* is an invalid grouping and just a form of *Australopithecus*.[25,26]

During the Lower Paleolithic, the development of the first stone tools by *Homo habilis* resulted in the stone tool industry known as Oldowan technology. These early tools were simple, usually made with one or a few flakes, chipped off another stone that resulted in a sharpened edge that could be used for cutting, chopping, or scraping.

The oldest Oldowan tools discovered are from Lomekwi 3, an archeological site in Kenya, where ancient stone tools were discovered dating to 3.3 mya. These tools are considered a major milestone in human evolutionary history because they represent the earliest evidence of cultural behavior. In order to migrate long distances, early humans needed to develop technology selected for long-distance transport and foraging in order to pursue large, dangerous terrestrial animals and cope with new environments. Increased stone tool production served as the primary way that early humans adapted to their ecological surroundings. The transition from Oldowan assemblages to Acheulean tools is representative of a significant stage for early human behavior development. The transition is believed by many to coincide with the origin of *Homo sapiens* and the first evidence of regional differentiation.[27]

Oldowan tools were very simple and are sometimes difficult to distinguish from naturally created objects. Oldowan tools were only about 7–8 cm of cutting edge, whereas Acheulean tools were often bifacial and could produce 30 cm of cutting edge.

Acheulean bifaces, such as stone hand axes and cleavers, were the main tool products of the Pleistocene period. These tools varied greatly in size, structure, weight, and quality depending on their maker. Acheulean flakes were much lighter than previous Oldowan technologies and could easily be reshaped or replaced when damaged, making them much more efficient than previous technology. Throughout most of Africa, Levallois cores and flakes developed from existing Acheulean manufacturing produced an array of diverse and sharper-edged tools.[6,27,28]

Stone tool technology was not the only evidence for complex symbolic behavior demonstrated by modern humans. The Levantine burial sites of Skhul and Qafzeh have uncovered more than just hominid fossils and tool kits. Eight of the skeletons were found in anatomical articulation, a position where the legs are flexed and the arms are extended, suggesting the bodies were deliberately buried. Additionally, two of the individuals at these sites were buried with grave goods such as boar mandibles and antlers of red deer. Personal adornment is another phenomenon of human behavior that represents complex conceptions of social value and material definition of self.[29,30] This form of modern human behavior is believed to have become popular in the very early stages of the Upper Paleolithic. Archeologists have recovered beads and pendants constructed from exotic fossil materials, bones, and animal teeth from excavation sites, indicating acknowledgment of human self-awareness. Scientists are finding that early *Homo sapiens* was far more civilized than they had previously believed. Structures of the brain associated with symbolic behavior have not evolved much from the Paleolithic time period to today.[31,32]

In order for humans to migrate and survive in new environments it was also important that early humans recognized the need for social structure. This implies a sense of cultural and social behavior. The development of social structures or lack of social structure created new criteria for selection. Biological fitness was still imperative; however, it was partially replaced by the need for cultural and social fitness. In other words, members of the group would need to have taken on certain social roles and develop organization skills in order to navigate the difficult journey. This would have resulted in both creative and maintenance roles allowing the group to function as a unit. These sociocultural skills would have been essential for the development of a primitive population with the same focus on survival and expansion.[33]

Scerri et al. presented a paleoenvironmental model of North Africa between 130 and 75 ka based on a multivariate analysis of stone tools from 17 corresponding archeological sites. The model explains the peopling of the Sahara and presents evidence for successive population budding and metapopulation structure, showing that *Homo sapiens* were strongly structured before they left Africa.[34,35]

The rise of the genus *Homo*

For more than 4 million years of hominin evolution, the fossil record indicates trends in canine reduction and cranial changes found in *Sahelanthropus*, *Orrorin*, *Ardipithecus*, *Australopithecus*, and *Paranthropus*. About 2.5 mya there is a change in the hominin fossil record that accompanied climate change. The increasingly cooler climate led to both anatomical and behavioral changes, resulting in the origin of the genus *Homo*.[15,36]

The genera *Homo*, *Paranthropus*, and *Kenyanthropus* are thought by many to have evolved from *Australopithecus*. The genus *Homo* is assumed to have emerged by around 2.8 million years ago, based on Homo habilis fossils found at Lake Turkana in Kenya. The explanation for the evolution of the genus, *Homo*, from *Australopithecus* is somewhat contentious, and for this reason the term hominin is often used to include both genera. Hominin technically includes chimpanzees and prehuman species as old as 10 million years ago. Although it is extremely difficult to determine the evolution of hominins, knowledge of *Paranthropus* and *Kenyanthropus* has grown over the years. The group were known to be some of the older hominins. *Kenyanthropus platyops* especially has made the evolutionary path of the hominins more confusing, because the specific species represents a new type. The discovery of *Kenyanthropus platyops* confirmed that it is one of the earlier species, living at the same time as *Australopithecus afarensis*.[25,26]

As mentioned earlier, the origin of the genus *Homo* in Africa is indicated by the shift from bipedal apes to primitive, large-brained, tool-making meat-eaters that moved across Africa and beyond. The evolution of *Homo* in Africa is represented by three species: *Homo habilis*, *Homo rudolfensis*, and *Homo erectus*, listed roughly in order of their appearance in the fossil record, although there is evidence for geographic and temporal overlap of all three species in East Africa.

H. habilis is from the Early Pleistocene of East and South Africa about 2.3–1.65 mya and is known for retaining some of the primitive features to Australopiths, including thick tooth enamel, long arms, and a moderately prognathic face (Fig. 7). It is also recognized for being the first stone tool maker in the *Homo* line. Less is known about *H. rudolfensis* that differs from *H. habilis* in its larger braincase, longer face, and larger molar and premolar teeth. *H erectus* paleobiology and evolution are well understood due to its rich fossil record. *H. erectus* was the first obligate, fully committed biped, with a body adapted for modern striding. It also signifies a major shift in hominin evolution, through its increased brain and body size and increasingly complex tools and behaviors. Finally, it was also the first in the *Homo* lineage to migrate outside of Africa[15] (Fig. 8).

Many researchers designate the earliest members of *H. erectus*, found in Africa, as *Homo ergaster*. Fossils of *Homo ergaster* tend to be smaller than *H. erectus* and are suggested by some to be ancestral to, or a sister group to, *H. erectus* in Asia.[17] However, most large-brained, large-bodied members of the genus *Homo* found across the Old World are not considered to display species-level variation, and are all described as *H. erectus*.[15,16] The morphological variations seen across the Old World are considered

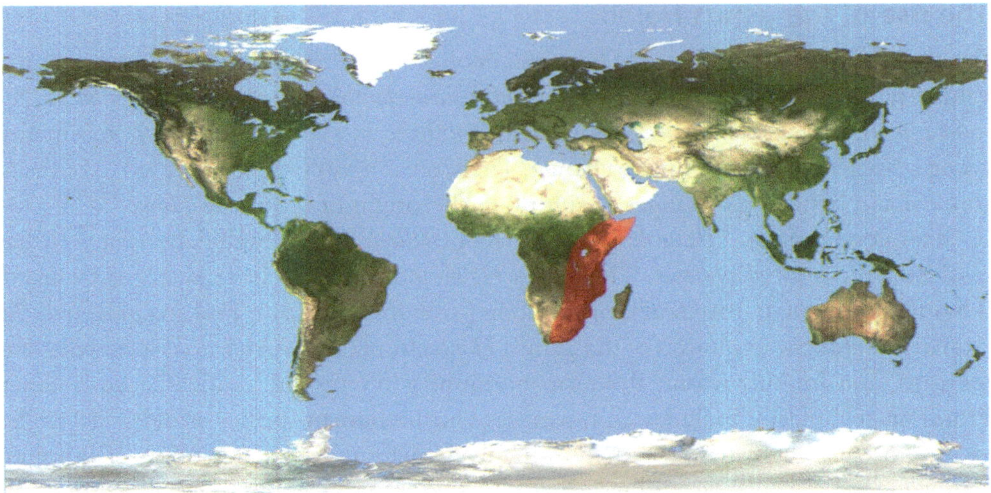

Fig. 7 Geographical range of *H. habilis* shown in *red* (http://www.abroadintheyard.com/wp-content/uploads/Homo-habilis-geographic-range.jpg).

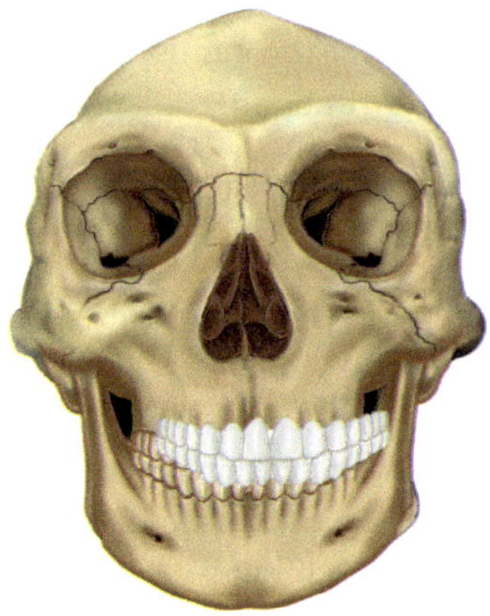

Fig. 8 This drawing represents the skull of *Homo erectus*. The drawing is based on the actual skull found in the link below. (https://en.wikipedia.org/wiki/Homo_erectus#/media/File:Peking_Man_Skull_(replica)_presented_at_Paleozoological_Museum_of_China.jpg).

to be the result of drift, gene flow, and selection working differently in the geographically dispersed populations (Fig. 9).

Outside of Africa, *H. erectus* fossils are some of the most primitive of the *Homo* line discovered, with remains retrieved from sites across Eurasia and Indonesia. Remains indicate that *H. erectus* left Africa as early as 1.8 mya. The skeletal and artefactual evidence for the genus *Homo erectus* in Asia from Dmanisi, Georgia, is dated to approximately 1.77–1.85 mya. Two incisors that belong to *Homo erectus* from Yuanmou, South

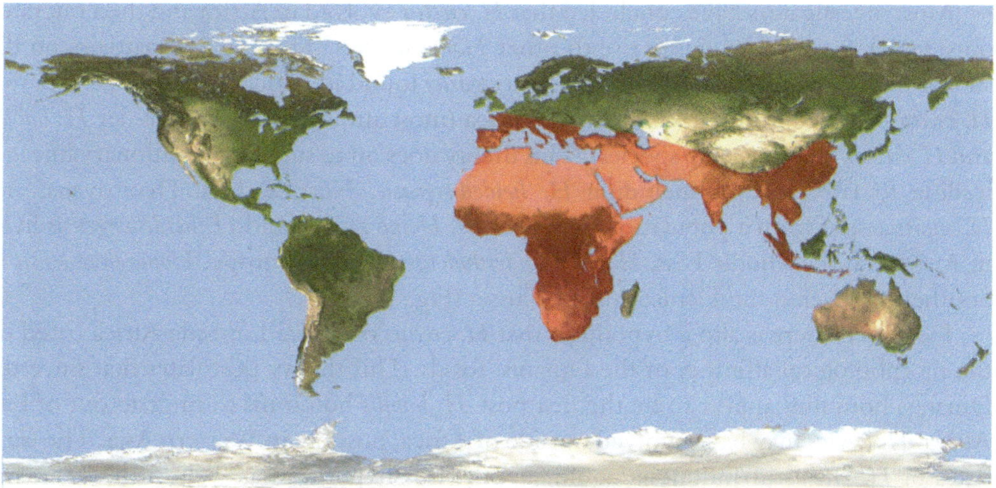

Fig. 9 Geographical range of *Homo erectus* shown in *red* (https://www.abroadintheyard.com/prehistoric-empires-geographic-ranges-human-species/).

Fig. 10 Regions of *H. erectus* fossil discoveries. *(From Herrera RJ, Garcia-Bertrand R.* Ancestral DNA, Human Origins, and Migrations. *London, United Kingdom: Academic Press.)*

China, are dated to 1.7 mya, and evidence for an *H. erectus* cranium from Lantian (Gongwangling) has been dated to 1.63 mya. The earliest hominin fossils from Sangiran Dome in Java are dated to about 1.5–1.6 mya (Fig. 10). Fossil evidence indicates they disappeared in Asia about 400 kya, and the last *H. erectus* may have survived in Indonesia about 550 kya.[37] Regardless of the various timelines for extinction, it is clear that *H. erectus* thrived and evolved across the tropical and temperate regions of Africa, Asia, and Europe with a much wider geographic distribution than any preceding hominins.[15,38]

Most investigators have concluded that *H. erectus* evolved in Africa and then migrated outside Africa, possibly giving rise to other *Homo* species. This is based partially on the similarity of *H. erectus* in Europe and *Homo ergaster* found in Africa as well as the idea that *H. erectus* evolved from *H. habilis* inside Africa (most authorities now consider *H. habilis* and *H. erectus* as sister group species). This theory goes on to suggest evolutionary lineages leading to later species, including *H. heidelbergensis*, Neanderthals, Denisovans, and *H. sapiens*, split from a population of *H. erectus*. *Homo sapiens* and *H. heidelbergensis* lived in Africa and the Middle East, *Homo neanderthalensis* lived in Europe, *Homo floresiensis* in Southern Asia, and the Denisovans in Russia (Fig. 11).

However, there is also a hypothesis that *H. erectus* originated outside Africa based on the morphological affinities of the Dmanisi fossils. This theory postulates that since the Dmanisi hominins appear to be the first post-*H. habilis* hominins to migrate out of East Africa, *H. erectus* may have evolved outside of Africa, possibly in Western Asia. The early dates of *H. erectus* in Indonesia could support this hypothesis as well as the *H. erectus* remains dated to 1.7 million years ago from the Yuanmou Basin in Southwest

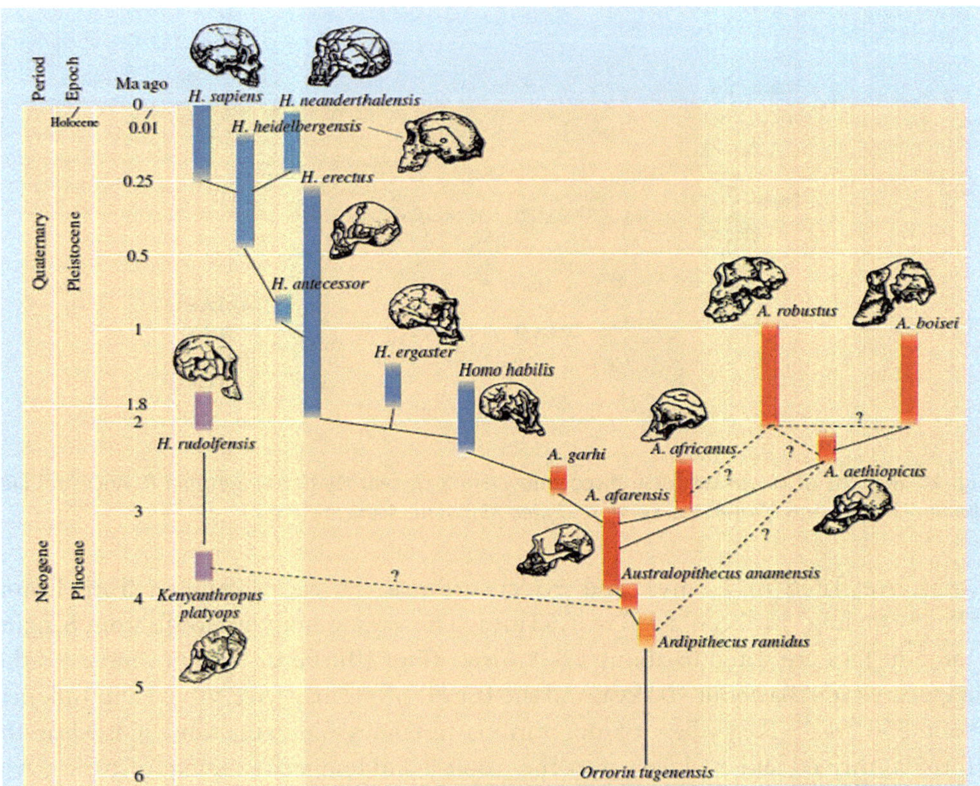

Fig. 11 Phylogenetic tree illustrating the evolution of hominins (http://imgur.com/gallery/UFc94).

China.[39] The evolution of *Homo erectus* outside of Africa may have been the result of gene flow between African and Eurasian hominin populations.[40]

Researchers who believe *erectus* evolved outside Africa point to evidence that indicates hominins migrated out of Africa before *Homo erectus* around 1.9–2 mya. These hominins may have been australopithecines or an unknown species of *Homo*. This theory suggests that the Dmanisi population may represent a missing link in the evolution of *Homo erectus*. Though their precise classification is controversial and disputed, the Dmanisi fossils are highly significant since they represent some of the oldest fossils from the early migrations out of Africa and into Western Eurasia.

For many years the Dmanisi fossils represented the oldest fossil record of hominins outside of Africa; however, a 2018 study by Zhu Zhaoyu et al. found evidence for the oldest human presence at Shangchen, Central China, dated 2.12 mya based on magnetostratigraphic dating of a layer containing stone artifacts. This discovery implied that hominins left Africa much earlier than indicated by the evidence from Dmanisi.[5]

Pulses of extreme dry and wet weather likely played a pivotal role in the dispersal of *erectus* out of Africa on more than one occasion since these climatic fluctuations created contraction and expansion episodes of population movements and ranges in gateway regions of migration out of Africa such as the Levant and the Horn of Africa. During dry conditions, populations contracted or migrated to other more livable terrains while in humid weather, groups expanded. It is likely that these environmental forces motivated *erectus* to leave Africa.

Considering that *erectus* originated in the tropical and permissive environment of sub-Saharan Africa and then rapidly migrated to areas with diverse and fluctuating climatic conditions, it is logical to think that *erectus* was capable of manufacturing some type of apparel for covering and protection against adverse weather soon after its origin. For instance, the Dmanisi site in a mountainous region of the Caucasus demonstrates that 200,000 years after its origins in Africa, *erectus* was able to survive the frigid winter conditions of the region. The weather at the time in the Caucasus was much colder and drier than now. Although no articles of clothing have been recovered to date, possibly due to the disintegration of the organic matter that they were made of, it is likely that *erectus* utilized skin and fur from the game they killed to make clothing, covers, bedding material, and hide for protection of some sort of dwelling.

It is also possible that *erectus* built huts. Primitive huts such as the ones in Terra Amata near Nice, France have been attributed to *erectus*. These shelters probably served as windbreakers and were fabricated out of hide extended over a wooden frame with foundations made up of stones and animal skins for floor. The Terra Amata huts were built around 400 kya as oval structures 26–50 feet long and 13–20 feet wide, reinforced with branches three inches in diameter and rocks at the bases. Other contemporary digs in Europe such as the one in Bilzingsleben, Germany exhibit similar circular structures 9–13 feet in diameter with a base made up of bones and stones. These rudimentary shelters attributed to *erectus* would have been of survival importance especially in regions experiencing extreme weather conditions.[41]

Erectus' ability to create and control fire probably means that they used it to keep predators away and keep warm at night. In addition, they may have employed the fire to cook. Cooking softens the meat and provides flavor. In relation to diet, it is noteworthy that cut marks created by tools have been detected in *erectus*' bones. This has been interpreted as evidence of some form of cannibalism, either ritualistic or as a mean of survival during lean times.

The debate about whether *Homo erectus* or another hominin was the first to leave Africa continues; however, it is agreed that whoever left Africa first needed to possess the physical and cultural attributes that would have aided dispersal through the arid environments of Northern Africa and the Middle East. In addition to bipedalism, the intelligence to cope with unfamiliar environments, and a diverse diet, they would have needed the technological skills to aid survival. Oldowan-style tools have been found at sites in Dmanisi, Georgia, and Northern China, dating to around 1.7 mya. Later waves of expansion are proposed around 1.4 mya showing Acheulean industries that are associated with *H. antecessor*, and 0.8 mya associated with *H. heidelbergensis*.[42,43]

Homo rhodesiensis is now considered a synonym for *H. heidelbergensis* or a possible African subspecies of *Homo heidelbergensis* that was present throughout Africa and Eurasia during the Middle Pleistocene. White et al. proposed *Homo rhodesiensis (H. heidelbergensis)* as ancestral to *Homo sapiens idaltu* (Herto Man); both early species have very similar morphologies to AMHs, and are thought by some to be the most recent ancestors to modern *Homo sapiens*. *Homo sapiens idaltu* fossils were discovered in the Middle Awash sites of Ethiopia and dated to 160 kya. Their skeletal morphology is distinct from that of late *Homo sapiens* but it is clear they share similar ancestry. Their globular braincase, facial features, and robust skulls resemble both primitive Africans and early modern humans. A long-standing source of controversy is whether humans replaced these early species or if they were similar enough to interbreed.[44,45]

Several investigators have suggested that Homo floresiensis was descended from an expansion, similar to *erectus*. It is not clear whether these earliest hominins leaving Africa should be considered *Homo habilis*, or late *Australopithecus* closely related to *Homo habilis*, or a very early form of *Homo erectus*. Regardless, the morphology of *H. floresiensis* has been found to show greatest similarity with *Australopithecus sediba*, *H. habilis*, and Dmanisi Man, raising the possibility that the ancestors of *H. floresiensis* left Africa before the appearance of *H. erectus*.[46] A phylogenetic analysis published in 2018 suggests that *H. floresiensis* was descended from a species (possibly australopithecine) ancestral to *H. habilis* to making it a "sister species" either to *H. habilis* or to a *habilis-erectus-ergaster-sapiens* clade. This would make *H. floresiensis* older than *H. erectus*. On this basis, *H. floresiensis* would represent an unknown very early migration out of Africa, dating to before 2.1 mya.[46,47]

Origin of modern humans

In 2017, human remains found in Jebel Irhoud, Morocco, were dated to 315 kya, making them the oldest specimens classified as *H. sapiens*. Because this date overlaps with that of *H. heidelbergensis*, the remains suggest that these transitional forms, including *H. idaltu*, represent a rapid evolution of the *sapiens* face, with gradual modifications to the braincase among populations distributed across Africa, beginning as early as 300,000 years ago.[48]

H. heidelbergensis is believed to have migrated out of Africa and into Europe, evolving into Neanderthals in Europe while separately evolving into *Homo sapiens* in Africa. These species are now considered to be an archaic *Homo sapiens* population from Europe, distinct from modern *Homo sapiens*.

Current evidence suggests that the lineage of *Homo sapiens* probably originated in Africa at least 500 kya with the earliest observed anatomically modern human morphology appearing about 300 kya. Nuclear DNA studies indicate that the highest level of genome diversity, when compared to populations outside Africa, is in sub-Saharan Africa, supporting an African origin of modern humans. The finding of greater genetic diversity is now supported by many genetic markers, making Africa the most likely geographical origin for modern human dispersal.

Studies of Y chromosome-specific haplogroups suggest that the paternal lineages of all modern humans had their origins in Africa,[49] while mtDNA studies support an African origin by revealing that the most common female ancestor lived in Africa about 200 kya.[50–52] Furthermore, anthropological, archeological, geological, and climate evidence all support an African origin.

Although most researchers believe an African origin is well established, one area of early human evolution still in question is where in Africa *Homo sapiens* emerged. The view that *Homo sapiens* evolved from a single region and/or population within Africa has been prescribed for many years, with the idea that humans originated most likely in modern-day Ethiopia. However, developments across multiple disciplines indicate new data are no longer consistent with the single region or single population view.[53]

Although migrations occurred in several waves and it is clear that evolution of *H. sapiens* took place in many regions of Africa, it was Eastern Africa where the first major demographic expansion occurred about 70–60 kya, which led to the modern-day peopling of the world. Rito et al. describe a possible explanation that important cultural traits were transmitted from South to East Africa around the time of migration. They identified a mitochondrial signal of the proposed dispersal soon after 70 kya (the only time in the last 200,000 years that humid climate conditions encompassed Southern and tropical Africa). This South to East African dispersal immediately preceded the out-of-Africa expansions, potentially providing the trigger for expansions by transmitting significant cultural elements from Southern Africa.[53]

Fossils from all over Africa have modern and ancient traits in varied combinations, including the 260 ky old Florisbad skull from South Africa,[54] and the 195 ky old remains from Omo Kibish[55] and the 160 ky old Herto skull, both from Ethiopia.[56] Some scientists have argued that these remains represent different subspecies of *Homo sapiens*, or different species altogether.[54]

New genetic and archeological data are consistently being acquired addressing the question of origin.[57,58] Based on recent data, Eastern, Southern, and Central Africa have all been proposed as the cradle of humanity. Southern Africa has been suggested based on the genome-wide variability of the Khoe-San people, indicating that ancient DNA extended as far north as Malawi in the Early Holocene.[59,60] Evidence for a Central African origin is based on mtDNA diversity, Y-chromosome data, and paleoclimatology, while an Eastern Africa origin is based mostly on paleontological evidence.[52,53]

A consistent feature in mtDNA and nuclear DNA genetic evidence is a deep split between hunter-gather "Khoe-San" groups in Southern Africa and all other present-day humans. Some of the genetic evidence supporting this split is seen in the mtDNA with haplogroup L0 in Southern Africa, and the L1′6 branch present in Central and Eastern Africa. This split cannot distinguish between a human origin in the north or the south, but does establish the existence of two groups of *H. sapiens*.[61,62]

While some scientists have placed *Homo sapiens*' origin in East Africa, others suggest a southern birthplace. The story always begins in a single region and moves on to describe how those ancestral hominids, most likely *Homo heidelbergensis*, slowly accumulated the characteristic morphological and cultural characteristics of our species.

In contrast to this, new data have challenged the view that *H. sapiens* was endemic to a single region or habitat, and imply a more complex African origin. This theory suggests that humans originated from several diverse populations that lived across Africa. Separated from each other by geographical barriers, they mostly evolved in isolation, with each group developing different traits. As a changing climate remodeled the African landscape, different populations were repeatedly drawn together and separated; however, when they did meet they mated, exchanging genes and ideas in a continent-wide melting pot that eventually coalesced into the features of AMHs.[34] This African multiregionalism theory (evolution within one continent) has gained popularity, especially since *Homo sapiens* fossils were found in Northwestern Africa, at Jebel Irhoud in Morocco, and dated to more than 300 kya. These fossils show similarities to *Homo sapiens* remains in both Eastern and Southern Africa.[48]

Other evidence supporting an African multiregionalism comes from early *H. sapiens* fossils which do not demonstrate a simple linear progression toward AMH morphology but instead exhibit remarkable amounts of morphological diversity and geographical dispersal. Recent archeological and genetic evidence are consistent with the idea that *H. sapiens* originated and diversified within subdivided populations, across Africa, that occasionally were connected by sporadic gene flow. These population subdivisions may have been shaped and sustained by shifts in cultural, ecological, and/or climatic boundaries. This concept of pan-African evolution not only may have included connections with other *H. sapiens* but also may have included hybridization between *H. sapiens* and more divergent hominins living in different regions.

The Pan African Evolution Group was established in 2018 at the Max Planck Institute for the Science of Human History to address problems of human origin through multidisciplinary research. The group uses large-scale fieldwork to generate multidisciplinary data from a transect across West Africa, one of the least understood regions of the continent. Some of the major goals are to recover human fossils, establish new chronometric dates, and isolate ancient DNA from the region. The group will also explore pan-African paleobiogeography, including animal genetics, in order to integrate different sources of information.[34]

The population genetics of small mammals, such as rodents, insects, rabbits, prairie dogs, and bats, are valuable to paleoenvironmental studies due to the animals' short life spans and fast reproductive rates, as well as narrow climatic and vegetative habitats. The presence or absence of these mammals indicate environmental change over short time periods which can be quantified and can help determine the environmental conditions under which humans lived.

The data supporting an African origin are solid; however, there are still questions regarding the African exodus, including: How many exits of AMHs are detected in the fossil or archeological record? Which of the exits are supported genetically rather than by interglacial spreads with other fauna in the fossil record? When were the routes taken and why?

Routes out of Africa

There is clear evidence that early humans migrated out of Africa into Eurasia on a number of occasions during the Upper Pleistocene (126 to 11.7 kya).[63] However, as mentioned earlier, debate still exists regarding the times and routes utilized by these early migrants. Two passageways have been delineated in connection with dispersals out of Africa: one is in Northeast Africa, the Levant, into what is today the Sinai Peninsula; and the other is the southern corridor from the Horn of Africa into modern-day Yemen.

Modern-day humans outside Africa derive most of their genetic ancestry from a single migration event 50–70 kya, according to genetic data from genome-wide mtDNA studies[64] and Y-chromosomal analyses.[65] It is likely that several dispersals took place and it has been proposed that at least four major migration events occurred, with three in the Upper Pleistocene and a fourth more recently.[66] Modern humans not only migrated out of Africa but also returned to Africa, possibly via the same corridors they used to leave.

The Horn of Africa

The southern path across the Strait of Bab-el-Mandeb on the Red Sea is thought by many researchers to be the gateway for the oldest known dispersal out of Africa (150–130 kya). This period corresponds closely to a glacial period (180–125 kya) which resulted in low sea levels or a land bridge during the glacial maxima. Estimates suggest the sea could have been as much as130 m lower than the current level.[63] The land, previously under the sea, would have provided easy passage for early modern humans to move into previously submerged coastal regions and occupy unexplored land.

Studies by Timmermann and Friedrich reconstructed conditions in Northeastern Africa and the Middle East, based on the astronomical cycles that drove the ice ages. They concluded that a wetter climate and lower sea levels could have enticed humans to cross from Africa into the Arabian Peninsula and the Middle East during four periods, around 100, 80, 55, and 37 kya.[67]

Across from the southern Red Sea corridor is the Jebel Faya site, which is an archeological site and limestone hill or escarpment near Al Madam in the United Arab Emirates. The site has been dated to about 127–95 kya.[63] Although the site does not contain human remains, it is associated with lithic tools manufactured using Acheulean technology which exhibit similarities to Northeast African tool technology, suggesting a connection between the two. The observation of similar Northeast African and Jebel Faya stone tool traditions dating to similar time periods may indicate an early settlement in the Arabian Peninsula and a possible long-term occupation at the Jebel Faya site.[63]

The Levant

Early modern human remains at the sites of Qafzeh and Skhul, across the northern corridor, in present-day Israel signal an occupation that likely penetrated Arabia via the Levant. Unlike the Jebel Faya site, where only lithic material was discovered, the Qafzeh and Skhul cave remains consist of fossils and artifacts dating to 130–90 kya.[68] These human remains were discovered in close proximity to Neanderthal fossils, yet it is not clear to what extent these two hominin species interacted at this site nor whether the disappearance of early modern humans from the Levant was the result of direct competition with Neanderthals, assimilation into a common gene pool, and/or retreat back to Africa.

Two routes have been proposed for the migration from Northeast Africa to the Levant corridor. One scenario suggests a coastal dispersal northward along the Red Sea, while a second one suggests a path across the central Sahara utilizing a river corridor that cut through the Libyan desert and led to the Mediterranean coast. Considering that this Libyan wet passageway existed about 130–117 kya, it is possible that the Qafzeh and Skhul settlements also resulted from movement along its basin. Lithic deposits of Abdur and Asef along the Red Sea coast, dating to about 125 kya both in present-day Eritrea, strengthen support for the coastal Red Sea route to the Nile corridor.[69]

This period of cohabitation seems unique since no modern human presence has been detected in the Levant after the period of the Qafzeh and Skhul occupation, until about 50 ky later, which resulted from the resettlement of Southwest Asia.[70] It is not clear whether the Levantine human population was able to migrate further east into Asia. However, as mentioned above some have speculated that this first wave of human expansion may have reached Southern China. Homo fossils found in the Chinese region of Zhirendong (Zhiren Cave) have been dated to over 100 kya and are the oldest fossils in East Asia.[71,72] In addition, there is some evidence of *Homo* in North America' as early as 125–130 kya.[73]

The Nile passageway was most likely used by modern humans from Northeast Africa into the Levant about 50,000 years ago.[70,74] This reoccupation of the Levant and Southwest Asia by modern humans subsequently led to the settlement of Southeast and Western Europe about 40–35 kya through the Danube River.[70]

Genetic data

Based on genetic data, the southern route is generally associated with the dispersal of modern humans from the Arabian Peninsula further eastward into South Asia across the modern-day Strait of Hormuz into the Indian subcontinent.[74,75] Today it is thought that all non-African lineages descend from a single African lineage, L3 for mtDNA or CT-M168 for the Y chromosome. Yet there has been a long-running debate about the early spread of Y-chromosomal lineages because their current distributions do not fit a simple phylogeographical model.[75]

mtDNA studies indicate that the L2 and L3 maternal lineages experienced an expansion during a period of 85–55 kya and suggested that a limited group of females carrying haplotype L3 migrated out of Africa approximately 70 kya through the Strait of Bab-el-Mandeb.[76]

Y-chromosome haplogroups suggest the most recent common ancestor (MRCA) of all modern human Y chromosomes arose between 142 and 338 kya.[77] The most ancient modern human Y haplogroups are A and B and are exclusively African in origin. Haplogroup A is mostly found among Khoisan populations, while haplogroup B is found among populations from the Congo. It is likely that haplogroups A and B never left Africa and that the migrants that dispersed out of Africa and populated the rest of the world carried an ancient point mutation on the Y chromosome known as M168, from which all non-African haplogroups originated outside of Africa. It is estimated that M168 individuals dispersed out of Africa about 70,000–50,000 years ago and differentiated into haplogroups C to T in Southwest Asia.[78] These date estimations suggest that the early modern human migrants' maternal and paternal lineages left Africa around the same time and possibly as part of the same migration. This parallelism in departure date for both uniparental set of markers favors the Horn of Africa as the putative route.

There is also evidence of a Y-chromosome haplogroup returning to Africa. Haplogroup J-M304, also known as J, is a Y-chromosome DNA haplogroup believed to have evolved in Western Asia. The clade spread from there during the Neolithic, primarily into North Africa, the Horn of Africa, Socotra, the Caucasus, Europe, West Asia, Central Asia, South Asia, and Southeast Asia. The haplogroup occurs in approximately 12% of native European populations. The average frequency of haplogroup J as a whole is today highest in the Near East (12%), followed by Europe (11%), the Caucasus (8%), and Northeast Africa (6%) (Fig. 12).

Hundreds of fully sequenced genomes from extant individuals living in Africa, Australia, Papua New Guinea, East Asia, and Oceania, as well as ancient DNA from the South Pacific, have contributed to a better understanding of early AMH migrations and their genetic contributions to modern-day humans. Each scientific team involved in these studies used complex computer algorithms to interpret genomic patterns of similarity and difference in the genomes of extant populations in order to infer genetic variability and population history of AMHs resulting from *H. sapiens* migration out of Africa.

Researchers at the University of Copenhagen examined Australia and New Guinea by comparing Aboriginal genomes to other groups. They concluded that Australian Aborigines diverged from Eurasians between 50 and 70 kya, after the group had already diverged from Africans. They also concluded that Aborigines and all other non-African people descend from the same out-of-Africa sweep, and that Australia was initially settled only once. Their DNA data suggest that a genetic bottleneck about 50 kya resulted in the

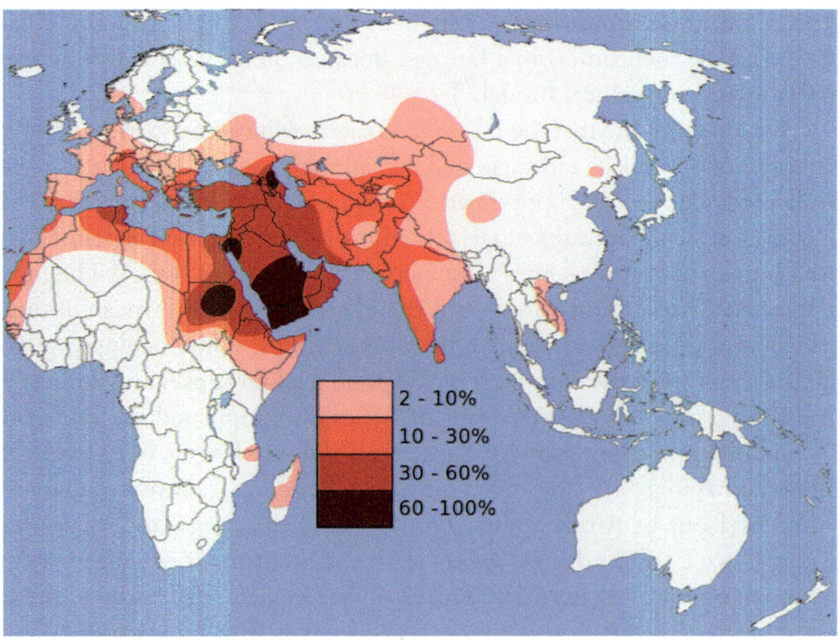

Fig. 12 Distribution of Y chromosome haplogroup J (https://en.wikipedia.org/wiki/Haplogroup_J_(Y-DNA)#/media/File:Haplogroup_J_(Y-DNA).svg).

small group that first colonized ancient Australia. The studies show Aboriginal ties to other Eurasians but a long period of isolation. This reaffirms unique genetic roots not seen anywhere else except Africa.[79]

Mallick et al. examined 300 genomes from 142 populations. The populations covered a broad range of human genetic, linguistic, and cultural diversity. The authors suggested that their results are similar to skeletal and archeological evidence that indicate early migrations of AMH outside of Africa, but their results suggest that modern humans, outside Africa, are almost completely descended from a single founding population. The authors concluded that early migrations occurred but did not contribute significantly to present-day populations. In other words, individuals who were involved in earlier migrations, 100–125 kya possibly through the southern route, died and contributed essentially nothing genetically to modern-day humans.[80]

In a study by Pagani et al.[64] that analyzed 379 genomes from 125 populations worldwide, the group concluded that at least 2% of the genomes of people from Papua New Guinea comes from an early dispersal of AMHs, who left Africa perhaps 120 kya. Their studies proposed that *Homo sapiens* left Africa in at least two waves, with the earliest migration from an ancestor that separated from Africans earlier than other Eurasians. This coincides with a multiple-dispersal model, as an early expansion of AMHs from Africa can be detected today in Australasia and indicates a minor genetic contribution to modern-day humans from the first AMH migrants to leave Africa.

Today most researchers agree that the most recent migration out of Africa explains more than 90% of the genetic ancestry of living people. However, researchers also agree that at least two migrations took place, with the evidence for early migrations coming from discoveries of early stone tools in India and Arabia, but those lineages mostly died out.

Migration of *Homo sapiens*

Various theories suggest that several expansions of archaic populations out of Africa occurred during the Lower and Middle Paleolithic. A widely accepted theory is that about 2 mya, the first of our direct ancestors moved northward from their homelands and out of Africa. The first human or human-like migrants out of Africa included hominins (either australopithecines or an unknown species of *Homo*), while *Homo erectus* left Africa around 1.7 mya. These early migrations occurred throughout Eurasia between 2.1 and 0.2 mya, and are collectively known as Out of Africa I. Around 200 kya, AMHs migrated into Eurasia in a movement known as Out of Africa II.

Some of the oldest known *Homo sapiens* fossils outside of Africa come from caves in Israel in Skhul (dated 90 kya), Qafzeh (dated 120 kya), and Misliya (dated 180 kya). These are thought to have been transient populations that intermittently occupied the region. There is no evidence of a direct evolutionary connection between the Misliya and the

Skhul and Qafzeh peoples. As mentioned previously, genetic studies also support the idea of earlier dispersals of modern humans out of Africa starting from about 220 kya[81] with the possibility of expansion into China.[71]

Southeast Europe is considered to be a major dispersal corridor and the human fossil record of this region appears to be more diverse than that of more isolated and less hospitable areas of Europe, reflecting the complexities of repeated migrations, higher survival rates, and the admixture of human groups. Two fossilized human crania (Apidima 1 and Apidima 2) from Apidima Cave, southern Greece, are among the most important finds from the region and were dated using U-series radiometric methods. Apidima 2 dates to more than 170 kya has a Neanderthal-like morphological pattern. By contrast, Apidima 1 dates to more than 210 kya and presents a mixture of modern human and primitive features. These results suggest that two late Middle Pleistocene human groups were present at this site—an early *Homo sapiens* population, followed by a Neanderthal population.[82]

Stone tools found in India dating to about 74 kya, in Yemen dating to between 70 and 80 kya, and in the United Arab Emirates dating to about 80 kya may indicate that earlier *sapiens* dispersals reached beyond the Levant. Some of these tools resemble African Middle Stone Age technology, while others are more like those used by *Homo sapiens* and Neanderthals in the Levant; however, no *Homo sapiens* or Neanderthal remains were found with the tools, although Neanderthal remains were found in the region. Most researchers have concluded from genetic and fossil evidence that mass migration of *Homo sapiens* occurred within the last 60–80 kya.[83]

After the first early dispersals out of Africa, various other groups left Africa in waves of migration which occurred during periods with favorable climatic and environmental conditions. These movements occurred eastward across Southern Asia more than 1 million years ago and into Western Europe around 900 kya. These migrations appear to have taken two forms—irregular short occupation of the Levant and nearby sites by small populations and later migration on a massive scale after 80–60 kya with larger waves of AMHs across Eurasia.[84]

Migrational motivation

Why AMHs decided to leave and move somewhere else is a difficult question to answer considering there are no written records and only artifactual clues. Migration could have resulted from unfavorable conditions and a search for something new. This could include things like unbearable changes in the climate, resulting in loss of food and other resources, natural disasters, competition with hostile neighbors, or following the movement of migrating animal food sources. The draw to new areas could have also been the possibilities of new rewards such as more productive lands with better climates and greater amounts of food and resources.[81]

As mentioned previously, there were prerequisites for successfully migrating to new regions that would have included the ability to deal with stressful and dangerous conditions. These conditions would have challenged any group's resourcefulness and ability to adapt. Adequate tools to hunt and gather successfully, or to protect against colder climates and cooperation in securing new resources, would have been essential.

Climate changes

The timing and extent of environmental and climatic conditions in northeast Africa, the Levant, and the Arabian Peninsula during the Middle and Late Pleistocene are essential to determining how and why hominins left Africa. Researchers have found evidence that during the Middle and Late Pleistocene, episodes of climatic change may have allowed passage between Northern Africa and Western Asia, and several models indicate the continued existence of an arid barrier between Northern Arabia and the Levant. Breeze et al.[85] evaluated the environmental setting for hominin dispersals between, and within, Northeast Africa and Southwest Asia during Marine Isotope Stages (MIS) 7–5 by determining surface freshwater availability to estimate environmental conditions at the time. They mapped paleohydrological features (lakes, wetlands, and rivers) in the region and integrated their results with paleoclimate models and paleoenvironmental data. Their analyses suggest limited potential for dispersals during MIS 7 and 6, but indicate the formation of a paleohydrological corridor (the "Tabuk Corridor") between the Levant and the Arabian interior during the MIS 6 to MIS 5e glacial–interglacial transition (191 and 123 kya, respectively). A slightly different route also occurred during MIS 5a (82 kya). They concluded that these paleohydrological and terrestrial data could be useful in establishing proposed routes for hominin dispersals and are consistent with archeological data.

As previously mentioned, Timmermann and Friedrich[67] presented results from spatiotemporal estimates of climate and sea level changes over the past 125 ky. The model simulates the overall dispersal of *H. sapiens* in close agreement with archeological and fossil evidence, suggesting that dramatic climate fluctuations created favorable environmental conditions that triggered periodic waves of human migration out of Africa every 20 ky or so, beginning just over 100 kya. Their data show prominent glacial migration waves across the Arabian Peninsula and the Levant region around 106–94, 89–73, 59–47, and 45–29 kya. The findings suggest that global climate swings played a key role in shaping Late Pleistocene global population distributions, and provide important clues about what drew early humans out of Africa and into Eurasia and beyond. The study further calls into question the deeply held dogma about a single prominent migration[86] out of Africa around 70–60 kya, and points out there were multiple opportunities for migration. The authors also concluded that millennial-scale abrupt climate changes, associated with Dansgaard-Oeschger events (rapid climate fluctuations that occurred 25 times during the last glacial period), had only limited regional effects.

In addition to cooling resulting from glaciation, widespread drying occurred from shutdown of the Atlantic meridional overturning circulation (AMOC). The AMOC is an injection of freshwater into the North Atlantic from the melting of the Laurentide ice sheet during Heinrich event H5 which led to development of a low-salinity cap, causing less warm water to flow northward. Shutdown of the AMOC forced colder air over the African continent by decreasing surface salinity and thermohaline circulation, which led to devastating cooling in the north and tropical Atlantic regions and drying in all regions except South Africa.[87]

As the environment and climate continued to change, hominid species were able to accumulate adaptations that increased their behavioral versatility in response to the fluctuations. Increasing aridity of subtropical lands and lowering of sea levels influenced migration from Central Africa to coastal areas. In the southern part of Africa, the coastline may have retreated by as much as 100 km. Coastal retraction was beneficial as the coast provided a more equable climate and a more constant and diverse supply of terrestrial and marine resources.[29]

Continual deterioration of the African environment into the Upper Pleistocene forced early humans living along the coasts to adjust to changing conditions by migrating out of Africa. Rapid resumption of the AMOC allowed European environments to flourish, quickly changing desert environments to open forests. It is likely during this time that hominids decided to cross the seas to access more desirable regions in Europe and Asia.[29]

Environmental changes

Overall, the theory of hominid dispersal is rooted in the evolutionary biology of fluctuating environments. An increase in aridity started in Africa around 1.5 mya and culminated around 1 mya, causing a shift in vegetation diversity and faunal distribution patterns. Paleoclimate data sets from the African continent suggest a shift toward a cooler and drier climate state with the onset of glaciation during the Late Pleistocene. A cold maximum occurred at about 75,000 years, greatly reducing rainforests and replacing them with montane forests and ecology.[29,32]

Hunter-gatherers most often move in response to resource availability, and once local foods are depleted, they migrate to other regions. It is estimate that contemporary hunter-gatherers have an annual foraging range of 5–15 km. If our ancient ancestors moved at a similar pace, they could have moved from East Africa to China (14,000 km) in just over 1000 years. This dispersal out of Africa took place during the glacial and interglacial conditions of the Pliocene and Pleistocene. Most likely movements generally occurred during the warmer intervals, and would have been more restricted during colder intervals.[88,89]

Once AMHs had migrated out of Africa they encountered a variety of primate species including Neanderthals, Denisovans, and possibly *H. erectus*. Sexual behavior between modern and archaic humans (Neanderthals and Denisovans) has been confirmed using DNA sequencing and statistical methods to discover where and how often humans, Neanderthals, and Denisovans mated. Investigators have determined that East Asians have three Neanderthal ancestors in their lineage, Europeans two, and Melanesians one. In addition, it is estimated that Denisovans mated with Melanesians at least once. Collectively, Denisovans, Neanderthals, and modern humans all bred together at some point, but it is difficult to tell how often, and matings that did not produce offspring cannot be traced. Neanderthal and Denisovan DNA sequences are concentrated in certain regions of the human genome and missing from others[90] (see Chapter 3).

In addition to Neanderthals and Denisovans, scientists have uncovered evidence that AMHs mated with another type of archaic human. Scientists speculate that this other archaic human was most likely *H. erectus*, but because they are unable to recover and sequence DNA from fossilized bones of *H. erectus*, it is unclear.

Conclusion

The story of human origin and early migration covers an immense area and time span and is further complicated by the fragmentary archeological record. There are numerous theories that attempt to explain the origin of anatomically modern humans (AMH). Thirty years ago, researchers thought the answers were straightforward and that AMHs originated and evolved in East Africa, then moved throughout Africa and into Eurasia and beyond in a single migration. Today it is clear that the story of human origin is much more complicated. As new technology was developed and further archeological discoveries were made, a clearer explanation of *Homo sapiens*' origin and evolution started to emerge. AMHs most likely evolved from the interaction of various human-like groups that existed throughout Africa. Those groups occasionally interacted and mated leading to different forms and selection pressures, eventually evolving into what we know today as *Homo sapiens*.

A similar level of confusion exists when trying to explain human migration, and to date there is no one explanation for how, when, or why the groups of hunter-gatherers went wandering within and out of Africa. For some groups, climate change may have pushed AMHs to seek more hospitable lands; others may have been looking for better or more abundant food sources, avoiding hostile or competing groups, or may have simply been curious adventurers wanting a change. The answers to the questions of human origin and migration are continuously changing as the technology advances and new discoveries are revealed. The field of genetics and ancient DNA analysis, fossil discoveries, and climatic and geological data are slowly attempting to piece together the story of human origin and migration, but the story will continue to evolve as new discoveries are made and new technology allows for deeper analysis into our past.

One thing is clear: once AMHs had migrated out of Africa they encountered a variety of primate species including Neanderthals, Denisovans, and possibly *H. erectus*. Sexual behavior between AMHs and Neanderthals and Denisovans has been confirmed using DNA sequencing and statistical methods to discover where and how often humans, Neanderthals, and Denisovans mated.

References

1. de Bonis L, Koufos GD. *Ouranopithecus* and dating the splitting of extant hominoids. *Comptes Rendus Palevol*. 2004;3(4):257–264.
2. Fuss J, et al. Potential hominin affinities of *Graecopithecus* from the Late Miocene of Europe. *PLoS One*. 2017;12(5):e0177127.
3. Langergraber KE, et al. Generation times in wild chimpanzees and gorillas suggest earlier divergence times in great ape and human evolution. *Proc Natl Acad Sci USA*. 2012;109:15716–15721.
4. Pozzi L, et al. Primate phylogenetic relationships and divergence dates inferred from complete mitochondrial genomes. *Mol Phylogenet Evol*. 2014;75:165–183. PMC4059600 https://doi.org/10.1016/j.ympev.2014.02.023.
5. Zhu Z, Dennell R, Huang W, et al. Hominin occupation of the Chinese Loess Plateau since about 2.1 million years ago. *Nature*. 2018;559:608–612. https://doi.org/10.1038/s41586-018-0299-4.
6. Grove M. Climatic variability, plasticity, and dispersal. *J Hum Evol*. 2014;87:32–47.
7. Russo GA, Christopher Kirk E. Foramen magnum position in bipedal mammals. *J Hum Evol*. 2013. https://doi.org/10.1016/j.jhevol.2013.07.007.
8. Biewener A, Daley MA. Unsteady locomotion: integrating muscle function with whole body dynamics and neuromuscular control. *J Exp Biol*. 2007;210:2949–2960. https://doi.org/10.1242/jeb.005801.
9. Lebatard AE, et al. Cosmogenic nuclide dating of Sahelanthropus tchadensis and Australopithecus bahrelghazali: mio-pliocene hominids from Chad. *Proc Natl Acad Sci USA*. 2008;105(9):3226–3231.
10. Haile-Selassie Y, Suwa G, White TD. Late miocene teeth from Middle Awash, Ethiopia, and early hominid dental evolution. *Science*. 2004;303:1503–1505. https://doi.org/10.1126/science.1092978.
11. Richmond BG, Jungers WL. Orrorin tugenensis femoral morphology and the evolution of hominin bipedalism. *Science*. 2008;319(5870):1662–1665. https://doi.org/10.1126/science.1154197 [PubMed] [Google Scholar].
12. Vignaud P, et al. Geology and palaeontology of the Upper Miocene Toros-Menalla hominid locality, Chad. *Nature*. 2002;418:152–155. https://doi.org/10.1038/nature00880.
13. Relethford JH. Genetic evidence and the modern human origins debate. *Heredity*. 2008;100:555–563.
14. Neubauer S, Gunz P, Hublin J-J. The pattern of endocranial ontogenetic shape changes in humans. *J Anat*. 2009;215:240–255.
15. Dunsworth HM. Origin of the Genus *Homo*. *Evo Edu Outreach*. 2010;3:353–366. https://doi.org/10.1007/s12052-010-0247-8.
16. Antón SC. Natural history of *Homo erectus*. *Yearb Phys Anthropol*. 2003;46:126–170.
17. Sano K, et al. A 1.4-million-year-old bone handaxe from Konso, Ethiopia, shows advanced tool technology in the early Acheulean. *Proc Natl Acad Sci USA*. 2020;117:202006370. https://doi.org/10.1073/pnas.2006370117.
18. Ponce de Leon MS, et al. Neanderthal brain size at birth provides insights into the evolution of human life history. *Proc Natl Acad Sci USA*. 2008;105(37):13764–13768. https://doi.org/10.1073/pnas.0803917105.
19. Leigh SR. Brain growth, life history, and cognition in primate and human evolution. *Am J Primatol*. 2004;62:139–164. https://doi.org/10.1002/ajp.20012.
20. Robson SL. Encephalization and life history: lessons from primate brain growth trajectories. *Am J Phys Anthropol*. 2008;S46:201.
21. Hofman M. Evolution of the human brain: when bigger is better. *Front Neuroanat*. 2014;8. https://doi.org/10.3389/fnana.2014.00015.

22. Nowinski WL. In: Miller K, ed. *Biomechanics of the Brain. Hofman M. Evolution of the Human Brain*. Berlin: Springer Science and Business Media; 2011:5–40.
23. Phillips H. Introduction: the human brain. *New Sci*. 2006;(September):2567. https://www.newscientist.com/article/dn9969-introduction-the-human-brain/.
24. Sylvester AD. Locomotor coupling and the origin of hominin bipedalism. *J Theor Biol*. 2006;242:581–590.
25. Spoor F, Leakey MG, O'Higgins P. Middle pliocene hominin diversity: *Australopithecus deyiremeda* and *Kenyanthropus platyops*. *Philos Trans R Soc B*. 2016;371:20150231. https://doi.org/10.1098/rstb.2015.0231.
26. Alemseged Z, Wynn JG, Geraads D, et al. Fossils from Mille-Logya, Afar, Ethiopia, elucidate the link between Pliocene environmental changes and *Homo* origins. *Nat Commun*. 2020;11:2480. https://doi.org/10.1038/s41467-020-16060-8.
27. Tyron CA, et al. Levallois lithic technology from the Kapthurin Formation, Kenya: Acheulian Origin and middle stone age diversity. *Afr Archaeol Rev*. 2006;22:199–229. https://doi.org/10.1007/s10437-006-9002-5.
28. Shipton C, Petraglia MD. Inter-continental variation in Acheulean bifaces. In: Norton C, Braun D, eds. *Asian Paleoanthropology. Vertebrate Paleobiology and Paleoanthropology*. Dordrecht: Springer; 2011. https://doi.org/10.1007/978-90-481-9094-2_5.
29. Shea JJ. Neanderthals meet modern humans. Modern human origins and Neanderthal extinctions in the Levant. *Athena Rev*. 1997;2:21–32.
30. Shea JJ, Stutz AJ, Nilsson-Stutz L. An early upper palaeolithic stone tool assemblage from Mughr El-Hamamah, Jordan: an interim report. *J Field Archaeol*. 2019;44(7):420–439.
31. Potts R, Shipman P. Cutmarks made by stone tools on bones from Olduvai Gorge, Tanzania. *Nature*. 1981;291:577–580.
32. Potts R. *Early Hominid Activities at Olduvai: Foundations of Human Behaviour*. New York, NY: Routledge; 2017:297–309.
33. Klüver J. The socio-cultural evolution of our species. *EMBO Rep*. 2008;9(Suppl. 1):S55–S58. https://doi.org/10.1038/embor.2008.35.
34. Scerri EML, et al. Did our species evolve in subdivided populations across Africa, and why does it matter? *Trends Ecol Evol*. 2018;33:582–594.
35. Scerri EML, et al. Earliest evidence for the structure of Homo sapiens populations in Africa. *Quart Sci Rev*. 2014;101:207–216.
36. deMenocal PB. African climate change and faunal evolution during the Pliocene–Pleistocene. *Earth Planet Sci Lett*. 2004;220(1–2):3–24.
37. Indriati E, et al. The age of the 20 meter Solo River Terrace, Ngandong (Java, Indonesia) reconsidered. *Am J Phys Anthropol*. 2010;(Suppl. 50):135.
38. Shen G, Gao X, Gao B, Granger DE. Age of Zhoukoudian *Homo erectus* determined with $^{26}Al/^{10}Be$ burial dating. *Nature*. 2009;458:198–200.
39. Rightmire G, Lordkipanidze D. Comparisons of Early Pleistocene Skulls From East Africa and the Georgian Caucasus: Evidence Bearing on the Origin and Systematics of Genus Homo. *Vertebrate Paleobiology and Paleoanthropology*; 2009:39–48. https://doi.org/10.1007/978-1-4020-9980-9_5.
40. Culotta E, Gibbons A. *Almost All Living People Outside of Africa Trace Back to a Single Migration More Than 50,000 Years Ago*; 2016. https://www.sciencemag.org/news/2016/09/almost-all-living-people-outside-africa-trace-back-single-migration-more-50000-years.
41. Kris Hirst K. *Terra Amata (France)—Neanderthal Life on the French Riviera*. ThoughtCo; August 25, 2020. thoughtco.com/terra-amata-france-neanderthal-life-173001.
42. Bar-Yosef O, et al. The excavations in Kebara Cave, Mt Carmel. *Curr Anthropol*. 1999;33:497–550.
43. Bar-Yosef O, Belfer-Cohen A. From Africa to Eurasia-early dispersals. *Quat Int*. 2001;75:19–28.
44. Hublin J-J. The middle pleistocene record. On the origin of Neandertals, modern humans and others. In: Begun RD, ed. *A Companion to Paleoanthropology*. John Wiley; 2013:517–537.
45. White TD, et al. Pleistocene Homo sapiens from Middle Awash, Ethiopia. *Nature*. 2003;23:742–747. https://doi.org/10.1038/nature01669.
46. Dembo M, et al. Baysian analysis of morphological supermatrix sheds light on controversial fossil hominin relationships. *Proc R Soc B Biol Sci*. 2015;282(1812):20150943. https://doi.org/10.1098/rspb.2015.0943. PMC 4528516. PMID 26202999.

47. Zhu ZY, et al. New dating of the *Homo erectus* cranium from Lantian (Gongwangling), China. *J Hum Evol.* 2015;78:144–157.
48. Hublin J-J, et al. New fossils from Jebel Irhoud, Morocco and the pan-African origin of *Homo sapiens*. *Nature.* 2017;546:289–292. https://doi.org/10.1038/nature22336.
49. Wang C-C, Li H. Inferring human history in East Asia from Y chromosomes. *Investig Genet.* 2013;4:11–21.
50. Soares P, et al. Correcting for purifying selection: an improved human mitochondrial molecular clock. *Am J Hum Genet.* 2009;84:740–759.
51. Cann R. DNA and human origins. *Annu Rev Anthropol.* 1988;17:127–143.
52. Soares P, Rito T, Pereira L, Richards MB. A genetic perspective on African prehistory. In: Jones SC, Stewart BA, eds. *Africa From MIS 6-2: Population Dynamics and Paleoenvironments.* Netherlands: Springer; 2016.
53. Rito T, et al. A dispersal of *Homo sapiens* from southern to eastern Africa immediately preceded the out-of-Africa migration. *Sci Rep.* 2019;9. Article number: 4728.
54. Stringer C. The origin and evolution of Homo sapiens. *Philos Trans R Soc Lond B Biol Sci.* 2016;371 (1698):20150237. https://doi.org/10.1098/rstb.2015.0237.
55. Fleagle J, et al. Paleoanthropology of the Kibish Formation, southern Ethiopia: introduction. *J Hum Evol.* 2008;55:360–365. https://doi.org/10.1016/j.jhevol.2008.05.007.
56. Sahle Y, Beyene Y, Defleur A, et al. Revisiting Herto: New Evidence of Homo Sapiens From Ethiopia. *Modern Human Origins and Dispersal.* Washington, DC: US Department of the Interior; 2019:73–104.
57. Chan EKF, et al. Human origins in a southern African palaeo-wetland and first migrations. *Nature.* 2019;575:185–189.
58. Batini C, Jobling MA. The jigsaw puzzle of our African ancestry: unsolved, or unsolvable? *Genome Biol.* 2011;12:118.
59. Henn BM, et al. Hunter-gatherer genomic diversity suggests a southern African origin for modern humans. *Proc Natl Acad Sci USA.* 2011;108:5154–5162.
60. Skoglund P, et al. Reconstructing prehistoric African population structure. *Cell.* 2017;171:59–71.e21.
61. Veeramah KR, et al. An early divergence of KhoeSan ancestors from those of other modern humans is supported by an ABC-based analysis of autosomal resequencing data. *Mol Biol Evol.* 2012;29:617–630.
62. Schlebusch CM, et al. Southern African ancient genomes estimate modern human divergence to 350,000 to 260,000 years ago. *Science.* 2017;358:652–655.
63. Beyin A. Upper Pleistocene human dispersals out of Africa: a review of the current state of the debate. *Int J Evol Biol.* 2011. https://doi.org/10.4061/2011/615094.
64. Pagani L, et al. Genomic analyses inform on migration events during the peopling of Eurasia. *Nature.* 2016;538:238–242. https://doi.org/10.1038/nature19792.
65. Poznik GD, et al. Sequencing Y chromosomes resolves discrepancy in time to common ancestor of males versus females. *Science.* 2013;341:562–565.
66. Armitage S, et al. The southern route "Out of Africa": evidence for an early expansion of modern humans into Arabia. *Science.* 2011;331:453–456.
67. Timmermann A, Friedrich T. Late Pleistocene climate drivers of early human migration. *Nature.* 2016;538:92–95. https://doi.org/10.1038/nature19365.
68. Mercier N, et al. Thermoluminescence date for the Mousterian burial site of Es-Skhul, Mt Carmel. *J Archaeol Sci.* 1993;20:169–174.
69. Walter RC, et al. Early human occupation of the Red Sea coast of Eritrea during the last interglacial. *Nature.* 2000;405:65–69.
70. Lahr MM, Foley R. Multiple dispersals and modern human origins. *Evol Anthropol.* 1994;3:48–60.
71. Lopez S, et al. Human dispersal out of Africa: a lasting debate. *Evol Bioinforma.* 2016;11s2:57–68.
72. Liu W, et al. Human remains from Zhirendong, South China, and modern human emergence in East Asia. *Proc Natl Acad Sci USA.* 2010;107:19201–19206.
73. Holen SR. A 130,000-year-old archaeological site in southern California, USA. *Nature.* 2017;544 (7651):479–483.
74. Mellars P. Archeology and the dispersal of modern humans in Europe: deconstructing the Aurignacian. *Evol Anthropol.* 2006;15:167–182.

75. Haber M, et al. A rare deep-rooting D0 African Y-chromosomal haplogroup and its implications for the expansion of modern humans out of Africa. *Genetics*. 2019;212:1421–1428.
76. Macaulay V, et al. Single, rapid coastal settlement of Asia revealed by analysis of complete mitochondrial genomes. *Science*. 2005;308:1034–1036.
77. Poznik GD, et al. Punctuated bursts in human male demography inferred from 1,244 worldwide Y-chromosome sequences. *Nat Genet*. 2016;48:593–599. https://doi.org/10.1038/ng.3559.
78. Underhill PA, et al. Y chromosome sequence variation and the history of human populations. *Nat Genet*. 2000;26:358–361.
79. Malaspinas A-S, et al. A genomic history of Aboriginal Australia. *Nature*. 2016. https://doi.org/10.1038/nature18299.
80. Mallick S, et al. The Simons genome diversity project: 300 genomes from 142 diverse populations. *Nature*. 2016;538:201–206. https://doi.org/10.1038/nature18964.
81. Groeneveld E. Early human migration. In: *World History Encyclopedia*; 2017, May 15. Retrieved from: https://www.ancient.eu/article/1070/early-human-migration/.
82. Harvati K, et al. Apidima Cave fossils provide earliest evidence of Homo sapiens in Eurasia. *Nature*. 2019;571:500–504. https://doi.org/10.1038/s41586-019-1376-z.
83. Lordkipanidze D. A complete skull from Dmanisi, Georgia, and the evolutionary biology of early Homo. *Science*. 2013;342:326–331.
84. Lawler A. In search of Green Arabia. *Science*. 2014;345:994–997. https://doi.org/10.1126/science.345.6200.994.
85. Breeze PS, et al. Palaeohydrological corridors for hominin dispersals in the Middle East ~250-70,000 years ago. *Quart Sci Rev*. 2016;144:155–185.
86. Oppenheimer S. Out-of-Africa, the peopling of continents and islands: tracing uniparental gene trees across the map. *Philos Trans R Soc B*. 2012;367:770–784. https://doi.org/10.1098/rstb.2011.0306.
87. Ziemen FA, et al. Heinrich events show two-stage climate response in transient glacial simulations. *Clim Past*. 2019;15:153–168.
88. Kappelman J, et al. First *Homo erectus* from Turkey and implications for migrations into temperate Eurasia. *Am J Phys Anthro*. 2008;135:110–116. https://doi.org/10.1002/ajpa.20739.
89. Shea J. Neandertals, competition, and the origin of modern human behavior in the Levant. *Evol Anthropol*. 2003;12:173–187.
90. Gibbons A. Rich sexual past between modern humans and Neanderthals revealed. *Science*. 2016. https://doi.org/10.1126/science.aaf4207.

CHAPTER 10

Neanderthal origin and fossil record

We are all cells in the same body of humanity.

Peace Pilgrim[1]

Origins of Neanderthals from early human species

Three decades ago, archeological evidence and genetic studies suggested that the evolution and origins of modern humans and their ancestors had been settled. The evidence pointed to modern human origin in East Africa around 4 million years ago (mya) evolving from *H. erectus*, followed by a single migration out of Africa into Europe and Eurasia, and the extinction of the Neanderthals, most likely through our superior intellect and ability to outcompete them for resources. Today, we have come to understand that the picture of our evolution and that of our ancestors is much more complex than previously thought.

From roughly 4 million until 2 million years ago, we see the hominin fossil record dominated by the genus *Australopithecus*. These hominins demonstrated anatomical changes in the lower limbs that provided improved walking ability and more time on the ground compared with the earlier genus *Ardipithecus*. Compared with Great Apes, *Australopithecus* no longer had a grasping foot, their legs were longer, and they had the same leg length to body mass ratio that we see in anatomically modern humans. Pelvic and fossilized footprint analyses indicate that these individuals had a modern gait. The upright and striding bipedal gait implies that these individuals could cover more ground and expand their range while consuming fewer calories than their predecessors. Fossils of *Australopithecus* reveal long arms and fingers, suggesting these hominins spent a lot of time in the trees, and dental records suggest a diet of plants. The plant-based diet and climbing adaptations tell us their foraging and daily activity remained mostly apelike.[2,3]

By 2 mya, fossil evidence reveals the genus *Homo* had evolved. The fossil record shows that *Homo* had modern human body proportions, different dietary needs indicated by cut-marked bones and unique stone tools, and unlike previous hominins, had expanded outside of Africa into the Levant, Europe, Eurasia, and Indonesia. It is thought that the use of stone tools, the consumption of meat, and the development of a hunting-and-gathering strategy separated the genus *Homo* from the other apes.[3]

The story of the genus *Homo* begins with what many believe to be our first human ancestor called *Homo habilis*. The *H. habilis* type specimen was found in 1960 in Olduvai Gorge, in Tanzania (East Africa) and represents some of the oldest remains ascribed to the

Sex and Cohabitation Among Early Humans
https://doi.org/10.1016/B978-0-12-819086-9.00014-2

genus *Homo* (2–1.5 million years old). Although *H. habilis* is regarded by most anthropologists as a representation of the first humans, there is an argument that *H. habilis* should be placed in the genus *Australopithecus* as *Australopithecus habilis*.[4,5]

H. habilis is typically regarded as the first human ancestor to consume meat, approximately 2 mya. Studies have shown that around the same time, hominid molars began to shrink, which facilitated chewing meat. However, our ancestors' molars began to shrink before their brains began to grow.[6] Encephalometric studies of *H. habilis* skulls revealed marked transverse expansion of the cerebrum, particularly the frontal and parieto-occipital parts, and increased bulk of the frontal and parietal lobes. *H. habilis* was the first hominid ever found to possess frontal and temporal brain regions that we know today are structural markers of the neurological basis of spoken language. Collectively, this indicates that the hominid brain had attained a new evolutionary level of organization, which can be determined as preceding meat consumption.[7]

H. erectus (African form referred to as *H. ergaster*) was the first human species to emigrate out of Africa and populate Southeast Asia and Europe and was phenotypically different from *H. habilis*, with a larger brain (approximately 900 cm^3) and taller body. *H. erectus* appears to have emerged around 1.8 mya, with fossil evidence discovered throughout Eurasia. *Homo erectus* is thought by some to be an ancestor to archaic *Homo sapiens*. Although *erectus* was found in Southeast Asia 1.8 mya, this hominin had not made its way into Europe until 1 mya. The existence of Asian *Homo erectus* at an earlier point in time has resulted in arguments suggesting that if *erectus* is the ancestor to anatomically modern humans, there must have been a multiregional evolution of modern humans in Africa and Asia (see Chapter 7).

Both Neanderthals and modern humans were initially thought to have evolved from *H. erectus* between 200 and 300 kya. However, around 600–700 kya, there is evidence of habitual use of fire and the presence of the larger-brained *H. heidelbergensis*. The fossils of *H. heidelbergensis* share features of both *H. erectus* and *H. sapiens*. Because *H. heidelbergensis* had a greater cranial volume (1250 cm^3) than *erectus*, unique dental features, more advanced tools than *erectus*, and were found in Africa, Europe, and West Asia, it is considered a separate species from *H. erectus*.[8]

The African form of *H. heidelbergensis*, sometimes referred to as *H. rhodesiensis*, was first discovered in Zambia in 1921 by Arthur Smith Woodward. Since then, *H. heidelbergensis* fossils have been found throughout Eastern and Southern Africa, in West Asia, and in Northern, Central, and Southern Europe. Early investigations determined that *H. heidelbergensis* lived 700–200 kya and is thought by some to have developed from *H. erectus* or *H. antecessor*, although its exact relation to its proposed predecessors and to Neanderthals, Denisovans, and modern humans is still debated.

It is not entirely clear when Neanderthals and anatomically modern humans diverged from one another; however, many scientists agree that their common ancestor was *Homo heidelbergensis*. One current estimate of divergence based on DNA evidence predicts that humans split from *Homo heidelbergensis* between 550,000 and 765,000 years ago.[1] Other estimates suggest that *H. heidelbergensis* inhabited Africa, Europe, and Asia from around 600,000 to 200,000 years ago. The later model assumes that after *Homo heidelbergensis* left Africa about 350,000–400,000 years ago, they evolved into *Homo neanderthalensis* in Europe, while *Homo heidelbergensis* populations that remained in Africa (*H. rhodesiensis*) gave rise to *Homo sapiens*, around 200,000 years ago.[9,10] The evidence for *H. heidelbergensis* giving rise to Neanderthals, Denisovans, and AMHs comes from the many fossils of *heidelbergensis* found in Europe, Eurasia, and *H. rhodesiensis* in Africa. However, paleoanthropologists disagree about how *H. heidelbergensis* relates to other human groups and some suggest they were a nonancestral species replaced by Neanderthals in Europe.[11]

To add to this confusion, recent investigations have overturned the understanding that modern humans and Neanderthals diverged a little over 550,000 years ago. New genetic data indicates that Neanderthals and Denisovans diverged from their human shared ancestor close to 744,000 years ago. Denisovans and Neanderthals then split approximately 300 generations after splitting from the ancestor of modern humans. These studies also recalculated the size of the global Neanderthal population and concluded that instead of several thousand, the number of individuals was closer to tens of thousands. The detected mutations still support the view that regionally, Neanderthals were likely small bands of individuals, which explains the genetic detection of inbreeding, but the global population was relatively larger than previously thought.[11,12] Archeological, paleontological, and genetic evidence suggest that Neanderthals were limited in their ability to adapt to all climatic changes and that they abandoned certain regions during extreme cold and in some cases recolonized those regions when conditions changed. However, many of these changes in geographical distribution resulted in local extinctions.[13]

Although the origin of Neanderthals is still being debated, there have been no fossils of Neanderthals or Denisovans found in Africa, suggesting they evolved outside Africa. The new dates of divergence (744 kya) proposed by Rogers et al. could suggest several of the fossils currently identified as *Homo antecessor* (*H. antecessor* is referred to as *H. heidelbergensis* by many investigators) found at Sima del Elefante dated at 1.2 mya[11,14] and those at Gran Dolina[15] dated at 800 kya could be reclassified as Neanderthal. Currently, the oldest Neanderthal fossils are thought to be 430,000 years old, found at Sima de Los Huesos. Alternatively, since the divergence of Neanderthals and modern humans has been pushed back before the emergence of *H. heidelbergensis* (600–700 kya), an alternative would be *H. sapiens* arising from *H. antecessor*, and that *H. heidelbergensis* was a pre-Neanderthal.

It should be clear that the origin of modern humans, Neanderthals, and Denisovans is still heavily debated and there are no known fossils that provide evidence of a direct ancestor to modern humans or Neanderthals or their speciation. The debate continues as to whether the Neanderthals and (AMHs) anatomically modern humans came from an *H. erectus* species or *H heidelbergensis*, and whether modern humans arose in Africa or parts of Asia. The idea that *H. heidelbergensis* was the sole species that gave rise to modern humans in East Africa has been heavily disputed recently, and there is mounting evidence that modern humans may have arisen from the interbreeding of a variety of hominin ancestors from East, West, North and Southern Africa. These different hominin groups could have been subspecies of *H. heidelbergensis* or a completely different *Homo* species. The resulting individuals from this interbreeding would have been what we describe today as anatomically modern humans before they left Africa. Although no "missing link" has been found, there are some fossils that have been identified as intermediates of modern humans and Neanderthals, and it is clear that there was interbreeding between these species.[16]

The origin of Neanderthals most likely relates to the recolonization of Western Eurasia by hominins of African origin carrying the Acheulean tool technology into Europe around 600 kya. Glacial and interglacial phases may have played a crucial role in Neanderthal evolution, along with an increase in encephalization, and genetic drift. The most likely scenario for the fixation of Neanderthal characteristics seems to be an accumulation of features during the second half of the Middle Pleistocene. Although a separation time for the African and Eurasian predecessors is difficult to determine, it certainly predates Marine Isotope Stage 11 (424 kya), as phenotypic Neanderthal features are documented as far back 430 kya.[13,17]

Overall, the Neanderthals appear to form part of a large endemic European clade like the Sima de los Huesos (see below) along with other specimens like those from Arago, Petralona, Kabwe, and Bodo, indicating that hominid European history is more complex than simply that of a single lineage evolving toward the terminal species *Homo neanderthalensis*.

Climatic conditions before and after the introduction of modern humans to Europe

Isotope studies of Greenland ice have shown that during the Neanderthal occupation of Europe, there were dramatic fluctuations in climate involving glacial and interglacial conditions. This unstable climate, 65–25 kya, referred to as Marine Isotope Stage 3 (MIS-3) was an interval between the two cold maxima—near-interglacial and peak-glacial conditions.[18] During this time, the climate fluctuated from extreme to mild cold for decades. MIS are the record of oxygen isotopes through time. This record of ^{18}O, or oxygen stable isotopes, comes from measuring oxygen in the microscopic skeletons of foraminifera (ancient amoeboid protists) that lived on the sea floor. This measure can be used as an indicator of changing temperature and glacial ice over time.

MIS-3 began with mild conditions and ended with ice covering Northern Europe.[18] Neanderthals were the only hominins in Europe at the beginning of MIS-3, with modern humans entering Europe toward the end. The MIS-3 studies point out that the climate was increasingly unstable toward the end of the last glacial maximum, which would have brought on dramatic ecological changes. Neanderthals would have needed to adapt to these conditions to survive.[19] Neanderthal populations adapted by adjusting their behavior and hunting cold-adapted animals during glacial periods and different animals during warmer periods. During extreme cold, they moved south toward warmer environments.

Large husky body structures are better at conserving heat, and because of the cold European ice age, the large husky build of Neanderthals could be a result of strong positive selection, allowing them to survive the extreme cold.[19,20] This theory tallies with Allen's rule of body shapes and proportions varying by climatic temperature and Bergmann's observations that species of larger size are found in colder climates. The idea is that large, stockier masses minimize the ratio of surface area to weight and reduce heat loss. Data indicating that Neanderthals may have contributed keratin genes to the modern human gene pool suggest that keratin may have also contributed to their ability to adapt to the cold.[21] Keratin provides toughness to skin, hair, and nails, and can be beneficial in colder environments by providing insulation. There was speculation that Neanderthals' large noses were the result of adaptations to the cold; however, studies of arctic animals and primates have shown that cold areas tend to favor sinus size reduction. This might be explained by Neanderthals' noses evolving in the warmer Middle East climate and then not changing when Neanderthals migrated into Europe, or that the larger nose provided a source for increased oxygen uptake, supporting high rates of physical activity and large muscle mass.[22]

The climate influences metabolic physiology and the level of physical activity required for subsistence, leading to variation in energy expenditure. In general, human foragers inhabiting colder climates have higher energy expenditure than those in warmer climates, in terms of both basal metabolic energy needs and the amount of energy spent obtaining resources. Results from studies using the fossil record indicate that Neanderthals would likely have had very high daily energy expenditure as part of their adaptive response to cold climates. Because anatomically modern humans entered Europe near the end of the last glacial cycle, studies imply that AMHs would have had substantially lower thermoregulatory and subsistence challenges, and therefore lower energy requirements, than Neanderthals in Europe.

Analysis of climatic effects on energy expenditure in Neanderthals and anatomically modern humans used mean annual temperature, along with body size, sex, and age, to predict basal metabolic rates. The study used fossil specimens of Neanderthals and Pleistocene modern humans, obtaining paleo-mean annual temperature values for fossil sites from data available from the Marine Isotope Stage 3 Project. The investigators then used basal metabolic rate values from climate-specific physical activity on extant human foragers living in different environments.[23] Their results indicate that Neanderthals would have had substantially higher energy needs than anatomically modern humans in similar climates, in the order of 100–350 kcal per day, which corroborated the results of previous studies using different methods. The investigators concluded that the difference was in large part due to the greater body mass in Neanderthals, which may have related mainly to a higher proportion of muscle mass compared to anatomically modern humans. Greater muscularity in Neanderthals would have provided them with greater thermoregulatory capability, may have served as insulation, and was probably part of a subsistence strategy of close range encounters with large mammalian prey. If greater muscle mass served these ends in Neanderthals, modern humans must have used other means of dealing with insulation and may have brought improved methods of insulation into Europe, along with various subsistence techniques. Lower adult energy needs could have provided modern humans with a competitive advantage over Neanderthals, who had higher, and thus harder to meet, energy demands.[24]

Neanderthal fossil discovery

Although Neanderthal specimens have been found scattered throughout Europe and Northern Asia, they have not been discovered in Africa. The closest region to Africa containing Neanderthal remains are the Levant and Gibraltar. Paleontologists do not agree when Neanderthals can first be recognized in the fossil record, but estimate that the first appearance of Neanderthals occurred outside Africa between 200,000 and 400,000 years ago. The first Neanderthal remains were discovered in Belgium in 1829 by Philippe-Charles Schmerling, but the specimen was not recognized by the scientific community as Neanderthal until 1936. The specimen found by Schmerling is referred to as Engis 2 and was initially thought to be the remains of a *Homo sapiens* similar to fossil remains called Engis 1. Engis 2 comprises fossils of a partial cranium and associated fragments of an upper and lower jaw of a 2- to 3-year-old Neanderthal. The specimen was found along with animal bones and Mousterian stone tools, and is estimated to be around 35 thousand years old (Fig. 1).

The type specimen used today was first discovered in 1856 near the city of Dusseldorf, Germany, in Feldhofer Grotto, Neander Valley (Fig. 2). While excavating in a limestone formation, miners discovered a skullcap, ribs, and part of a pelvis. When the remains were examined, the bones were recognized as coming from a human-like source, although the skull had unusual eyebrow ridges, and a retreating forehead that made the specimen look pathological. For many years, most scientists regarded the specimen as a pathological specimen from a contemporary human. In 1863, William King, a professor at Queen's College in Ireland, suggested the remains be recognized as a different species, *Homo neanderthalensis*, thereby acknowledging the fossils' link to humanity, yet distinguishing it from *H. sapiens*. The word *Homo* is Latin for "human" or "man" and the word *neanderthalensis* is based on the location where the type specimen was found (Neander Valley). The German word for valley is "Tal," but in the 1800s it was spelled "Thal." The names

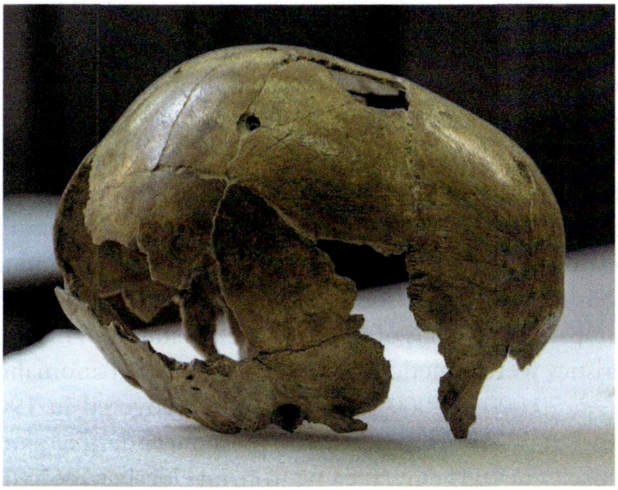

Fig. 1 The fossil Engis 2, discovered in Belgium in 1829 by Philippe-Charles Schmerling. This specimen represents a partial cranium from a 2- to 3-year-old Neanderthal (https://upload.wikimedia.org/wikipedia/commons/4/40/Engis_2_Schaedel_1.jpg).

Fig. 2 The drawing in figure 2 represents the bones of the original Neanderthal type specimen discovered in the Neander valley in 1856. The link below takes you to the original bones. https://www.sciencephoto.com/media/170932/view/original-type-specimen-of-neanderthal-man.

Neanderthal and Neandertal reflect the original and modern German spelling, respectively. More recently, in 1997, archeologists returning to the Neander Valley site found the remains of three Neanderthal individuals, stone tools, and animal remains. Radioactive dating of the material indicated that the fossils were from Neanderthals that had lived 42,000 years ago.[25,26]

In 1886, two Neanderthal skulls were found in Spy, Belgium, along with the remains of ancient mammoths, woolly rhinoceroses, and stone tools. These fossils, along with ensuing discoveries, began to convince 19th-century scientists that the Neanderthal bones found in the Neander Valley were ancient and probably represented robustly built humans or human forebears.

By the early 1900s, a large number of Neanderthal fossils had been found in the Dordogne region of southwestern France. The region contained a number of almost complete Neanderthal skeletons that were reconstructed to form the first impression of a Neanderthal.[27] Some of the key findings in France include the following:

- Le Moustier 1 and 2—Le Moustier 1 is a 45,000-year-old adolescent skull and several long-bones discovered in 1908 in Le Moustier. Le Moustier 2, discovered in 1914, is an almost complete skeleton of a newborn. The distinctive features of Neanderthals are already apparent in both the newborn and adolescent individuals. These remains helped to establish different developmental stages of the Neanderthals and the differences when compared to AMHs of the same age. In addition, the remains indicate that these characteristics were genetic and not developmental anomalies.
- La Ferrassie 1—a 70,000–50,000-year-old skull discovered in 1909 in La Ferrassie. This skull is the most complete specimen of a Neanderthal skull ever found and has a cranial capacity of 1641 cm^3. It came from an elderly male and has a large sloping forehead and large nasal openings, features that are associated with what is now described as "classic" European Neanderthals. The leg and foot bones indicate he walked upright (Fig. 3).

Fig. 3 One of the largest assemblages of Neanderthal bones ever found was from seven individuals discovered over a century ago at La Ferrassie (a limestone cave) in southwestern France. La Ferrassie is in the Dordogne, not far from La Chapelle. Hundreds of fragments of animal bone, and hundreds of flakes of flint from stone tools were found to be about 70,000 years old. The figure shows the La Ferrassie vulva. Among the thousands of Neanderthal artifacts, almost none represent unambiguous attempts at art or adornment, and those which have been found in La Ferrassie are disputed as being created by Neanderthals (http://factsanddetails.com/world/cat56/sub400/entry-5970.html).

- La Chapelle-aux-Saints—a 50,000–60,000-year-old skull discovered in 1908 in La Chapelle-aux-Saints. This male individual is estimated to have been around 40 years old when he died. The remains indicate that his mandibular molars were absent and his skeleton showed evidence of major injuries including a healed broken hip, and advanced arthritis of the lower neck, back, hip, and shoulders. He survived for quite some time with these debilitating injuries, which may indicate that Neanderthals showed some altruistic tendencies toward the sick and elderly.

Since the initial discoveries, thousands of skeletal remains of more than 500 Neanderthals have been uncovered from sites across Europe, Western Asia, and the Middle East. The remains include babies, children, and adults up to the age of 40, with approximately half of the discoveries being children. Fossil evidence indicates that Neanderthals had a high incidence of fractures, possibly due to their interaction with large mammals while hunting. Additional signs of trauma have been found that may indicate fighting among individuals.[28]

Neanderthal remains suggests that Neanderthals inhabited Western and Central Europe, Ukraine and Western Russia, and Western, Central, and Northern Asia up to the Altai mountains.[27,29] Some important examples include the following:

- Shanidar 1—found in Iraqi Kurdistan in 1957, the remains are of an almost complete skeleton with an upper jaw and teeth that show heavy wear, a characteristic that is even found in young Neanderthals. The Shanidar Cave has yielded 10 Neanderthal individuals. Shanidar 1, known as "Nandy," was 40–50 years old and displayed trauma-related abnormalities that resulted in severe deformities. The deformities indicate that life would have been painful and very difficult without assistance.
- Amud 1—a 55,000-year-old skull discovered in 1961 by Hisashi Suzuki in Amud, Israel. This individual is unique for its unusual size and brain capacity. The remains indicate it was more than 180 cm tall and had the largest brain of any fossil human (1740 cm^3). Neanderthals probably migrated to the Middle East during periods of harsh European winters. Like other Neanderthal specimens in the Levant, Amud 1's skull is long, and broad, with an especially large nose and a large face, with pronounced midfacial prognathism (protruding jaw) and a large palate. Overall, its body is stocky, robust, and has short limbs, similar to the cold-adapted bodies of West European Neanderthals.[30,31]
- Kebara 2—also known as "Moshe," Kebara 2 is a 60,000-year-old partial skeleton discovered in 1983 in Kebara Cave, in Northern Israel's Carmel mountain range. This relatively complete skeleton (missing the cranium) belonged to an adult male. It appeared to be deliberately buried, but no grave goods were found, so no ritualistic behavior can be inferred.
- Lagar Velho—refers to a 24,000-year-old skeleton of a boy discovered in 1998 in Abrigo do Lagar Velho, Central Western Portugal. The remains are of a complete skeleton of an approximately 4-year-old child, buried with a pierced shell and red

> ochre. The discovery is important because it appears to be a *Homo sapiens-Homo neanderthalensis* hybrid. This interpretation was based on Neanderthal-like knee and leg proportions but the specimen had head, pelvis, and forearms that are human-like. This morphological mosaic indicates admixture between late archaic and early modern humans in Iberia, refuting the hypotheses of complete replacement of the Neanderthals by early modern humans without intermixing. In addition, it emphasizes the complexities of the cultural and biological processes that were involved in modern human emergence and Neanderthal extinction.[32]

Some have argued that it is more likely that the robustness of the Lagar Velho specimen is a climatic adaptation, and that it is not a hybrid but a modern human. Comparisons to other humans of this period are difficult due to lack of knowledge on variations within child populations.[33]

The easternmost evidence for Neanderthal occupation is from a skull specimen called "Maba Man" that was discovered in 1958 in a Chinese cave near the town called Maba, in the north part of Guangdong Province. The Maba Man partial skull, dated to about 120,000–140,000 years old, is comparable to European Neanderthals and its discovery in Southern China suggests the possibility that Neanderthals traveled further east than previously thought.

The evidence for *H. heidelbergensis* giving rise to Neanderthals, Denisovans, and AMHs comes from the many fossils found in Europe, Eurasia, and Africa. Over the years, the focus on human evolution and technological advances has come from European and North American institutions. However, recent interest from East Asia has revealed several unique transitional fossils found in modern-day China that are different from European and African *H. heidelbergensis*. The findings suggest that other intermediate forms of *Homo* may have existed, or that another group of hominins that descended from *H. erectus* gave rise to East Asian populations that differed from the European types. This proposed scenario would have involved a species like *H. erectus* interbreeding with groups from Africa and Eurasia, ultimately resulting in modern East Asians. This would assume that there was a multiregional origin of *H. sapiens*. Because there are so few remains from East Asia, it is purely speculation at this point.[34]

Use of stone tools

Neanderthals persisted in Eurasia through multiple glacial-interglacial cycles in the Middle Paleolithic (MP), which most scholars place between 300 and 35 kya. One main characteristic of the MP is the stone tools, which show a decrease in the frequency of large hand-axes and cleavers and the appearance of the Levallois technique (Fig. 4). The Levallois technique is a distinctive type of stone knapping (shaping of flint, chert, obsidian, or other fracturing stone) developed by Neanderthals and precursors to modern humans. It is named after tools found in the 19th century in Levallois-Perret, a suburb of Paris,

Fig. 4 *Arrowheads* showing the Levallois technique of flaking (https://upload.wikimedia.org/wikipedia/commons/c/cf/Pointe_levallois_Beuzeville_MHNT_PRE.2009.0.203.2.fond.jpg).

France. Studies have suggested that these tools were used for a wide variety of tasks, including butchery and food processing. The use of Levallois technology demonstrates that Neanderthals had the cognitive abilities necessary to design and execute complex knapping products[35] (further discussion of tools and behavior can be found in Chapter 4).

Growth and development of Neanderthals

Since the discovery of Neanderthal fossils, investigators have asked how Neanderthals grew and developed, and whether their growth and development was different from that of anatomically modern humans. Paleontology uses the comparative anatomy of fossils as evidence of species development through evolution. Although the study of fossils has allowed scientists to point out clear anatomical differences between Neanderthals and anatomically modern humans of the time, it is important to remember that both *H. sapiens* and *H. neanderthalensis* had diverse body types, and could have lived under different environmental conditions, and that the ranges provided by investigators are those of individual remains. Just as today we would not say all human males are between 1.7 and 1.9 m tall, we cannot say all species of these prehistoric times were identical and represent all types whose remains we have not uncovered.[36,37] However, morphologically, Neanderthals postdating MIS-7 (243 kya) display a unique combination of features and morphometric architecture that makes them easily distinguishable from anatomically modern humans. Post-MIS-7 Neanderthal fossils show a very small number of differences in overall anatomic variability, while AMHs of the same period displayed high levels of variability. This suggests that different evolutionary processes were at work in AMHs and Neanderthals. Drift and the influence of culture might have played a role in human variability, while stabilizing selection may have played a major role in Neanderthals' anatomic stability.[13]

Studying skeletal remains has provided some insight into Neanderthal birth and maturation rates. For example, the Neanderthal superior pubic ramus (pubic bone located at the front and base of the pelvis) was longer than that of modern humans; researchers have suggested that this was because of a larger birth canal in Neanderthals needed to accommodate the relatively large head size, or, alternatively, that it was due to a difference in Neanderthal locomotion and posture.[6,37] It is unclear what the larger pubic ramus bone indicates, but reconstruction studies on infant skeletons have indicated that the gestation period and head size of Neanderthals were probably similar to those of modern humans.[38]

In one study, skeletal remains of a 49 thousand years old Neanderthal child recovered from the El Sidron cave in Asturias, Spain, in 1994, has shed some light on the early development of Neanderthals. The skeletal remains were of a boy (determined by dental analysis) approximately 7–8 years old weighing approximately 26 kg and 110 cm tall. In 2017, Rosas and others used the skeletal remains to determine that AMHs and Neanderthals regulated their growth differently in order to adapt to differences in energy consumption and their physical characteristics. It was determined that AMHs and Neanderthals had a similar overall growth rate, which makes sense because Neanderthals and AMHs are thought to have evolved from a recent common ancestor, *H. heidelbergensis*.[39]

The El Sidron J1 child developed in a similar fashion to modern-day humans; however, the divergent aspect in the growth of AMHs versus Neanderthals was maturation of the vertebral column and extended brain development after birth. The cartilaginous joints of the middle thoracic vertebrae and the atlas are the last to fuse in hominids, but in the El Sidron Neanderthal, fusion occurred about 2 years later when compared to modern humans. The delay of this fusion may indicate that Neanderthals had a disconnection between certain aspects of development, compared to AMHs, in the transition from infancy to puberty. The investigators speculate that this feature could be related to the characteristic enlarged shape of the Neanderthal torso, or slower brain growth that may have resulted from energy constraints during development.[39,40]

Brain development in humans has a high energy cost and results in the slow development of the rest of the body. The growth and development of the El Sidron child appears to match typical human ontogeny, and slow body growth ensued between birth and puberty, most likely compensating for the large brain development in Neanderthals.

Dental characteristics

Teeth are the best-preserved portions of skeletal remains and are the only fossil remains showing no effect of age as opposed to other skeletal remains. Because fossil tooth morphology is only affected by dental disease and attrition, both of which are easily recognized, teeth can be a marker to study life history (developmental history and the timing of crown and root formation), age of individuals, their health, and nutrition.[41] Analyses of *Australopithecus* teeth exhibit signs of plant consumption similar to those in earlier hominins and extant apes. The lack of large sharp canines and thick enameled molars suggest that *Australopithecus* diets contained mostly fibrous foods. By the time *H. habilis* (typically regarded as the first human ancestor) had evolved, their molars were much smaller, possibly to make chewing meat easier. Similarly, one of the distinguishing features of *H. erectus* from more advanced species is the size difference in teeth. *H. erectus* had relatively large teeth in comparison to those of *H. sapiens* and Neanderthals, possibly due to eating raw meat instead of cooked meat like *H. sapiens* and Neanderthals. Recent analyses of the food trapped inside Neanderthal fossilized dental calculus show that Neanderthals ate a balanced diet of meat with plenty of plants and grains.[3]

Enamel thickness figures prominently in studies of human evolution, particularly for phylogeny and dietary reconstruction. Most attention has focused on molars, using advanced imaging technologies. However, Buti et al. examined 3D enamel thickness of canines in Neanderthals and anatomically modern humans. Overall, they found that crown size was significantly greater in Neanderthals than in Upper Paleolithic and recent modern humans. In addition, their study supported the general findings that Neanderthals have relatively thinner enamel than modern humans, similar to molars, indicating that unworn or slightly worn canines can be used to discriminate between the two groups. They were unable to determine if the differences in enamel thickness are functionally related or are the result of pleiotropic or genetic drift effects.[42]

Changes to the dental morphology and jaw development are major elements of hominid evolution, driven by the types and processing of food, and it is no surprise that they distinguish AMHs from our nearest ancestors. Discoveries of Neanderthal jaws and teeth have resulted in some broad conclusions about Neanderthal anatomical and physiological features. Compared to modern humans, Neanderthals have a distinctive set of dental features, including large anterior crowns, long roots, molars with enlarged pulp chambers, large, shovel-shaped incisors, and canines that grew more quickly in Neanderthals compared to AMHs. The evolution of the jaw and teeth in modern humans are thought to have facilitated encephalization, speech, and the formation of the human chin, which is absent in all other hominins.[43,44]

Neanderthal molars are typically taurodont (a condition found in the molar teeth whereby the body of the tooth and pulp chamber is enlarged vertically at the expense of the roots, moving the tooth toward the tip of the tongue and down—referred to as bull-like teeth). Studies using high-resolution microcomputed tomography have shown that the timings of molar crown and root completion in Neanderthals match those of modern humans. The data suggest similar timings of tooth initiation relative to birth in Neanderthals and modern humans, leading to a predictable extended period between birth and first molar (M1) emergence, by which time about 90% of brain volume would have been attained. The results also indicate a more complex enamel-dentine junction morphology and a late peak in root extension rate in Neanderthals compared to AMHs. Other investigators have measured permanent mandibular molar root morphology in Middle and Late Pleistocene *Homo neanderthalensis*, and Late Pleistocene and recent *Homo sapiens*, to assess if variation in root form could be related to differences in tooth function. They used microtomographic imaging to visualize and quantify the external and internal dental morphologies of both isolated molars and molars embedded in the mandible. They concluded that the molars of recent *H. sapiens* are smaller than both Pleistocene *H. sapiens* and Neanderthals. They also showed that the first molar had the largest average root surface area in modern *H. sapiens* and Neanderthals. The results indicate that because root surface area is linked to masticatory function, the data suggests a distinct occlusal (overbite) loading regime in Neanderthals compared with both recent and Pleistocene *H. sapiens* (i.e., an increase in the pressure applied on the molars of Neanderthals during chewing, which may be the result of dietary adaptations).[45,46]

The unique Neanderthal craniofacial morphology (facial and dental anatomy) has been proposed to be an adaptive response to the application of heavy forces resulting from both the masticatory and cultural use of the anterior teeth. The evidence for this hypothesis, known as the anterior dental loading hypothesis (ADL), comes from observations of heavy anterior tooth wear (indicating heavy chewing forces) in Neanderthal specimens. If the ADL hypothesis is correct, it might be predicted that the wear shown on the anterior teeth of Neanderthals would greatly exceed that of the posterior teeth and that the difference would be greater than in other hominins with different facial morphologies.

Numerous metric and nonmetric features typically distinguish the crania and mandibles of Neanderthals from those of modern humans in relation to ADL. The appearance of derived features in the fossil record can point to the action of directional natural selection or genetic drift, and the retention of primitive features can indicate stabilizing natural selection (Fig. 5).

Clement et al.[47] examined tooth wear patterns to test the ADL hypothesis in a large group of Late Pleistocene hominins and a group of Inuit hunter-gatherers from Igloolik, Canada. The results showed that all Late Pleistocene hominins, including Neanderthals, had heavily worn anterior teeth relative to their posterior teeth, but the Inuit had heavier anterior tooth wear relative to their posterior teeth than Neanderthals. There was, however, a characteristic Neanderthal pattern showing even wear spread between anterior teeth and posterior teeth, unlike that seen in modern humans. The evidence suggests that all Late Pleistocene hominins habitually applied heavy forces between their anterior teeth and that Neanderthals were not exceptional, thus these results appear to discredit the ADL hypothesis.

Evidence from comparisons of tooth enamel development in AMHs and Neanderthal fossils suggests that Neanderthal enamel development was about 15% faster after fetal

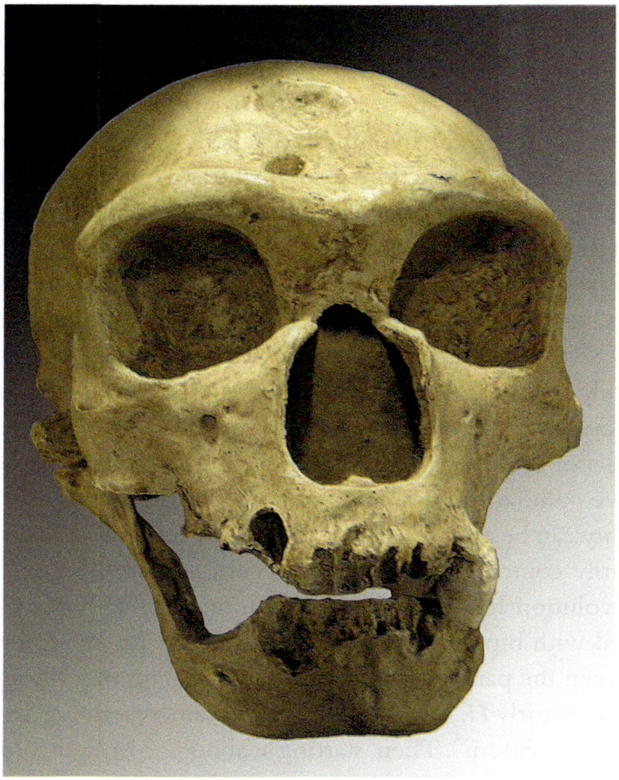

Fig. 5 Neanderthal skull showing distinctive features of the Neanderthal face (https://en.wikipedia.org/wiki/Neanderthal#/media/File:Homo_sapiens_neanderthalensis.jpg).

development than that of modern humans.[30,36] This could suggest that Neanderthals sexually matured at a faster rate, leading to earlier reproduction. Faster maturation could have evolved as a way to overcome high mortality rates; however, it is not clear if Neanderthals sexually matured at a faster rate than humans.

The disruption of dental growth can also indicate disease, stress, and malnutrition. Since teeth do not grow after they form and do not produce new enamel, pits and grooves found in the enamel of fossil teeth can indicate enamel hypoplasia. This can be caused by a variety of factors, such as premature birth, malnutrition, bacterial and viral infections, or trauma to newly developing teeth and the mouth. An investigation of Neanderthal fossil crowns compared to other prehistoric humans indicated that 75% of Neanderthal individuals suffered from fluctuating asymmetry (random deviations from bilateral symmetry due to environmental pressures during development) and enamel hypoplasia. The authors suggest that the results indicated stress lasting from several weeks to 3 months, and that Neanderthals were under greater developmental stress than all other prehistoric AMH samples.[48,49]

Cranial anatomy

Three explanations have dominated debates about the evolution of Neanderthal cranial features: cold climatic adaptation, adaptation to anterior dental loading (ADL), and genetic drift. Weaver[50] tested the hypothesis that genetic drift was the major contributor to craniofacial features using predictions from quantitative and population genetics with a sample of 37 cranial measurements collected on 20 Neanderthal specimens and 2524 recent humans. His data indicated strong support for genetic drift as the major contributor. Weaver also suggested that support for the genetic drift hypothesis comes from the fossil record where gradual changes in Neanderthal features accumulated over a period of 300,000 years. Weaver concluded that climatic adaptation was the most reasonable explanation for the wide bodies and short limbs of Neanderthals, that some additional postcranial features could be secondary consequences of the body's skeletal proportions, and that changes in diet or activity patterns may explain some mild effects on postcranial features. Others have come to similar conclusions about the contribution of ADL.[47]

Cranial capacity can be directly related to brain development, although a larger brain does not necessarily coincide with an increase in cognitive ability. Brain enlargement during human evolution has been dramatic. During the early stages in the evolution of traits associated with bipedalism in the hominin lineage, the brain started enlarging slowly.[51,52] Between the partitioning of the great ape-hominin lineages, approximately 5 mya, and the time of early *H. habilis*, 2.5 mya, the brain experienced a modest increase in size from about 400 to 650 cm^3. Then, starting at about 2.5 mya, the hominin brain experienced a faster rate of growth (Fig. 6).

Subsequently, after a growth plateau lasting about 800,000 years, during the period between *H. erectus* (1.5 mya) and *Homo heidelbergensis* (700 kya), encephalization relative to body size was especially pronounced and coincided with a period of strong climate fluctuations and an increased range of environments. This encephalization culminated with humans reaching an average cranial capacity of 1250–1400 cm^3 and Neanderthals 1350–1700 cm^3.[3,53] It has been calculated that during the 2.5-million-year period of hominin brain growth, the number of neurons increased at an average rate of about 100,000 neurons per generation.[54] During this time, the brain also experienced a number of significant morphological changes including expansion and convolutions of the cerebral cortex and greater myelination of neurons, which is associated with faster neuron signal transmission. By the end of encephalization, larger brains allowed hominins to process and store information, to plan ahead, to solve abstract problems, and to formulate solutions to diverse survival challenges.

Studies of fossil remains from the Sima de los Huesos cave site in the Atapuerca Mountains of Northern Spain dated to 430,000 years ago represent some of the oldest fossils with Neanderthal and *H. heidelbergensis* features. Meyer et al.[55] examined the mitochondrial genome sequence of a hominin from Sima de los Huesos and showed that it

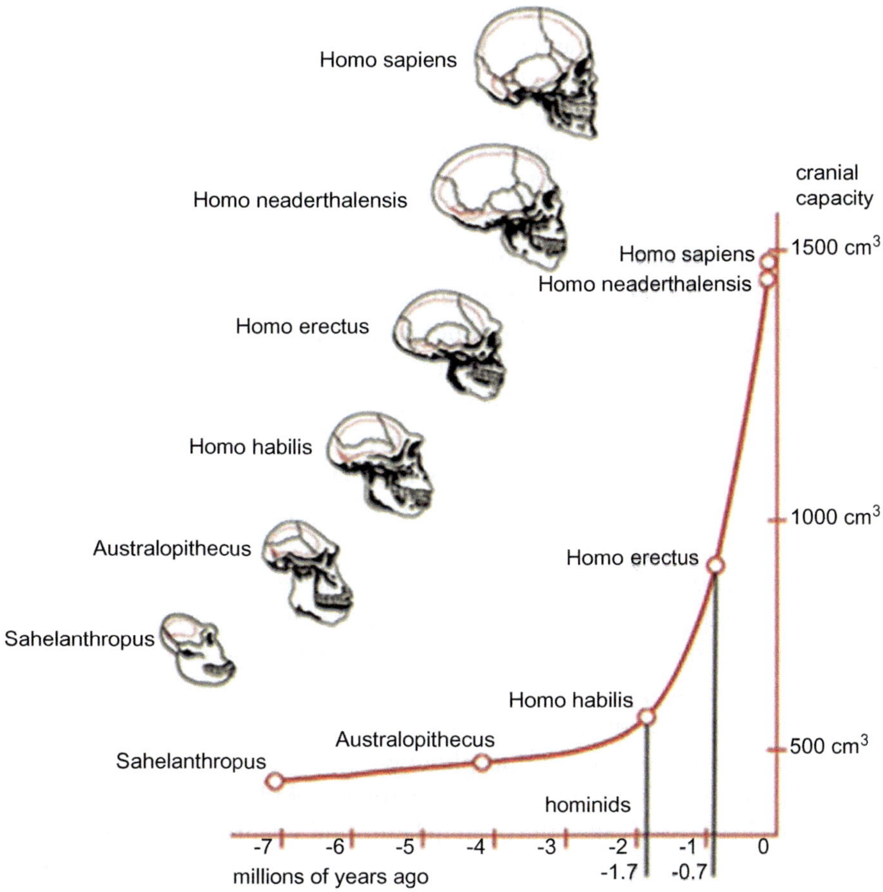

Fig. 6 Evolution of the human brain. The growth rate of hominid cranial capacity began to increase rapidly around 2 million years ago. *(From Evolution du volume cerebral des Hominides.* Le Journal du Net *2010. http://www.linternaute.com/science/biologie/dossiers/06/0608-memoire/8.shtml. Copyright 2011 Benchmark Group—69-71 avenue Pierre Grenier 92517 Boulogne Billancourt Cedex, France. http://www.linternaute.com/science/biologie/dossiers/06/0608-memoire/8.shtml. Also in Herrera RJ, Garcia Bertrand R.* Ancestral DNA, Human Origins and Migrations. *Academic Press; 2018:67.)*

was closely related to the lineage leading to mitochondrial genomes of Denisovans. Because of the fossils' age and anatomical features, the remains were initially thought to be from *H. heidelbergensis*[56]; however, further anatomical and DNA sequencing studies by Meyer et al. in 2016, using two of the 28 individuals recovered, indicated that the remains are from early Neanderthals.[17] Measurements of 17 skulls from the site indicated that they are much smaller than those of most Neanderthals and their average brain size is smaller than that of modern humans. The contrasts compared to other European fossils in the Middle Pleistocene could be explained by age, regional variations, and interbreeding with other archaic populations, and possibly demonstrate early examples of Neanderthal skull and brain evolution. These studies indicate that later Neanderthals developed larger skulls than their earlier ancestors, and that there was room for other neural developments during the evolution of the Neanderthal brain.[51]

Three-dimensional reconstruction studies of juvenile and adult skull shape in AMHs and Neanderthals from Russia and Syria indicate that early brain development of Neanderthals and humans started with a similar pattern in the womb but may have diverged after birth. However, it has been argued that humans who lived during the same period as Neanderthals had roughly the same cranial capacity as Neanderthals and that our modern cranial capacity evolved to accommodate our smaller stature and our different nutrient uptake.[57] Studies indicate that softer agricultural diets may have resulted in a less robust craniofacial morphology in early farmers (the mastication of softer foods is mechanically less demanding in terms of chewing than forager diets). The studies indicate that although the effects of agriculture, softer foods, and mastication are modest, the effects are reliably directional and most pronounced in craniofacial features. In general, these studies have shown a relatively potent effect of dietary change on the shape and size of the mandible and, in some cases, a commensurate effect on the form of the skull.[58] Others have argued that cranial differences could be related to diet, but they could also reflect differences in the specific genetic population histories of farmers and foragers, and additional factors, such as migration of new peoples into the area, could also have an impact on the observed patterns of morphology.[59]

Distinctive features of skull shape in Neanderthals include a rounded brain case, a skull that had a bulge in the back called the occipital bun, and a depression (the suprainiac fossa) for the attachment of thick neck muscles. The face showed thick, rounded brow ridges and large, rounded eye sockets. The larger eye sockets and larger areas of the brain devoted to vision suggest that Neanderthals' eyesight may have been better than that of modern humans.[57] Frontal features show a relatively flat and receding forehead, a large broad nose, a chinless lower jaw, and a midface that looked as if it was pulled forward by the nose (facial prognathism—a condition where the jaw protrudes, resulting in the top and lower teeth misaligning).[43] The jaws and teeth of Neanderthals were larger than those of modern humans and had a gap (the retromolar space) behind the wisdom teeth. Dental remains from two Italian sites, Fontana Ranuccio (Latium) and Visogliano (Friuli-Venezia Giulia), indicate that Neanderthal dental features had evolved by around 450,000 years ago.[60]

Three-dimensional reconstructions of nasal cavities and computational fluid dynamics indicate that Neanderthals and modern humans adapted their noses independently and through convergent evolution to accommodate breathing in cold and dry conditions.[61] The large nose of Neanderthals and *Homo heidelbergensis* also affected the shape of their skull and the muscle attachments, and gave them a weaker bite than modern humans.[62]

Wroe et al.[63] examined several Neanderthals' unique craniofacial features, including their wide, tall nasal aperture, a wide, projecting nasal bridge, their retromolar gap, "swept back" zygomatic arches, and a depressed nasal floor. They used computer simulations to determine if these craniofacial features represent adaptations to using their teeth as tools, better conditioning to cold, dry air, increased ventilatory flows to sustain higher energetic demands, genetic drift, or just retained traits shared by their common ancestor. They tested their hypotheses using three-dimensional models of Neanderthals, modern humans, and *Homo heidelbergensis*, applying finite-element analysis (FEA) and computational fluid dynamics (CFD). These studies extended those of de Azevedo et al., who did not use CFD in their analysis.[61] The FEA analysis of Wroe et al. revealed few differences between *H. heidelbergensis*, modern humans, and Neanderthals in their capacities to sustain high anterior tooth loadings (heavy biting). The CFD analysis showed that the nasal cavities of Neanderthals, and especially modern humans, condition air more efficiently than those of *H. heidelbergensis* do, suggesting that both evolved to better withstand cold and/or dry climates. They also determined that Neanderthals could move considerably more air through the nasal pathway than *H. heidelbergensis* or modern humans could. They concluded that Neanderthal facial morphology evolved to reflect improved capacities to better condition cold, dry air, and to move greater air volumes in response to higher energetic requirements, but not heavy biting.[63]

Body mass and development

The limbs of Neanderthals were thick, with large joints, and shin bones and forearms were relatively short, which are typical features of people living in cold environments. The pelvis was wider than that of modern humans, which led to the idea that Neanderthals lacked developed spinal curvatures, and therefore had bad posture. However, recent studies using virtual reconstruction of the La Chapelle-aux-Saints 1 Neanderthal skeletal remains have shown that the angle of the pelvis and the way the vertebrae stacked produced an S-shaped spine that curved at the neck and inward from the lower back toward the waist, resulting in the usual human lower back and neck curvature (lordosis) and a fully upright and balanced posture.[6,64] Neanderthal ankle and heel bones suggest that Neanderthals could not run as well as modern humans over long distances. Their heel bones are longer than those of modern humans, resulting in a longer Achilles tendon. Shorter Achilles tendons like those seen in modern humans are more efficient for running. Because of the cold climate and ambush tactics used in hunting, Neanderthals would not generally need to be good long-distance runners.

A virtual 3D reconstruction of the thorax of the 60,000-year-old adult Neanderthal male known as Kebara 2 revealed that the thorax is significantly different but not larger than that of modern humans. The thorax is wider in its lower segment, which parallels his wide bi-iliac width (referring to the widest measure of the pelvis between the outer edges of the upper iliac bones), and with a more invaginated vertebral column. The analyses showed that rib cages that are wider in their lower segment produce greater overall respiratory capacity during inspiration. This may have helped with high energy demands and could have had biomechanical consequences for stronger muscle attachment. Thus, the investigators hypothesize that Neanderthals may have had a subtly different breathing mechanism to modern humans.[65,66]

Although height and weight varies considerably with time, geographic region, nutritional source, social status, and prevalence of disease in a population, skeletal reconstructions can be used to estimate average height and weight. These measurements indicate that the height of male Neanderthals was approximately 165–168, and 150–156 cm for females. Average weight was estimated from 26 specimens of Neanderthals recovered from various European regions and different glacial periods. Body mass was estimated using femoral and head dimensions, and stature. Average weight for males was determined to be 77.6 and 66.4 kg for females. *Homo Heidelbergensis* males stood at an average of 174–176 cm, while females were shorter, with an average height of 156–158 cm.[24]

Anatomically modern European humans at the time were approximately the same height as *H. heidelbergensis*.[27] Studies of Neanderthals' genes, in contemporary humans, indicate that Neanderthal gene regulatory variants contributed to the genomic complexity of modern humans. Some examples include rs950169, an SNP in an extracellular matrix protein *ADAMTSL3* that is significantly associated with height and schizophrenia risk[67] (see Chapter 9 for more detail on Neanderthal DNA).

Overall, Neanderthals had many common features, and it is typical when describing Neanderthal characteristics to list the classical traits that are indicative of the species (physically stronger than modern humans with a stocky build, a wide short pelvis, heavily muscled limbs, prominent brow, large head, and distinctive skull morphology); however, the level of Neanderthal-derived traits over their 400,000-year timeline varied between evolutionary periods and geographic regions (Fig. 7). Clear differences between individual fossil remains suggest that Neanderthals evolved to have distinguishing temporal and regional characteristics. This sequence of physical, morphological, and genetic alterations eventually produced populations that were distinct from the original ancestor (investigators sometimes describe these stages of development as early pre-Neanderthals, pre-Neanderthals, and early Neanderthals). Two models have been proposed to explain the variation seen between the European Middle Pleistocene (MP) forms and the later Neanderthals. The models, known as the "two-phases model" and the "accretion model," attempt to explain the differences, and both have implications for the taxonomy used to describe Neanderthal European fossils.

The two-phases model suggests a rapid anatomical change between MP specimens assigned to *H. heidelbergensis*, conceived either from a sequential change in the developmental pattern of the Western Eurasian clade or as an Afro-European root group. This model proposed that the first phase (early MP) was characterized by an increase in body

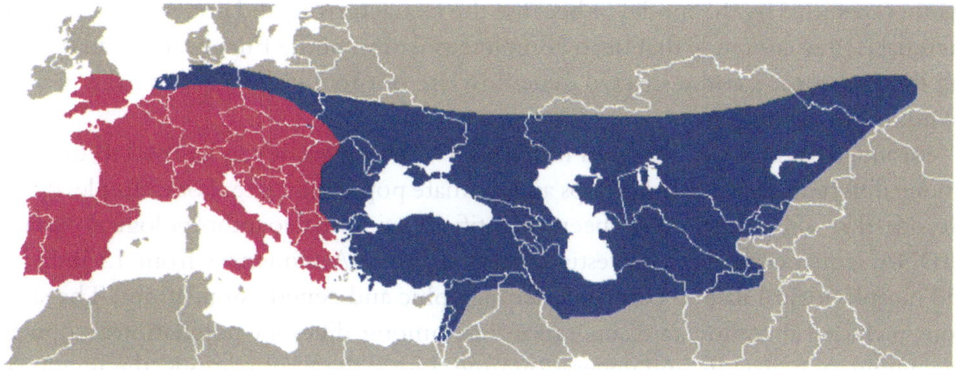

Fig. 7 The geographical range of Neanderthals over their existence in Europe. The environmental differences and genetic drift may have led to some of the morphological variation seen in the pre-Neanderthal, early Neanderthal, and late Neanderthal. Approximate Neanderthal range: pre-Neanderthal and early Neanderthal range shown in purple, late Neanderthal range in *blue* (https://en.wikipedia.org/wiki/Neanderthal#/media/File:Neanderthal_Range.png).

size, postcranial robustness, and midfacial prognathism. Speciation from an ancestral *H. heidelbergensis* population occurred in the late MP, and resulted in a major reorganization of the cranial architecture, leading to relative increases in the size of the occipital and temporal neural areas. Their results indicated a strong association of facial shape with climate, particularly temperature, and that facial centroid size was associated with genetic distances. The investigators concluded that Neanderthals represent a distinct evolutionary lineage, allowing easy identification of the MP *H. heidelbergensis* and *H. neanderthalensis*.[13]

The accretion model suggests that Neanderthals of the Late Pleistocene evolved in partial or complete genetic isolation from the rest of human hominids through the gradual accumulation of distinctive morphological traits in Western European populations. As the traits became more common, they also became less variable. The model's supporters propose that genetic drift and random fixation caused this evolution, resulting from an initial small Western European population and either complete isolation or drastic reduction in gene flow between it and human populations elsewhere. Changes in the cranial architecture occurred in relation to a growing degree of encephalization, likely resulting from selective pressures similar to those experienced by *H. sapiens*. However, they followed a different pathway that did not result in a more globular brain overhanging a retracted face, but maintained a more primitive pattern of craniofacial structure. In this model, there would be no clear divide between *H. heidelbergensis* and early *H. neanderthalensis*.[13] However, in 2001, Hawks and Wolpoff[68] disputed this model, indicating that the pattern of trait changes over time within Europe does not support the accretion model, either because the characters did not change in the manner specified by the model or because the characteristics did not change at all. From their data, they conclude that near-complete isolation of the European population during the Pleistocene and restricted gene flow are not required to explain the pattern of evolution in the region.

Neanderthal DNA is discussed in Chapter 9; however, it can be used to distinguish among different Neanderthal groups and estimate population size. For example, different Neanderthal subgroups have been identified using paleoanthropological data and mtDNA sequence analysis. Investigators used mtDNA sequences from 12 individuals and morphological data to construct demographic and genetic simulations. The simulations included mutation rate, migration rate among different populations, population size, population growth, and generation time. Using these simulations, the investigators concluded that Neanderthals can be divided into at least three groups: one in Western Europe, a second in Southern Europe, and a third in Western Asia. Results from these studies also indicate that the size of the Neanderthal population was not constant and that some migration occurred among the subgroups. The study was limited to Neanderthals before the arrival of humans in the regions.[69]

Researchers can use genetic diversity as a clock to measure how long groups of individuals have lived apart. When the investigators compared the genetic diversity of the mitochondrial DNA (mtDNA) from six Neanderthal individuals, they found much lower diversity than they would expect compared with six mtDNA sequences from modern Europeans. Based on mitochondrial DNA sequencing of the Neanderthals, the investigators estimated that fewer than 3500 breeding Neanderthal females lived in Europe at any one time between about 38,000 and 70,000 years ago.[12]

More recent studies using DNA from the 1000 Genomes Project and Neanderthal and Denisovan samples from the Max Planck Institute for Evolutionary Anthropology analyzed a few million nucleotide sites that shared a gene mutation in two or three human groups and established 10 distinct nucleotide site patterns. The analysis revealed that 20% of nucleotide sites bore a mutation shared by Neanderthals and Denisovans, but not seen in modern humans. The researchers' results indicate that the Neanderthals numbered in the tens of thousands.[11]

The discovery of a female skeleton during the early 1930s in Israel at Mugharet et-Tabun, and a second find at the Mount Carmel site, Mugharet es-Skhul, produced remains of 10 individuals that resembled Neanderthals but with some significantly modern human features (delicate jaw, smaller brow ridge, more pronounced chin, a more modern-shaped cranium). These individuals clearly represented a transitional middle ground between humans and previously discovered Neanderthals. Later analyses of some of these individuals suggested that these specimens represent early *H. sapiens*.[27] It has also been argued that Amud-1 (described above) also represents a transitional fossil.[70] Other investigators have documented that Levantine Neanderthals had phenotypes significantly more similar to modern humans than European Neanderthals had to modern humans[71].

References

1. Peace Pilgrim. *We Are All Cells in the Same Body of Humanity*. https://www.brainyquote.com/quotes/peace_pilgrim_183390.
2. Pontzer H. The crown joules: energetics, ecology, and evolution in humans and other primates. *Evol Anthropol*. 2017;26(January/February):12–24.
3. Pontzer H. Evolved to exercise. *Sci Am*. 2019;320:22–29. https://doi.org/10.1038/scientificamerican0119-22.
4. Wood RL, Richmond BG. Human evolution: taxonomy and paleobiology. *J Anat*. 2000;2000 (197):19–60.
5. Gabunia I, et al. Earliest Pleistocene hominid cranial remains from Dmanisi republic of Georgia: taxonomy, geological setting, and age. *Science*. 2000;228:1019–1025.
6. Rak Y, Arensburg B. Kebara 2 Neanderthal pelvis: first look at a complete inlet. *Am J Phys Anthro*. 1987;73:227–231.
7. Tattersall I. New twist added to the role of culture in human evolution. *Sci Am*. 2014;311 (September):54–59.
8. Manzi G. Humans of the Middle Pleistocene: the controversial calvarium from Ceprano (Italy) and its significance for the origin and variability of *Homo heidelbergensis*. *Quat Int*. 2016;411:254–261.
9. Stringer C. *The Origin of Our Species*. London: Penguin Books; 2011:26–29.
10. McDougall I, et al. Stratigraphic placement and age of modern humans from Kibish, Ethiopia. *Nature*. 2005;2005(433):733–736.
11. Rogers AL, et al. Early history of Neanderthals and Denisovans. *Proc Natl Acad Sci USA*. 2017;214:9859–9863.
12. Laursen L. Neanderthal census reveals diversity. *Nature News*. 2009. https://doi.org/10.1038/news.2009.701.
13. Hublin JJ. The origin of Neandertals. *Proc Natl Acad Sci*. 2009;106:16022–16027.
14. Carbonell E, et al. The first hominin of Europe. *Nature*. 2008;2008(452):465–469.
15. Campana I, et al. New interpretation of the Gran Dolina-TD6 bearing *Homo antecessor* deposits through sedimentological analysis. *Sci Rep*. 2016;6:34799.
16. Green RE, et al. A draft sequence of the Neandertal genome. *Science*. 2010;328:710–722.
17. Meyer M, et al. 2016 Nuclear DNA sequences from the Middle Pleistocene Sima de los Huesos hominins. *Nature*. 2016;531:504–507.
18. Dansgaard W, et al. Evidence for general instability of past climate from a 250-yr ice-core record. *Nature*. 1993;364:218–220.
19. Wong K. Our inner neanderthal. *Sci Am*. 2012;22:76–81.
20. Mayr E. Geographical character gradients and climatic adaptation. *Evolution*. 1956;10:105–108.
21. Sankararaman S, et al. The genomic landscape of Neanderthal ancestry in present-day humans. *Nature*. 2014;507:354–357.
22. Rae TC, et al. The Neanderthal face is not cold adapted. *J Hum Evol*. 2011;60:234–239.
23. Froehle AW, et al. Energetics and the origin of modern humans. In: JCM A, Smith FH, eds. *The Origins of Modern Humans: Biology Reconsidered*. Hoboken, NJ: Wiley-Blackwell; 2013:285–320.
24. Froehle AW, Churchill SE. Energetic competition between neandertals and anatomically modern humans. *PaleoAnthropology*. 2009;96–116.
25. Schmitz R, et al. The Neanderthal type site revisited: interdisciplinary investigations of skeletal remains from the Neander Valley. *Proc Natl Acad Sci USA*. 2002;99:13342–13347.
26. Seddon C. *Humans: From the Beginning*. San Bernardino, CA: Glanville Publications; 2015.
27. Campbell BG, et al. *Humankind Emerging*. 9th ed. Boston, MA: Pearson; 2006:358–362.
28. Berger TD, Trinkaus E. Patterns of trauma among Neandertals. *J Archaeol Sci*. 1995;22:841–852. https://doi.org/10.1016/0305-4403(95)90013-6.
29. Prüfer K, et al. The complete genome sequence of a Neanderthal from the Altai Mountains. *Nature*. 2014;505:43–49.
30. Neubauer F. A brief overview of the last 10 years of major late Pleistocene discoveries in the old world: *Homo floresiensis*, Neanderthal and Denisovan. *J Anthropol*. 2014. https://doi.org/10.1155/2014/581689. 7 pages, Article ID 581689.
31. Stringer C, Gamble C. In search of the Neanderthals. *Nature*. 1993;376:100–101.

32. Duarte C, et al. The early Upper Paleolithic human skeleton from the Abrigo do Lagar Velho (Portugal) and modern human emergence in Iberia. *Proc Natl Acad Sci USA*. 1999;96:7604–7609.
33. Tattersall I. Neanderthals, Homo sapiens, and the question of species in paleoanthropology. *J Anthropol Sci*. 2007;85:139–146.
34. Qiu J. The forgotten continent: fossil finds in China are changing ideas about the evolution of modern humans and our closest relatives. *Nature*. 2016;535:218–220.
35. Hardy BL. Neanderthal behavior and stone tool function at the Middle Palaeolithic site of La Quina, France. *Antiquity*. 2004;78:547–565.
36. Gomez-Olivencia A, et al. La Ferrassie 8 Neanderthal child reloaded: new remains and re-assessment of the original collection. *J Hum Evol*. 2015;82:107–126.
37. Will M, Stock JT. Spatial and temporal variation of body size among early Homo. *J Hum Evol*. 2015;82:15–33.
38. Ponce de Leon MS, et al. Neanderthal brain size at birth provides insights into the evolution of human life history. *Proc Natl Acad Sci USA*. 2008;105:13764–13768.
39. Rosas A, et al. The growth pattern of Neandertals, reconstructed from a juvenile skeleton from El Sidrón (Spain). *Science*. 2017;357:1282–1287.
40. Rozzi FVR, Bermudez de Castro JM. Surprisingly rapid growth in Neanderthals. *Nature*. 2004;428:936–939.
41. El Zaatari S, et al. Ecogeographic variation in Neandertal dietary habits: evidence from occlusal molar microwear texture analysis. *J Hum Evol*. 2011;61:411–424.
42. Buti L, et al. 3D enamel thickness in Neandertal and modern human permanent canines. *J Hum Evol*. 2017;113:162–172. https://doi.org/10.1016/j.jhevol.2017.08.009.
43. Schwartz JH, Tattersall I. The human chin revisited: what is it and who has it. *J Hum Evol*. 2000;38:367–409.
44. Zilberman U, Smith P. A comparison of tooth structure in Neanderthals and early Homo sapiens sapiens: a radiographic study. *J Anat*. 1992;180:387–393.
45. Kupczik K, Hublin JJ. Mandibular molar root morphology in Neanderthals and Late Pleistocene and recent Homo sapiens. *J Hum Evol*. 2010;59:525–541.
46. Macchiarelli R, et al. How Neanderthal molar teeth grew. *Nature*. 2006;444:748–751.
47. Clement AF, et al. Tooth wear, Neanderthal facial morphology and the anterior dental loading hypothesis. *J Hum Evol*. 2012;62:367–376.
48. Barrettt CK, et al. Revisiting dental fluctuating asymmetry in Neandertals and modern humans. *Am J Phys Anthropol*. 2012;149:193–204.
49. Guatelli-Steinberg D, et al. Prevalence and the duration of linear enamel hypoplasia: a comparative study of Neandertals and Inuit foragers. *J Hum Evol*. 2004;47:65–84.
50. Weaver TD. The meaning of Neandertal skeletal morphology. *Proc Natl Acad Sci USA*. 2009;106:16028–16033. https://doi.org/10.1073/pnas.0903864106.
51. Wong K. Neanderthal minds: analysis of anatomy, DNA and cultural remains have yielded tantalizing insights into the inner lives of our mysterious extinct cousins. *Sci Am*. 2015;312:36–43.
52. Sylvester AD. Locomotor coupling and the origin of hominin bipedalism. *J Theor Biol*. 2006;242:581–590.
53. Campbell BG, et al. *2006 Humankind Emerging*. 9th ed. Boston, MA: Pearson; 2006:367–380.
54. Edelman GM. *Second Nature: Brain Science and Human Knowledge*. New Haven, CT, USA: Yale University Press; 2006.
55. Meyer M, et al. A mitochondrial genome sequence of a hominin from Sima de los Huesos. *Nature*. 2014;505:403–406.
56. Stringer CB. The status of Homo heidelbergensis (Schoetensack 1908). *Evol Anthropol*. 2012; 21:101–107.
57. Pearce E, et al. New insights into the differences in brain organization between Neanderthals and anatomically modern humans. *Proc R Soc B*. 2013;280:1758.
58. Katz DC, et al. Changes in human skull morphology across the agricultural transition are consistent with softer diets in preindustrial farming groups. *Proc Natl Acad Sci USA*. 2017;114:9050–9055.
59. von Cramon-Taubadel N. Measuring the effects of farming on human skull morphology. *Proc Natl Acad Sci USA*. 2017;114:8917–8919.

60. Zanolli C, et al. The Middle Pleistocene (MIS 12) human dental remains from Fontana Ranuccio (Latium) and Visogliano (Friuli-Venezia Giulia), Italy. A comparative high resolution endostructural assessment. *PLoS One*. 2018;13:e0189773.
61. de Azevedo S, et al. Nasal airflow simulations suggest convergent adaptation in Neanderthals and modern humans. *Proc Natl Acad Sci USA*. 2017;114:12442–12447.
62. Toro-Ibacache V, et al. The relationship between skull morphology, masticatory muscle force and cranial skeletal deformation during biting. *Ann Anat*. 2016;203:59–68.
63. Wroe S, et al. Computer simulations show that Neanderthal facial morphology represents adaptation to cold and high energy demands, but not heavy biting. *Proc Biol Sci*. 2018. https://doi.org/10.1098/rspb.2018.0085.
64. Haeusler M, et al. Morphology, pathology, and the vertebral posture of the La Chapelle-aux-Saints Neandertal. *Proc Natl Acad Sci USA*. 2019;116:4923–4927.
65. Bastir M, et al. The relevance of the first ribs of the El Sidron site for the understanding of the Neanderthal thorax. *J Hum Evol*. 2015;80:64–73.
66. Gomez-Oliencia A, et al. 3D virtual reconstruction of the Kebara 2 Neandertal thorax. *Nat Commun. 2018*;9. Article number: 4387.
67. McCoy RC, et al. Impacts of Neanderthal-introgressed sequences on the landscape of human gene expression. *Cell*. 2017;168:916–927.
68. Hawks JD, Wolpoff MH. The accretion model of Neandertal evolution. *J Hum Evol*. 2001;40:A10.
69. Fabre V, et al. Genetic evidence of geographical groups among Neanderthals. *PLoS One*. 2009. https://doi.org/10.1371/journal.pone.0005151.
70. Arensburg B, Belfer-Cohen A. Sapiens and Neandertals. In: *Neandertals and Modern Humans in Western Asia*; 2002:311–322.
71. Ayala FJ, Cela-Conde CJ. *Processes in Human Evolution: The Journey From Early Hominins to Neanderthals and Modern Humans*. New York: Oxford University Press; 2017:220–221.

CHAPTER 11

Fossil and genetic evidence for Neanderthal introgression

> *We don't meet people by accident.*
> *Everyone is meant to cross our path for a reason.*
>
> **Kathryn Perez**[1]

The early evolution of humans and Neanderthals

Before looking at the introgression between Neanderthals and anatomically modern humans (AMHs), it is important to provide a perspective of how each species evolved from early ancestors. The most relevant source of ancient data comes from fossil discoveries, especially those of early *Homo* species. Current data points to the evolution of *Homo erectus* around 2 million years ago which spread to east and southern Africa. Then, approximately 1.8 million years ago, *H. erectus* migrated out of Africa by way of the Levantine corridor and the Horn of Africa into Eurasia, reaching Southeast Asia by about 1.7 million years ago and populating Europe about 1.2 million years ago.

Another archaic species of the Lower Paleolithic, *Homo antecessor*, was present in Western Europe between 1.2 million and 0.8 million years ago and is thought by some to be descended from *Homo erectus*. Fossils of *Homo antecessor* are restricted to Western Europe, specifically Spain, France, and England. Some experts believe *Homo antecessor* to be the ancestral group from which *Homo heidelbergensis* derived, although others suggest that *H. antecessor* is simply an early form of *Homo heidelbergensis* or a Western European variety of *H. erectus*.

Homo heidelbergensis inhabited Africa, Europe, and Asia from about 600 to 200 kya (thousand years ago). *Homo heidelbergensis* is distinct from *Homo erectus* primarily due to its large brain size and body proportions, which are comparable to contemporary humans. It is theorized that after *H. heidelbergensis* left Africa, about 300 kya, they evolved into *Homo neanderthalensis* in Europe and Denisovans in Asia, while in Africa, they gave rise to *Homo sapiens* around 200–300 kya.[2]

Although most investigators agree that *H. heidelbergensis* is the ancestor of Neanderthals and AMHs, there is still debate over the timing when AMHs evolved and the identification of *H. heidelbergensis* in Eastern and Western Europe. Some investigators like Roksandic et al.[3] and others argue that the fossil record of the early Middle Pleistocene

https://doi.org/10.1016/B978-0-12-819086-9.00006-3

in Eastern and Western Europe suggests that two different populations of *H. heidelbergensis* inhabited Europe: a population in the west of the continent with derived Neanderthal morphology; and a more variable population in the east which can be characterized by a combination of ancestral traits shared by members of the group and others shared by two or more taxa. Some investigators suggest that in order to continue using the nomenclature of *Homo heidelbergensis*, the sample from which the characters of a population are to be inferred must be revised to include only the specimens from the eastern group.[3]

The phylogenetic relationships between more archaic hominins like *Homo antecessor*, *Homo heidelbergensis*, and hominins that appear later in the fossil record (*Homo sapiens*, *Homo neanderthalensis*, Denisovans) are still heavily debated. This is because it is difficult to determine relationships based on the interpretation of individual fossil samples from various geographic regions and the lack of fossils from West Africa. The degradation of ancient DNA further hampers studies. Recent research has demonstrated that the analysis of ancient proteins can help address some of these issues by examining dental enamel proteomes. Welker et al.[4] examined the proteome of *H. antecessor* from Atapuerca, Spain, and *Homo erectus* from Dmanisi, Georgia. They provided evidence that *H. antecessor* is a close sister lineage to subsequent Middle and Late Pleistocene hominins, including modern humans, Neanderthals, and Denisovans. Their results implied that the AMH-like face of *H. antecessor* may reflect deep ancestry in the genus *Homo* and that the cranial morphology of Neanderthals represents a derived form. Their results have provided insights into the evolutionary relationships between *H. antecessor* and other hominin groups, and have demonstrated the usefulness of enamel proteomes to investigate hominin biology across the genus *Homo*.[4]

Synopsis of *Homo sapiens* emergence from Africa

Research from the 1990s and early 2000s concluded that *H. sapiens* evolved in East Africa around 200 kya, then moved out of Africa in their first successful migration around 60–70 kya, spreading along the southern coast of Asia and into Oceania by about 50 kya. Fossil evidence also suggests that AMHs had not entered Europe before 40–50 kya where they encountered Neanderthals and shared the continent until the Neanderthals became extinct about 27 kya.[5] Early investigations determined that although humans and Neanderthals occupied the region, AMHs did not admix with Neanderthals during their range expansion into Europe.[6]

From early studies of AMH fossils, two models were proposed for the evolution of *H. sapiens*: (1) the Multiregional Model of AMH origins, which implied the gradual evolution of large-scale archaic hominin populations evolving modern human morphology over the course of 2 million years[7]; and (2) the Out of Africa event, also known as the Recent African Origin model, which implied a single African origin (200–100 kya) of modern humans and subsequent replacement of other populations.[8]

As investigations continued into the 21st century, it became obvious that there had been intermixing between AMHs and other ancient species. Because AMH genomes contain DNA through interbreeding with ancient populations in more than one region of the world outside of Africa, some conclude that our origins were multiregional. However, because the amount of archaic genetic material in AMHs is modest, and fossil evidence outside of Africa shows little sign of the long-term morphological continuity in AMHs predicted by the multiregional model, the Recent African Origin (RAO) model for modern humans is still supported by some of the data.[9] However, new fossil discoveries, new technologies, and improved methods of ancient DNA analysis have enhanced our understanding of the evolution of *H. sapiens* and have caused investigators to move away from both the Multiregional Model and the RAO model.[8]

New approaches to understanding human evolution have shifted the emphasis from replacement to the varied demographic and genetic outcomes of recent hominin interactions. Subsequent models to address the complex evolutionary geography of hominins have included: multiple dispersals out of Africa and the role of population structure in Africa; models based on understanding the evolutionary mechanisms that gave rise to the anatomically different Neanderthals in various regions of Europe; and an ever-increasing number of genetic studies that examine ancient and extant human populations to follow the changing genome structure that led to modern-day humans.[10–15]

Recent discoveries of *Homo sapiens* fossils in West and South Africa suggest that humans evolved earlier and in several regions in Africa. Examples include fossils from Morocco dated 315 kya, the Florisbad skull from South Africa dated to 259 kya, remains of *H. naledi* in South Africa dated to 236–335 kya, and remains from the Omo National Park in Ethiopia dated to 200 kya.[16–19]

The finding of a 315-kya *sapiens* in Northwest Africa indicates that the original theory that *sapiens* evolved only in East Africa is not correct, and the date estimate is at least 100 ky older than previously thought. Furthermore, a number of reports on mtDNA diversity outside of Africa argue for AMHs out of Africa about 270 kya. These new mtDNA dispersal estimations are based on the fact that the elevated DNA variation observed outside Africa could not be accounted for by the previously predicted 70 kya exodus because more evolutionary time is needed to generate the mtDNA diversity observed in Eurasia. Overall, the recent archeological and genetic data suggests that the origins and dispersal of modern humans out of Africa are more ancient than previously thought. This paradigm shift also accommodates the very old Chinese *sapiens* fossils.[14]

Other recent studies have also concluded that *H. sapiens* occurred earlier in Africa than previously thought. Mounier and Marta[15] used a phylogenetic modeling method to statistically predict possible morphologies of a last common ancestor of all modern humans, which they compared to Late Middle Pleistocene African fossils. Computerized determination of a virtual skull (based on computerized tomography scans) representative of the earliest modern humans suggested that modern humans arose between 260 and 350 kya. Mounier and Marta's results support a complex process for the evolution of *H. sapiens*, with the recognition of different geographic populations and lineages in Africa, not all of which contributed to the origin of *Homo sapiens*. Based on the data of the scanned fossils, *H. sapiens* appears to have originated from the merging of South and East African source populations, while Mounier and Marta conclude that North African fossils may represent a population which introgressed into Neanderthals during the Late Middle Pleistocene.[15]

A new timeline for the emergence of AMHs sets the dates between 250 and 350 kya and suggests that our species may have evolved in subdivided populations within Africa.[19] There is still controversy about how many times *H. sapiens* left Africa and when these migrations were successful. Although AMH fossils dating to 180 kya have been found in Israel, and in China dated 80–120 kya, many investigators assumed that these were from rare migrations that left no survivors and that the first successful migration of *H. sapiens* out of Africa occurred 60–70 kya.[20] If the fossils from Israel and China represent earlier travelers who survived, they may have encountered other *Homo* species including *H. heidelbergensis*, *H. neanderthalensis*, and Denisovans much earlier than previously thought.

Synopsis of Neanderthal evolutionary history and appearance in Europe

Neanderthals represent an extinct hominid lineage that existed in Europe and Asia for nearly 400,000 years. They thrived in these regions for much of this time, but declined in numbers and went extinct around 30 kya. Many researchers believe their disappearance occurred after the arrival of modern humans into these areas, which has prompted some to argue that Neanderthals were displaced by better suited and more adaptable modern humans, or that Neanderthals were assimilated into the gene pool of modern humans by admixture. Initial conclusions about the relationships between Neanderthals and contemporary humans were based solely upon fossil and archeological records. However, genomic analysis has provided a powerful tool for studying the levels of genetic interactions between Neanderthals and modern humans, and has provided different timelines for the interaction between the species.

Considerable debate still exists regarding the time of divergence of the lineages leading to modern humans and Neanderthals, and arguments continue regarding the identity and geographic distribution of the species that gave rise to humans and Neanderthals. One of the reasons for this debate is that the fossil record is much more complete from 130 kya onwards, and makes up the bulk of known Neanderthal skeletons. Other causes include the scarcity, fragmentary preservation, often unclear chronology, and the questionable taxonomic classification of fossil remains.

One major source of evolutionary information is cranial fossils. Daura et al.[21] recognize three different cranial morphologies in the European Middle Pleistocene hominin record. First, there are fully Neanderthal crania, with nearly the entire group of derived traits, dated less than 200 kya. Second is a cranial configuration with many Neanderthal traits in the face such as supraorbital torus, temporal bone, and mandible, while the general shape of the neurocranium is not Neanderthal-like, but instead a mosaic indicative of Neanderthal cranial evolution. Daura et al. recognize this second type as resembling the Sima de los Huesos fossils.[22] Third, there are crania that do not show Neanderthal-derived traits in the preserved regions or in which the features are more ambiguous. Daura et al. attribute these crania to an incipient stage of Neanderthal evolution, to *Homo heidelbergensis*, or to a subspecies of *Homo erectus*. This third description represents the 400 kya Aroeira 3 crania, which could represent a transitional phase in Neanderthal evolution where ancestral and derived morphs could have lived concurrently. The presence of all of these cranial morphology differences suggests that intra- or interdeme hominin diversity and complex population dynamics characterized the Middle Pleistocene, including variable population replacement with fluctuating levels of isolation and admixture.[21]

The timing and the identity of the last common ancestor (LCA) of *Homo neanderthalensis* and *Homo sapiens* are still undetermined.[15] Studies of ancient DNA and cranial morphology generally point to a divergence time of 500–300 kya.[23,24] In addition, typically discussed evolutionary scenarios tend to assume that at least some Middle Pleistocene hominins dated to 600–400 kya, or even younger, were part of the last common ancestral species to Neanderthals and modern humans. Multiple anatomical studies of the fossil evidence, however, have indicated that some Middle Pleistocene European hominins, particularly those belonging to the Sima de los Huesos sample, show clear affinities with Neanderthals,[22] while a more variable population in the east can be characterized by a combination of ancestral traits shared by members of the group and others shared by two or more taxa.[3] After some early inconsistent results, the Sima de los Huesos sample is now dated to 430 kya. The analyses of the nuclear DNA of this population demonstrated an evolutionary affinity with classic Neanderthals, indicating that divergence between Neanderthals and modern humans is necessarily older than the age of the Sima de los Huesos fossils.[22,25,26] With this in mind, Gomez-Robles[27] used quantitative analyses of dental evolutionary rates and Bayesian analyses of hominin phylogenetic relationships to show that any divergence time between Neanderthals and modern humans younger than 800 kya ago would necessitate unexpectedly rapid dental evolution in early Neanderthals from Sima de los Huesos. The results support a divergence time of the last common ancestor for Neanderthals and modern humans of 800 kya. Although the molecular data and morphological data do not always agree, it is important to ensure that all data are presented. Since we have not unearthed all the fossils and have limited data on those we have unearthed, the time lines are still undetermined.

Studies of nuclear DNA from Neanderthals, humans, and Denisovans concluded that Denisovans were a sister group of Neanderthals that evolved after the divergence of Neanderthals and humans. mtDNA from the various groups provided evidence for at least two distinct mtDNA branches, which implied a sister group relationship between (1) modern humans and (2) Neanderthals and Denisovans as an mtDNA outgroup. Based on Y chromosome data from Neanderthals and AMHs[28] and the genome-wide human mutation rates,[29] the humans-Neanderthals split was calculated to 806–447 kya and 765–550 kya, respectively.

Evidence for introgression between Neanderthals and AMHs

One of the major questions in human evolution is whether introgression occurred between humans and archaic hominids inside and/or outside of Africa. Evidence for human evolution inside Africa suggests that humans evolved in various regions of Africa and most likely introgressed with other hominids, leading to AMHs.[19] Early evidence for introgression outside of Africa was suggested by several fossils that appeared as intermediate forms between AMHs and Neanderthals or some other hominid species.[30] Then, when the Neanderthal genome was sequenced, comparisons of Neanderthal and non-African genomes suggested that there was introgression with Neanderthals in non-African humans. However, just the appearance of Neanderthal DNA in modern-day humans outside of Africa was not enough to demonstrate introgression. An alternative explanation for Neanderthal DNA sequences in modern humans, known as ancient population structure, is that Neanderthal DNA evidence is linked to ancient population structure. This implies that the subdivision in the ancestral population that gave rise to Neanderthals and AMHs may have resulted in some non-African modern humans being more closely related to Neanderthals than other groups that stayed in Africa.

In order to show introgression, investigators had to look more closely at the genomic structure of Neanderthals and non-African AMHs to see what genomic changes had taken place over time (comparing genomes since Neanderthals and humans diverged and since the proposed time of introgression). One way to do this is by looking at genetic linkage and linkage disequilibrium, which can help to distinguish ancient population structure from introgression.

When chromosomes are transferred during fertilization, they carry sets of genes attached to the chromosome like beads on a string (linkage). These genes occupy specific positions on the chromosome (loci) and have different forms (alleles). When members of two different populations interbred (for example, Neanderthals and humans), they each contributed very different genes and alleles that had evolved since the time of their split (in the case of Neanderthals and humans, 550–800 kya). Each parent contributed half their genome to their offspring (known as a haplotype). Thus, offspring from

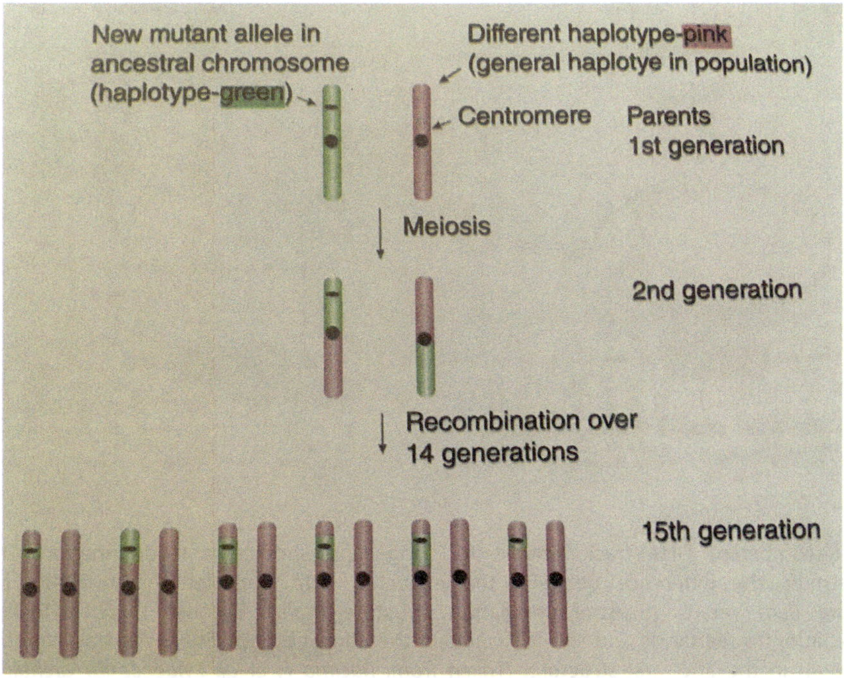

Fig. 1 Recombination with human genomes over 14 generations. Chromosome recombination from two different haplotypes. The figure shows two homologous chromosomes from the original parents (representing two different haplotypes: *pink* and *green*). The *green* haplotype is from a Neanderthal; the *pink* is from an anatomically modern human. A mutation *(black line)* is in the *green* ancestral haplotype. After several generations and multiple meiotic recombination events with human parents, the mutant allele is still present with segments of the ancestral Neanderthal chromosome (*green* haplotype). *Note*: The *green* chromosome represents the contribution from the Neanderthal parent with mutant allele *(black)*. The *pink* chromosome represents the contribution from the human parent. The bottom of the figure shows the 15th generation and the continued genetic contribution of Neanderthals and the mutant allele in some of the chromsomes.

Neanderthal-human breeding contained two different haplotypes representing an assortment of different chromosomes and alleles contributed by each parent to the resulting hybrid (Fig. 1). Each parent chromosome contributed to the hybrid offspring could only be broken up by recombination in future generations. As recombination occurred in subsequent generations, the length of Neanderthal DNA tracts in the chromosomes would diminish because recombination would break down the Neanderthal haplotypes into shorter and shorter fragments (Fig. 2). The human contribution would have dominated when Neanderthals began to go extinct, and what would presumably be observed is the shortening by recombination of Neanderthal haplotypes being distributed in the human genomes of the human gene pool.

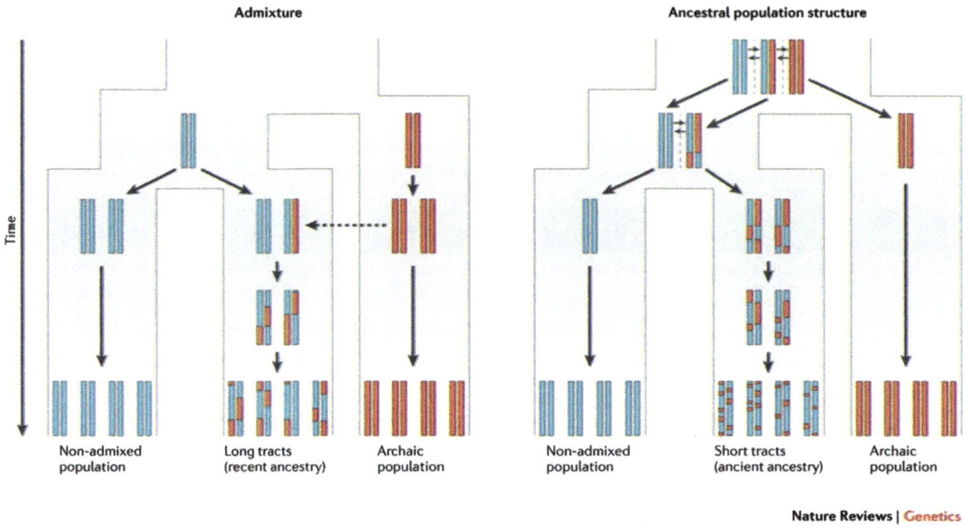

Fig. 2 Recombinant DNA track lengths for long- and short-term recombination. The figure demonstrates the difference between the expected DNA tract lengths in AMHs based on admixture *(left)* versus ancestral population structure *(right)*. Because there is less time for recombination to rearrange and shorten tracts in the admixed population, the tract lengths will be longer than in the ancestral structure. *(Taken from: Racimo F, et al. Evidence for archaic adaptive introgression in humans.* Nat Rev. Genet *2015;16:359–371.)*

Investigators can see these combinations of Neanderthal genes and alleles when they look at the genomes of modern-day humans, because gene flow between genetically distinct populations creates a condition known as linkage disequilibrium (LD) among all loci that have different alleles and different allele frequencies in the founding populations. Linkage disequilibrium—the nonrandom association of alleles at different loci—is a good indicator of the population genetic forces that structure a genome.

Linkage and linkage disequilibrium

Linkage equilibrium (LE) occurs when linked alleles associate randomly with each other in a population. LE assumes random association of alleles over time; however, LE does not occur in a single generation but is dependent on the rate of recombination and the number of generations. If there is no selection, LE increases over time and with an increasing number of generations (Fig. 3). Suppose we look at a simple example where two genes (*A* and *B*) each with two alleles (*A a* and *B b*) are linked. If the alleles are in linkage equilibrium, then the arrangement of alleles in coupling (*AB*; *ab*) and in repulsion (*Ab*; *aB*) will be random.

Linkage Equilibrium and Disequilibrium

If I know the first allele in a haplotype, can I predict the second allele?

If answer is Yes then it is in Linkage Disequilibrium
If answer is No then it is in Linkage Equilibrium

Fig. 3 Linkage equilibrium and disequilibrium simplified.

Linkage disequilibrium (LD) occurs when there is a nonrandom distribution of linked alleles in a population (i.e., when alleles travel together much more often than expected). For example, aB and Ab travel together most of the time and you rarely see AB and ab in the population. The classic measure of linkage disequilibrium (D) is the observed allele frequency minus the expected frequency (if the alleles were randomly segregating). For two loci, each with two alleles (A and a, and B and b), the observed frequency (P) is the number of individuals in the population that have the genotype $AB = P_{AB}$. The expected frequency (P') is the product of the two allele frequencies in the population: $P_A \times P_B = (P_A P_B)$. Therefore, the measure of the magnitude of LD is: $D = P_{AB} - P_A P_B$.

For example, assume loci A and B are linked. Taking into account only two alleles at each locus, alleles A and a are linked to alleles B and b; this would lead to various linkage combinations, AB, Ab, aB, and ab.

The observed genotypic frequency of A and B alleles in the population are:

$$AB = 0.6; Ab = 0.1; aB = 0.2; ab = 0.1.$$

Given these genotypic frequencies in the population, you can determine the allele frequency by adding up the frequencies of A, B, a, and b.

$$\text{Allele frequency}: A = 0.7; B = 0.8; a = 0.3; b = 0.2.$$

The LD can be computed by taking the observed frequency minus the expected frequency:

$$D = P_{AB} - P_A P_B = 0.6 - (0.7 \times 0.8) = 0.04.$$

During the first few generations, admixture can especially lead to LD. When two diverse genotypes come together, each chromosome will vary by allele composition, and genes along those chromosomes will be linked by ancestry. Because each parent contributes a different set of chromosomes with varying allele combinations, admixture leads to LD over very large chromosomal segments. For example, if a population with homozygous genotype *AABBDDEE* is admixed with a population whose homozygous genotype is *aabbddee*, the offspring will all show complete disequilibrium in the first generation, where the heterozygous genotype *AaBbDdEe* is really composed of two distinct haplotypes in one diploid genome, namely *ABDE*/*abde*. The higher the number and size of LD regions on the chromosomes, the earlier the migration event occurred.

When recombination occurs in future generations, LD declines and ancestral alleles become fragmented with only very tightly linked alleles remaining in LD.[31]

Because LD decreases over time, admixed populations can be used to determine the time of the migration event by looking at the decay of LD. The use of modern data from admixed populations along with appropriate statistical methods can help determine the level of admixture remaining and the number of generations since the initial event (e.g., migration).

Introgression versus ancient population structure

The difficulty in identifying introgression is to distinguish it from shared ancestral genetic variation. Any two populations with a common ancestor will always share some segments of DNA, because both populations descended from the same population and might therefore have inherited some of the same DNA segments from the ancestral population. Various statistical methods are available to identify introgression using sequencing, genome-wide data, single nucleotide polymorphisms (SNPs), identity by descent, and linkage disequilibrium.[32] For our purposes, we will only discuss identity by descent, tract lengths, and linkage disequilibrium.

The presence of Neanderthal DNA sequences in non-African modern-day humans is generally accepted as evidence of archaic introgression estimated to have occurred several times between 47 and 65 kya. An alternative explanation for Neanderthal DNA sequences in modern humans, known as ancient population structure, is that no hybridization occurred, and Neanderthal DNA evidence is strongly linked to the strength of population structure in the ancient populations that gave rise to both groups. This implies that the subdivision in the ancestral population that gave rise to Neanderthals and AMHs may have resulted in some modern human groups (like the ancestors of Eurasians) being more closely related and carrying similar DNA sequences to Neanderthals than other groups that stayed in Africa.[32,33]

However, the date of the last Neanderthal gene exchange into modern-day Eurasians, determined by examining the distribution of the length of Neanderthal DNA tracts in modern-day humans, favors the admixture scenario. As noted earlier in the chapter, recombination breaks down haplotypes into shorter and shorter fragments (Fig. 1). DNA from recent introgression events should, therefore, fall into longer contiguous tracts than DNA resulting from old introgression episodes or ancient population structure (Fig. 2). Measurements of DNA tract lengths in modern humans indicate that the last gene flow event occurred 50–60 kya.[32,34] This time frame is too recent to support a scenario of ancestral structure, as the divergence time between humans and Neanderthals has been estimated to have occurred 550–800 kya.[27]

As previously described, the length of an introgressed DNA sequence depends on the time since introgression. Under certain assumptions, it has been determined that the lengths of introgressed DNA tracts should follow an approximately exponential distribution with mean length (L) the expected length of a shared sequence. $L = ([1-m]r[t-1])^{-1}$, or approximately $[1/(r\ t)]$, where t is the number of generations separating the two populations, r is the recombination rate per base pair, and m is the proportion of the population that was replaced by migrants from another population (admixture fraction).[32,35]

For example, a haplotype of 40,000 base pairs (bp) was shared between two populations that split 8000 generations ago. If this region has a recombination rate of 2.5^{-8}, you can calculate the expected length of shared sequence:

$$1/\left(2.5^{-8} \times 8000\right) = L = 5000\,\text{years bp}$$

To determine the probability of the block (L) being due to ancient population structure and not introgression, use the formula $P = \exp(-k/L)$ where k is the original block size, $\exp(-40{,}000/5000) = \exp(-8)$, so the haplotype has a probability of 0.0003 of having persisted due to ancestral shared polymorphism and not introgression.

The result of introgression should generate longer tracts than ancient population structure. Using the length of the introgressed haplotype provides a strong method for distinguishing between introgression and shared ancestral variation. One weakness of this method is that it assumes that haplotypes could not be joined and rebuilt in small inbred populations as are found during bottleneck events which clearly occurred as humans ventured into Eurasia. Another drawback is that introgressed haplotypes are not directly observable, but instead must be inferred from genetic data or statistics that summarize haplotype length. The S* statistic provides a method for extracting this information.[36] S* was originally derived to identify genome-wide evidence of introgression similar to Patterson's *D* statistic (which uses genome-wide data to measure the different allele sharing between a Neanderthal compared to two modern human populations). S* differs from Patterson's *D* by using genome-wide data to find SNPs at a given locus, instead of looking at alleles. S* statistics discover subsets of SNPs within a window that are in high linkage disequilibrium. The higher the S* values, the higher the LD. S* has been used to identify highly divergent haplotypes that contain variants that show strong linkage disequilibrium introgressed from Neanderthals into non-Africans.[32,36]

Nature of Neanderthal introgression

Scientists have argued over whether *H. sapiens* and *H. neanderthalensis* are truly separate species since the early discovery of their ancient fossils (see Chapter 2). Neanderthal fossils resemble ours and are clearly members of our hominin family tree as well as members of the genus *Homo*. Early debates centered around whether Neanderthals were just a weird variant or more primitive form of *sapiens*, or whether they were really something different. Close examination of skeletal remains indicated that Neanderthal features are distinct from those of *H. sapiens*. When it was clear that Neanderthals and humans introgressed, the debate started again, and the discussions over the concept of different species continue.

Although it is now clear that Neanderthals and humans introgressed, Neanderthal genes may have been a burden on humans' health, made some individuals sterile, or human introgression overcame the small number of Neanderthals. As a result, Neanderthal genes became rarer, and many of their genes and DNA sequences disappeared from our gene pool. Today, Neanderthal DNA has been found in the genomes of all contemporary human populations, accounting for 1%–4% of genomes for people outside Sub-Saharan Africa and 0%–0.3% for people in Africa.[37] The amount of Neanderthal DNA is highest in East Asians, intermediate in Europeans, and lowest in Southeast Asians and Melanesians.[38]

The amounts of Neanderthal DNA, the specific sequences, and the origin of the Neanderthals that contributed their DNA vary between AMH populations. In a recent study by Skov et al.,[39] 14.4 million putative archaic chromosome fragments were recovered from 27,566 Icelanders, which covered 38.0%–48.2% of the callable genome. On the basis of the similarity with known archaic genomes, Skov et al. assigned 84.5% of fragments to an Altai or Vindija Neanderthal origin and 3.3% to Denisovan origin; 12.2% of fragments were of unknown origin. They found that Icelanders had more Denisovan-like fragments than expected; this was best explained by Denisovan gene flow, either into ancestors of introgressing Neanderthals or directly into humans. In addition, they found that although the overall rate of mutation was similar in humans and Neanderthals during the 500 ky that their lineages were separate, there were differences in the relative frequencies of mutation types. Skov et al. speculated that these mutation differences could be due to different generation intervals for males and females.[39]

In the case of Neanderthal human introgression, the initial force leading to LD was admixture. However, some of the forces that maintain LD are positive selection or balancing selection (Figs. 4 and 5), or inbreeding/bottlenecks (Fig. 6), or introgressed DNA was not removed from the population by negative selection or genetic drift (Figs. 7 and 8). Distinguishing between these different forces requires additional statistical analyses, because when a genomic region contains introgressed DNA, the pattern of polymorphism may be different than that expected under a neutral model with no introgression. These patterns of selection can be distinguished by a variety of statistical methods.[32]

Selection and adaptive introgression

When AMHs mated with Neanderthals, they picked up genes that we still carry today. Some of these genes were beneficial and may have accelerated AMH adaptation and improved survival in the new environments, in a process known as adaptive

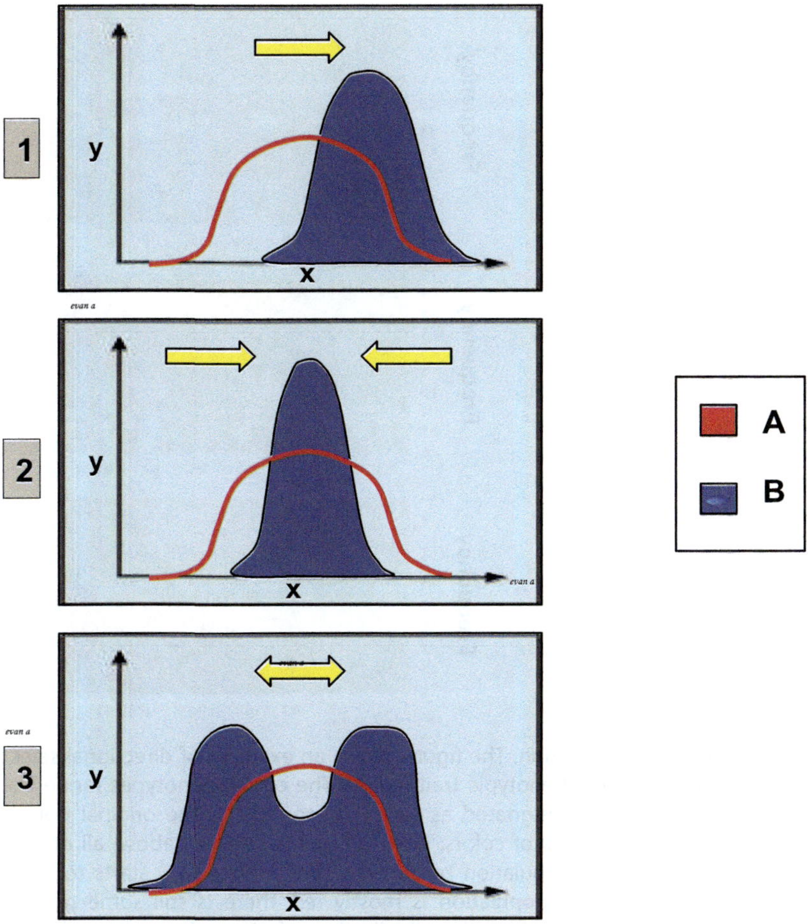

Fig. 4 Types of genetic selection. The figure shows three different types of genetic selection (directional, stabilizing, and disruptive or diversifying). The *x*-axis indicates the type of phenotypic trait and the *y*-axis indicates the number of organisms showing the trait. Group A *(red)* is the original population and Group B *(blue)* is the population after selection. Graph 1 shows directional selection, in which an extreme phenotype is favored. Graph 2 depicts stabilizing selection, where the intermediate phenotype is favored. Graph 3 shows disruptive selection, in which the extreme phenotypes are favored over the intermediate. *(Source: https://en.wikipedia.org/wiki/Directional_selection.)*

introgression. Over the last several years, scientists have revealed genomic regions of DNA that signal adaptive introgression. An indicator of adaptive introgression is the presence of mutations in strong LD that exist at high frequency in a particular population and that are only present in the archaic source population, while being absent or at very low frequencies in other present-day human populations.[32]

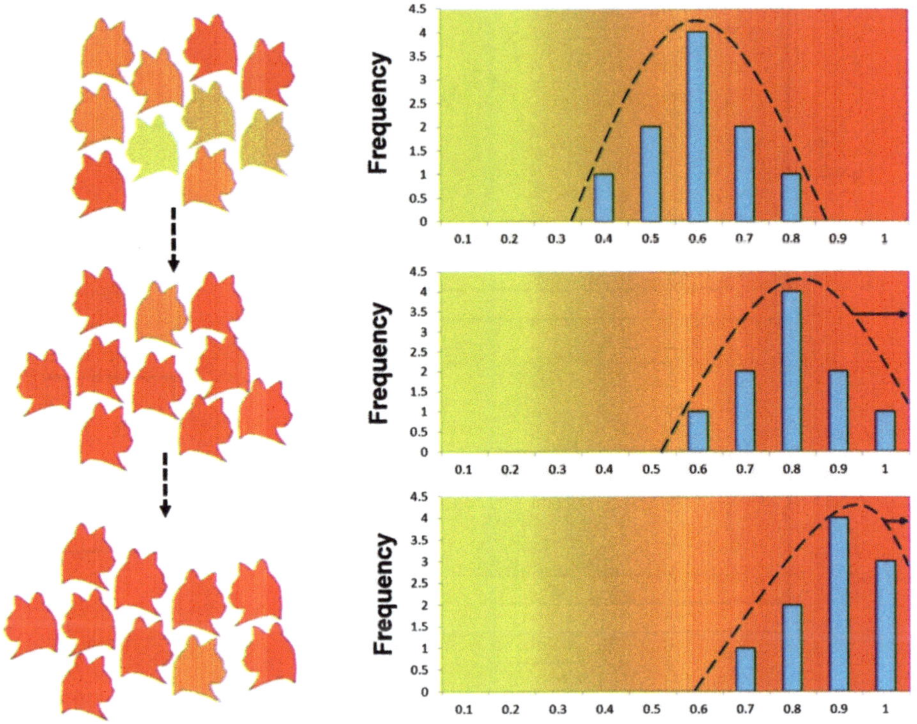

Fig. 5 Example of balancing selection. The figure shows an example of directional selection using a feline population with a single phenotypic trait (color). The color phenotypes range from *yellow* to *orange* to *red* along the x-axis, designated as *yellow* (0) to *red* (1). The original population at the top of the figure has a diverse mix of colors; however, *red* is selected above all others. Over time, the average color of the entire population becomes *redder* while *yellow* starts to disappear. Note, however, that although the final population is mostly *red*, there is still some minor variation in colors. These changes are reflected in the distribution of the color-coding alleles *(blue bars)*, showing movement toward the *red* end of the spectrum. *(Source: https://theg-cat.com/tag/balancing-selection.)*

Selection can favor certain allelic combinations that impart a higher level of fitness. Positive selection produces an increase in selected allele frequencies over time. A direct way of inferring selection is the detection of significantly high frequencies of an introgressed fragment in a specific population, relative to other populations. Selection may act on a favorable gene that carries a whole series of tightly linked alleles (that may or may not be favorable), maintaining disequilibrium. A favorable allele linked to neutral or unfavorable alleles is the classic example of genetic hitchhiking. The neutral or unfavorable allele increases in frequency due to the selection of the favorable allele and LD.[40] The nonrandom association of alleles at different loci, due to genetic coadaptation, may exist with some alleles within a population. This may be due to specific allelic combinations imparting an adaptive advantage to the organism. Disequilibrium will only occur if

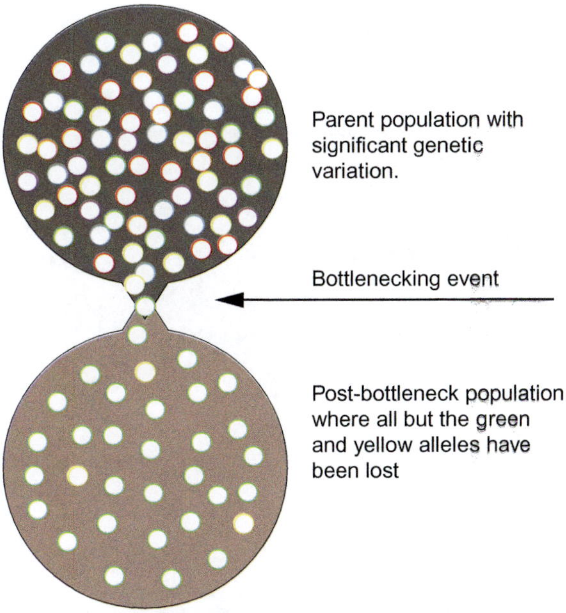

Fig. 6 Bottlenecks. The figure shows the effects of a genetic bottleneck where a small representative group of individuals survives from a disaster, disease, or some other mechanism of selection. In this example, mostly those individuals with the *green* phenotype survive. As the remaining surviving individuals go on to reproduce, the future population will consist mostly of those with the *green* phenotype. This is one method by which new populations can show different allele frequencies and produce a founder effect by genetic drift. *(Source: https://upload.wikimedia.org/wikipedia/commons/2/24/Bottleneck_effect.jpg.)*

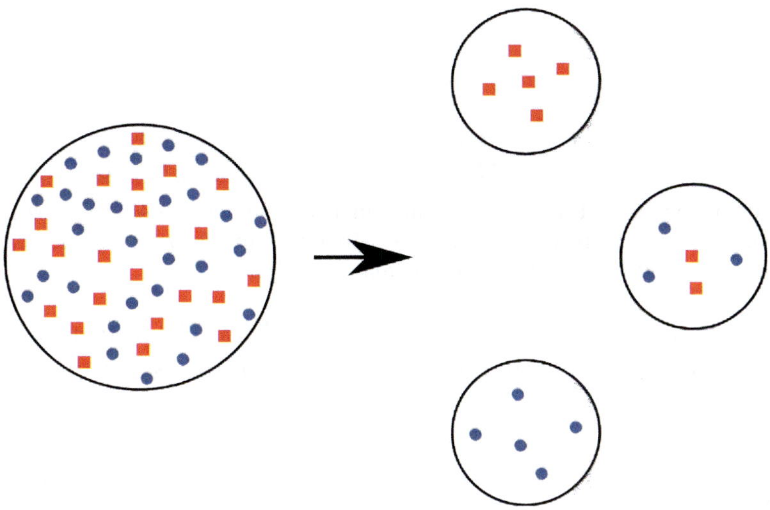

Fig. 7 Founder effect. The founder effect reduces genetic diversity, and results when a population is descended from a small number of colonizing ancestors. This can occur through a bottleneck or when a new population is established by a very small number of individuals migrating away from a larger population. This is another example of genetic drift.

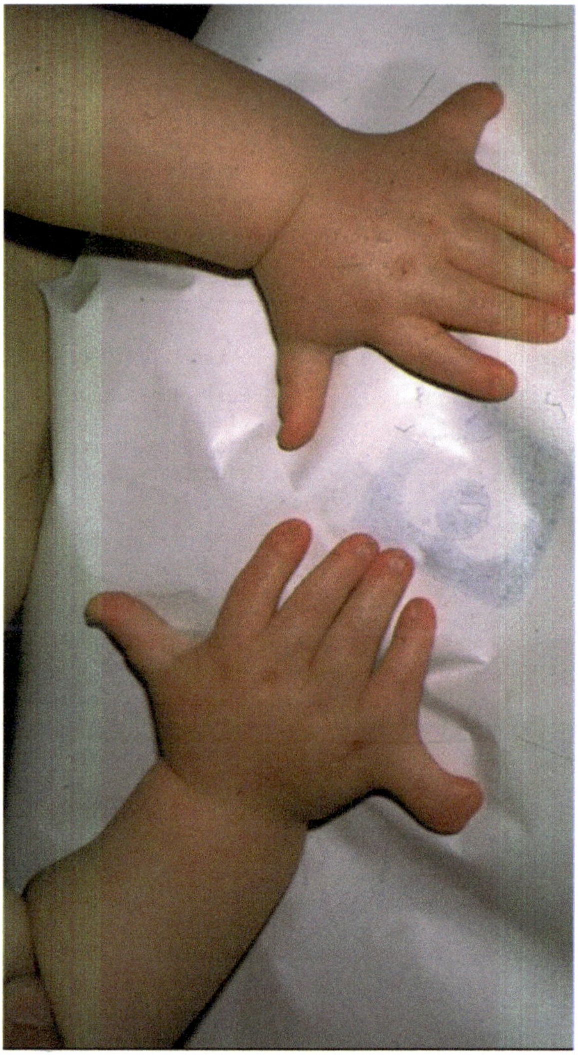

Fig. 8 A result of genetic drift. The founder effect in action: among the Amish, babies with Ellis-Van Creveld Syndrome are born with six fingers. *(Source: https://socratic.org/questions/what-is-the-difference-between-genetic-drift-founder-effect-and-bottleneck-effec.)*

natural selection favors a particular combination of alleles. The further the linked alleles are from each other on the chromosome, the stronger the forces of natural selection need to be in order to maintain disequilibrium.

Vernot and Akey[36] used whole-genome sequencing and S* statistics to identify admixture between extant European and Asian populations and Neanderthals. They recovered a substantial number of Neanderthal sequences, which revealed that positive and purifying selection occurred, and found that admixture occurred both before and after the divergence of non-African modern humans. Some of the Neanderthal genes that revealed positive selection include those that impart immunity,[41] affect skin cell response to ultraviolet-B irradiation, cause red hair color, produce keratin found in skin, hair, and nails,[42] and regulate sugar metabolism, muscle contraction, and oocyte meiosis. All of these genes and gene regions involved with these processes were inherited from Neanderthals and have been involved in archaic human adaptive introgression.[43]

Another example of adaptive introgression is the progesterone receptor gene (PGR). Progesterone is important for preparing the uterine lining for egg implantation and the early stages of pregnancy. The gene encoding *PGR* carries introgressed Neanderthal haplotypes with two missense mutations and a transposable *Alu* element. These Neanderthal gene variants constitute nearly 20% frequency in non-Africans and have been associated with preterm birth. Researchers have shown that one of the missense substitutions appears to be fixed in Neanderthals, while the other substitution along with the *Alu* insertion were polymorphic. The two Neanderthal haplotypes carrying the *PGR* gene entered AMHs, and present-day carriers of the Neanderthal haplotypes express higher levels of the PGR. This likely resulted in more siblings, fewer miscarriages, and less bleeding during early pregnancy, suggesting that the Neanderthal progesterone receptor alleles promote fertility.[44]

Studies by Enard and Petrov[45] found evidence that modern humans inherited new viruses during introgression, some of which were related to influenza, herpes, and HIV. They hypothesized that during the several million years since our ancestors split from other primates, a third of the adaptive changes in our proteins had occurred among those proteins that interact with viruses. They determined that Neanderthals and humans were exposed to novel viruses, and the exchange of adaptive alleles provided resistance against these viruses. They found adaptive segments of Neanderthal DNA in modern Europeans enriched for proteins that interact specifically with RNA viruses. These genes inherited from Neanderthals through interbreeding protected our ancestors from infections just as they protected Neanderthals. Enard and Petrov's studies also demonstrated that retained segments of Neanderthal ancestry could be used to detect ancient epidemics.[46]

While less is known about the adaptive nature of introgressed DNA sequences retained in humans, most introgressed DNA segments from Neanderthals were removed by purifying selection, which pushed introgressed variants to low frequencies resulting in a depletion of introgressed tracts in functional genomic regions, possibly due to hybrid sterility or negative selection. This is because while some Neanderthal gene variants may have been optimal for active lives outdoors in prehistoric Europe, some are no longer beneficial because modern lifestyles and environments are very different. For example, a gene that makes blood clot faster might have been life-saving for individuals hunting with primitive weapons, and who had a short life span. But today that variant boosts the risk of stroke. Overall, Neanderthal gene variants that may have been optimal for active outdoor lives in prehistoric Eurasia may be problematic for contemporary humans, who live mostly indoors and get less sunlight and exercise.[47–49] Not all Neanderthal gene variants were removed by purifying selection, and some detrimental genes remain in AMHs.

By scanning health records of 28,000 Americans, investigators have shown that Neanderthal gene variants may boost the risk of sunburn, depression, skin lesions, and blood clots. Their work reveals a dozen Neanderthal genes that could raise the risk of disease in modern humans.[47,50]

In another study, Dannemann and Kelso[51] at the Max Planck Institute for Evolutionary Anthropology analyzed data from 112,338 Britons from the UK Biobank. They discovered Neanderthal versions of genes that affect how often AMHs get sunburned, what color their hair is, what color their eyes are, and variants that boost the odds that a person smokes (CHRN3 allele), whether they are an evening person rather than a morning person, and whether they are prone to sunburn and depression. These results do not mean that Neanderthal genetic mutations cause these traits in modern humans, or that the genes imparted the same traits in Neanderthals. For example, while one of the Neanderthal mutations is linked with an increased likelihood of smoking, it does not mean that Neanderthals had a cigarette habit.

Kelso and colleagues also found another intriguing pattern. Genetic variants that lead to early puberty also became rare in older age groups, implying that natural selection may have preserved those variants even though they shorten life span because they also boosted fertility.[51]

Studies identified a gene cluster on human chromosome 3, inherited from Neanderthals, as a risk locus for respiratory failure after infection with SARS-CoV-2, which causes the COVID-19 disease. Data collected from 3199 hospitalized COVID-19 patients and controls showed that it is the major genetic risk factor for severe SARS-CoV-2 infection and hospitalization. Investigators found that the version that raises people's risk of severe COVID-19 is the same version found in a Neanderthal who lived in Croatia 50 kya. The investigators showed that the risk is conferred by a genomic segment of approximately 50 kb and occurs at a frequency of about 50% in South Asia, 16% in Europe, and 4% in East Asia.[52] The haplotype harbors variants across its entire 50,000-nucleotide span that are inherited together more than 98% of the time. Long haplotypes can be a result of positive selection, maintained in our genomes because they confer or conferred increased chances of survival and reproductive fitness. In this case, the haplotype was introduced as a result of interbreeding with Neanderthals.

The data indicates that in Bangladesh, individuals of Bangladeshi origin (i.e., native to Bangladesh) were twice as likely to die from COVID-19 as non-native members of the population, although other risk factors could contribute to their statistics. In Bangladesh, 63% of people carried at least one copy of the Neanderthal variant. Studies that analyzed data from 40% of all people admitted to the hospital with COVID-19 in England, Scotland, and Wales between February and May 2020 concluded that people from a South Asian background had a mortality rate 20% higher than the white population. Ethnic minorities in hospital with COVID-19 were more likely to be admitted to critical care and receive ventilation than whites, despite similar disease severity on admission, similar duration of symptoms, and being younger with fewer comorbidities. South Asians are at greater risk of dying, due at least in part to a higher prevalence of preexisting diabetes.[53]

Harrison et al.[53] speculate that the Neanderthal SARS CoV-2 variant haplotype might have been protective against other ancient pathogens, and therefore positively selected for in certain populations. One possibility is that the Neanderthal version originally provided a benefit, perhaps by providing an immune response that worked against ancient viruses. But when individuals are infected with the SARS-CoV-2 coronavirus, the protective immune response mediated by these ancient genes might be overly aggressive, leading to uncontrolled attacks by an individual's own immune system that end up scarring their lungs, causing inflammation and leading to a potentially fatal immune response observed in people who develop severe COVID-19 symptoms. As a result, a haplotype that at times in our past might have been beneficial for survival could now be having an adverse effect.[54]

Drift

Introgression introduces new genetic variation, but as long as the level of introgression is low, a rare allele introduced by introgression will not usually segregate at high frequencies. To increase the frequency, genetic drift or selection must act on the allele to increase its frequency. Thus, allele frequency differentiation methods can distinguish between selection and pure genetic drift.

Drift can constantly remove haplotypes by chance from large populations, and subsequently recombination will decrease LD. In small populations, drift can lead to LD by parents passing on blocks of linked alleles that, by chance, result in more offspring due to favorable combinations being passed to subsequent generations. Additionally, LD can reach an equilibrium value in these small populations. Through drift, a block of DNA may increase in frequency in the population.[55] In addition, it is sometimes impossible to distinguish between an older, larger haplotype and one which has been reassembled through recombination, which is one reason why the use of autosomal haplotypes to reconstruct ancestry is difficult using autosomal "haplotypes" which can be broken and reassembled. In small populations, genetic drift can cause rates of recombination between highly related haplotypes to rise. As bottleneck events are often found associated with human migrations, such reconstructions may have become more common at times when modern humans encountered archaic populations as they migrated from Africa throughout Eurasia and beyond.

Balancing selection

It is challenging to identify introgressed sequences that persist in humans due to balancing selection. Admixture and balancing selection both generate increased within-population variability and deep coalescent genealogies, which make it difficult to determine if one or both events influence a particular region of the genome. One study that tried to look for adaptive introgression via balancing selection looked at highly polymorphic human leukocyte antigen (HLA) class I, which are vital immune system components subject to strong balancing selection. The data from Abi-Rached et al. showed how modern humans acquired the HLA-B*73 allele that is in strong linkage disequilibrium with variants that are present in the Denisovan genome and currently present in West Asia through admixture. They identified the introgressed allele by looking for long *HLA* haplotypes that showed a large number of differences with other haplotypes at the same locus, that had high sequence homogeneity in individuals with the haplotypes, and that showed extreme LD in non-African populations. The investigators argued that these are unusual features that are unlikely to be predicted exclusively by balancing selection: even though this type of selection generally acts to preserve sequence diversity, it should not produce long homogenous haplotypes that are deeply divergent from other haplotypes, especially in the HLA region, which is known to undergo rapid diversification by recombination. Therefore, the adaptive introgression may not have been due to balancing selection.[32,56]

Kelso and others looked for links between Neanderthal DNA and traits in genotyped data from 500,000 people from the UK Biobank data set, which was released in 2017. They discovered Neanderthal alleles that help cause baldness, help cause mental illness, boost certain immune functions, and help explain why AMH heads are round and not shaped like Neanderthal skulls.[57]

Comparing three Neanderthal genomes, Mafessoni et al.[58] found that the number of changes in genes expressed in the striatum of the brain during adolescence was remarkable. They speculated that these changes accumulated in Neanderthals due to their small population size, possibly in combination with a relaxation of selection on genes expressed in the striatum. These genes expressed in the striatum overlap more frequently than expected with genomic regions where Neanderthal introgressed fragments in modern human genomes are rare. Mafessoni et al. speculated that striatal genes may carry Neanderthal-specific changes that were disadvantageous when introduced into modern humans. This, in addition to positive selection for derived changes in the modern human lineage, may underlie regions that are depleted of Neanderthal ancestry in modern-day humans, so-called Neanderthal deserts.[58]

How humans both dominated and interacted with other species

Since the beginning of recorded history, human groups have often had an aversion to dissimilarities in other human cultures, languages, and religious beliefs. These aversions target particular populations, social groups, or individuals. In many cultures, these prejudices can include skin color, gender, race, age, sexual orientation, nationality, socioeconomic status, weight, height, and intellect. These discriminatory practices have led to violent behavior, major and minor conflicts, and the suppression and/or domination of many groups.

Looking back at the first encounters between Neanderthals and anatomically modern humans (AMHs), there must have been episodes of intolerance. Based on torso and facial features alone (prominent brow ridges, a large wide nose, prognathic jaw, barrel-shaped ribcage, large hands and feet, short stature, no prominent waistline), it would have been clear that Neanderthals were not members of an AMH group. How then did humans interact with other human groups and eventually interbreed with Neanderthals? Why did other human groups disappear while AMHs continued to thrive and ultimately dominate?

Over the last 300 kya, *Homo sapiens* shared the planet with at least four other human species; however, we were the only ones to survive. Some reasons for our perceived dominance include: we were the best hunters, the smartest, and the most technologically savvy. However, it has become apparent over the last several decades that some of the other human species were more or equally technologically advanced, had been around for much longer, and had brains as big as or larger than ours. For example, archeological evidence indicates that Neanderthals were stronger, highly skilled with weapons, capable of hunting very large mammals, adapted to harsher weather conditions, acquired the ability for speech,[59] and had evolved a striatum with unique functions involved in cognition and reward processes in the brain.[58] There is increasing evidence that they buried their dead, cared for their sick and injured, and adorned themselves with makeup and jewelry. Given this data, some predictions of which human species would have survived favor Neanderthals.[60,61]

Many investigators now believe that what allowed *Homo sapiens* to thrive was our ability to interact peacefully with each other—our ability to interact and work together with people that we know well or with strangers. Some researchers believe this ability develops before we can walk or talk, based on child observations, and was the gateway to a sophisticated social and cultural world, allowing us to inherit the knowledge of prior generations. In evolutionary terms, the definition of friendliness relates to positive behaviors, either intentional or unintentional, toward others.[61]

The idea is that this "friendliness" may have evolved in *H. sapiens* through natural selection, by a process known as self-domestication. Although we typically perceive domestication as something we achieved by interacting with animals and artificial selection, domestication can also occur through natural selection. It is difficult to study how friendliness or self-domestication might have evolved, because it is a behavior that is impossible to measure quantitatively. However, some investigators believe that domestication in humans can be visualized in a similar fashion as with other animals—that is, over time, animal domestication results in physiological and morphological changes (shape of the face, hormone levels, reproductive cycles, behavior, and the nervous system).

The self-domestication hypothesis predicts that if *H. sapiens* were self-domesticated, we should find evidence of selection for friendliness in the Pleistocene. Since behavior cannot be determined by speculating on human interactions, investigators have turned their attention to the study of neurohormones that regulate behavior and that shape our skeletons. This allows scientists to trace changes through paleoanthropological specimens. For example, the more testosterone you have during puberty, the thicker your brow ridge and the longer your face, so that men typically have longer faces than women. In addition to shaping skeletal features, testosterone and its interactions with other hormones modulate aggressive behavior.[61]

Investigators found that *H. sapiens*, prior to 80 kya, had longer faces and much larger brow ridges than in the Late Pleistocene. On average, more recent skulls had a 40% reduction in how far their brow ridges projected from the face, and faces were 10% shorter and 5% narrower than older skulls. This pattern continued so that faces of modern hunter-gatherers and agriculturalists grew more delicate in appearance, suggesting a decrease in testosterone. In addition, serotonin may have promoted a set of changes that led to smaller brains and less aggression. Suggestive increases in serotonin appear early on during domestication and may also have been involved in skull development.[61,62]

Signs of *H. sapiens* technology and progress appeared before 80 kya; however, it was not until after that time that the fossil record suggests complex cultural traditions and technologies. These milestones started to spread across long distances, leading to cultural innovations that could be shared at greater speed. From 50 kya onward, AMHs began to leave evidence of expanding social networks and cultural expertise that seem to be related to cooperative communication and social tolerance. The benefits of social interactions would have included the ability to solve problems better than individuals could on their own, resulting in the ability to share knowledge across generations that produced the technology and culture that allowed *H. sapiens* to survive. It may have also given AMHs the edge needed to outlast other hominins.[61]

This does not assume that all interactions with other human groups were friendly. Some of the same hormonal changes that underlie friendliness also support violence and intolerance. For example, oxytocin is crucial to compassionate parental behavior; however, the same hormone increases in a mother at birth and sustains the rage she feels when someone threatens her newborn. Oxytocin is also implicated in some forms of male aggression and compassion.

This suggests that as our species was shaped by self-domestication, our increased friendliness also brought new forms of aggression. Group members bonded with one another, and the bonds among them were so strong that the individuals often came with a willingness to defend unrelated group members aggressively. Humans became more violent when members of the group were threatened.

AMH interactions with Neanderthals and other human groups may have become violent due to aversion to dissimilarities in appearance or cultural practices, or disputes over resources. In these instances, introgression of AMHs and Neanderthals may have occurred as a result of rape and/or kidnapping as one group dominated the other and acquired captives. These captives may have been turned into slaves or, in the case of young captives, integrated into the group where they later introgressed into the general population.

In an alternative scenario, AMH peaceful interactions with Neanderthals and other human groups may have come when groups were developing, and there was an acceptance of others with different physical or cultural characteristics. In this case, individuals must have perceived that those outside the group did not pose a threat. This form of self-domestication would have involved not only close physical proximity while group size expanded but also an ability to read rapidly the intentions of those outside the group as they were integrated into the group.

Whether introgression occurred forcefully or peacefully, it most likely occurred in rare instances. Neanderthals lived in relatively small groups depending on their location. Siberian Neanderthals lived in relatively isolated populations of fewer than 60 individuals, whereas Neanderthals from Europe and AMHs seem to have lived in populations of more than 100 individuals. This is reflected in the analysis of introgressed Neanderthal DNA, where European Neanderthal populations contributed more DNA to present-day populations than Siberian Neanderthals.[58]

Rare instances of introgression may have resulted in some hybrid individuals that may have been shunned, abused, or killed because of their differences. Some hybrids may have been sterile or otherwise unable to reproduce. Today, DNA evidence suggests that only hybrid females from Neanderthal male-human female mating may have survived, since there is no evidence of Neanderthal Y chromosomes or Neanderthal mtDNA in AMHs. In support of rare Neanderthal-human mating encounters, studies by Currat and Excoffier[63] used a model of admixture during a spatial expansion to study the hybridization of Neanderthals with modern humans during their spread out of Africa. They found that observed low levels of Neanderthal ancestry in Eurasians are compatible with the rate of interbreeding less than 2%, potentially attributable to a very strong avoidance of interspecific mating, low fitness of hybrids, or both. Their results suggest the presence of very effective barriers to gene flow between the two species. Although the exact demography of the Paleolithic populations is uncertain, their results are compatible with the observed lack of Neanderthal mtDNA and Y chromosome introgression.[63]

Stone age people avoided incest

Early modern humans were not just having sex with Neanderthals; they were also reproducing with each other. A study by Sikora et al.[64] suggests that ancient humans lived in large, complex social networks which helped them swap genes and ideas, but did not practice incest.

The study analyzed the DNA of one man, two children, and a hollowed-out thigh bone, all buried at a 34 kyo site called Sunghir, in Russia. The man was buried by himself, the children buried together, with the hollowed-out thigh bone placed next to their bodies.

Sikora et al.[64] found that none of the individuals were related, and they were genetically distinct, revealing they were at most second cousins. Their genomes showed the same, or even more, genetic diversity than present-day populations of hunter-gatherers in Africa and Amazonia. This suggests that they had pretty similar social networks 34 kya as we do today: limited family relationships within a group, avoiding incest, and mating outside that group.

Given the genetic diversity, Sikora et al. calculated that about 300 people would need to be having sex with each other to produce so much genetic variation. That is a much bigger population than most researchers anticipate would have aggregated during those times; however, it is possible that smaller groups were embedded in a larger social network spread out geographically where people had sex and could exchange resources and ideas.

For several years, investigators believed that Neanderthals were incestuous most likely due to their small groups found in caves and burial sites. Initial studies in 2013 presented a high-quality genome sequence of a Neanderthal woman from Siberia. Studies showed that her parents were related at the level of half-siblings and that mating among close relatives was common among her recent ancestors.[65]

The bones found in the El Sidrón cave located in the limestone hills of the Asturias region of northern Spain reveal something of the sex life of the cave's inhabitants. Analysis of the skeletal remains and mtDNA of a group of 13 Neanderthals that date to between 50 and 47 kya supports the hypothesis that the 13 individuals represented a family group. Seven of the 13 individuals share the same mtDNA haplotype, and 3 of the 4 adult females have different mtDNA lineages. The younger juvenile and the infant share mtDNA with one of the adult females, and thus they were likely her children. Genetically, the three adult males in the group were closely related enough to be brothers, cousins, or uncles, while the four adult females in the group came from three distinct genetic lines. Thus, the men were all closely related, but the women were from outside the group. That suggests this Neanderthal family practiced a patrilocal residence pattern. Other evidence of close relatedness includes dental anomalies and congenital anomalies represented by structural malformations of various body parts that occur while an individual is developing in the womb.[66]

Genome analyses of 19 bones from a cave in Croatia, dating back to 34 kya, indicated that Neanderthals, who would have bred with modern humans, showed no signs of incestuous relationships. Matings between close relatives most likely occurred in small isolated groups but were probably rare.[65]

Introgression inside Africa

Introgression from Neanderthals and Denisovans in modern humans outside Africa has been well established. However, until recently the contribution of archaic hominins to the genetic variation of present-day Africans was poorly understood. Durvasula and Sankararaman[67] provided evidence for archaic introgression into four West African populations. Their analyses indicated that these populations derive 2%–19% of their genetic ancestry from an archaic population that diverged before the split of Neanderthals and modern humans. They accomplished this by building genome-wide maps of archaic ancestry in the Yoruba and the Mende populations that revealed segments of archaic ancestry at high frequency in these populations, representing potential targets of adaptive introgression. Their results show the substantial contribution of archaic ancestry that helped shape the gene pool of present-day West African populations.[67]

In another study, investigators detected Neanderthal ancestry in Africans[68] through a method known as IBDmix, which utilizes identity by descent (IBD). IBD identifies a section of DNA from two individuals that is identical because those individuals once shared a common ancestor. The length of the IBD segment depends on how long ago those individuals shared a common ancestor. For example, siblings share about 50% of their DNA, resulting in long IBD segments because their shared ancestor (a parent) is only one generation removed. Alternatively, fourth cousins share shorter IBD segments because their shared ancestor is several generations removed.

The investigators used IBD to identify Neanderthal DNA in the human genome by distinguishing sequences that look similar to those of Neanderthals because we once shared a common ancestor approximately 500 kya. They distinguished DNA segments from those that look similar because we interbred with Neanderthals more recently, around 50–60 kya. Previously African populations were believed to carry little or no Neanderthal DNA, but studies were hindered by using a reference population to determine ancestry. IBDmix does not use an African reference population but instead uses characteristics of the Neanderthal sequence itself, like the frequency of mutations or the length of the IBD segments, to distinguish shared ancestry from recent interbreeding. The researchers were therefore able to identify 0%–0.3% Neanderthal ancestry in Africans for the first time. In addition, their results refined the understanding of Neanderthal ancestry in non-African populations and demonstrate that remnants of Neanderthal genomes survive in every modern-day human population studied.[37]

Were there early interactions between Neanderthals and anatomically modern humans?

There have been conflicting genome studies of global populations with some suggesting one AMH migration from Africa, while other studies have concluded that there were multiple waves of migration out of Africa. Studies also differ with regard to the dates when AMHs migrated and settled in regions outside of Africa.

Some of the evidence for a single migration comes from the Simmons Genome Diversity Project, which looked at 300 individuals from 142 diverse populations. The data generated in these studies indicated that the ancestors of some pairs of present-day human populations were substantially separated by 100 kya. Additionally, indigenous

Australians, New Guineans, and Andamanese do not derive substantial ancestry from an early dispersal of AMHs; instead, their ancestry is consistent with coming from the same source as other non-Africans.[11,69]

Studies by Pagani et al.[20] used the Estonian Biocentre Human Genome Diversity Panel, a dataset of 483 high-coverage human genomes from 148 worldwide populations, including 379 new genomes from 125 populations. Using this extensive data, researchers examined continent-wide patterns of heterozygosity, long- and short-distance gene flow, archaic admixture, and changes in effective population size through time, as well as looking for signals of positive or balancing selection. The investigators found a genetic signature in present-day Papuans that suggested at least 2% of their genome originated from an early and largely extinct expansion of AMHs out of Africa, contributing to the evidence for the presence of AMHs out of Africa earlier than 75 kya.

Early modern human remains at the sites of Qafzeh and Skhul in present-day Israel signal an occupation that likely penetrated Arabia via the Levant. Unlike the Jebel Faya site where only lithic material was discovered, the Qafzeh and Skhul cave digs are made up of fossils and artifacts dating to 130–90 kya.[70] Significantly, these early human remains were discovered in close proximity to Neanderthal fossils. Yet it is not clear to what extent these two species were interacting at these locations. It is not evident, for example, whether the disappearance of these early modern humans from the Levant was the result of direct competition with Neanderthals, assimilation into a common gene pool, and/or retreat back to Africa. At these Levant sites, the Neanderthals and early modern humans shared the Levallois technique for making tools as well as cave habitats. It is not possible to determine whether individual blades and triangular flakes were made by Neanderthals or early humans, thus preventing meaningful comparisons between modern human artifacts in the Levant and Northeast Africa. What is clear about this cohabitation is that no human presence is seen in the Levant after the period of the Qafzeh and Skhul occupation. Fossil evidence of modern humans is not seen again in the region until 50 kya due to the resettlement of Southwest Asia.[71] In addition, as with the Jebel Faya settlement, it is not clear whether the Levantine human population was able to migrate further east into Asia. The Nile passageway was possibly used by modern humans from Northeast Africa into the Levant about 50 kya, and could have led to the settlement of Southeast Europe and eventually Western Europe about 35–40 kya.[72] This gap in the fossil record between 90 and 50 kya is assumed to reflect the demise of humans in the region or their relocation, but may also be due to the limited number of fossil remains found in the region.

Another one of the first encounters between AMHs and Neanderthals may have come in Israel's Misliya Cave, where investigators found animal bones, tools, and human jaw and teeth bones dated between 177 and 194 kya. These are some of the oldest AMH fossils ever found outside Africa and suggest that *Homo sapiens* might have spread to the Arabian Peninsula around 180 kya. It is unclear if the fossils represented a brief incursion and were part of a population that ultimately died out or a more lasting expansion of the species that reproduced and spread to other parts of the world. If these early AMHs did survive, it might indicate that people were moving through this land corridor from one continent to another, consistently occupying the region, and they most likely encountered and interbred with Neanderthals before 200 kya.[73]

In addition to AMH fossils found in Israel and China, recent studies reexamined European fossilized human crania uncovered in the 1970s from Apidima Cave in southern Greece. New dates of origin were determined using U-series radiometric methods and determined that the Apidima #2 fossil was 170 kya, and the Apidima #1 fossil was 210 kya. Reconstructions indicated that Apidima 2 had a Neanderthal-like morphological pattern, while Apidima 1 showed a mixture of modern human and primitive features. These results insinuate that an early *Homo sapiens* population, followed by a Neanderthal population, were present at the site. These findings support the idea of multiple AMH dispersals out of Africa and possible Neanderthal-AMH interactions very early during AMH evolution. The researchers concluded that the findings highlight the complexity of populations that characterized Pleistocene human evolution and AMHs in Southeast Europe.[74]

Studies from Kuhlwilm et al.[75] also support early AMH migration out of Africa. They examined genomes of a Neanderthal and a Denisovan from the Altai Mountains in Siberia together with sequences of chromosome 21 of two Neanderthals from Spain and Croatia. Their results indicate that a population which diverged early from other modern humans in Africa contributed genetically to the ancestors of Neanderthals from the Altai Mountains. They determined that the genetic contribution of AMHs to Altai Mountain Neanderthals occurred around 100 kya. They did not detect a similar genetic contribution in the Denisovan or the two European Neanderthals, which led them to conclude that in addition to later interbreeding events, Neanderthals from the Altai Mountains and early modern humans met and interbred, possibly in the Near East, thousands of years earlier than in Europe.[75]

Other evidence that AMHs and Neanderthals encountered each other earlier than previously thought comes from studies of mtDNA. The closer affinity of the Neanderthal mtDNA to modern humans than to Denisovans suggests that it is a result of gene flow from an African source into Neanderthals more than 100 kya. Posth et al.[14] reported the complete mtDNA sequence of a Neanderthal femur from the Hohlenstein-Stadel cave in southwestern Germany that carried the deepest divergent mtDNA lineage that split from other Neanderthals about 270 kya. This finding provided a lower boundary for the time of the putative mtDNA introgression event. Their data suggested that a complete Neanderthal mtDNA replacement is feasible over the proposed 270 ky interval, even with minimal hominin introgression. They suggested that the highly divergent branch is indicative of greater mtDNA diversity during the Middle Pleistocene.[14,75]

Conclusion

One of the major questions in human evolution is whether introgression occurred between humans and archaic hominids inside and/or outside of Africa. Evidence for human evolution inside Africa suggests that humans evolved in various regions of Africa and most likely introgressed with other hominids, leading to AMHs. Although the origin and migration of humans out of Africa are still debated, it is clear that humans and Neanderthals shared a common African ancestor that once split and evolved in parallel with humans in Africa, and Neanderthals in Eurasia. Evidence for introgression outside of Africa was suggested by several fossils that appeared as intermediate forms between AMHs and other hominid species. When the Neanderthal genome was sequenced, comparisons of Neanderthal and non-African genomes suggested there was introgression with Neanderthals in non-African AMHs. Studies concluded that the Neanderthal DNA evidence was not linked to ancient population structure but instead was the result of human-Neanderthal introgression. Studies have since revealed that 2%–4% of the genome of non-Africans comes from Neanderthals, suggesting a more complex scenario of modern human evolution than previously thought. Most data points to introgression occurring 50–60 kya, although there is clear evidence that AMHs occupied regions in the Levant 150 kya where there is evidence of Neanderthal occupation. Studies have suggested that the relatively low levels of Neanderthal ancestry in Europeans and Eurasians are compatible with a low occurrence of interbreeding. This may be attributable to a strong avoidance of interspecific mating, a low fitness of hybrids, or both, suggesting the presence of effective barriers to gene flow between the two species. Once introgression had occurred, Neanderthal genes may have been a burden on modern humans' health or made it harder to reproduce. As a result, Neanderthal genes became rarer, and most disappeared from our gene pool, including Neanderthal mtDNA and the Neanderthal Y chromosome.

Throughout recorded human history, there have been numerous examples of intolerance among humans leading to discrimination as well as minor and major conflicts. Looking back at the first encounters between Neanderthals and AMHs, there must have been episodes of intolerance. Based on torso and facial features alone, it would have been clear that Neanderthals were not members of an AMH group. These encounters may have resulted in violence and rape, or they may have been sociable. While introgression from Neanderthals has been documented in AMHs outside of Africa, the contribution of archaic hominins to the genetic variation of modern Africans was unclear. Today, data has shown that archaic introgression occurred in at least four West African populations that diverged before the split of Neanderthals and modern humans. DNA evidence indicates that Neanderthal ancestry in modern-day Africans is due to migrations back to Africa, predominately from ancestral Europeans, or gene flow into Neanderthals from an early dispersing group of humans out of Africa.

References

1. Perez K. https://www.goodreads.com/quotes/1091791.
2. McDougall I, et al. Stratigraphic placement and age of modern humans from Kibish, Ethiopia. *Nature.* 2005;433:733–736.
3. Roksandic M, et al. Revising the hypodigm of Homo heidelbergensis: a view from the Eastern Mediterranean. *Quant Intern.* 2018;466:66–81.
4. Welker F, Ramos-Madrigal J, Gutenbrunner P, et al. The dental proteome of Homo antecessor. *Nature.* 2020;580:235–238. https://doi.org/10.1038/s41586-020-2153-8.
5. Villa P, Roebroeks W. Neandertal demise: an archaeological analysis of the modern human superiority complex. *PLoS ONE.* 2014;9, e96424.
6. Currat M, Excoffier L. Modern humans did not admix with Neanderthals during their range expansion into Europe. *PLoS Biol.* 2004;2, e421.
7. Bräuer G. The origin of modern anatomy: by speciation or intraspecific evolution? *Evol Anthropol.* 2008;17:22–37. https://doi.org/10.1002/evan.20157.
8. Stringer CB. Modern human origins: progress and prospects. *Philos Trans R Soc Lond Ser B Biol Sci.* 2002;357:563–579. https://doi.org/10.1098/rstb.2001.1057.
9. Stringer C. Why we are not all multiregionalists now. *Trends Ecol Evol.* 2014;29:248–251.
10. Lachance J, et al. Evolutionary history and adaptation from high-coverage whole-genome sequences of diverse African hunter-gatherers. *Cell.* 2012;150:457–469. https://doi.org/10.1016/j.cell.2012.07.009.
11. Malaspinas A-S, et al. A genomic history of Aboriginal Australia. *Nature.* 2016;538:207–214.
12. Marta ML. The shaping of human diversity: filters, boundaries and transitions. *Philos Trans R Soc B Biol Sci.* 2016;371(1698). https://doi.org/10.1098/rstb.2015.0241, 20150241.
13. Chris S. The origin and evolution of Homo sapiens. *Philos Trans R Soc B Biol Sci.* 2016;371(1698). https://doi.org/10.1098/rstb.2015.0237, 20150237.
14. Posth C, et al. Deeply divergent archaic mitochondrial genome provides lower time boundary for African gene flow into Neanderthals. *Nat Commun.* 2017;8:16046.
15. Mounier A, Lahr M. Deciphering African Late Middle Pleistocene Hominin diversity and the origin of our species. *Nat Commun.* 2019;10(1):3406. https://doi.org/10.1038/s41467-019-11213-w. 31506422.
16. Liu W, Martinón-Torres M, Cai Y, et al. The earliest unequivocally modern humans in southern China. *Nature.* 2015;526:696–699. https://doi.org/10.1038/nature15696.
17. Smith TM, et al. Earliest evidence of modern human life history in North African early Homo sapiens. *Proc Natl Acad Sci U S A.* 2007;104:6128–6133.
18. Hublin J-J. New fossils from Jebel Irhoud, Morocco and the pan-African origin of Homosapiens. *Nature.* 2017;546:289–292.
19. Scerri EML, et al. Did our species evolve in subdivided populations across Africa, and why does it matter? *Trends Ecol Evol.* 2018;33:582–594. https://doi.org/10.1016/j.tree.2018.05.005. PMC 6092560. PMID 30007846.
20. Pagani L, et al. Genomic analyses inform on migration events during the peopling of Eurasia. *Nature.* 2016;538:238–242. https://doi.org/10.1038/nature19792.
21. Daura J, et al. New Middle Pleistocene hominin cranium from Gruta da Aroeira (Portugal). *Proc Natl Acad Sci U S A.* 2017;114:3397–3402. https://doi.org/10.1073/pnas.1619040114.
22. Meyer M, et al. Nuclear DNA sequences from the Middle Pleistocene Sima de los Huesos hominins. *Nature.* 2016;531. https://doi.org/10.1038/nature17405.
23. Weaver TD, et al. Stringer close correspondence between quantitative- and molecular-genetic divergence times for Neandertals and modern humans. *Proc Natl Acad Sci U S A.* 2008;105:4645–4649.
24. Endicott P, et al. Using genetic evidence to evaluate four palaeoanthropological hypotheses for the timing of Neanderthal and modern human origins. *J Hum Evol.* 2010;59:87–95.
25. Arnold LJ, et al. Luminescence dating and palaeomagnetic age constraint on hominins from Sima de los Huesos, Atapuerca, Spain. *J Hum Evol.* 2014;67:85–107.
26. Meyer M, et al. A mitochondrial genome sequence of a hominin from Sima de los Huesos. *Nature.* 2014;505:403–406.

27. Gomez-Robles A. Dental evolutionary rates and its implications for the Neanderthal–modern human divergence. *Sci Adv*. 2019;5, eaaw1268.
28. Mendez FL, et al. The divergence of Neanderthal and modern human Y chromosomes. *Am J Hum Genet*. 2016;98:728–734.
29. Prufer K, et al. The complete genome sequence of a Neanderthal from the Altai Mountains. *Nature*. 2014;505:43–49.
30. Fu Q, et al. An early modern human from Romania with a recent Neanderthal ancestor. *Nature*. 2015;524:216–219. https://doi.org/10.1038/nature14558.
31. Moorjani P, et al. The history of African gene flow into Southern Europeans, Levantines, and Jews. *PLoS Genet*. 2011;7, e1001373.
32. Racimo F, et al. Evidence for archaic adaptive introgression in humans. *Nat Rev Genet*. 2015;16:359–371.
33. Eriksson A, Manica A. Effect of ancient population structure on the degree of polymorphism shared between modern human populations and ancient hominins. *Proc Natl Acad Sci U S A*. 2012;109:13956–13960. https://doi.org/10.1073/pnas.1200567109.
34. Fu Q, et al. Genome sequence of a 45,000-year-old modern human from western Siberia. *Nature*. 2014;514:445–449.
35. Liang M, Nielsen R. The lengths of admixture tracts. *Genetics*. 2014;197:953–967.
36. Vernot B, Akey JM. Resurrecting surviving Neanderthal lineages from modern human genomes. *Science*. 2014;343:1017–1021.
37. Chen L, et al. Identifying and interpreting apparent Neanderthal ancestry in African individuals. *Cell*. 2020. https://doi.org/10.1016/j.cell.2020.01.012.
38. Wolf AB, Akey JM. Outstanding questions in the study of archaic hominin admixture. *PLoS Genet*. 2018;14, e1007349.
39. Skov L, et al. The nature of Neanderthal introgression revealed by 27,566 Icelandic genomes. *Nature*. 2020;582:78–83.
40. Fay JC, Wu CI. Hitchhiking under positive Darwinian selection. *Genetics*. 2000;155:1405–1413.
41. Mendez FL, Watkins JC, Hammer MF. A haplotype at STAT2 Introgressed from Neanderthals and serves as a candidate of positive selection in Papua New Guinea. *Am J Hum Genet*. 2012;91:265–274.
42. Ding Q, et al. Neanderthal introgression at chromosome 3p21.31 was under positive natural selection in East Asians. *Mol Biol Evol*. 2014;31:683–695.
43. Sankararaman S, et al. The genomic landscape of Neanderthal ancestry in present day humans. *Nature*. 2014;507:354–357.
44. Zeberg H, et al. The Neandertal progesterone receptor. *Mol Biol Evol*. 2020. https://doi.org/10.1093/molbev/msaa119, msaa119.
45. Enard, et al. Viruses are a dominant driver of protein adaptation in mammals. *elife*. 2016. https://doi.org/10.7554/eLife.12469.001.
46. Enard D, Petrov DA. Evidence that RNA viruses drove adaptive introgression between Neanderthals and modern humans. *Cell*. 2018. https://doi.org/10.1016/j.cell.2018.08.034.
47. Gibbons A. Neandertal genes linked to modern disease. *Science*. 2016;351:648–649. https://doi.org/10.1126/science.351.6274.648.
48. Barton NH. Genetic linkage and natural selection. *Philos Trans R Soc B*. 2010;365:2559–2569.
49. Ennis S. Linkage disequilibrium as a tool for detecting signatures of natural selection. *Methods Mol Biol*. 2007;376:59–70.
50. Simonti CN, et al. The phenotypic legacy of admixture between modern humans and Neanderthals. *Science*. 2016;351:737–741.
51. Dannemann M, Kelso J. The contribution of Neanderthals to phenotypic variation in modern humans. *Am J Hum Genet*. 2017;101:578–589.
52. Zeberg H, Pääbo S. The major genetic risk factor for severe COVID-19 is inherited from Neanderthals. *Nature*. 2020. https://doi.org/10.1038/s41586-020-2818-3.
53. Harrison EM, et al. Ethnicity and outcomes from COVID-19: the ISARIC CCP-UK prospective observational cohort study of hospitalised patients. *SSRN*. 2020. https://doi.org/10.2139/ssrn.3618215. https://ssrn.com/abstract=3618215.

54. Luo Y. Neanderthal DNA raises risk of severe COVID. *Nat News*. 2020. https://doi.org/10.1038/d41586-020-02957-3.
55. Ohta T. Linkage disequilibrium due to random genetic drift in finite subdivided populations. *Proc Natl Acad Sci U S A*. 1982;79:1940–1944.
56. Abi-Rached L, et al. The shaping of modern human immune systems by multiregional admixture with archaic humans. *Science*. 2011;334(6052):89–94. https://doi.org/10.1126/science.1209202.
57. Gibbons A. Genetic data on half a million Brits reveal ongoing evolution and Neanderthal legacy. *Science*. 2019. https://doi.org/10.1126/science.aaw5042.
58. Mafessoni F, et al. A high-coverage Neandertal genome from Chagyrskaya Cave. *Proc Natl Acad Sci U S A*. 2020;117:15132–15136.
59. Wu SS, et al. Functional heterogeneity of inferior parietal cortex during mathematical cognition assessed with cytoarchitectonic probability maps. *Cereb Cortex*. 2009;19:2930–2945.
60. Hare B. Survival of the friendliest: *Homo sapiens* evolved via selection for Prosociality. *Annu Rev Psychol*. 2017;68:155–186.
61. Hare B, Woods V. Humans evolved to be friendly. *Sci Am*. 2020;58–63.
62. Cieri RL, et al. Craniofacial feminization, social tolerance, and the origins of behavioral modernity. *Curr Anthropol*. 2014;55(4):419. https://doi.org/10.1086/677209.
63. Currat M, Excoffier L. Strong reproductive isolation between humans and Neanderthals inferred from observed patterns of introgression. *Proc Natl Acad Sci U S A*. 2011;108:15129–15134. https://doi.org/10.1073/pnas.1107450108.
64. Sikora M, et al. Ancient genomes show social and reproductive behavior of early Upper Paleolithic foragers. *Science*. 2017;358:659–662. https://doi.org/10.1126/science.aao1807.
65. Prufer K, et al. A high-coverage Neandertal genome from Vindija Cave in Croatia. *Science*. 2017;358:655–658. https://doi.org/10.1126/science.aao1887.
66. Weyrich LS, et al. Neanderthal behavior, diet, and disease inferred from ancient DNA in dental calculus. *Nature*. 2017;544:357–361.
67. Durvasula A, Sankararaman S. Recovering signals of ghost archaic introgression in African populations. *Sci Adv*. 2020;6(7). https://doi.org/10.1126/sciadv.aax5097, eaax5097.
68. Willson J. There and back again—ancient genes reveal early migrations. *Nat Rev Genet*. 2020;21:205. https://doi.org/10.1038/s41576-020-0222-3.
69. Mallick S, et al. The Simons genome diversity project: 300 genomes from 142 diverse populations. *Nature*. 2016;538:201–206.
70. Mercier N, et al. Thermoluminescence date for the Mousterian burial site of Es-Skhul, Mt Carmel. *J Archaeol Sci*. 1993;20:169–174.
71. Lahr MM, Foley R. Multiple dispersals and modern human origins. *Evol Anthropol*. 1994;3:48–60.
72. Mellars P. Archeology and the dispersal of modern humans in Europe: deconstructing the Aurignacian. *Evol Anthropol*. 2006;15:167–182.
73. Callaway E. Israeli fossils are the oldest modern humans ever found outside of Africa. *Nature*. 2018;554:15–16.
74. Harvati K, et al. Apidima cave fossils provide earliest evidence of Homo sapiens in Eurasia. *Nature*. 2019;571:500–504. https://doi.org/10.1038/s41586-019-1376-z.
75. Kuhlwilm M, et al. Ancient gene flow from early modern humans into eastern Neanderthals. *Nature*. 2016;530:429–433.

CHAPTER 12

Neanderthal habitat, culture, language, and intellect

The creation of something new is not accomplished by the intellect but by the play instinct acting from inner necessity. The creative mind plays with the objects it loves.

Carl Jung[1]

Human origins and the development of culture

Fossil discoveries over the last 30 years indicate that 1–4 mya there were periods when our ancestors and several close relatives were present at the same time in Africa. During this period, it was not uncommon for different human ancestors to meet and mate in various regions of Africa. Because of these interactions, recent data indicate that the story of *Homo* evolution is far more complex than most researchers realized even 20 years ago.

Genetic and fossil evidence of multiple lineages in the past 100 ky show that Neanderthals, Denisovans, and *Homo floresiensis* are our direct ancestors who shared regions of the earth with several other *Homo* species. Archeological evidence indicates that *H. erectus* probably survived later than was originally thought and interacted with other lineages. In this scenario, introgression most likely occurred between Neanderthals, Denisovans, and some archaic group such as *H. erectus* or one of its contemporaries similar to the fossil that was recently recovered in Harbin city in northeastern China.[2] The DNA from the archaic group(s) was then passed on to modern humans through introgression with Neanderthals and Denisovans.[3]

In addition to Neanderthal and Denisovan introgression with AMHs, evidence has emerged for at least one more "ghost lineage" as evidenced by ancient DNA found in contemporary modern humans from West Africa. This ancient ghost lineage diverged from those of modern humans before the split between Neanderthals and modern humans. These different lineage interactions, and possibly others, not only altered evolutionary pathways through the mixture of genetic material but also likely led to new developments in technology, behavior, and cultural practices during their coexistence.[4]

Evidence from archeological sites and DNA tell a story of Neanderthals and *Homo sapiens* in Europe, Eurasia, and the Middle East who underwent relocation, introgression, migration, and important technological developments. Although many aspects of archaic early lifestyle remain a mystery, through archeological evidence and dating of artifacts it is known

Sex and Cohabitation Among Early Humans
https://doi.org/10.1016/B978-0-12-819086-9.00003-8

that the development of technologies and culture during the Pleistocene were additive: they built on prior trials and knowledge and became increasingly complex over time.[5]

Neanderthal behavior

The types of behavior and cultural practices associated with Neanderthals have been the subject of an ongoing debate. One issue is the term "behavior," which encompasses numerous, sometimes complex topics including but not limited to: art, language, spiritual beliefs, cognition, technology, food acquisition and consumption, introgression, genetics, reproduction, health and safety, and (in the case of Neanderthals) extinction. All of these topics include some aspect of behavior and are difficult to ascertain, because all of the evidence for Neanderthal behavior lies in their skeletal remains, the artifacts they left behind, and their genomes. Some of these sources provide direct evidence for behaviors, whereas others provide indirect evidence, or are open to speculation. Skeletal remains and artifacts can provide direct evidence for use of fire, shelters, diet, environment, use of tools, indications of violence, some diseases, and possible hunting strategies. However, the genome can provide evidence for disease, some external phenotypes, and some physiological characteristics. What are less evident are the group and/or community interactions, water and land navigation, family structure, communication, and day-to-day activities. The cognitive, innovative, and emotional abilities of Neanderthals are almost purely speculative because of the inability of those traits to be fossilized and our lack of understanding of the genetics of these complex traits.

A number of behavioral differences have been noted when comparing modern humans with Neanderthals. With regard to lifestyle, fossil evidence suggests that Neanderthal males, females, and juveniles participated in a narrow range of activities that centered on obtaining large prey. The apparent absence of role differentiation in the Neanderthals is consistent with the distinctive anatomical features of Neanderthal populations, the small groups found at archeological sites, and their perceived lifestyle.

The absence of role differentiation and division of labor could have decreased the competitiveness of Neanderthals. The ability to share decision-making, especially in a democratic process, can be useful in forming social bonds, strengthening group interactions and cooperation, and developing new cultural practices and technologies that can be passed on to future generations.[6]

The cooperative economic systems of AMHs may have given them a demographic advantage over Neanderthals, facilitating the rapid expansion of human culture throughout Eurasia. Kuhn and Stiner[7] argue that cooperative systems are more likely to have evolved first in the tropics or subtropics. They argue that although we take for granted today the division of labor by sex and age, it is not clear when it evolved; however, there are definite advantages to division of labor, and the advantages may have made modern humans more competitive than Neanderthals.

The small Neanderthal groups may have resulted in, or may have been the result of, interbreeding among close relatives, or matrilocal or patrilocal behavior. Sequencing of the hypervariable regions of mtDNA and Y chromosome DNA analyzed from 12 Neanderthal individuals from the El Sidron site in Asturias, Spain revealed low genetic diversity and suggested patrilocal mating behavior.[8] Genetic analysis of a female toe bone excavated from the Altai Mountains revealed that the parents of the woman were closely related, possibly the result of a mating between half-siblings, or a mating between an uncle or an aunt. It is not clear whether inbreeding or patrilocal practices were common among Neanderthals or whether these were special cases that were unavoidable because of the small number of Neanderthals who lived in the area.[9]

Several Neanderthal remains show signs of violence. These include deeply grooved ribs showing signs of healing, misshapen bones, and broken hands and arms.[10] However, it is not clear if the damage was done by humans, by other Neanderthals, by accident, or by animal attacks. There is evidence that some Neanderthals may have practiced cannibalism. The evidence comes from 100 kyo remains from a cave at Moula-Guercy, near the Rhône River. The scattered distribution of Neanderthal bones, burned bones, marks on bones showing de-fleshing, hacked off arms and feet, and splintered bones for marrow extraction indicates consumption of the dead.[6,10] It is not clear if the individuals were killed or just consumed after death, and/or if this was an isolated incident because of limited food resources.[11,12]

Some authors have speculated that child mortality rates may have been high among Neanderthal children because of the number of juvenile remains uncovered. They have associated these high mortality rates with the aggressive lifestyle of Neanderthals and the focus on collecting large prey.[13] Comparisons of Neanderthal remains with modern groups with comparable lifestyles (hunter-gatherers, nomads, etc.) found no evidence of excessive or unusual trauma and no evidence that trauma played a more significant role in the lives of Neanderthals.[5,14] This is not to say that Neanderthals lived without trauma, but it indicates that the trauma they faced may not have been any more severe than that seen in ancient human skeletal remains. Research shows that throughout AMH history, cannibalism was widespread; mass killings, homicides, and assaults have been clearly documented.[15]

Neanderthal tools

The Middle Paleolithic is associated with the evolution of *Homo Neanderthalensis* from *H. heidelbergensis* about 300 kya and their subsequent extinction about 40 kya.[16] Neanderthals occupied Europe for approximately 200 ky, and the most recent radiocarbon dates indicate that the Neanderthals went extinct about 5000 years after the first modern humans arrived in Europe, around 45 kya.[16] The Middle Paleolithic brought about advanced tool-manufacturing techniques and the creation of more elaborate devices,

Fig. 1 Representations of tools used during the Paleolithic (https://www.quora.com/What-are-the-differences-between-the-tools-of-the-Palaeolithic-the-Mesolithic-and-the-Neolithic-ages).

such as the stone-tipped spear, which allowed hunters to access a greater variety of food sources[17,18] (Fig. 1).

Sites where hominin fossils are directly associated with stone tools and other artifacts are rare but critical to determine hominin occupation, group size, organization, and their level of technology, behavior, artifact transport, cultural characteristics, rituals, and beliefs. Fossil remains and extensive archeological discoveries reveal a large variety of innovations belonging to the Neanderthals, including hunting tools, weapons, and leather-burnishing tools made of animal bone. With a few exceptions such as the Schöningen spears and wooden tools at Pogetti Vecchi, almost all of our knowledge about the Middle Paleolithic comes from bones and stone tools; however, new discoveries and methods of analysis have increased our level of understanding regarding the technologies imparted by Neanderthals.[19]

The style of flint implements used during the Middle Paleolithic are associated primarily with Neanderthals, and are referred to as Mousterian (or the Mousterian industry or tradition). The Mousterian is an archeological industry of stone tools, associated primarily with the Neanderthals in Europe, and to the earliest anatomically modern humans in North Africa and West Asia. The name Mousterian comes from the site of Le Moustier, in the Dordogne region of southwestern France, where Neanderthal remains and artifacts were discovered (Fig. 2).

In 1961, Francois Bordes divided the tools into Mousterian points, denticulates (those with multiple notches), backed knives, bifaces, and convex and convergent scrapers.[20] Materials used for stone tools were mostly local, but there is evidence that some of these materials were transported into Western Europe, within 5 km of the source, indicating some foresight and planning, and simultaneously providing a clue to the range of Neanderthal mobility. Since the original discovery, Mousterian tools have been found throughout Europe and parts of Northwest Africa.[21]

Fig. 2 Example of Mousterian points associated with Neanderthals (https://en.wikipedia.org/wiki/Mousterian).

Many investigators believe that Neanderthals are also associated with an early Upper Paleolithic and the Châtelperronian industry which occurred around 35 kya. The evidence comes from two sites in France: Arcy-sur-Cure[22] and Saint-Césaire,[23] where Châtelperronian tools were discovered with Neanderthal remains. The Châtelperronian is similar to the Aurignacian, and since the rest of the Upper Paleolithic is associated with modern humans, the connection with Neanderthals to the Châtelperronian has generated controversy. The controversy is over some evidence that the sites in France may have been contaminated by mixing archeological sediments and/or that Neanderthals acquired or mimicked the technology developed by AMHs.[12]

The first definitive archeological evidence of human settlement on the European continent was discovered in 1964, in the Grotta del Cavallo, in Apulia, Italy. Investigators discovered infant teeth that were later determined to be 43–45 kyo, and because of their unusual hybrid-like structure they were some of the first evidence to suggest that Neanderthals and humans coexisted over thousands of years and interbreeding occurred between the two species. In addition, these interactions may have led to shared tool technology and/or shared development of new technology.

The appearance of anatomically modern humans in Europe and the nature of the transition from the Middle to Upper Paleolithic are matters of intense debate. Most researchers accept that before the arrival of AMHs, Neanderthals had adopted several "transitional" techno-complexes. Two of these are the Uluzzian, a prehistoric stone tool technology used by Neanderthals and adopted by early modern humans from southern Europe, and the Châtelperronian, a blend of tool types from the Middle and Upper Paleolithic from Central and Southwestern Europe (Fig. 3). These industries are key to current interpretations regarding the timing of arrival of AMHs in Europe, their potential interaction with Neanderthals, the cognitive abilities of Neanderthals, and the reasons behind Neanderthal extinction. However, the actual fossil evidence

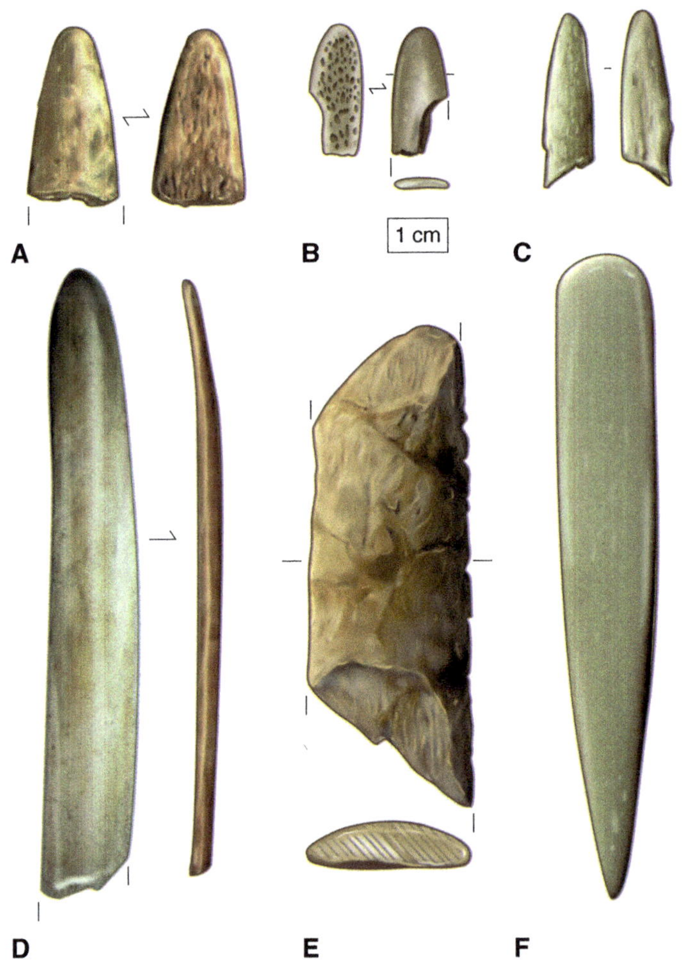

Fig. 3 Tools from the Châtelperronian industry made by Neanderthals representing the earliest Upper Paleolithic industry in central France, southwestern France, and northern Spain (https://www.researchgate.net/publication/255790943_Neandertals_Made_the_First_Specialized_Bone_Tools_in_Europe).

associated with the Uluzzian and Châtelperronian is limited, and recent work has questioned attributing the Châtelperronian industry to Neanderthals on the basis of mixed fossils analysis.[24] Uluzzian artifacts have been unearthed from caves in Italy and Greece and are considered the most likely transitional tool types following the Mousterian, which then spread across Europe before the Aurignacian. Bennazi and colleagues[24] reanalyzed the deciduous molars from the Grotta del Cavallo in Apulia and associated them with the Uluzzian. Using two independent morphometric methods based on microtomographic data, they showed that the specimens can be attributed to AMHs. The teeth provided crucial evidence that the makers of the Uluzzian technology were not Neanderthals. In addition, new chronometric data for the Uluzzian layers of the Grotta del Cavallo obtained from associated shell beads and included in a Bayesian age model show that the teeth must date to 43–45 kya.[25]

However, in the debate over tool technology, later studies by Zilhao J. et al.[26] concluded that the Uluzzian technology could not be attributed to modern humans and that Neanderthals were responsible for the technology. Further excavations and data may eventually resolve the debate.

Neanderthals created tools for domestic uses that were distinct from hunting tools. These tools included scrapers for tanning hides, awls for punching holes in hides to make loose-fitting clothes, and burins for cutting into wood and bone. Tools also included those used to sharpen spears, process animals, and prepare foods. Bone tools shaped from deer ribs, referred to as lissoirs (used today by leather workers to treat animal hides to make them softer, tougher, and waterproof), were discovered at two different sites in Dordogne. The tools were pressed against animal hides judging from the wear marks on the bone similar to modern-day wear. The site where the tools were found dates to around 51 kya, suggesting that they were an independent invention of the Neanderthals or indicating that modern humans influenced Neanderthals earlier than previously believed.[27] Some have speculated that Neanderthals were able to plan their subsistence strategies and adapt their toolmaking to these strategies, indicating that they had the ability to be flexible and solve problems. Although the evidence for Neanderthal use of symbols, ornaments, and sophisticated tools provides some indication of their potential ability for higher-order thought, some argue that Neanderthals might have acquired the technology from their interaction with modern humans.[28] If this were the case, it would indicate that the Neanderthals were capable of mimicry rather than complex thought.

In support of the mimicry theory, recent investigations of artifacts from excavations at Bacho Kiro Cave (Bulgaria) determined that tooth and mitochondrial DNA from bone fragments were from *H. sapiens*. These findings link the expansion of Initial Upper Paleolithic technologies with the spread of *H. sapiens* into the mid-latitudes of Eurasia before 45 kya. The bone artifacts included pendants made from cave bear teeth that are indicative of those later produced by the Neanderthals of Western Europe. These data are consistent with models based on the arrival of multiple waves of *H. sapiens* into Europe interacting with Neanderthal populations and spreading cultural and technological practices.[22]

However, other investigators have argued that Neanderthal tool production was much more sophisticated than previously thought and not merely a case of mimicry. These investigators have argued that Neanderthal tools were much more difficult to produce than Acheulean tools (associated with *H. erectus* and *H. heidelbergensis*), requiring a much higher degree of preparation and planning. Neanderthals used a technique called Levallois flaking whereby a stone core is first prepared and then a larger oval flake is removed and further refined.

Neanderthals, as well as some of their relatives, were creating stone blanks of specific size and shape through a multistage process of core preparation and flake or blade production known as Levallois. Many of the blanks were then subject to edge retouching with additional production steps. The discovery that some blanks were attached to wooden handles or shafts with a binding agent to render hafted or composite artifacts is inferred from traces of hafting wear and adhesives on the stone blanks. Neanderthals made both stone-tipped wooden spears and hafted cutting or scraping tools, and used a variety of adhesives.[29] The total level of complexity, inferred by the discovery of artifacts and the process of tool development, suggests that the Neanderthals were making food-getting artifacts more complex than those of some modern-day hunter-gatherers.[30]

In addition to adhesive technology, recent investigations have uncovered individual twisted fibers on stone tools from the Abri du Maras (located in a valley near the Ardèche River, a tributary of the Rhône River), leading to the conclusion that Neanderthals acquired fiber technology and string production (Fig. 4). The technology was in the form of a three-ply cord fragment made from inner bark fibers on a stone tool recovered from the site. These twisted fibers most likely provided the basis for clothing, rope, bags, nets, mats, boats, etc., which would have become an indispensable part of daily life. The site dates to 90,000 years ago and has also been associated with the consumption of birds, aquatic mammals, and fish.[31] The investigators concluded that the ability to understand and use twisted fibers implies the use of complex multicomponent technology as well as a mathematical understanding of pairs, sets, and numbers. This discovery, in addition to evidence of birch bark tar, art, and shell beads, continues to discredit the idea that Neanderthals were cognitively inferior to their modern human counterparts.[32,33]

For many years, archeological excavations in the Levant produced fossils and stone tool assemblages that revealed that both Neanderthals and *Homo sapiens* occupied the region. This suggested a potential mixing site for the two populations. Until recently, the distinct Nubian Levallois method of tool assemblages found in the region was argued to have been produced only by *Homo sapiens*.

In a recent study by Blinkhorn et al.,[34] investigators re-examined the fossil and archeological record of the Israeli Shukbah Cave. This study marked the first time the human

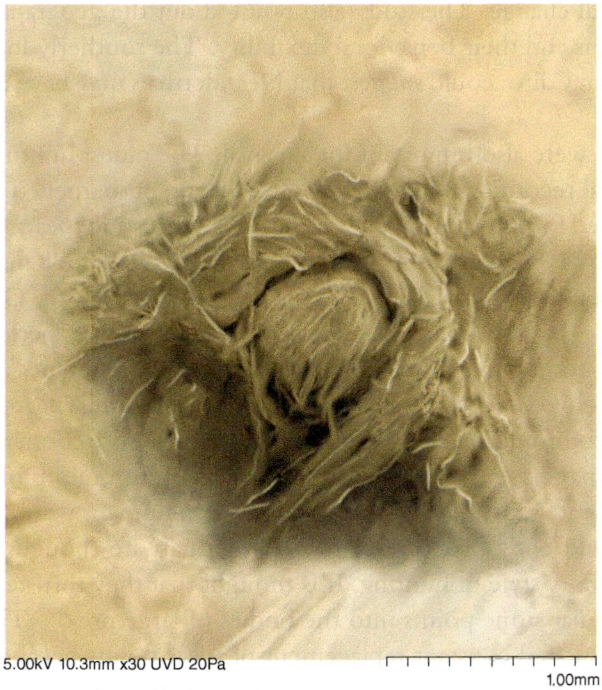

Fig. 4 A scanning electron micrograph of a 50,000-year-old string associated with Neanderthal site in France (https://www.npr.org/2020/04/10/828400733/the-oldest-string-ever-found-may-have-been-made-by-neanderthals).

tooth from the site was studied in combination with a comparative study of the stone tool assemblage.

Shukbah Cave was first excavated in 1928 by Dorothy Garrod, who discovered an assemblage of animal bones, Mousterian-style stone tools cemented in breccia deposits, and what was then thought to be a human molar. Until recently, the specimen was kept in a private collection prohibiting studies using modern methodology. The recent re-examination of the tooth measured the size, shape, and both the external and internal 3D structure. A comparison of the data to Holocene and Pleistocene *Homo sapiens* and Neanderthal specimens concluded that the tooth belonged to an approximately 9-year-old Neanderthal child, extending the southernmost range of Neanderthals.

This study was important because the argument has been the Nubian Levallois technology belonged exclusively to *Homo sapiens* and Nubian Levallois tools have been used to track human dispersals in the absence of fossils. This study suggests that researchers can no longer make a simple link between this technology and *Homo sapiens*. In addition, Southwest Asia could be particularly important to examine interactions between Neanderthals and *Homo sapiens* due to its importance in hominin demography, behavior, and environmental change. The study also pointed out the geographic range of Neanderthal populations and their behavioral flexibility. The southerly location of Shukbah, about 400 km from Cairo, could suggest that Neanderthals may have even entered Africa at times.[34]

Hunting tools were also important for survival. The oldest hunting weapons known from archeological records are 300–400 kyo, consisting of sharpened wooden staves that may have been used for throwing and/or close-range thrusting spears. Similar spears were most likely used by Neanderthals for hunting. Gaudzinski-Windheuser et al.[23] reported that 120 kyo skeletons of two fallow deer, from Neumark-Nord, Germany, contained perforations made by Neanderthals. Detailed studies of the perforations, including micro-computed tomography imaging and ballistic experiments, demonstrated that the perforations resulted from the close-range use of thrusting spears. The authors concluded that such confrontational ways of hunting require close cooperation between participants, which may have shaped important aspects of hominin biology and behavior.

The hunting strategies of Neanderthal and modern humans were different. The more robust anatomy and body proportions of Neanderthals are thought to have allowed a more close-proximity style of hunting. It is thought that they thrusted and forced heavy spears with triangular stone points into the bodies of large prey.[35] This type of hunting would have necessitated constant replacement of spear points because of the force used during the penetration of the prey. Therefore, the expectation was that a number of point replacements would be found at Neanderthal caves. This was indeed the case where stone flakes shaped into various types of tools and spear points are found in caves occupied by Neanderthals. Many of the spearheads and flint tools from this type of archeological culture are Mousterian.

The evidence that Neanderthals used stone-tipped spears to hunt comes from impact scars on stone tips. Retouched Mousterian points from several sites in Western Europe have impact scars that suggest they were probably used at close range.[15] The weapons associated with modern human remains are more fragile wooden spears made for throwing at prey as projectiles from a distance. Neither the stone points found in Neanderthal caves nor the features of Neanderthal anatomy support the notion that they used long-distance projectiles.[36,37]

As with all animals, human individuals act aggressively to attain goals that will maximize their survival and reproduction. Natural selection favors aggression when it facilitates food security and protection of other resources (shelter, tools, etc.). As *H. sapiens* moved into Europe, they encountered Neanderthals and Denisovans, who had occupied the territory for thousands of years. This may have initially led to cooperation but would have ultimately led to competition as AMH populations increased and resources became scarce. Research shows that throughout AMH history, mass killings, homicides, and assaults have occurred.[15]

Early on, when AMHs first left Africa, there would have been a balance of power between Neanderthals and early *H. sapiens* because of similar tools and possibly shared technology (hand axes and spears used for hunting). Until the advent of AMH projectile technology, Neanderthals may have even had a slight advantage in combat because of their extensive knowledge of the region, their muscular physique, and their close-range hunting skills. However, as *H. sapiens* became more gregarious and developed new weapons, the advantage would have swung dramatically in their favor. AMH groups that did not band together and pass on the technology for new weapons may not have succeeded.[35,38]

Neanderthal language

Language is a social invention, meant to serve social or public purposes and convey ideas through sounds or symbols. Language can be thought of as complex and multifaceted when used as a mechanism for discourse, or as a simple method of categorizing objects and basic ideas (similar to what we see in very young children or some mammals). Modern human language is thought to have slowly evolved into a mechanism for discourse through both biological and culture evolution, so any initial linguistic abilities of Neanderthals and early modern humans must have been a very simplistic method of communication. A study that looked at genetics, paleontology, and linguistics concluded that humans and Neanderthals may have shared some components of speech, and proposed not only that Neanderthals used language but also that *H. heidelbergensis* may have also attained some form of speech.[39]

Studies and discussions claiming that Neanderthals engaged in a range of symbolic behaviors, including personal ornaments,[40] body esthetics,[41] funerary practices,[42] and visual arts,[43] have become a foundation to infer that Neanderthals engaged in symbol use and thus possessed some form of language. A critical overview by Botha[44] of the archeological hypotheses and speculations on behavior concluded that even if these behaviors were symbolic, inferring that these symbolic behaviors led to language would be an arbitrary conclusion.

Instead, Botha[44] proposes that specific hunting behaviors may be an indicator of linguistic communication in Neanderthals. For example, he suggests that ambush hunting of large prey may indicate that Neanderthals possessed some form of language because such behaviors required close cooperation between individuals and thus some form of communication.

Stout and Chaminade[45] propose that there are cognitive processes which underlie both stone-tool-related and linguistic behaviors. Further, they suggest that language might rely on processes and neural networks which were reused from processes and neural networks originally involved in stone-tool-related behaviors. In other words, early stone toolmaking may have favored adaptations later incorporated into an evolving language capacity. Studies by Gabric et al.[46] also concluded that toolmaking could have led to the evolution of cognition and language.

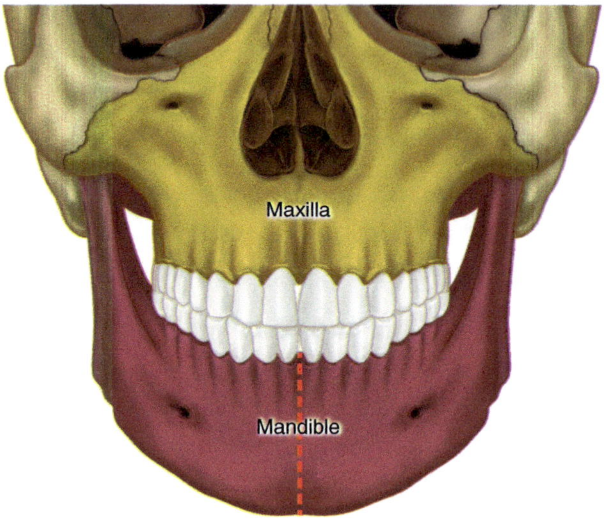

Fig. 5 The bottom of the red stripe (base of the chin) indicates the region of the mental protuberance (https://en.wikipedia.org/wiki/Mandibular_symphysis#/media/File:Symphysis_menti_(Gray190_edit).png).

Initial investigations of Neanderthal remains led several authors to argue that Neanderthals did not have the physical characteristics suitable for speech. However, several anatomical features of Neanderthals suggest that they had the ability for speech. An anatomical feature that may be related to speech is the mental protuberance, which makes up the point of the chin and is responsible for the shape and size of the chin (Fig. 5). The mental protuberance contains the mentalis muscle, which helps to move the lower lip. The mental protuberance has been seen in several Neanderthal fossils.[47] The large Neanderthal brain reflected by endocranial casts resembles that of modern humans in an area important for speech, suggesting Neanderthals had the neural development necessary for language.[48] In addition to brain size, a 3D X-ray imaging of a Neanderthal hyoid bone (horseshoe-shaped bone in the neck) revealed that the bone was indistinguishable from that of modern humans (Fig. 6). The hyoid allows for the proper interactions between the tongue and larynx to articulate sounds, which supports the notion that Neanderthals possessed the ability to develop speech.[49]

A study by Conde-Valverde et al.[50] examined the hearing abilities of humans and Neanderthals. The authors used computerized tomography scans and a comprehensive model from auditory bioengineering to establish sound power transmission through the outer and middle ear, and calculated the occupied bandwidth in Neanderthals. The high-resolution CT scans were used to create virtual 3D models of the ear structures of both humans and Neanderthals. They used this data to model the frequencies that each species could hear the best (Fig. 7).

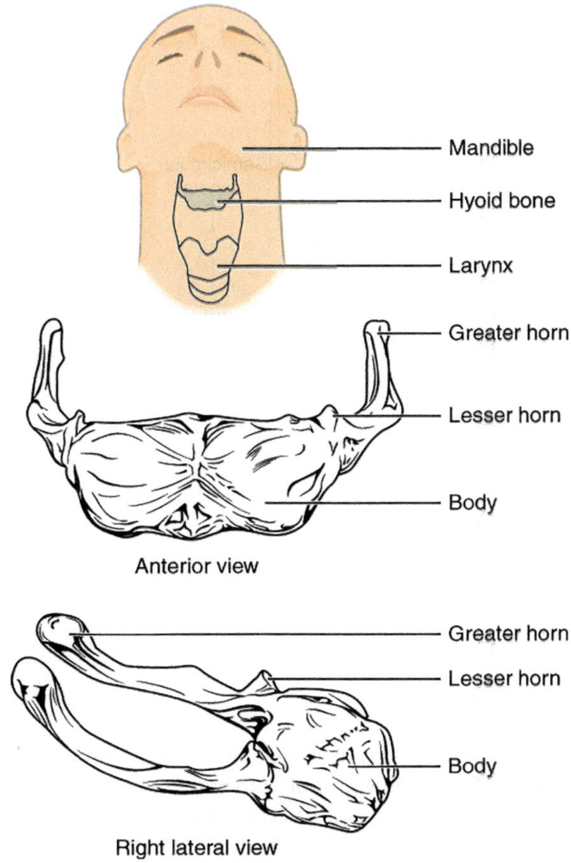

Fig. 6 The hyoid bone, a horseshoe-shaped bone situated in the anterior mid-line of the neck between the chin and the thyroid cartilage. The bone allows for interactions between the tongue and larynx to articulate sound. Neanderthal fossils contain a hyoid bone which supports the notion that Neanderthals possessed the ability to develop speech (https://en.wikipedia.org/wiki/Hyoid_bone#/media/File:712_Hyoid_Bone.jpg).

The human ear can hear sounds between frequencies of 20 Hz and 20 kHz; the majority of human speech sounds are up to 5 kHz. Neanderthals showed a greater sensitivity in the frequency range of 4–5 kHz, similar to *Homo sapiens*. The occupied bandwidth is directly related to the efficiency of the vocal communication system of a species, i.e., they can distinguish a wider range of sounds. The fact that the authors' results showed that Neanderthals' occupied bandwidth was wider than that of their ancestors and similar to that of *Homo sapiens* implied that Neanderthals evolved the auditory capacities to support a vocal communication system as efficient as modern human speech. The authors suggest that Neanderthals could even have had a form of language, but believe that does

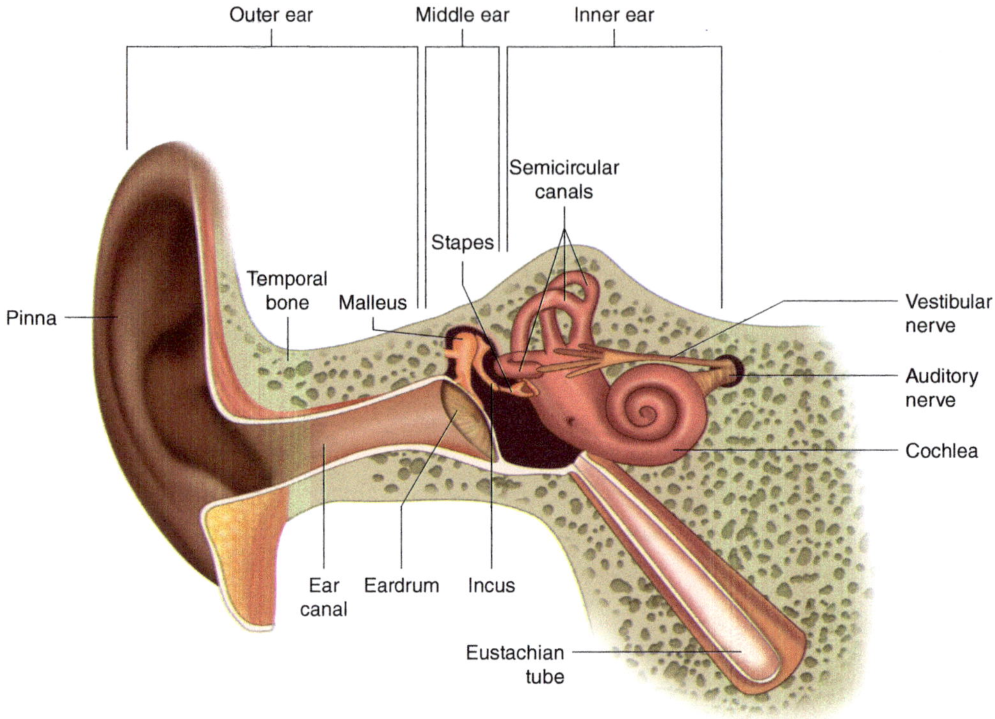

Fig. 7 The anatomy of the human ear. Recent studies, employing high-resolution CT scans, were used to create virtual 3D models of the ear structures of both humans and Neanderthals. The data was used to model the sound frequencies that each species could hear. The results showed that Neanderthals' occupied bandwidth was wider than that of their ancestors and similar to that of *Homo sapiens*, implying that Neanderthals evolved the auditory capacities to support a vocal communication (https://www.nidcd.nih.gov/health/how-do-we-hear).

not necessarily mean they had the mental capabilities to speak the same language as ancient humans.[50]

Although language acquisition is a multifactorial trait (i.e., no single gene is responsible for language, and it is dependent on environment), scientists have tried to identify single genes that contribute to linguistic ability. In 2001, the FOXP2 gene was implicated in the development of language externalization and speech, and mutant forms of FOXP2 were found to lead to defects in orofacial movements, language processing, and/or reading.[51] The gene codes for a protein that is common in mammals and is expressed more in females. The form of FoxP2 protein unique to humans has two small changes in the building blocks of the protein (amino acids) that appear to have been fixed in the human lineage since our split with chimpanzees (i.e., FoxP2 protein found in humans is unique among mammals and is not found in nonhuman primates).

Studies of ancient DNA have shown that Neanderthals had the same FOXP2 gene that is found in modern humans. Krause et al.[52] found these same two amino acid substitutions in two Neanderthal samples. The findings of Krause et al. were challenged as possibly being the result of modern human contamination[53]; however, since the initial sequence was published, it has been confirmed that the FOXP2 genes from Neanderthals and humans are the same, but the genes are not located on the same chromosome and not regulated in the same way.[54]

Further studies by Vernot and Akey[55] confirmed that the Neanderthal version of FOXP2 is not found in modern human genomes, suggesting that the Neanderthal version of FOXP2 was lost in Neanderthal-human hybrids by natural selection. Vernot and Akey also suggest that FOXP2 may not have functioned as an adequate component for modern language in Neanderthals. Most scientists believe that although FOXP2 is a possible contributor to speech, it is not the only source of linguistic ability.

The CNTNAP2 gene has been associated with language development and has been implicated in multiple neurodevelopmental disorders, including Gilles de la Tourette syndrome, schizophrenia, epilepsy, autism, ADHD, and intellectual disability in children.[56]

Mutations in the ASPM gene important for the division of cells in the developing brain have been associated with tone deafness. The CNTNAP and ASPM genes have both been associated with brain development, and neither is found in the Neanderthal and Denisovan genomes, suggesting that if Neanderthals had language, it may have differed from human language.[57]

Although there is speculation on many fronts regarding the evolution of language and whether Neanderthals could speak, the truth is that all the conclusions are based on assumptions and, in some instances, weak archeological and genetic evidence. There must have been some form of communication that would allow Neanderthals to hunt, reproduce tools, and pass on some forms of knowledge; however, studies of nonhuman animals, archeological evidence, genetics, and modeling have not provided sufficient data to come to any concrete conclusions. This lack of solid evidence should not be surprising because currently we know very little about the genetics and evolution of modern human language, let alone that of Neanderthals.[58]

Neanderthal diet

Diet is a significant component of any cultural analysis, and food-centered activities influence even modern human cultures. A group's diet can determine its activity, group size, predation risk, social structure, and potential cognitive skills.[59] In addition, researchers have credited changes in technology and diet as forces that led to the brains of our ancestors doubling in size between 2 million and 700 kya. As early humans got better at killing animals and processing meat, they ate a higher-quality diet, which gave them more energy to fuel the growth of their hungrier brains.

To identify the source of ancient diets, scientists use a variety of methods, including habitat reconstruction, floral and faunal collections at fossil sites, dental wear, dental tartar, isotope analysis, and identification of oral bacteria. Habitat reconstructions can provide limited information about the climate, the variety of flora and fauna at the time, and the overall environment at the time, because there may be no modern analogs for comparisons. Even when there is ample evidence of flora and fauna in the region during a specific period (through pollen and bone analysis), the potential food resources are not necessarily those used in the diet.[59] Most analysis on diet is the result of findings at a fossil site. The accumulations of various animal bones, the distribution of bones, and cut marks on the bones have all been used to determine diet; however, there is often little consideration as to how the animal remains got to the site (washed in, brought in by predators, already at the site when hominins arrived, etc.) and little consideration that food may have been prepared at the scavenger or hunting site and bones left behind, and/or meat brought back to the living area. The same can apply for plant remains. Plant material, mostly pollen and seeds, found at fossil sites may be the result of windblown activity or animals that deposited the material. Plant material found in coprolites (fossilized feces) is a better indication of ingestion. However, even coprolites can contain pollen and phytoliths (plant microscopic silica structures) blown in and attached to coprolites, and it is difficult to determine the age of coprolites.[60]

Tooth wear can also be used to predict diet, but only specific types of dental wear can be used to infer diet because teeth can be worn when used as tools, and wear can result from dust during food ingestion[61] (Fig. 8). Optical 3D topometry is a noninvasive technique for 3D scanning of the tooth surface to detect roughness or other characteristics. The characteristics of the scans can be compared with those of modern people whose diets are known. This technology has been used to determine that Neanderthal diets varied between warmer and colder climates with colder climates being high in protein.[59] Looking at dental micro-wear, El Zaatari et al.[62] found that most Neanderthals had micro-wear similar to populations that eat mostly meat, but that some populations found in forest habitats had a mixed diet (both plant and animal).

Dental tartar (a hardened deposit of biofilm and oral fluids, i.e., mineralized plaque) can also be used to study mineral deposits left by food consumption. These studies have suggested that Neanderthals had a very broad diet that included meat and a variety of plants. Mineral deposits left on tooth tartar suggest that plant material was a source of food or possibly medicine.[63] These studies show various lipids and protein products of a variable diet. However, it has not been clearly shown that these plant deposits were the result of direct consumption. An alternative explanation could be that the plant material came come from chyme (the stomach contents of prey) that may have consumed the plant material. Chyme is considered a delicacy in several cultures.[59]

The stable isotope ratios 13C/12C and 15N/14N in collagen recovered from fossils are used for diet reconstruction. 13C and 18O values from tooth enamel can sometimes

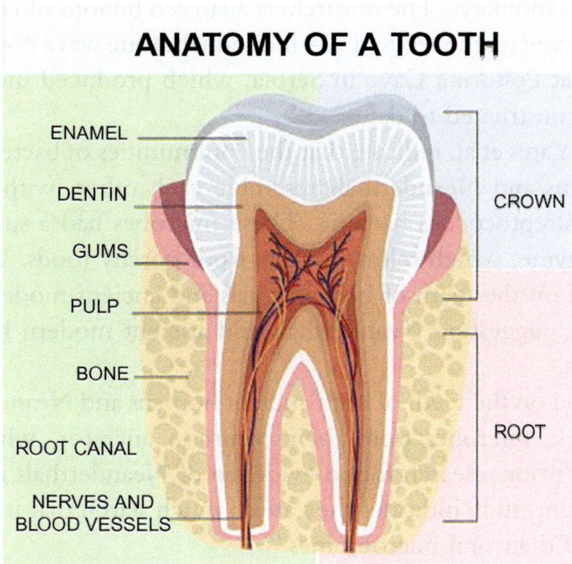

Fig. 8 The detailed anatomy of a tooth (https://depositphotos.com/262814436/stock-illustration-tooth-anatomy-vector-dental-infographics.html).

discriminate among terrestrial and marine foods. To make predictions, the stable isotope ratio from two different isotope pairs in a sample is measured relative to a standard producing an accurate estimate of food source. Bone chemistry from Neanderthal sites and the predicted cold habitat of the time suggests that large mammals were a primary source of food.[64] Fossil evidence from Gorham's Cave and Vanguard Cave reveal that Neanderthals consumed dolphins, seals, shellfish, birds, and rabbits. The isotopic analysis suggests that Neanderthals were at the top of the food chain consuming large mammals.[31,65] The problem with isotope analysis is that it only measures protein; therefore, it is challenging to measure plant consumption. The isotopic ratios can be used to measure plant proteins, but it is not clear if the plant proteins were directly consumed by the Neanderthals, or if plants were consumed by animals, followed by subsequent consumption of the animals.

Most anthropologists agree that the primary food source for Neanderthals was most likely large mammals. These would have supplied a high protein diet needed to survive in the cold climate. A healthy diet, however, would have required the consumption of carbohydrates in addition to animal protein (because high-protein diets can cause ketosis and ammonia toxicity). Carbohydrates could be derived from seasonal fruits and other plant material or through the stomach contents of the plant-eating prey.[66]

Recent studies by Yates and others[67] confirmed that Neanderthals ate large amounts of plant material. The investigators looked at the oral bacteria stuck to the teeth of Neanderthals, preagrarian modern humans who lived more than 10,000 years ago, chimps, gorillas, and howler monkeys. The researchers analyzed billions of DNA fragments from dead bacteria preserved on the teeth of 124 individuals. One was a Neanderthal who lived 100,000 years ago at Pešturina Cave in Serbia, which produced the oldest oral microbiome genome reconstructed to date.

The studies by Yates et al. indicate that the communities of bacteria in the mouths of the modern humans and Neanderthals resemble each other, with both harboring an unusual group of streptococcus bacteria. These microbes had a special ability to bind to the amylase enzyme, which releases sugars from starchy foods. The presence of the bacteria was found on the teeth of Neanderthals and ancient modern humans, but not the other primates, suggesting Neanderthals and ancient modern humans were eating more starchy foods.

The streptococci on the teeth of both ancient humans and Neanderthals also suggests they inherited these microbes from their common ancestor, who lived more than 600 kya. Although prior research found evidence of Neanderthals eating grasses tubers and barley, the recent study indicates they ate so much starch that it dramatically altered the composition of their oral microbiomes.[67]

Neanderthal symbolism and cognition

The measure of Neanderthal cognition, including their cultural and linguistic capabilities, has been debated for decades. This debate revolves around inherent uncertainties in reconstructing prehistory out of indirect evidence and current analytical limitations. In addition, interpretation difficulties between disciplines, various data analysis of artifacts, research biases, and the considerable amount of data complicate any attempt to move toward answers. Consistent and trustworthy answers may only lie in a careful comprehensive interdisciplinary synthesis of findings by considering diverse perspectives within the literature. To date, while some uncertainties remain, the current approach paints the picture of an extensive shared humanity between anatomically modern humans and Neanderthals.[68]

Studies of brain fossils suggest that Neanderthals invested more neural tissue to their visual and somatic systems, leaving less neural tissue for areas associated with problem solving and critical thinking like the prefrontal cortex and parietal lobe. On the other hand, Cro-Magnon brains were organized differently, allotting more neural tissue to regions associated with social adaptations and problem solving. This data supports the notion that AMHs had increasingly more cognitive abilities and social networks than Neanderthals.[69]

However, discoveries of sophisticated art, ornamentation, and an awareness of time reveal that Neanderthals were far more technologically advanced than previously thought (Fig. 9). These studies show that Neanderthals engaged in a range of symbolic behaviors once believed to be unique to *H. sapiens*. These behaviors included decorating their

Fig. 9 Ornaments and ivory ring from the Châtelperronian levels at the Grotte du Renne, France (https://www.pnas.org/content/107/47/20147).

bodies with paint, jewelry, and feathers,[40] funerary practices,[42] visual arts,[70] and body esthetics.[41]

Although most researchers believe these data provide evidence of symbolism, others argue that from certain philosophical perspectives, there is no logical foundation to conclude that jewelry, cave art, body esthetics, and funerary practices reflect or constitute symbolism, especially when these behaviors are used to support the evolution of Neanderthal language.[44]

For most investigators, the clearest examples of cognition are innovation, symbolism, and the development of culture; for many years, there was very little evidence that Neanderthals had anything more than simple tools and hunting strategies. Recent evidence is beginning to point to a more sophisticated lifestyle than previously thought. Polished eagle talons with markings, found in Krapina in modern-day Croatia, show that they were pieces of jewelry. The talons were dated 80,000 years before modern humans entered Europe.[40] Similarly, anthropologists at sites in France and Italy have uncovered eagle talons dating from 40,000 to 90,000 years ago. Across Europe, decorative feathers, seashells bearing traces of pigment (red ochre), and seashells with holes (suggesting they were used as jewelry) are indications that Neanderthals had symbolic thinking.[71] In addition to decorative items, there are indications that Neanderthals also used art as a method of expression. Geometric-like engravings estimated to be more than 39,000 years old in a Gibraltar cave indicate that at least one Neanderthal tried to use art to convey an idea or message.[72,73]

Although the evidence for Neanderthal use of symbols, ornaments, and sophisticated tools provides some indication of their potential ability for higher-order thought, some argue that Neanderthals might have acquired the technology from modern humans.[28] If this were the case, the use of these symbols, ornaments, and tools would merely indicate

Fig. 10 Wall paintings from the Cueva de La Pasiega, or Cave of La Pasiega (https://en.wikipedia.org/wiki/Cave_of_La_Pasiega#Cave_paintings).

that the Neanderthals who interacted with modern humans were capable of mimicry rather than complex thought.

Cave art is important because many scientists believe that symbolic thinking is a precursor to language. Evidence of the world's oldest cave art suggests that it was produced by Neanderthals (Fig. 10). Investigators discovered more than a dozen paintings in a Spanish cave, dated to 65 kya, which predates AMHs in Europe. The investigators also discovered perforated seashell beads and pigments at Cueva de los Aviones, a cave in southeastern Spain, that are at least 115 kyo.

The seashell beads and ochre pigments are the oldest ornaments ever found and predate by 20,000–40,000 years anything found in Africa. The authors of the study suggest that these findings indicate that Neanderthals had a cultural competence that mimics the ancient artifacts of modern humans.[43]

A key feature of modern humans and their capacity for caring and emotion is exemplified in their burial of the dead. Archeologists have argued that several Neanderthal sites show the signs of ritual burial (Fig. 11). However, others have disputed those claims as speculation. Some of the evidence presented for burial is as follows: tools and animal corpses found with remains, red ochre pigment in Middle Eastern graves, and Neanderthal corpses found in the fetal position.[42]

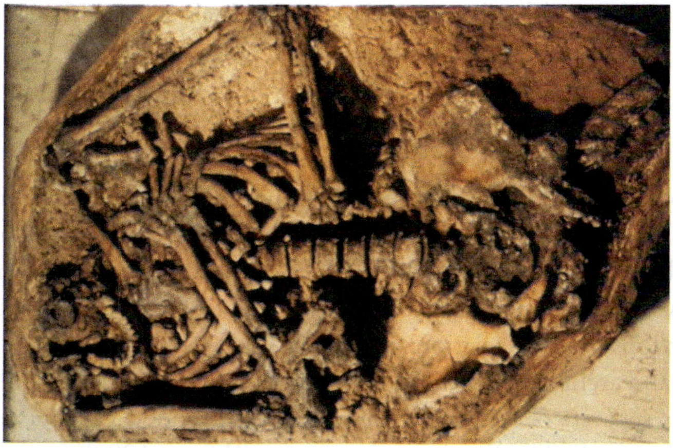

Fig. 11 A Neanderthal burial found at Kebara Cave in Israel. The site was dated to 60,000 years ago and the position of the body and mineral pigments suggest complex behavior and symbolic thought (https://www.nature.com/scitable/knowledge/library/neanderthal-behavior-59267999).

Skeptics claim that the evidence probably represents nonsymbolic and nonreligious corpse disposal. One study found large amounts of pollen within the grave, suggesting flower or medicinal plant deposits.[74] The pollen was identified as coming from yarrow, bachelor's button, ragwort, grape hyacinth, etc., all of which have showy flowers and medicinal properties.[75] Although this represented strong evidence for ritual burial, rodent deposits of large numbers of flowers, and not human activity, later explained the pollen at the grave site.[76]

In 2013, new evidence for burial came from a 12-year re-excavation of the La Chapelle-aux-Saints site. Rendu et al.[77] found features that suggest the remains of an individual were modified for burial, a grave was dug, and the remains were buried quickly to avoid it being eaten by carnivores. The site dates back 50,000 years, before any interaction with modern humans could have taken place, suggesting that the practice was original and not copied from modern human practices.

In addition to burial, there are a couple of examples of caring for the sick that may have been practiced by Neanderthals. The remains of an elderly man were found at Shanidar Cave in Iraqi, Kurdistan. The remains, referred to as Shanidar 1, were one of four almost complete skeletons. Shanidar 1 was a 40- to 50-year-old whose remains displayed debilitating abnormalities, including a crushing fracture to the left side of his face, a withered right arm that caused the loss of his right hand, and deformities in his lower legs and foot that would have resulted in painful walking. All injuries showed signs of healing long before death, implying that he was supported during his lifetime. In addition, the man of La Chapelle-aux-Saints was older than average, and his skeleton revealed that he was severely bent over by arthritis and lost all but two teeth. The age and disabilities of the Shanidar 1 and La Chapelle-aux-Saints' individuals suggest that Neanderthals may have cared for the sick and aged.[78]

Neanderthal and humans interacting

The specifics of Neanderthal and modern human interactions are still unclear; however, genetic and archeological evidence has revealed they interacted within the 5000 years that they both inhabited Europe. Details of their sexual interactions are covered in a previous chapter; however, we will cover some of the evidence here to emphasize the importance of their interactions and the relation to lifestyle.

Geological and archeological evidence suggests that Neanderthals inhabited the Near East during two separate occasions, being pushed south due to glaciers and cold climatic conditions in the north. The first incursion occurred from 120,000 to 100,000 years ago and a second one 61,000 to 48,000 years ago, with Neanderthals going extinct between 39,000 and 41,000 years ago.[79] The timing of modern humans leaving Africa and Neanderthal migration suggests that humans and Neanderthals could have cohabitated twice in parts of their ranges, in Europe and the Near East. The specific nature of the interactions between humans and Neanderthals is not clear, yet considering the options of conflict or collaboration, it is difficult to imagine them living side by side for the estimated amount of time without some social interaction.

Indications of shared living quarters in caves in the Near East have been documented in present-day Israel. This cohabitation between Neanderthals and humans could have occurred during the emergence of the first modern humans out of Africa. The Mount Carmel range near the Sea of Galilee in northern Israel includes the caves of Tabun, Jamal, el-Wad, and Skhul. A series of discoveries of early modern human remains were made in the western slopes of the range. The remains were from the Early Middle Paleolithic age with dates ranging between 100 and 120 kya. Of specific interest, these caves contain the some of the earliest modern human remains outside Africa, and the samples belong to the same strata as Neanderthal bones in the adjacent Tabun Cave.[80] A human jawbone dated to 177 kya was found at Mount Carmel, Israel.[81] Dating techniques indicated that the Tabun Neanderthal samples were coexistent with the modern human remains found in the Mount Carmel range, with the Neanderthal samples dated to about 120 kya. This proximity in time and space of modern humans and Neanderthal fossils suggests a possible cohabitation of the two groups.

Another example is a study that exposed evidence of shared living spaces in caves of present-day Israel, depicting cohabitation and shared technology of Neanderthals and modern humans during the first modern human migration from Africa into the Near East. The modern human remains found at this site displayed skeletal features that seemed to represent a mix of Neanderthal and Early modern human anatomy. This evidence and numerous genetic studies support the admixture theory, in which Neanderthals mated and produce viable and fertile offspring with modern humans. The admixture of Neanderthals and early humans played a significant role in the evolution of modern-day Europeans and Asians, and Neanderthals may have contributed both useful and detrimental genes to the human gene pool.[79,82–84]

Behavioral differences between humans and Neanderthals have been noted in the Levant. In spite of potential coexistence within the same region or even the same cave, modern humans and Neanderthals may have differed in habitat lifestyle. In one example, it appears that the two groups used different patterns of occupying caves. Animal remains (winter prey, mainly herbivores, such as gazelles) found in association with modern human fossils at the Skhul, Qafzeh, and Tabun caves suggest single-season occupation, and that modern humans shifted their living quarters seasonally, including camping outdoors when the weather was favorable. Neanderthal fossils, from about the same time, are found in the context of local animals that were abundant during the entire year, suggesting Neanderthals practiced yearlong occupation.[83]

Genetic data

Although geological data, art, and relics are very useful for distinguishing different populations and cultures, and providing information on living conditions, intellect and cultural sophistication, genome data can reveal subtle differences that may not be obvious in geologic and archeological studies. Unraveling the genetic history of humans and Neanderthals is a difficult task, due to the huge amount of genetic data; however, the ability to sequence nuclear DNA, mitochondrial DNA, and Y-chromosomal DNA from prehistoric human remains has provided vital information about human-Neanderthal interactions, phenotypes, and genetic contributions to the evolution of modern-day humans. A variety of modern studies indicate interaction between humans and Neanderthals that resulted in anatomical hybrids and genetic modifications still seen in modern-day humans.[84,85]

Homo sapiens interbred with Neanderthals in Europe and the Middle East, and these introgressions occurred frequently enough that genomes of non-African people today are up to 3% Neanderthal while African people contain an average of 0.03% Neanderthal DNA. Because different people carry different bits of Neanderthal DNA, the sum total of Neanderthal genetic material that persists in modern-day humans is around 20%.

The Neanderthals were not the only archaic humans with whom *H. sapiens* interacted. Denisovans also interbred with our ancestors. Moreover, sex with archaic hominins seems to have benefited some *H. sapiens*, allowing them to acquire genes that aided their survival: DNA inherited from Neanderthals seems to have boosted immunity, and a gene variant from Denisovans that allows red blood cells to cope with low oxygen levels helps Tibetans live at high altitudes.

However, many modern-day genetic illnesses likely came from Neanderthal DNA. Genes associated with a higher risk of blood clots and strokes, depression, skin lesions, a propensity for nicotine addiction, and malnutrition due to imbalanced thiamine (vitamin B1) have been attributed to Neanderthal genes. These traits were most likely related to adaptations that would have been beneficial to Neanderthals and AMHs in prehistoric times when their bodies were regulated by circadian rhythms, a different diet, and an increased need for an immediate immune response. Unfortunately for modern humans, those traits now lead to health concerns.[82]

Because of human-Neanderthal introgression, hybrid intermediate forms have been recovered. For example, in 2002 Erik Trinkaus[86] retrieved specimens of three individuals from the Cave of Bones (Peştera cu Oase) in Romania. The findings included a lower jawbone, the partial skull of a 15-year-old adolescent, and a left temporal bone, dated to about 37 to 42 ky old, exhibiting a mixture of Neanderthal and anatomically modern human features. Recently, geneticists determined that the genome of one of the Oase men contained approximately 6%–9% Neanderthal DNA and that the individual had a Neanderthal ancestor only four to six generations before.[79]

Another recent study based on a tooth and fragmentary remains from Bacho Kiro Cave in Bulgaria, with individuals dated to between 45.9 and 42.6 kya, all had recent Neanderthal forebears. The genomes of modern non-Africans usually harbor about 2% Neanderthal ancestry, but the Bacho Kiro individuals had slightly more at 3.4%–3.8%, and the chromosome segments—which shorten in successive generations—were considerably longer. By measuring these segments, the researchers estimated that the Bacho Kiro individuals had Neanderthal ancestors as recently as the past six or seven generations—and probably in Europe, not the Middle East.[87]

In other reports, Fu and colleagues[87,88] presented genome-wide data from 51 Eurasians that dated from 45 to 7 kya. The data confirmed that modern humans and Neanderthals had mated but the amount of Neanderthal DNA decreased from 6% in the oldest specimens down to 2% in the younger specimens, suggesting that random drift or natural selection may have eliminated some detrimental Neanderthal DNA In addition, a comparison of the Oase man's genome to that of modern-day Europeans led the researchers to conclude that the Oase individual did not contribute a substantial portion of his genetics to later humans in Europe and that the population to which the Oase man belonged went extinct. In other words, the data suggest that the earliest humans in Europe (those older than 37 kya) did not contribute genetically to later populations.

Others have found that as time passed and modern humans continued to breed, fewer genetic remnants of the Neanderthals remained. For example, the remains of a 35 kyo male discovered by Edouard Depont, in Goyet, Belgium was used to extract DNA. The DNA sequence revealed that this European ancestor contained the earliest evidence of a genetic contribution to modern-day Europeans, and although the sequenced DNA did contain Neanderthal DNA, it was much less than that found in the remains of individuals from the Châtelperronian culture. This indicated Neanderthal DNA had been diluted out of the individual through human-to-human breeding and the extinction of Neanderthals by this time.[16]

Conclusion

Genetic and fossil evidence of multiple lineages in the past 100 ky shows that Neanderthals, Denisovans, and *Homo floresiensis* are our direct ancestors who shared regions of the earth with several other *Homo* species. Although many aspects of archaic early lifestyle remain a mystery, through archeological evidence and dating of artifacts, it is known that the development of technology became more complex over time and that Neanderthals were not the doltish figures that they were portrayed as when first discovered. Studies of the types of behavior and cultural practices associated with Neanderthals have revealed the existence of art, language, spiritual beliefs, cognition, technology, food acquisition and consumption, introgression, and eventual extinction. The evidence for Neanderthal behavior lies in their skeletal remains, the artifacts they left behind, and their genomes. Some of these sources provide direct evidence for behaviors, whereas others provide indirect evidence, or are open to speculation. Neanderthals and humans were very similar but distinct in many ways. The more robust anatomy and body proportions of Neanderthals are thought to have allowed a more close-proximity style of hunting. The large Neanderthal brain reflected by endocranial casts resembles that of modern humans in an area important for speech, suggesting that Neanderthals had the neural development necessary for language. Interactions between humans and Neanderthals would have required some form of communication to hunt large animals. Most anthropologists agree that the primary food source for Neanderthals was mostly large mammals. These would have supplied a high-protein diet needed to survive in the cold climate. Studies have shown that Neanderthals engaged in a range of symbolic behaviors once believed to be unique to *H. sapiens*. These behaviors included decorating their bodies with paint, jewelry, and feathers, funerary practices, visual arts, and body esthetics. Discoveries of sophisticated art, ornamentation, and an awareness of time reveal that Neanderthals were far more technologically advanced than previously thought. Sex with archaic hominins seems to have benefited some *H. sapiens*, allowing them to acquire genes that aided their survival while other genes appear detrimental to the health of modern-day humans. It is now evident that AMH interaction with Neanderthals was a big part of our history and that Neanderthals were more closely related to humans than previously thought.

References

1. Jung C. https://www.brainyquote.com/quotes/carl_jung_125713?src=t_intellect.
2. Ni X, et al. Massive cranium from Harbin in northeastern China establishes a new Middle Pleistocene human lineage. *Innovation (N Y)*. 2021. https://doi.org/10.1016/j.xinn.2021.100130.
3. Bernard W. Welcome to the family. *Sci Am*. 2014;(September):43–47.
4. Durvasula A, Sankararaman S. Recovering signals of ghost archaic introgression in African populations. *Sci Adv*. 2020;6(7). https://doi.org/10.1126/sciadv.aax5097, eaax5097.
5. Neubauer F. A brief overview of the last 10 years of major late Pleistocene discoveries in the old world: *Homo floresiensis*, Neanderthal and Denisovan. *J Anthropol*. 2014;7. https://doi.org/10.1155/2014/581689, 581689.
6. Schmitz R, et al. The Neanderthal type site revisited: interdisciplinary investigations of skeletal remains from the Neander Valley. *Proc Natl Acad Sci U S A*. 2002;99:13342–13347.
7. Kuhn SL, Stiner MC. The division of labor among Neandertals and modern humans in Eurasia. *Curr Anthropol*. 2006;47:953–981.
8. Lalueza-Fox C, et al. Genetic evidence for patrilocal mating behavior among Neanderthal groups. *Proc Natl Acad Sci U S A*. 2011;108:250–253.
9. Prüfer K, et al. The complete genome sequence of a Neanderthal from the Altai Mountains. *Nature*. 2014;505:43–49.
10. Campbell BG, et al. *Humankind Emerging*. 9th ed. Boston, MA: Pearson; 2006:358–380.
11. Defleur A, et al. Neanderthal cannibalism at Moula-Guercy, Ardeche, France. *Science*. 1999;286:128–131.
12. Monnier G. Neanderthal behavior. *Nat Educ Knowl*. 2012;3:11.
13. Seddon C. *Humans: From the Beginning*. San Bernardino, CA: Glanville Publications; 2015.
14. Estabrook VH. Is trauma a Krapina like all other Neanderthals trauma? A statistical comparison of trauma patterns in Neandertal skeletal remains. *Period Biol*. 2007;109:393–400.
15. Walker PL. A bioarcheological perspective on the history of violence. *Annu Rev Anthropol*. 2001;30:573–596.
16. Longo L, et al. Did Neandertals and anatomically modern humans coexist in northern Italy during the late MIS 3? *Quat Int*. 2012;259:102–112.
17. Marlowe FW. Hunter-gatherers and human evolution. *Evol Anthropol*. 2005;14(2):15294.
18. Allington-Jones L. The Clacton spear: the last one hundred years. *Archaeol J*. 2015;J172:273–296.
19. Aranguren B, et al. Wooden tools and fire technology in the early Neanderthal site of Poggetti Vecchi (Italy). *Proc Natl Acad Sci USA*. 2018;115:2054–2059.
20. Campbell BG, et al. *Humankind emerging*. 9th ed. Boston, MA: Pearson; 2006:376–379.
21. Delagnes A, et al. *Middle Palaeolithic Raw Material Procurement, Transport and Exploitation Patterns in the Charente River Basin*; 2006.
22. Hublin JJ, Sirakov N, Aldeias V, et al. Initial Upper Palaeolithic *Homo sapiens* from Bacho Kiro Cave, Bulgaria. *Nature*. 2020;581:299–302. https://doi.org/10.1038/s41586-020-2259-z.
23. Léveque Gaudzinski-Windheuser S, Noack ES, Pop E, et al. Evidence for close-range hunting by last interglacial Neanderthals. *Nat Ecol Evol*. 2018;2:1087–1092. https://doi.org/10.1038/s41559-018-0596-1.
24. Benazzi S, et al. Early dispersal of modern humans in Europe and implications for Neanderthal behavior. *Nature*. 2011;479:525–528.
25. Higham T, et al. The earliest evidence for anatomically modern humans in northwestern Europe. *Nature*. 2011;479:521–524.
26. Zilhao J, et al. Analysis of site formation and assemblage integrity does not support attribution of the uluzzian to modern humans at Grotta del Cavallo. *PLoS ONE*. 2015;, 0131181.
27. Soressi M, et al. Neandertals made the first specialized bone tools in Europe. *PNAS*. 2013;110(35):14186–14190. https://doi.org/10.1073/pnas.1302730110.
28. Appenzeller T. Neanderthal culture: old masters. *Nature News*. 2013;(may).
29. Kozowyk PRB, et al. Experimental methods for the Palaeolithic dry distillation of birch bark: implications for the origin and development of Neandertal adhesive technology. *Sci Rep*. 2017;7:8033.
30. Hoffecker JF. The complexity of Neanderthal technology. *PNAS*. 2018;115(9):1959–1961. https://doi.org/10.1073/pnas.1800461115.

31. Hardy BL, Moncel M-H. Neanderthal use of fish, mammals, birds, starchy plants and wood 125-250,000 years ago. *PLoS ONE*. 2011;6, e23768.
32. Hardy BL, Moncel MH, Kerfant C, et al. Direct evidence of Neanderthal fibre technology and its cognitive and behavioral implications. *Sci Rep*. 2020;10:4889. https://doi.org/10.1038/s41598-020-61839-w.
33. Moncel MH, Chacón MG, Vettese D, et al. Late Neanderthal short-term and specialized occupations at the Abri du Maras (South-East France, level 4.1, MIS 3). *Archaeol Anthropol Sci*. 2021;13:45. https://doi.org/10.1007/s12520-021-01285-5.
34. Blinkhorn J, et al. Nubian Levallois technology associated with southernmost Neanderthals. *Sci Rep*. 2021. https://doi.org/10.1038/s41598-021-82257-6.
35. Villa P, Soriano S. Hunting weapons of Neanderthals and early modern humans in South Africa: similarities and differences. *J Anthropol Res*. 2010;66:5–38.
36. Rhodes JA, Churchill SE. Throwing in the middle and upper Paleolithic: inferences from an analysis of humeral retroversion. *J Hum Evol*. 2008;56:1–10.
37. Shea JJ, Sisk ML. Complex projectile technology and Homo sapiens dispersal into western Eurasia. *PaleoAnthropol*. 2010;2010:100–122.
38. Marean CW. The most invasive species of all. In: *The Rise of Humans*. New York, NY: Scientific American; 2020. ISBN:978-1-948933-13-1.
39. Dediu D, Levinson SC. On the antiquity of language: the reinterpretation of Neandertal linguistic capacities and its consequences. *Front Psychol*. 2013;4:397.
40. Radovčić D, et al. Evidence for Neandertal jewelry: modified white-tailed eagle claws at Krapina. *PLoS ONE*. 2015;10. https://doi.org/10.1371/journal.pone.0119802, e0119802.
41. Roebroeks W, et al. Use of red ochre by early Neandertals. *Proc Natl Acad Sci USA*. 2012;109:1889–1894. https://doi.org/10.1073/pnas.1112261109.
42. Balzeau A, et al. Pluridisciplinary evidence for burial for the La Ferrassie 8 Neandertal child. *Sci Rep*. 2020;10:21230. https://doi.org/10.1038/s41598-020-77611-z.
43. Hoffmann DL, et al. U-Th dating of carbonate crusts reveals Neandertal origin of Iberian cave art. *Science*. 2018;359:912–915. https://doi.org/10.1126/science.aap7778.
44. Botha R. *Neanderthal Language: Demystifying the Linguistic Powers of Our Extinct Cousins*. Cambridge University Press; 2020. ISBN:978-1-108-49132-7.
45. Stout D, Chaminade T. The evolutionary neuroscience of tool making. *Neuropsychologia*. 2007;45:1091–1100. https://doi.org/10.1016/j.neuropsychologia.2006.09.014.
46. Gabrić P, et al. Palaeolithic toolmaking and the evolution of cognition and language. In: Strle T, Markič O, eds. *Cognitive Science*. Ljubljan: Institut "Jožef Stefan"; 2018:12–16. https://doi.org/10.5281/zenodo.4674332.
47. Schwartz JH, Tattersall I. The human chin revisited: what is it and who has it. *J Hum Evol*. 2000;38:367–409.
48. Le May M. The language capability of Neanderthal man. *Am J Phys Anthropol*. 1975;42:9–14.
49. D'Anastasio R, et al. Micro-biomechanics of the Kebara 2 hyoid and its implications for speech in Neanderthals. *PLoS ONE*. 2013;8, e82261.
50. Conde-Valverde M, et al. Neanderthals and *Homo sapiens* had similar auditory and speech capacities. *Nat Ecol Evol*. 2021;5:609–615. https://doi.org/10.1038/s41559-021-01391-6.
51. Fisher SE, Scharff C. FOXP2 as a molecular window into speech and language. *Trends Genet*. 2009;25:166–177.
52. Krause J, et al. The derived FOXP2 variant of modern humans was shared with Neanderthals. *Curr Biol*. 2007;17:1908–1912.
53. Coop G, et al. The timing of selection at the human FOXP2 gene. *Mol Biol Evol*. 2008;25:1257–1259.
54. Maricic T, et al. A recent evolutionary change affects a regulatory element in the human FOXP2 gene. *Mol Biol Evol*. 2013;30:844–852.
55. Vernot B, Akey JM. Resurrecting surviving Neanderthal lineages from modern human genomes. *Science*. 2014;343:1017–1021.
56. Falsaperia R, et al. Intronic variant in *CNTNAP2* gene in a boy with remarkable conduct disorder, minor facial features, mild intellectual disability, and seizures. *Front Pediatr*. 2020;11. https://doi.org/10.3389/fped.2020.00550.

57. Berwick RC, et al. Neanderthal language? Just-so stories take center stage. *Front Psychol.* 2013;4:1–2.
58. Hauser MD, et al. The mystery of language evolution. *Front Psychol.* 2014;5:401. https://doi.org/10.3389/fpsyg.2014.00401.
59. Buck LT, Stringer CB. Having the stomach for it: a contribution to Neanderthal diets? *Quat Sci Rev.* 2014;96:161–167.
60. Kullmer O, et al. Hominid tooth pattern database (HOTPAD) derived from optical 3D topomentry. In: Marfat B, Delingette H, eds. *Three-Dimensional Imaging in Paleoanthropology and Prehistoric Archeology. Liege Acts of the XIV UISPP Congress*; 2002:71–82. BAR International Series; vol. 1049.
61. Lucas PW, et al. Mechanisms and causes of wear in tooth enamel: implications for hominin diets. *J R Soc Interface.* 2013. https://doi.org/10.1098/rsif.2012.0923.
62. El Zaatari S, et al. Ecogeographic variation in Neandertal dietary habits: evidence from occlusal molar microwear texture analysis. *J Hum Evol.* 2011;61:411–424.
63. Hardy K, et al. Neanderthal medics? Evidence for food, cooking, and medicinal plants entrapped in dental calculus. *Naturwissenschaften.* 2012;99:617–626.
64. Richards MP, Trinkaus E. Isotopic evidence for the diets of European Neanderthals and early modern humans. *Proc Natl Acad Sci U S A.* 2009;106:16034–16039.
65. Stringer CB, et al. Neanderthal exploitation of marine mammals in Gibraltar. *Proc Natl Acad Sci U S A.* 2008;105:14319–14324.
66. Florenza L, et al. To meat or not to meat? New perspectives on Neanderthal ecology. *Am J Phys Anthropol.* 2014;156:43–71.
67. Yates JAF, et al. The evolution and changing ecology of the African hominid oral microbiome. *PNAS.* 2021;118(20). https://doi.org/10.1073/pnas.2021655118.
68. Breyl M. Triangulating Neanderthal cognition: a tale of not seeing the forest for the trees. *Wiley Interdiscip Rev Cogn Sci.* 2021. https://doi.org/10.1002/wcs.1545.
69. Pearce E, et al. New insights into the differences in brain organization between Neanderthals and anatomically modern humans. *Proc R Soc B.* 2013;280:1758. https://doi.org/10.1098/rspb.2013.0168.
70. Hoffmann DL, et al. U-Th dating of carbonate crusts reveals Neandertalorigin of Iberian cave art. *Science.* 2018;359:912–915. https://doi.org/10.1126/science.aap7778.
71. Wong K. Neanderthal minds: analysis of anatomy, DNA and cultural remains have yielded tantalizing insights into the inner lives of our mysterious extinct cousins. *Sci Am.* 2015;312:36–43.
72. Rodriguez-Vidal R, et al. A rock engraving made by Neanderthals in Gibraltar. *Proc Natl Acad Sci U S A.* 2014;111:13301–13306.
73. Pike AWG, et al. U-series dating of Paleolithic art in 11 caves in Spain. *Science.* 2012;336:1409–1413.
74. Solecki RS. Shanidar IV, a Neanderthal flower burial in northern Iraq. *Science.* 1975;190:880–881.
75. Lietava J. Medicinal plants in a Middle Paleolithic grave Shanidar IV? *J Ethnopharmacol.* 1992;35:263–266.
76. Sommer JD. The Shanidar IV 'flower burial': a re-evaluation of Neanderthal burial ritual. *Camb Archaeol J.* 1999;9:127–129.
77. Rendu W, et al. Evidence supporting an intentional Neanderthal burial at La Chapelle-aux-saints. *Proc Natl Acad Sci U S A.* 2013;111:81–86.
78. Trinkaus E, Villotte S. External auditory exostoses and hearing loss in the Shanidar 1 Neandertal. *PLoS ONE.* 2017;12(10), e0186684.
79. Fu Q, et al. An early modern human from Romania with a recent Neanderthal ancestor. *Nature.* 2015. https://doi.org/10.1038/nature14558.
80. Coppa A, et al. Newly recognized Pleistocene human teeth from Tabun cave Israel. *J Hum Evol.* 2005;49:301–315.
81. Hershkovitz, et al. The earliest modern humans outside Africa. *Science.* 2018;359:456–459.
82. Simonti CN, et al. The phenotypic legacy of admixture between modern humans and Neandertals. *Science.* 2016;351:737–741. https://doi.org/10.1126/science.aad2149.
83. Lieberman DE. *Neandertal and Early Modern Human Mobility Patterns.* Cham, Switzerland: Springer International publishing; 2002.
84. Sankararaman S, et al. The genomic landscape of Neanderthal ancestry in present-day humans. *Nature.* 2014;507:354–357.
85. Compton T, et al. The morphology of the Late Pleistocene hominin remains from the site of La Cotte de St Brelade, Jersey (Channel Islands). *J Hum Evol.* 2021;152. https://doi.org/10.1016/j.jhevol.2020.102939.

86. Trinkaus E. An early modern human from the Peştera cu Oase, Romania. *Proc Natl Acad Sci U S A*. 2003;100:11231–11236.
87. Callaway E. Oldest DNA from a *Homo sapiens* reveals surprisingly recent Neanderthal ancestry. *Nature*. 2021;592:339. https://doi.org/10.1038/d41586-021-00916-0.
88. Fu Q, et al. The genetic history of ice age Europe. *Nature*. 2016;534:200–205.

CHAPTER 13

To what extent did Neanderthals and modern humans interact?

> *We have only indirect means of knowing the courage and activity of the Neanderthals in the chase, through the bones of animals hunted for food which are found intermingled with the flints around their ancient hearths.*
>
> ***Henry Fairfield Osborn***[1]

Historical precedence

Historical accounts of modern humans' behavior in situations of first encounters involving populations with different levels of technological sophistication fall into two categories: (1) *sapiens* at times empathized with the less technologically advanced population; or (2) alternatively, *sapiens* behaved aggressively. Furthermore, in instances of peaceful first encounters, there was a tendency for relationships to deteriorate with time as the more advanced population eventually tried to take advantage of the other. Historical accounts, such as the colonization of the Americas and Africa, the European expansion into the American West, and the European settlement of the Pacific Islands, describe only a few of the instances in which contemporary humans have been compassionate at times, yet extremely aggressive and brutal in others. Belligerent behavior was not a monopoly of Europeans toward the natives. Intertribal warfare among Native American groups is a good example of pre-Columbian brutality.[2] Although a highly contentious issue among historians,[3] the Carib tribes of the Antilles were notorious for the practice of institutionalized slavery and captive mutilation. They routinely captured Taino (native inhabitants of the Greater Antilles at the time of the first contact by Christopher Columbus) women, raped them, and even ate the resulting children.[4]

Independent of behavioral modes among human populations in the recent past was the recurrent impact of infectious diseases transmitted as a consequence of contact and subsequent cohabitation. In the Americas, European infectious diseases totally devastated entire groups of people. At times this decimation occurred unintentionally, resulting from casual contact. At other times, purposeful infection took place as a form of biological warfare. A notorious example of intentional infestation was attempted by the British, who contaminated Native Americans with smallpox by providing peace emissaries with

Sex and Cohabitation Among Early Humans
https://doi.org/10.1016/B978-0-12-819086-9.00001-4

laced blankets and a handkerchief from the nearby hospital. This was done in the hopes that the items would infect the natives and then spread the deadly disease to nearby tribes.[5] Smallpox was not the only disease that had fatal consequences subsequent to contacts in different parts of the world. Other illnesses include bubonic plague, chickenpox, cholera, the common cold, diphtheria, influenza, malaria, measles, scarlet fever, gonorrhea, typhoid, typhus, tuberculosis, and pertussis. Not all infectious diseases traveled with the Europeans all over the world; some like syphilis may have reached the Old World from the Americas.[6]

There are also many examples of empathy involved in historical encounters between different groups of people. One such case is illustrated by the actions of Bartolomé De Las Casas (1474–1566), a Spanish priest, social reformer, and historian.[7] He was a principal figure and defender of the rights of Native Americans in 16th-century Spain and Spanish America. As a young man, De Las Casas sailed for Española (today Dominican Republic/Haiti) where he initially battled the natives, secured land, enslaved the Tainos, and felt no regrets for his actions. In the conquest of the Americas, aggression went in both directions: it was directed at the natives as they were enslaved and at the invaders as the natives retaliated (Fig. 1). At the age of 40, De Las Casas' attitude toward the natives changed dramatically. He initiated a movement within the Spanish legal system to secure the establishment of model colonies in which Spanish farmers would live and labor side by side with the natives in a peaceful coexistence that would lead the latter to Christianity. De Las Casas won an impressive victory in the Spanish courts when in 1542 laws were

Fig. 1 The Natives of Cumaná attack the mission after Gonzalez de Ocampo's slaving raid. Colored copperplate by Theodor de Bry, published in the "Relación brevissima" (https://en.wikipedia.org/wiki/Bartolom%C3%A9_de_las_Casas).

adopted that outlawed Indian slavery, prohibited Indian forced labor, and provided for gradual abolition of the *encomienda* system (natives living on agricultural lands in serfdom). The life story of De Las Casas illustrates how humans are capable both of acts of brutal aggression and of kindness and empathy, even within the lifetime of a single individual.

Self or different?

A central question that comes to mind when pondering the topic of the initial encounters between Neanderthals and anatomical modern humans (AMHs) is: "How did these two groups of people regard each other?" (see Fig. 2). In other words, did the encounter trigger a reaction of recognition as belonging to the same group or a different type of beings; that they did not belong together? History shows that *sapiens* are capable of becoming confused about the identity of people whom they meet. Consider, for instance, how incorrect original Europeans were when they encountered Native Americans. Columbus and his crew thought that they were facing Asians. Part of their confusion resulted from the expectation that their trek would take them to Asia. Ironically, knowing what we know today about the Asian ancestry of Native Americans, they were not wrong.

Fig. 2 Comparison of faces of early European *Homo sapiens (left)* and *Homo neanderthalensis (right)* based on forensic facial reconstructions (https://en.wikipedia.org/wiki/Neanderthal_anatomy; https://en.wikipedia.org/wiki/Neanderthal_anatomy#/media/File:Comparison_of_faces_of_Homo_sapiens_and_Neanderthal.jpg).

Part of the answer to the question of *self or nonself* depends on basic issues, such as how different, physically and behaviorally, these two hominins were from each other. For example, were they able to articulate words and subsequently able to learn each other's language and communicate? Although some scholars think that both were anatomically and genetically capable of vocalizing, there are no data indicating that they actually did, or if they did, how effective they were.[8]

To a group of AMHs encountering Neanderthals for the first time, they would have looked stronger and more muscular than AMHs. They would have noticed that they looked relatively peculiar, with a projecting face and jaw, no protruding chin, and a broad nose. Maybe the most observant members of the *sapiens* group would have noted that the Neanderthals' hands and feet were larger than their own, and they had broad shoulders, chests, and pelvises, and that their waists were hardly different from the rest of the trunk of the body. AMHs may have also noticed that Neanderthals walked in a different way from them. Other less obvious characteristics of Neanderthals included skulls that were long and low with prominent brow ridges above their big eyes. In general, they were shorter than AMHs with bulky torsos and short lower legs and arms. Adults grew to about 1.50–1.75 m tall and weighed about 64–82 kg.[9,10] Although most AMHs differ from Neanderthals with respect to these features, some of the abovementioned characteristics were present in both Paleolithic species and are even seen in some contemporary *sapiens*. The retention of some of these features characteristic of Neanderthals in contemporary humans may result from ancestral hominin traits (i.e., inherited from a common ancestor) shared with Neanderthals.

In terms of skin color, DNA analysis of 40,000- and 50,000-year-old Neanderthal fossils from Spain and Italy suggest that at least some Neanderthals had light-colored skin and reddish hair.[11] On the other hand, an independent study discovered that some Neanderthals from Croatia had dark skin and brown hair.[12] Several genes determine skin pigmentation in AMHs, and likely similar genes controlled skin color in Neanderthals.[13] Thus, as in AMHs, it is likely that Neanderthals' skin color was variable and polymorphic, as might be expected for a population that inhabited a large territorial expanse.[14] Therefore, it is not possible to assess the skin color of the Levantine Neanderthals who met AMHs for the first time. On the other hand, DNA analysis indicates that AMHs did not gain their pale complexion until 11,000 to 19,000 ya.[14] The decline in pigmentation came about rather suddenly by way of what geneticists call "selective sweeps" (strong selective pressure) that favored genes for light skin color. It is theorized that at high latitudes (e.g., Northern Europe), genes for fair skin are strongly favored by selection, since pale skin allows a greater amount of sunlight to reach the epidermis of the skin and produce sufficient amounts of vitamin D, promoting bone growth. Conversely, people with dark skin would be at a disadvantage in northern latitudes due to vitamin D deficiency, since the larger amounts of the skin pigment, melanin, would absorb the sunlight, preventing it from reaching the epidermis. On the other hand, since vitamin D in high levels

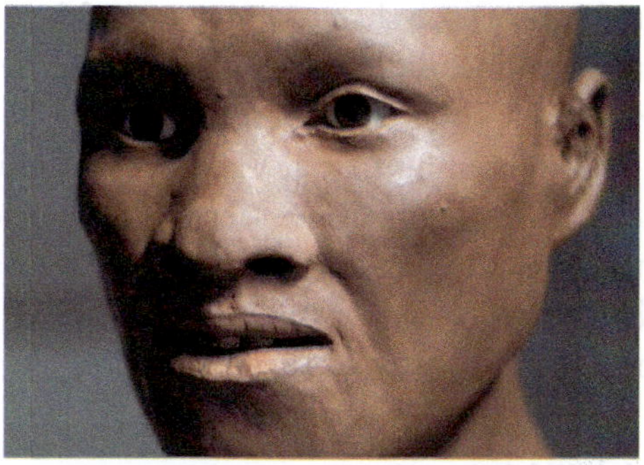

Fig. 3 Reconstruction of an ancient modern human found in Romania 43,000 ya gives us a glimpse of what the Siberian man might have looked like (https://www.zmescience.com/science/anthropology/humans-first-interbred-neanderthals-least-50000-years-ago).

is toxic, in populations living in lower latitudes close to the equator, genes for dark pigmentation are favored. Taking all of this into account, it is likely that the AMHs who invaded the Near East had dark complexions, as depicted in Fig. 3.

A violent past

It is likely AMHs 120,000 years ago (ya), when they first encountered Neanderthals in the Levant, were behaviorally different from historical or contemporary humans. Although it has been postulated that in the earliest hunter-gatherer groups, population density was low enough to prevent frequent conflicts,[15] fieldwork and archeological research on primitive cultures indicate that conflicts in primitive cultures have been intense, cruel, pervasive, and lethal.[16] Therefore, it has been theorized that as population density increased, after the origin of AMHs and sedentism (living in one place for an extended amount of time) was adopted, competition for resources intensified. It has been proposed that intra- and interspecies competition was partially responsible for the various waves of migrations out of Africa into Eurasia. These migrations into territories not occupied by AMHs (e.g., the Levant) may have been motivated by competition for resources. It is thought that under these conditions, Paleolithic *sapiens* began to practice organized raiding of competitors' settlements using novel tactics, such as spear-throwing and ambush warfighting techniques.[17]

These notions are not totally hypothetical. Cave paintings approximately 30,000 years old depict Paleolithic people being wounded or killed by different types of projectiles. These representations may represent violence over game and/or resources, although they may also illustrate human sacrifice, assassination, or systemic warfare.[18] In addition, there are cases of points, darts, or arrows lodged in the fossilized bones of AMHs. The cave art and points embedded in bones provide compelling evidence of *sapiens* killing *sapiens* during the

Upper Paleolithic. Painted figures with shafts coming out of their bodies have been discovered in several European caves including Cougnac and Pech Merle, both in France. At Cougnac, various lanced AMHs have been identified. Some had spears protruding from the back and buttocks. It is interesting that at Cougnac, some images of speared people are superimposed over the painted animals. It seems that the cave artists were depicting an event in which animals were moving into the foreground of an area that the observers were in possession of, having wounded or killed the intruders. These paintings seem to memorialize an event in which the artist's clan prevailed.

Violence among Paleolithic sapiens

Cioclovina

Cases of interpersonal violence using a number of different weapons are found throughout the prehistory of AMHs.[19–22] During the Paleolithic, violence was pervasive and often deadly. Individuals of all ages and both sexes are represented. A number of cases clearly illustrate the point. A recent description of injuries observed in a calvarium (cover of the cranial cavity) from Cioclovina in west central Romania provides indisputable evidence of interpersonal conflict.[23] The calvarium was found in the Pestera Cioclovina cave; it was dated to about 33,000 ya and represents one of the earliest modern AMHs in Europe. This individual was associated with the Aurignacian lithic industry. The calvaria shows two perimortem (damage occurring at or near the time of death, without any evidence of healing) fractures in the occipital and parietal bones (Fig. 4). It is theorized that the first impact caused a linear fracture at the base of the skull, and the second one generated a depressed fracture on the right side of the cranial vault. Since only part of the skull was recovered, it is possible that additional injuries were inflicted postcranially as

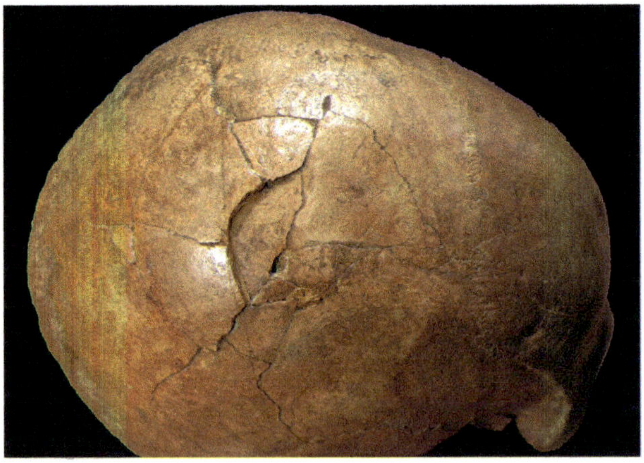

Fig. 4 Right lateral view of the Cioclovina calvaria exhibiting a large depressed fracture (https://doi.org/10.1371/journal.pone.0216718.g001).

part of the attack. The fractures were caused by a club-like object and the attack was so violent that it must have caused severe internal brain injuries, likely leading to the quick death of the victim. The lack of any signs of healing in the areas of the injuries indicates that the Cioclovina individual did not survive this violent attack. It seems that the assault was a face-to-face confrontation between two standing individuals, and the aggressor was left-handed.

Sunghir 1

Another example of a violent incident involving AMHs took place about 30,000–32,000 ya. It ended the life of an older adult male in his late 50s or early 60s in a location near the town of Vladimir in northern European Russia about 200 km northeast of Moscow.[24] Although the victim was of advanced age, he was not geriatric, and physically he was a robust and generally healthy individual. The fossil goes by the name of Sunghir 1 and he was given one of the most elaborate burials known from the Paleolithic era. As part of the interment, he was thoroughly covered in red ochre and decorated with about 3000 mammoth ivory beads distributed over the body, a labor-intensive task of several thousand hours for the community. Significantly, the Sunghir 1 fossil exhibits evidence of a fatal perimortem sharp trauma in the first thoracic vertebra (Fig. 5) that never had a chance to heal, thus indicating that the individual died shortly after the incident. The wound and incision were caused by a sharp object about 10 mm wide, 1–2 mm thick, and at least 6–7 mm long that penetrated the left ventral-lateral body of the T1 vertebra. The entrance of the sharp object occurred through the anterior-lateral left side of the neck, penetrating the body of the vertebrate. It is likely that a spear point or a knife pierced through the lower neck, entering between the left clavicle and left first rib, and eventually reaching and creating a hole in the vertebra (Fig. 5). This lesion is not a clear-cut case of lethal interpersonal violence, since it is possible, for example, that a clan member may have generated the wound accidentally during a hunting accident. Yet considering the age of the individual and his supposed high status within the clan (as indicated by the funeral decorations), it is unlikely that he was out hunting with the younger members of the group. A potential scenario for his demise may therefore be some type of interpersonal violence.

Skhul IX

The Skhul site is located in Mount Carmel about 20 km south of the city of Haifa in Israel and about 3 km (1.9 mi) from the Mediterranean Sea. The cave was originally explored and excavated by British archeologist Dorothy Garrod (1892–1968) during the summer of 1928. From the Skhul Cave, microlith stone tools and burial artifacts of AMHs have been recovered. The cave also indicates that Neanderthals may have cohabitated with AMHs about 100,000 ya.[25] During the 1929–35 time period, the remains of 10 AMHs were recovered. These included seven adults and three children;

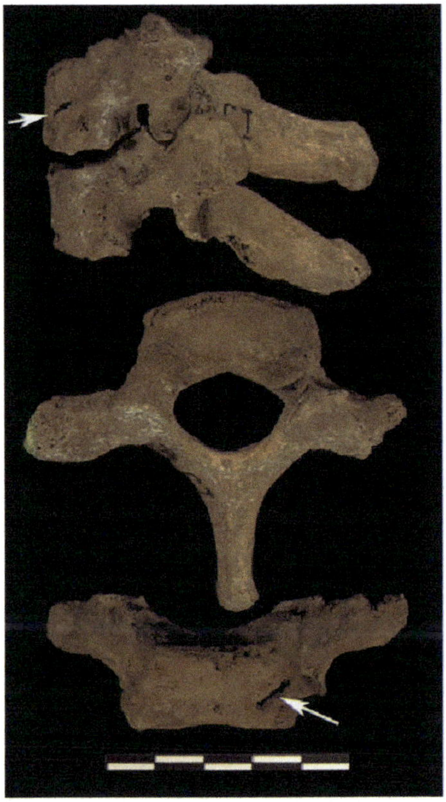

Fig. 5 The first (T1) and second (T2) thoracic vertebrae of Sunghir 1: (A) T1 and T2 articulated in left lateral view; (B) T1 in cranial view; and (C) T1 in ventral view. Scale in centimeters. The lesion is indicated by the *arrows* in the lateral and ventral views. *From Trinkaus E., Buzhilova A.P. The death and burial of sunghir 1.* Int J Osteoarchaeol. *2012;22:655-666.*

three of them were identified as burials. These *sapiens* fossils have been dated to about 80,000–120,000 ya using electron paramagnetic resonance and thermoluminescence dating techniques (see section on "Absolute dating" in Chapter 4).[26] Interestingly, the AMHs at the Skhul Cave exhibit a mosaic of ancestral and derived traits. Although the skulls are rounded as in AMHs, they possess evident brow ridges and a projecting facial profile reminiscent of Neanderthals. This combination of traits has prompted some to suggest that they represent *sapiens*/Neanderthal hybrids, but most scholars attribute the mosaicism to the retention of archaic ancestral *sapiens* characters rather than Neanderthal features.[27]

Among the Skhul individuals, two exhibit signs of trauma that may have been caused by acts of interpersonal violence. The least convincing of the two is the one of a child whose skull has a depressed area in the mid-line of the frontal bone. Examination of the fossil led investigators to conclude that the injury was the result of an impact by a blunt object. In addition, this skull displays a hole and fracture of the right temporal bone in the roof of the ear, also resulting from an impact. The investigators determined that the wounds were inflicted at the time of death.[28] Based on these conclusions, the likelihood of two massive consecutive impacts in different parts of the head leading to death occurring by accident is improbable.

The second and more convincing case of interpersonal attack coming from the Skhul Cave is that of Skhul IX, an adult. Dated at 91,000 ya, this skull exhibits injuries that appear to have been caused by a spear-like weapon. Investigators indicated that the injuries were inflicted at the time of death. The spear was thrust through the upper leg and into the pelvic cavity.[29] In the process, the weapon severed the whole thickness of the head of the femur and the floor of the acetabulum (socket into which the head of the femur fits), and penetrated into the pelvic cavity. To generate such an extensive injury, the spear must have had a hard tip and the person who inflicted the wound must have been strong. Skhul IX also shows signs of trauma to the head caused by a strike that impacted at an angle. The two lesions happened at the time of death and the victim did not survive the attack. Together, the two injuries causing the death of the individual are compatible with a premeditated act of interpersonal violence.[29] It is possible that the unfortunate victim was first immobilized by the wound to the pelvis and was then finished by the mortal blow to the head.

Violence among Neanderthals

Cranium 17

Several cases involving violence among Neanderthals have been documented throughout their geographic range and time period. One clear event is that of Cranium 17 from Sima de los Huesos, located in Atapuerca, northeastern Spain. The site is made up of a series of limestone caves near the city of Burgos. Within the caves, archeologists have recovered approximately 1600 hominin fossils corresponding to 28 individuals. The fossils date to a time period ranging from 600,000 to 300,000 (Middle Pleistocene), and most scholars consider them Neanderthals due to their projecting mid-face, long and narrow pubic bones, and thick finger bones. A number of theories have been proposed to account for the accumulation of remains at Sima de los Huesos, with the currently most widely accepted view being that bodies were dropped from an opening to the cave down a 13 m vertical chimney. It is likely that the bodies were thrown into the small cavity during the active period of the Neanderthal morgue. Given the mode of deposition and the free fall, there is no surprise that the remains are not arranged in any particular manner, and they simply rested in a mass on top of each other.

However, one skull—Cranium 17—is of particular interest because it represents a 430,000-year-old fractured skull, which is the earliest clear evidence of interpersonal

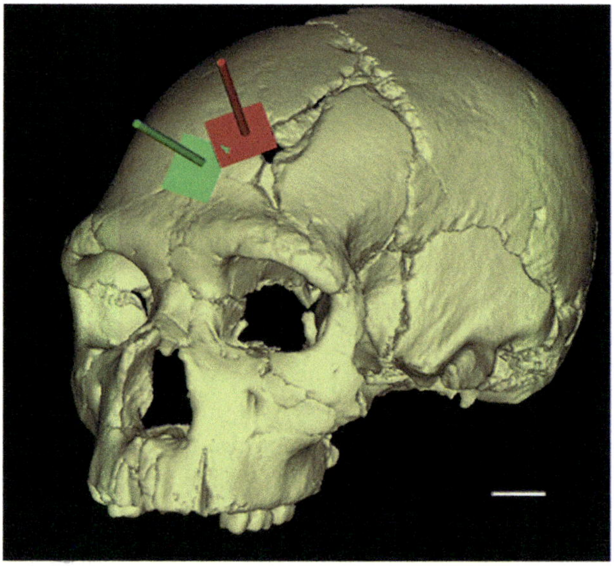

Fig. 6 Reconstruction of the trajectory of the impacts relying on the planar orientations of the external outlines of the fractures (*green*=T1 and *red*=T2) (https://www.researchgate.net/publication/277326376).

violence in hominins.[30] Cranium 17 exhibits two clear perimortem depression fractures on the frontal bone (Fig. 6), which are likely to have been caused by a blunt object. Both fractures compromised the bone-brain barrier and either individually would have caused the death of the individual. It was a frontal attack by a right-handed attacker using a club. Considering the lethality of both impacts, it is reasonable to think that the intention was to inflict impacts resulting in death. The severity of the traumas and the lack of healing and bone remodeling in the injured area suggest that the individual did not survive the attack. Based on the forensic evidence, the best explanation for the fractures is an intentional act of vicious interpersonal violence.

Shanidar 3

The Shanidar site is located in the Zagros Mountains of northeastern Iraq. Initial excavations were carried between 1957 and 1961 by a team of archeologists from Columbia University led by American anthropologist Ralph Stefan Solecki (1917–2019). In the Shanidar Cave, the remains of 10 Neanderthal individuals dating from 65,000 to 35,000 ya have been discovered, including seven adults and three infants.[31] Shanidar is of especial interest since four out of the seven adult individuals exhibit signs of trauma. This collection represents the largest group of Neanderthals showing injuries from a single site.

One of the most highly incapacitated individuals found at the site is Shanidar 3, the skeleton of an adult male who died between the ages of 35 and 50. Shanidar 3 exhibits an antemortem injury on the superior edge of the left ninth rib.[32] The injury created an incision about 1.5 mm long that was partially healed (some bone remodeling took place after the lesion was made while the lithic was still embedded) at the time of death (Fig. 7), several weeks after the incident. The angle of the injury has suggested to some scholars that the wound was not self-inflicted and likely was the result of an interpersonal confrontation involving a right-handed individual who stabbed Shanidar 3 while they were standing face to face.[32]

Due to the nature of the injury, it is likely that the point punctured the pleura (membranes lining the thorax and enveloping the lungs) of the left lung, which led to a partial or complete pneumothorax or a collapsed lung. Assuming only a partial collapse of the lung, the probability of posttraumatic survival would have been fairly good.[33] In the case of Shanidar 3, no evidence of infection is present. It is not clear how Shanidar 3 died several weeks after the incident.

More recent controlled stabbing experiments using replicas of Mousterian (cultural tradition associated with Neanderthals in Europe) and Levallois (lithic tradition made by AMHs) points directed against the thoraces of pig and goat carcasses suggest that the lesion may have been caused by a long-range projectile.[34] The experiments were designed to mimic lesions made by the two types of points moving with low or high kinetic energy (energy possessed by an object in motion). In other words, the experiments were conducted to ascertain the effects of two lithic-type weapons moving at different speeds. The two experimental conditions simulate the damage resulting from the impact of a thrusting spear moving with high kinetic energy compared to a long-range projectile weapon moving with low kinetic energy. The damage patterns of the two were

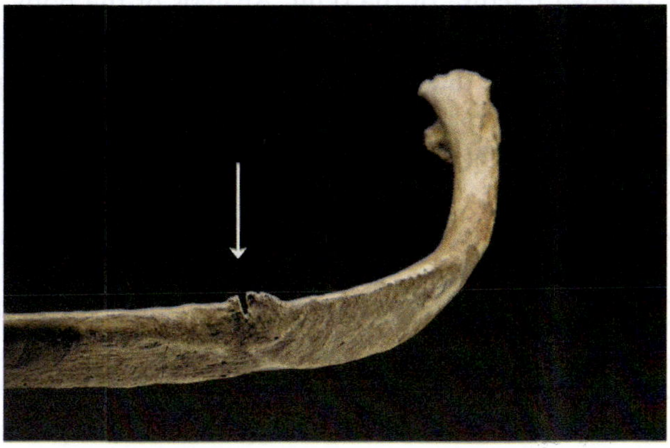

Fig. 7 Partly healed stab wound in Shanidar 3 (https://humanorigins.si.edu/evidence/human-fossils/shanidar-3-neanderthal-skeleton).

found to be different. Impacts resulting from high kinetic energy and momentum (generated by a thrusting spear) produce greater damage at the wound site involving multiple ribs while low kinetic energy and momentum stabs (generated by long-range projectiles) tend to cause more restricted lesions involving single ribs. The lesion seen in the rib of Shanidar 3, being limited and restricted only to part of a rib, is most consistent with injury from a low kinetic energy and low momentum weapon, such as a long-range projectile. And although it does not rule out accidental injury, the track, placement, and orientation of the rib lesion are more congruent with a projectile traveling along a ballistic trajectory.

Given that Shanidar 3 lived 45,000–35,000 ya, it is possible that his time period overlapped with the beginning of AMH presence in West Asia. Thus, it is likely that the two species coexisted in the region. Since Neanderthals made spears and used them by thrusting them to create massive wounds while AMHs used their spears as projectiles that created smaller lesions, it is likely that the perpetrator of the violent attack on Shanidar 3 was a *sapiens*. Alternatively, the attacker may have been a weapon-acculturated Neanderthal.

St Césaire 1

In 1979 the crushed remains of a young adult male Neanderthal were discovered in the Roche-à-Pierrot rock shelter, near the village of Saint-Césaire in southwest France.[35] The fossil was associated with Châtelperronian artifacts and dated at 36,000 years old. It thus represents one of the last members of its species in Europe. Although clearly a Neanderthal, this specimen possesses a number of traits characteristic of late members (more recent) of the species, such as smaller teeth, modest eyebrow ridges, less massive jaws, and limb bones that resemble those of early AMHs. This last feature indicates a shift in the patterns of locomotion and arm use.

Computer-tomographic imaging and computer-assisted reconstruction of the skull brought to light a healed fracture of the skull. Analysis of the bony scar indicates that the trauma was caused by the impact of a sharp object. It also suggests that an implement was used to strike the skull of the individual during an act of interpersonal violence. Although it is possible that the injury resulted from an accident, the location of the fracture is more compatible with interpersonal violence. Interspecies violence involving an AMH attacker cannot be ruled out.

Assuming the victim and attacker were in an upright position, the direction of the fracture, as well as its top position on the skull, is diagnostic of a hit from the front or from the back. The immediate outcome of such an injury would have been heavy bleeding, confusion, and impairment. Considering that St Césaire 1 survived the attack and lived for a while, as indicated by the observed degree of healing, it is likely that members of his clan assisted him back to health. This suggests that compassion was part of the Neanderthal behavioral repertoire. Other members of the group likely nursed this severely impaired individual over an extended period of time. If the attack was carried out by a member of the same clan, the care and support provided to the impaired individual probably was done in spite of (or assisted by) the perpetrator.

El Sidrón

El Sidrón Cave is located in the province of Asturias in northwest Spain about 50 km from the city of Oviedo, the provincial capital. At the site, the remains of at least 13 Neanderthal individuals were accidentally discovered in 1994. Three were males, four females, three adolescent boys, and three children, including one infant. The fossils were found in one of the chambers of the cave, the Ossuary Gallery, which measures 28 m long and 12 m wide. The fossils were found by cave explorers when they came across two jaws that they thought belonged to contemporary humans. Initially, it was believed that the bones were the remains of Republican fighters who had died in the cave while hiding from Franco's forces during the Civil War. However, after the police arrived and more bone fragments were identified and collected, forensic scientists were summoned to the site and it was eventually determined that the bones were not of Republican militia members who died hiding in the cave, but instead were the remains of Neanderthals who died approximately 48,000–49,000 ya.

The bones were found mixed with gravel and mud, suggesting that the individuals had died not at the site, but somewhere outside the cave, and that subsequently the bones had been carried by a flood into the chamber. Clues as to what actually happened at El Sidrón were evident when researchers examined the bone fragments more closely and began to notice cut marks, flakes, percussion pittings, conchoidal (smooth shell-shaped concave surface) scars, and adhering flakes.[36] The researchers realized that the 13 Neanderthals had been killed approximately 50,000 ya and had been eaten by another clan of the same species.[37] Indeed, this area seems to have been the site of a gruesome massacre. No evidence of fire or burns in bones was detected, suggesting that the flesh was eaten raw, possibly in one sitting. The hungry Neanderthals not only ate the meat, but cracked open the long bones to eat the marrow and smashed the skulls for the brain tissue and the tongues. The lithics found with the fossils were crafted from rocks in a quarry located just a few kilometers away from the cave. These implements were likely used to slice muscle from the bone of the assassinated victims. A scenario could be envisioned in which the victims might have wandered into the territory of another clan of Neanderthals and, for their indiscretion, they were mass-murdered and eaten.

In addition to the potential of Neanderthals to exhibit extreme violence toward other members of the same species, El Sidrón has provided scholars with a view of the species' family structure. For example, all three men had the same mtDNA (the entire mtDNA sequences were identical), which could mean they were brothers, cousins, or uncles. However, the four females possessed different mtDNA haplgroups, indicating they came from different families. The genetic make-up of this clan of Neanderthals suggests that they lived in a small extended family made up of close relatives who practiced a patriarchic family system. It is possible that this family structure may have led to some degree of endogamy and inbreeding among the species. It is likely that when two bands met and they were not too hungry, they sometimes exchanged daughters. Other information derived from the genetic analysis is that two individuals possessed a gene variant for red hair.

AMH violence directed at Neanderthals?

Les Rois

Les Rois, in southwestern France, is an archeological site where AMH fossils have been found in association with Aurignacian technology. Les Rois is a cave located 2 km south of the village of Mouthier-sur-Boëme in Charente. Two archeological layers have been identified and dated to 28,000–30,000 ya. The site was discovered in the late 1920s and reported in the literature for the first time in 1937.[38] At that time, a few teeth, now lost, were collected from inside the cave. Subsequently, in 1958, 2 mandibles (A and B) and 37 teeth were identified.[39] Most scholars consider the human fossils of Les Rois to be AMHs of the Early Upper Paleolithic. Yet during the 1958 study, it was noticed that both mandibles, but especially mandible B, and the teeth exhibited a number of Neanderthal-like traits. Furthermore, in mandible B a number of cut marks were observed. Since then, it has been proposed that mandible B belonged to a juvenile exhibiting Neanderthal traits, whereas mandible A is attributed to an AMH. Of particular interest are the enamel development profiles seen in mandible B, which resemble the state of those of Neanderthals and are unlike the enamel development profiles of AMHs from the Upper Paleolithic of Europe.

Several interpretations are possible for this scenario with regard to the cut marks. One theory, provided by Ramirez Rozzi and colleagues,[40] is that the juvenile mandible may represent the consumption of a Neanderthal individual by Aurignacian AMHs, possibly resulting from an act of interpersonal violence and/or symbolic ritualistic death. This interpretation portrays AMHs hunting down the last Neanderthals in the area. Part of the problem with this interpretation is that the proposed date of this putative encounter (the date of the site) was 28,000–30,000 ya. This age is quite recent to be compatible with most of the estimations for the extinction of Neanderthals in Europe. Other interpretations have been proposed, including the possibility that the Aurignacian utensils at Les Rois were made by AMHs/Neanderthal hybrids possessing both AMH and Neanderthal traits. In addition, since many European AMH fossils from 40,000 to 25,000 ya show different degrees and combinations of Neanderthal and AMH features, possibly due to the retention of ancestral traits in common or introgression, it is possible that the Les Rois remains belonged to ancient AMHs with ancestral traits or hybridization. These interpretations would suggest that ancient AMHs possessed considerable variation of ancestral and derived (modern) characteristics during the Upper Paleolithic. Yet Rozzi and collaborators believe the jawbone with cut marks is compatible with AMHs attacking

Neanderthals, and sometimes killing them, bringing back their bodies to caves to eat and/or to use their skulls or teeth as trophies.[40]

Increase in interpersonal violence since the Middle Paleolithic

Table 1 summarizes documented cases of trauma related to interpersonal violence among hominins during the Middle Paleolithic. A review of the table shows that the traumas represent different degrees of severities and body locations, affecting both sexes and various age categories. Additional data that compare injuries caused by blunt objects versus projectiles during the period of 300,000–50,000 ya (Middle Paleolithic) indicate a ratio of 20/1, respectively, while 50,000–10,000 ya (Upper Paleolithic) was 13/3, and 10,000–5000 ya (Mesolithic) was 60/17. This indicates not only an increase in the number of injuries, but also an escalation in the use of projectiles or spears as weapons. The increase in violence during the Mesolithic has been well documented.[43] This era has been identified as an epoch in which violence became far more common compared to earlier periods of human history. The underlying reasons for this increase are unclear. Causes such as sedentism, conflicts between foragers, pastoralists, and agricultural societies, accumulation of goods and wealth by certain sectors of the community, prestige, or other elements associated with societal complexities have been proposed.[44] Alternatively, it has been postulated that the Mesolithic spike is simply the result of sampling error resulting from the greater recovery of skeletal remains compared to earlier periods. The augmentation in the use of projectiles is likely a function of the fact that spear-throwing technology was a *sapiens* innovation, while earlier hominins, including Neanderthals, lacked the know-how. The minimal number of injuries during the Middle Paleolithic in Europe corroborates the notion that Neanderthals did not possess the projectile technology that *sapiens* imported from Africa to Europe and that made them so deadly.

Hard-wired to violence

In addition to the stereotypic perception of Neanderthals as being mentally inferior to humans, traditionally they have been portrayed as harsh, clumsy, and dangerous. This reputation is, in part, the result of how early scholars described their fossils and interpreted their anatomy as deformed versions of *sapiens*. Furthermore, since the early years after their discoveries, scholars began to notice in Neanderthals' skeletons an unusually high incidence of traumatic injuries, especially in the head and neck region.[45] Investigators have theorized that the injuries were due to Neanderthals' dangerous lifestyles, which included violent social behavior, a mobile hunter-gatherer subsistence, attacks by predators, and a close-range hunting style that required stabbing or thrusting spears into large dangerous game, such as rhinoceri.[46] AMHs practiced the less risky hunting style of using spears as projectiles in which the game was killed from a safe distance. This important

Table 1 Summary of Middle Paleolithic trauma possibly related to interpersonal violence.

Degree of severity	No. individuals	Area of trauma	By age class	By sex
First degree	4			
Nonlethal trauma to postcranial skeleton	Feldhofer 1	Ulna	Old	Male
	Krapina 180	Ulna	Adult	Unknown
	Krapina 188.8	Ulna	Adult	Unknown
	Le Moustier 1	Mandible	Adolescent	Male
Second degree	10			
Minor cranial trauma	Feldhofer 1	Occipital	Old	Male
	Krapina 4	Frontal	Adult	Male
	Krapina 5	Parietal	Adult	Male
	Krapina 20	Frontal	Adult	Female
	Krapina 31	Frontal	Adult	Female
	Šala 1	Frontal	Adolescent	Female
	Shanidar 1	Frontal	Old	Male
	Shanidar 5	Frontal	Old	Male
	Skhul 1	Frontal	Child	Unknown
	Skhul 1	Temporal/ Mandible	Child	Unknown
Third degree	5			
Major cranial trauma, probably involving a loss of consciousness	Krapina 34.7	Parietal	Adult	Unknown
	Shanidar 1	Frontal/ Parietal	Old	Male
	St. Césaire	Parietal	Adolescent	Male
	Cova Negra 1	Parietal	Unknown	Unknown
	Qafzeh 11	Frontal	Adolescent	Male?
Fourth degree	2			
Fatal trauma	Shanidar 3	Rib	Old	Male
	Skhul IX	Innominate	Prime	Male

Note: For references to site publications, see Estabrook[41] and Wu et al.[42]
From: Mark W. Allen, Terry L. Jones. *Violence and Warfare among Hunter-Gatherers*. Routledge, Abingdon-on-Thames, UK; 2016.

hunting innovation and *sapiens*'s more "civilized" nature explained the fewer incidents of traumatic injuries observed in AMH fossils. For the most part, the studies that led to these inferences were based on case-by-case reports of injuries on specific individual skeletons, rather than on systematic, population-wide statistical analyses. Scholars typically would discover bones exhibiting injuries and signs of trauma and then report their findings in the literature as a result of accidents, predator attacks, or violence. However, in a recent study, the investigators took a different approach. They performed a systematic survey of injuries by using quantitative and population-wide analyses of a sizeable number of Neanderthals and AMHs.[47] The advantage of this methodology is the capacity to evaluate large numbers of samples, which allows rigorous statistical analyses of the data. Statistical analyses were used to assess if the observed differences between groups (i.e., Neanderthals and AMHs) are greater than we would expect to observe by random chance. For example, the investigators identified 21 specimens with 1 or several cranial lesions (nine Neanderthals and 12 Upper Paleolithic AMHs) in a sample of 114 Neanderthals and 90 AMHs. When these numbers are statistically evaluated, they indicate that Neanderthals and Upper Paleolithic AMHs incurred comparable incidences of cranial trauma. In addition to showing equivalent numbers of injuries in both species, the data demonstrated that male individuals incurred a higher incidence of trauma in both species, likely reflecting sex-specific behaviors and activities. Division of labor, in which males hunted and females took care of infants and nursed injured clan members, could have generated such a sex bias in the number of injuries. In addition, the injured Neanderthals, as a group, were younger. This age-dependent difference between the taxa may be a reflection of the age of the individuals at the time of injury. It is possible that Neanderthals were more likely to be injured at a younger age. Alternatively, the mortality risk of trauma survivors was higher among Neanderthals (long-term survival after a nonlethal injury among AMHs).

Comparative empirical data also confirms that primates, and specifically Paleolithic hominins, are violent by nature. Employing a unique phylogenetic approach comparing the incidence of violence among mammals, it was concluded that lethal aggression is deeply rooted in the primate lineage and is particularly prevalent among hominins.[48] In other words, the propensity for violence is not only encoded in the DNA of AMHs, but can also be considered an ancestral trait that goes back to early primate evolution. With this approach, Gómez and colleagues estimated AMHs' rates of lethal violence at the origin of *sapiens* approximately 160,000–200,000 ya. It was concluded that early AMHs were about as violent as expected based on other primate species. Specifically, according to their calculations, these early AMHs were about six times more violent than the average mammal.

In order to estimate the levels of violence in specific taxa, Gómez and collaborators compared data on more than 4 million remains of 1024 mammal species from 137 (or 80%) taxonomic mammalian families. The AMH samples were from different time periods, including the Paleolithic, Mesolithic, New and Old World Neolithic, Bronze Age, and Iron Age, and from the past few centuries. For each species, the authors assessed the proportion of deaths out of the total resulting from interpersonal violence involving member(s) of the same species. From the incidence of violence in present-day species, the investigators extrapolated that 1 in 300 (0.30%) deaths were violent at the time of the origin of mammals about 200 million years ago. The rate of violence increased continuously during mammalian evolution. By the time of the ancestors of the Great Apes, the rate of mortality by violence was 1.8%. Among the mammalian species, the increase in violent deaths was found to correlate with communal living and territoriality. Examination of the Paleolithic specimens (12–50 kya) generated 2% of violent deaths, similar to the rate at the origin of our species. Subsequently, during the time period between 3000 and 500 ya, violent death jumps from 15% to 30%, only to decline in contemporary populations.

The comprehensive comparative phylogenetic approach utilized by Gómez and collaborators strongly suggests an evolutionary tendency since the origin of mammals to kill members of their own species. The data suggest that this propensity for lethal violence has culminated in *sapiens*.

Violence was not ubiquitous

From the above discussion of the archeological record, it is clear that both *sapiens* and Neanderthals used violence as part of their behavioral repertoire. Yet it is likely that at least during the early stages of the migration of AMHs into the Levant and Europe, when population densities of both Neanderthals and AMHs were low, avoidance by dispersal or just moving away from each other functioned efficiently as a mechanism for conflict resolution. In those early years, sharing the same general territory may not have led to frequent confrontations, as Neanderthals were decreasing in numbers and modern humans were moving into extensive territories. It was probably not difficult to avoid each other as well as members of opposing clans of the same species.

Although for many years scholars thought of the disappearance of Neanderthals and their replacement by AMHs as a simple process of competition and extinction, today it is clear that this was a much more complex population phenomenon. The available data on interbreeding clearly demonstrate that it was not a simple replacement of Neanderthals. It is even possible that different combinations of hominins, including Neanderthals, Denisovans, and AMHs (and possibly other species), became quite friendly with each other when they did meet.

Neanderthals were not brutes. They were a sophisticated species. They built tools, made jewelry, created cave paintings, buried their dead, were stronger than us, had bigger brains, and may have been just as smart or smarter. Considering the overall physical and behavioral similarities between Neanderthals and AMHs (Fig. 2), it is not clear to what degree each species thought of each other as "self." Did Neanderthals and AMHs find each other repugnant or attractive? Were they able to express affection or even love for each other? Today we still know very little to help us answer these fascinating questions.

We must have seen something in each other. We mated with each other repeatedly during different time periods. We mated with Neanderthals about 100,000 ya in Siberia, then 50,000–60,000 ya in the Levant and likely more recently while in Europe. Clearly, our rendezvous were not one-time events. Estimating the number of mating events is difficult, since such calculations require knowledge of the number of individuals of each species in the area at the time of the encounters. It could have been a couple of hundred matings or thousands of matings. Although population size is unknown, most scholars think both Neanderthals and AMHs lived at low population density in extended family groups of one to a few dozen individuals. Taking this into consideration, the encounters were probably not frequent, and the fact that various gene flow events are detected suggests that they were willing to mate when they did meet. This indicates that both species shared the same mate recognition system. In other words, since animals mate with partners that they recognize and are comfortable with, Neanderthals and AMHs must have overlapped in their mate recognition system, to the extent that they were attracted to each other.

The direction of gene flow from these sexual encounters is revealing. Were Neanderthal males impregnating *sapiens* females or AMH males fertilizing Neanderthal females or both? The genetic data indicate that gene flow went in both directions.[49] Children of AMHs and Neanderthals were eventually incorporated into *sapiens* as well as Neanderthal populations. Since children usually stay with their mother, that means that a *sapiens* male likely mated with a Neanderthal female to produce a hybrid child that introduced *sapiens* DNA into the Neanderthal gene pool. Conversely, a Neanderthal male mated with an AMH female in instances of Neanderthal DNA in the *sapiens* gene pool. Of course these two scenarios are only valid if the females (Neanderthal or *sapiens*) were not abducted into clans of the other species. Some may argue that a Neanderthal father and *sapiens* mother may be more compatible with a conceptual intercourse. What is significant, in the context of this discussion, is that independent of which species provided the sperm or the ovum, the hybrid children would have grown up in a community of individuals who were somewhat different to them. The implication of this is that hybrid children were accepted and became integrated into the *sapiens* or Neanderthal communities. In other words, they were able to survive and pass their genes to the next generation and so on. Were these hybrid children discriminated against? Clearly, any discrimination was not enough to prevent them from surviving, mating, and starting their own families. Another scenario, although less likely, is that some interspecies families existed in Eurasia and this was common and accepted. In other words, families were made up of different combinations and/or proportions of Neanderthal and AMH DNA. It is not clear how such mixed family units could have interacted with single-species family units. Would they have engaged in daughter exchanges with mixed communities?

The fact that gene flow occurred in both directions provides answers to some of the questions posed earlier in this chapter. It suggests that some degree of kinship existed between the two species. It also indicates that they saw each other sufficiently similar to be sexually attracted to each other and to mate. Therefore, it is possible that the interactions between humans and Neanderthals were not totally aggressive and some harmony occurred at times. It is likely that the interspecies interactions and relationships were complex, variable, and dependent on the circumstances and the specific individuals and clans involved, and thus difficult to generalize. It is also possible that Neanderthals and AMHs did not realize they were different.

Other signs of intimacy

At times, very valuable scientific information is discovered in unexpected places. This was definitely the case when scientists decided to examine the fossilized dental plaques of Neanderthals who lived in Europe about 48,000–49,000 ya.[50] They were not looking for Neanderthal DNA; they were searching for the DNA of microbes that lived in the Neanderthal's mouths. It turns out that a number of microorganisms (microbiota) live by attaching themselves to the surface of animals' dentures, including ours, close to the gums and feeding from any residual food left behind after meals. This is a benefit offered to microbes by not only humans, but also other animals, including hominins. Microorganisms thrive in these areas, and the greater the amount of residual food, the more they like it. The accumulation of decaying food has a negative impact on the health of dentures as well as the gums, and for that reason dentists urge their patients to brush and floss their teeth regularly, especially after meals. Fortunately for these molecular biologists interested in human evolution, Neanderthals did not brush their teeth after dinner, and they thus accumulated tartar loaded with microbes and their DNA.

The fauna that the investigators found within the plaques provided a treasure trove of information on the lifestyle of Neanderthals from two locations: El Sidrón in Spain and Spy in Belgium. This information included the type of diet consumed by Neanderthals. It turns out that the calculus residues recovered by the investigators contained not only DNA from microbes, but also DNA from the food material that they ate. Thus, it was possible to ascertain the species of plants and animals that Neanderthals consumed in the two different locations. Currently, scientists have DNA data on most groups of organisms, allowing the identification of most living things, including the species that were alive during the Paleolithic. This allows investigators to ascertain the type of microbes that were alive in tartar 50,000 ya. Scientists, for example, found out that the inhabitants of El Sidrón were surviving mainly on a vegetarian diet while those in Spy were primarily carnivores. This finding made sense, since the ecology in Spy at the time consisted of plains and grassland frequented by various fauna, including rhinoceri, wild sheep, mammoths, horses, and reindeer. El Sidrón, on the other hand, was a forested area where mushrooms and plants were abundant. Although it is likely that not all foods have an equal chance of getting incorporated into plaque and that not everything has a chance of being fossilized and preserved, these findings still tell us that Neanderthals were very adaptable and flexible. Not only were they omnivores, but also they were able to survive by eating very different foods.

Of particular interest were the types of microbes that the investigators found. One of them is the archaeal single-cell organism *Methanobrevibacter oralis*. Similar to bacteria, this microbe thrives in low-oxygen environments, such as the mouths of carnivores. It turns out that the genetic material of *M. oralis* started differentiating (becoming different) in the teeth of Neanderthals and AMHs as soon as their lineages separated about 500,000 ya in Africa. The DNA of these two microbial evolutionary lines continued evolving their separate ways, increasingly becoming more and more distinct. Since the rate of change of DNA is known, the amount of DNA differences between Neanderthals and contemporary AMHs' *M. oralis* DNA was used as a proxy to construct a molecular clock to estimate the date at which the two lineages of microbes separated in the past. It was assessed that the two lineages of *M. oralis* diverged from each other about 112,000–143,000 ya. In other words, the DNA of the strain of *M. oralis* in humans today indicates that approximately 112,000–143,000 years have passed since it separated from the strain of *M. oralis* that lived within Neanderthals' tartar. Since this time is long after AMHs and Neanderthals separated in Africa, the acquisition of the microbes by the latter must have happen during subsequent interactions between Neanderthals and AMHs when *sapiens* migrated out of Africa, potentially in the Near East. If this timeline is accurate, it may indicate that AMHs came in contact with Neanderthals earlier than 70,000 ya and more in line with more current, older AMH fossils dating between 120,000 and 177,000 ya.[51]

The question remains how the microbes were transferred. Could it have been via sharing food while eating a friendly interspecies meal? Or by Neanderthals scavenging food previously partially consumed by AMHs? Acts of cannibalism in which parts of the vocal cavity, including the tongue, are consumed raw may have facilitated microbial transfer (see section on "El Sidrón" above). This last mode of transfer is an act not of intimacy, but of violence. Other scenarios include some degree of physical contact, involving spit transfer during kissing. It is not possible to discriminate among these possibilities with the current data. Although it is not known if Neanderthals engaged in lip-to-lip kissing, chimpanzees and bonobos not only kiss but the latter are known to practice tongue-kissing, oral sex, and genital massaging.[52] Yet it is not known if ancient *sapiens* and/or Neanderthals were inclined to kiss. What is clear is that the two species had some form of close interaction, and the observations suggest that instead of aggressive behavior, the two species may have interacted sensually.

An effort to explain Neanderthals' extinction

Although AMHs and Neanderthals have a common origin in *Homo heidelbergensis*, sometime between 520,000 and 630,000 ya in Africa, the two species began to evolve independently after they split. After the separation, Neanderthals migrated to Eurasia where most of their evolution took place. The *sapiens* branch stayed in Africa where it evolved until they too migrated to Eurasia. Thus, after about half a million years of separation, the two lineages became neighbors, now somewhat different from each other.

Over the years since Neanderthal fossils have been recognized as a distinct species, a number of theories have been put forward to explain their extinction. Theories have ranged from Neanderthals actively being killed by *sapiens* to having limited capacity to develop complex social networks and/or inability to adapt to climate changes. It is likely there was no single cause to account for the extinction of Neanderthals, but a combination of factors contributed to their demise. In addition, it is likely that they became extinct in different regions at different times and for different reasons. In some areas where populations of both species were in close contact, it could have been the result of competition, while in other regions where their numbers were small, it may have been due to inbreeding and/or interbreeding. The speculations fall into a number of general categories including violence from encroaching AMHs, competition, technological advances, better clothing/shelter, improved hunting techniques, more diverse subsistence strategies, larger group sizes, more elaborate social networks, differences in birth/mortality rates, longer interbirth intervals, shorter life expectancy, interbreeding, inbreeding, and natural catastrophes, among others. Some of these putative causes have been elaborated earlier in this chapter and additional ones are discussed as follows.

Competition

Generally speaking, competition could be direct/active or indirect/passive. In a direct competition scenario, AMHs and Neanderthals may have been involved in physical confrontation. In such encounters, AMHs may have been favored due to numerical or technological superiority. Alternatively, passive struggles for resources, such as game, shelter, water, etc., may have been a factor. These passive rivalries may have taken the form of intimidation by occupying a territory or hunting area with a large number of clan members.

Bigger eyes and visual brain centers

The separate evolutionary paths of Neanderthals and AMHs included the structure and function of the brain. Both lineages experienced an increase in overall brain size, and both lineages ended up with similar-sized brains. Yet Neanderthals enlarged their brains differently, and their morphology consequently differed. Neanderthals expanded their crania lengthwise, which resulted in skulls shaped more like oval footballs, while

AMHs' skulls are more globular-looking, resembling round soccer balls. These overall topological differences of the two skulls as well as other anatomical postcranial features are no surprise considering the environmental differences between the two locales in which they lived and the resulting different selection pressures. Specifically, one of the differences noticed between Neanderthals and AMHs is an enlargement of the visual processing areas of the brain in the former and the relatively large olfaction regions in the later. Larger eyes and visual centers in the brain are thought to be adaptations to the long, dark nights and winters of Europe where Neanderthals lived.[53] Relatively large olfaction regions contributed to a greater sense of smell, which was beneficial for consciously or subconsciously identifying safe foods or detecting social information in AMHs.[54]

One thesis proposes that Neanderthals evolved a large visual processing system, including eyes and visual cortex, as an adaptation to living in high latitudes.[53] In other words, it has been suggested that the limited amount of light in higher latitudes selected for an enhanced visual system in Neanderthals. AMHs, on the other hand, living in the tropics where light is more plentiful, evolved in a different direction. They instead evolved bigger frontal lobes, allowing them to develop more complex social lives. Since Neanderthals did not possess significantly larger brains than AMHs, this meant that the two species' brains were organized differently and that a greater proportion of the overall brain tissue in Neanderthals was invested in visual and somatic systems, while proportionally less neural tissue was left for other functions, such as higher-level cognitive functions, compared to AMHs. Therefore, it is argued that Neanderthals became specialized for receiving and processing visual information, but when confronted with challenges requiring teamwork, for example, they were at a disadvantage. This limitation is thought to have contributed to their extinction.

Inbreeding

Inbreeding comes about when genetically related individuals mate. It occurs when family members or related individuals within small groups are isolated from outsiders and reproduce. Genetically, inbreeding results in homozygosity (having two identical forms of a gene in an individual) of deleterious genes. In higher organisms, genes exist in duplicate and when both copies are defective and/or compromised, the normal function performed by that sequence of DNA is impaired. Generally speaking, defective genes that do not perform their function, or that are less efficient, reduce the biological fitness of individuals and of the population at large. This is because defective genes lower the ability of the affected individual to reproduce and reduce the probabilities of the population to survive.

It has been suggested that late during Neanderthals' existence, they lived in small family groups made up of extended family members.[55] These small clans were spread out across Eurasia. These living conditions provided for intensive inbreeding as close family members mated repeatedly, generation after generation.[56] Under these conditions, genetic variability rapidly decreased, homozygosity increased, and deleterious genes combined within individuals, thus decreasing their individual fitness and the genetic health of the clan and the population. With time, it is thought that these isolated groups died out as members became afflicted with genetic diseases caused by those deleterious/lethal genes. This scenario is congruent with studies reporting a high incidence of congenital defects in Neanderthal vertebra fossils recovered from El Sidrón.[57] Fig. 8 illustrates the 17 congenital anomalies observed within the El Sidrón Neanderthal family group. As discussed earlier in this chapter, the family unit at the El Sidrón site was made up of an extended family of 13 individuals. The three males in the group had the same mtDNA, which could mean they were brothers, cousins, or uncles. This suggests that this family practiced some degree of endogamy and inbreeding. In addition to the negative impact of deleterious/lethal genes, populations made up of small groups of individuals are more prone to random changes in birth rate and mortality, which can bring about their extinction.

Technological advances and improved hunting techniques

Approximately 100,000 ya, early AMHs in Africa and the Levant were making tools that were similar to those made by Neanderthals living in Eurasia about the same time. These utensils were simple flake and core lithics known as Mousterian tools (see Chapter 1 for description of Mousterian tools). The arsenal of Mousterian tools consisted mainly of flakes and small hand axes made from disk-shaped cores (Fig. 9). The flakes took the shape of points that likely were used for cutting. In addition, some of the lithics had notches, which were probably used for carving. Tool making did not change much until 90,000–75,000 ya, when some AMH groups began producing more diverse and specialized lithics. Neanderthals, on the other hand, continued making their tools using Mousterian technology, which eventually died with them when they became extinct.

More sophisticated types of utensils were produced starting 70,000–60,000 ya when some *sapiens* groups began to heat-treat stones prior to flaking. This preparation step facilitated flaking and allowed stones to be shaped with finer cutting edges. At this time, blades were struck off from conical cores and used to create long, narrow flakes called blades (Fig. 10). These blades were retouched to make scrapers, and some were set in hafts or handles. In addition to lithics, bones and antlers were made into points. All of this new technology was unknown to the Neanderthals.[58] This new tool tradition goes by the name of Aurignacian culture and was marked by the invention of the burin, or tool engraving, which was a form of portable art. Some scholars think of burin as a way to personalize sets of tools used by specific individuals, a way of saying, "These are my tools." During the Aurignacian epoch, artistic expressions reached new heights. This was the period when AMHs began to create sculptures in the form of highly stylized statuettes of pregnant women, the so-called Venus figures—presumably fertility figures—as well as the cave paintings of Lascaux and Altamira.

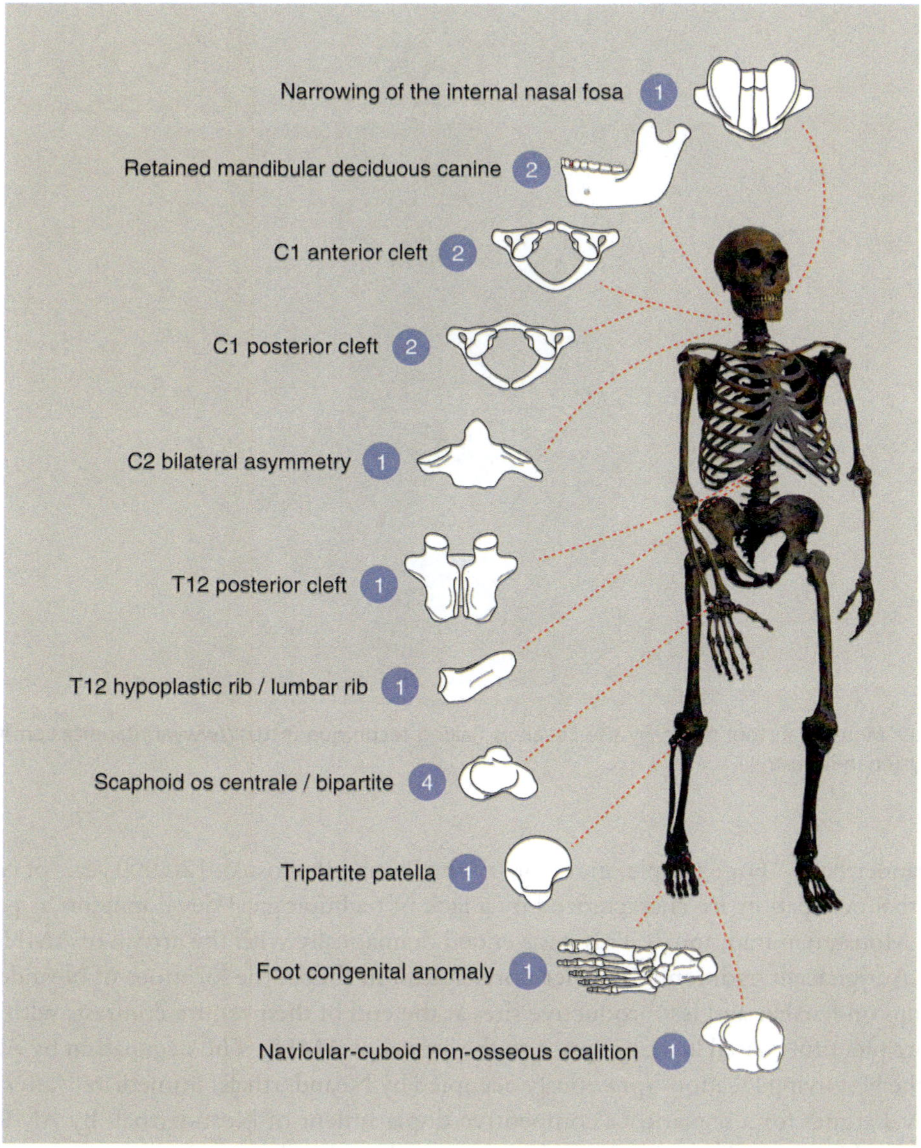

Fig. 8 Summary of the 17 congenital anomalies observed within the El Sidrón Neanderthal family group. The number of observations for each condition is shown in the *blue circles*, together with a schematic representation of the condition (https://www.nature.com/articles/s41598-019-38571-1/figures/2).

The beginning of the Aurignacian tradition coincides sharply with Neanderthals' disappearance. In fact, the onset of the Aurignacian cannot be detached from the demise of the Neanderthals.[59] This correlation has prompted many scholars to postulate that the technological turnover was linked to population replacement and the extinction of

Fig. 9 Mousterian tool made by the Levallois flaking technique (https://www.britannica.com/topic/Mousterian-industry).

Neanderthals.[60] For example, in regions of the Iberian Peninsula, 120,000 years of Neanderthal occupation are characterized by a lack of technological developments as part of the Mousterian tradition.[61] This stasis ended dramatically with the arrival of AMHs and the Aurignacian tradition.[59] Furthermore, at least in Iberia, the locations of Neanderthal camps on harsher and less productive sites at the end of their tenure contrasts with their more plentiful coastal locations prior to the arrival of AMHs. The occupation by AMHs of the best living locations (previously occupied by Neanderthals) immediately after their arrival argues for a scenario of competitive displacement of Neanderthals by AMHs.

In addition to the technological advances ushered in by the Aurignacian tradition, AMHs also used some of their more advanced tools in different ways. Take, for instance, their hunting styles and strategies. Based on the angles of penetration and cracks left behind as part of the perforation, it has been surmised that Neanderthals thrust their weapons into animals rather than throwing them. The available data suggest that Neanderthals used wooden spears to stave.[62] These spears were not as sophisticated as those made by AMHs. They were plain staves and lacked stone points. By reconstructing the scene of the hunt, it seems that they ambushed the game at specific sites—places that

Fig. 10 Aurignacian blades (http://www.lithiccastinglab.com/gallery-pages/2002marchaurignacian page3.htm).

gave them an advantage, such as cul-de-sacs, bottlenecks, depressions, precipices, and blind corners, places where they had the element of surprise in their favor. They likely approached their prey by slowly surrounding it, in collaboration with other clan members. When they were close enough, the game was staved with the wooden spear, usually from an underhand angle.[63] The animal was generally in an upright posture when staved by the low-velocity thrust of the spear. This hunting technique requires cooperation, careful planning, and concealment. It was sophisticated and involved cooperative ambush tactics. This type of hunting is quite different from that employed by AMHs, which involved running, chasing down the prey, and throwing projectiles from a distance at vulnerable parts of the body. Neanderthals, being stocky and heavily built, were likely slower and could not engage their prey in such a dynamic way. Some scholars have even theorized that Neanderthals were anatomically and physiologically incapable of throwing spears for long distances.

Neanderthals tended to use the same sites repeatedly and kill as many of the animals as possible indiscriminately and then selectively butcher the biggest, fattest ones. Every kill was skinned, but the ones not butchered were left to rot.[64] This kind of wasteful approach to hunting would have been destructive to the survival of the game population, possibly contributing to the depletion or extinction of the game in specific areas and negatively impacting the livelihood of the Neanderthal population.

Although hunting by thrusting was efficient enough to allow Neanderthals to survive in Eurasia for about 500,000 years, it came at a price. This style of hunting required coming in close proximity to the prey, and in some instances the prey was big, powerful, and endowed with weapons in the form of antlers or horns capable of killing hunters with a single blow. The risk to Neanderthals is evident in the number of healed bone fractures seen in the fossil record. On the other hand, the ability to throw spears provided greater safety to AMHs. Throwing a spear also increased the types of hunting habitats and the variety of species that could be targeted. Thus, the hunting technologies and practices employed by *sapiens* likely played a role in the disappearance of Neanderthals.

Assimilation

It has been proposed that Neanderthals did not become extinct in the sense that species often do, which is by a decrease in numbers due to lack of resources, for example. Rather, this hypothesis posits that Neanderthals did not become extinct, but instead were assimilated into the AMHs' gene pool. It has been suggested that they bred extensively with other hominin species, such as *sapiens* and Denisovans, to the point that they were incorporated into the other hominin lineage. In other words, they ceased to exist as a "pure" species as their DNA was diluted into the gene pool of another lineage. In this way, AMHs and Neanderthal gene pools coalesced after repeated episodes of introgression. Thus, the identity of Neanderthals as a distinct entity gradually disappeared. This hypothesis has some support, considering the existing evidence of gene flow from AMHs to Neanderthal groups and vice versa. Yet it is likely that the level of interbreeding detected so far was not sufficient by itself to account for the disappearance of Neanderthals[65] but occurred in combination with other factors, such as different forms of competition, climatic stress, inbreeding, etc. to contribute to their demise. In relation to this last point, it is important to keep in mind that the level of interbreeding reflected today in fossils and contemporary humans may not necessarily indicate the actual level of introgression that took place during the different periods of interaction. In other words, the amounts of introgressed DNA seen in a limited number of fossils, from a small number of locations, may not represent an accurate measurement of assimilation. Similarly, today's estimates of Neanderthal DNA may be only a fraction of the gene flow that occurred then.

References

1. Osborn HF. *Men of the Old Stone Age: Their Environment, Life and Art*. New York, USA: Charles Scribner's Sons; 1915.
2. LeBlanc S. *Constant Battles: The Myth of the Peaceful, Noble Savage*. New York, USA: Saint Martin's Press; 2003.
3. Zinn H. *A People's History of the United States*. New York, USA: Harper Collins; 1980.

4. Morrison S. *Journals and Other Documents on the Life and Voyages of Christopher Columbus*. New York, USA: Heritage Press; 1963.
5. Barras V, Greub G. History of biological warfare and bioterrorism. *Clin Microbiol Infect*. 2014;20:497–502.
6. Farhi D, Dupin N. Origins of syphilis and management in the immunocompetent patient: facts and controversies. *Clin Dermatol*. 2010;28:533–538.
7. Griffin N. *Bartolomé De Las Casas Short, Account of the Destruction of the Indies*. London, UK: Penguin; 1999.
8. Lieberman P. The evolution of human speech. *Curr Anthropol*. 2007;48:39–66.
9. University of Washington. *Study Reconstructs Neandertal Ribcage, Offers New Clues to Ancient Human Anatomy*. Phys. org; 2018. Retrieved 31 October 2018.
10. Gómez-Olivencia A, et al. 3D virtual reconstruction of the Kebara 2 Neandertal thorax. *Nat Commun*. 2018;9:4387.
11. Green RE, et al. A draft sequence of the Neanderthal genome. *Science*. 2010;328:710–722.
12. Racimo F, et al. Evidence for archaic adaptive introgression in humans. *Nat Rev Genet*. 2015;16:359–371.
13. Cerqueira CCS. Predicting *homo* pigmentation phenotype through genomic data: from neanderthal to James Watson. *Am J Hum Biol*. 2012;24:705–709.
14. Beleza S, et al. The timing of pigmentation lightening in Europeans. *Mol Biol Evol*. 2013;30:24–35.
15. Kelly R. The evolution of lethal inter-group violence. *PNAS*. 2005;102:24–29.
16. Keeley LH. *War Before Civilization: The Myth of the Peaceful Savage*. Oxford, UK: Oxford University Press; 1996.
17. Kelly RC. *Warless Societies and the Origin of War*. Ann Arbor, USA: University of Michigan Press; 2000.
18. Otterbein KF. *How War Began*. College Station, USA: Texas A&M University Press; 2004.
19. Dyer M, Fibiger L. Understanding blunt force trauma and violence in Neolithic Europe: the first experiments using a skin-skull-brain model and the Thames Beater. *Antiquity*. 2017;91:1515–1528.
20. Messina AD, et al. Fatal cranial injury in an individual from Messina (Sicily) during the times of the Roman Empire. *J Forensic Legal Med*. 2013;20:1018–1023.
21. Pasini A, et al. Violence in the Early Bronze Age. Diagnosis of skull lesions using anthropological, taphonomic and scanning electron microscopy techniques. *Forensic Sci Med Pathol*. 2019;15:324–328.
22. Schulting R, Fibiger L. *Skeletal Evidence for Interpersonal Violence in Neolithic Europe*. Oxford, UK: Oxford Scholarship; 2012.
23. Kranioti EF, Grigorescu D, Harvati K. State of the art forensic techniques reveal evidence of interpersonal violence *ca*. 30,000 years ago. *PLoS ONE*. 2019;14, e0216718.
24. Trinkaus E, Buzhilova AP. The death and burial of sunghir 1. *Int J Osteoarchaeol*. 2012;22:655–666.
25. Olson S. *Mapping human history*. New York, USA: Houghton Mifflin; 2003.
26. Lewin R, Foley RA. *Principles of Human Evolution*. Malden, USA: Blackwell Publishing; 2004.
27. Oppenheimer S. *Out of Eden: The Peopling of the World*. London, UK: Brown Book Group; 2003.
28. Coqueugniot H. Earliest cranio-encephalic trauma from the Levantine Middle Palaeolithic: 3D reappraisal of the Qafzeh 11 skull, consequences of pediatric brain damage on individual life condition and social care. *PLoS ONE*. 2014;9, e102822.
29. McCown TD, Keith A. *The Stone Age Men of Mount Carmel: The Fossil Human Remains from the Levalloiso-Mousterian*. Oxford, UK: Oxford University Press; 1939.
30. Sala N, et al. Lethal interpersonal violence in the Middle Pleistocene. *PLoS ONE*. 2015;10:1–12.
31. Trinkaus E, Zimmerman MR. Trauma among the Shanidar Neandertals. *Am J Phys Anthropol*. 1982;57:61–76.
32. Trinkaus E. *The Shanidar Neandertals*. New York, USA: Academic Press; 1983.
33. Weissberg D, Refaely Y. Pneumothorax: experience with 1,199 patients. *Chest*. 2000;117:1279–1285.
34. Churchill SE, et al. Shanidar 3 Neandertal rib puncture wound and paleolithic weaponry. *J Hum Evol*. 2009;57:163–178.
35. Zollikofer CPE, et al. Evidence for interpersonal violence in the St. Césaire Neanderthal. *PNAS*. 2002;99:6444–6448.
36. Rosas A, et al. Paleobiology and comparative morphology of a late Neandertal sample from El Sidrón, Asturias, Spain. *Proc Natl Acad Sci*. 2006;103:19266–19271.

37. Lalueza-Fox C, et al. Genetic evidence for patrilocal mating behavior among Neandertal groups. *Proc Natl Acad Sci*. 2011;108:250–253.
38. Coiffard J. L'Aurignacien en Charente. *Bullet Mém Soc Archèol Hist Charente*. 1937;16:113–128.
39. Vallois HV. Les restes humains d'âge aurignacien de la grotte des Rois, Charente. *Bull Mém Soc Anthropol Paris*. 1958;9:138–159.
40. Ramirez Rozzi FV, et al. Cutmarked human remains bearing Neandertal features and modern human remains associated with the Aurignacian at Les Rois. *J Anthropol Sci*. 2009;87:153–185.
41. Estabrook VH. *Sampling Biases and New Ways of Addressing the Significance of Trauma in Neandertals* [Unpublished Ph.D. dissertation]. Department of Anthropology, University of Michigan, Ann Arbor; 2009.
42. Wu X-J, Schepartz LA, Liu W, Trinkaus E. Antemortem trauma and survival in the Late Middle Pleistocene human cranium from Maba, South China. *Proc Natl Acad Sci USA*. 2011;108:19558–19562.
43. Thorpe N. Origins of war. Mesolithic conflict in Europe. *Br Archaeol*. 2000;52:9–12.
44. Pospisil L. I am very sorry I cannot kill you anymore: war and peace among the Kapauku. In: Reyna SP, Downs RE, eds. *Studying War. Anthropological Perspectives*. Amsterdam, The Nederland: Gordon and Breach Publishers; 1994.
45. Underdown SA. Comparative approach to understanding Neanderthal trauma. *Period Biol*. 2006;108:485–493.
46. Berger TD, Trinkaus E. Patterns of trauma among the Neandertals. *J Archaeol Sci*. 1995;22:841–852.
47. Beier J, et al. Similar cranial trauma prevalence among Neanderthals and Upper Palaeolithic modern humans. *Nature*. 2018;563:686.
48. Gómez J, et al. The phylogenetic roots of human lethal violence. *Nature*. 2016;538:233–237.
49. Kuhlwilm M. Ancient gene flow from early modern humans into eastern Neanderthals. *Nature*. 2016;530:429–433.
50. Weyrich L, et al. Neanderthal behaviour, diet, and disease inferred from ancient DNA in dental calculus. *Nature*. 2017;544:357–361.
51. Hershkovitz I, et al. The earliest modern humans outside Africa. *Science*. 2018;359:456–459.
52. Manson JH, Perry S, Parish AR. Nonconceptive sexual behavior in bonobos and capuchins. *Int J Primatol*. 1997;18:767–786.
53. Pearce E, Stringer C, Dunbar RIM. New insights into differences in brain organization between Neanderthals and anatomically modern humans. *Proc R Soc B*. 2013;280:0168.
54. Bastir M, et al. Evolution of the base of the brain in highly encephalized human species. *Nat Commun*. 2011;2:588.
55. Rogers AR, et al. Early history of Neanderthals and Denisovans. *Proc Natl Acad Sci U S A*. 2017;114:9859–9863.
56. Vaesen K, et al. Inbreeding, allee effects and stochasticity might be sufficient to account for Neanderthal extinction. *PLoS ONE*. 2019;14, e0225117.
57. Ríos L, et al. Skeletal anomalies in the Neandertal family of El Sidrón (Spain) support a role of inbreeding in Neandertal extinction. *Sci Rep*. 2019;9:1697.
58. Watson W. *Flint Implements: An Account of Stone Age Techniques and Cultures*. London, UK: The Trustees of the British Museum; 1968.
59. Cortés-Sánchez M, et al. An early Aurignacian arrival in southwestern Europe. *Nat Ecol Evol*. 2019;3:207–212.
60. Higham T, et al. The timing and spatiotemporal patterning of Neanderthal disappearance. *Nature*. 2014;512:306–309.
61. Haws JA, et al. Coastal wetlands and the Neanderthal settlement of Portuguese Estremadura. *Geoarchaeology*. 2010;25:709–744.
62. Gaudzinski-Windheuser S, et al. Evidence for close-range hunting by last interglacial Neanderthals. *Nat Ecol Evol*. 2018;2:1087–1092.
63. Milks A. Making an impact. *Nat Ecol Evol*. 2018;2:1057–1058.
64. White M. Shoot first, ask questions later: interpretative narratives of Neanderthal hunting. *Quat Sci Rev*. 2016;140:1–20.
65. Currat M, Excoffier L. Strong reproductive isolation between humans and Neanderthals inferred from observed patterns of introgression. *Proc Natl Acad Sci U S A*. 2011;108:15129–15134.

CHAPTER 14

The extinction of Neanderthals

Extinction is the rule. Survival is the exception.

Carl Sagan

What happened?

The enigma of Neanderthal extinction has been a topic for many conversations and speculations.[1–4] Neanderthals had survived for hundreds of thousands of years under varying climatic conditions and geological events, and then suddenly disappeared after seeking out isolated refuges in the southernmost regions of Europe.[5] For many years, it was thought that Neanderthals became extinct in southern Iberia about 24–28 kya; however, it is now accepted that Neanderthals went extinct around 39–41 kya.[6,7]

One reason for the interest in Neanderthal extinction is its relation to our own demise. Therefore, knowing what factors may have contributed to Neanderthal extinction and what might have been avoided could provide lessons for our future.

There are several theories regarding the extinction of the Neanderthals, as follows:

(1) Many investigators believe that ancient, anatomically modern humans (AMHs), known as Cro-Magnons in Europe, had a competitive advantage over the Neanderthals. Their advancements in tool manufacturing and weapons led to a larger variety of food sources. Also, their social structure placed the European early AMHs a step ahead of the Neanderthals who were competing for the same prey and other resources. Some suggest that competition led to the slaughter of Neanderthals by AMHs.

(2) Humans migrating out of Africa brought diseases (pathogens or parasites) that were deadly to Neanderthals.

(3) Another theory is that the relatively small population of Neanderthals and interbreeding with the relatively large population of AMHs led to some genetically unfit hybrids, and selection was more efficient in removing deleterious genes from the AMH population.

(4) Lastly, some investigators suggest that the extinction of the Neanderthals was due to their inability to adapt to the ecological changes that came with the fluctuating climates in southern Europe. The harsh, cold environments may have led to significant changes and a decrease in food sources, potentially affecting Neanderthal survival, or

https://doi.org/10.1016/B978-0-12-819086-9.00004-X

an environmental catastrophe, like a volcanic eruption, may have killed off many plants and animals used by Neanderthals.

All but one of the proposed theories for the extinction of Neanderthals suggest the involvement of AMHs. However, if AMHs were involved in Neanderthal extinction, then what is the evidence for the presence of both species in the same region at the same time, and did AMHs have one or more chances to interact with Neanderthals?

Evidence for Neanderthal and anatomically modern human interactions

There are several time periods where AMH-Neanderthal interactions could have occurred. Fossil and archeological data suggest that the exodus of *Homo sapiens* out of Africa and into Eurasia between ~50 and 120 kya occurred in several migration episodes. Individuals crossed vegetated corridors from Northeastern Africa into the Arabian Peninsula and the Levant, and expanded further into Eurasia, Australia, and the Americas, during time-varying climate and sea level conditions.[8]

To substantiate these claims, Timmermann and Friedrich[9] developed a numerical model that estimated climate and sea-level changes over the past 125 ky, and proposed AMH dispersal. Their results were in close agreement with archeological and fossil data, and suggested prominent glacial migration waves across the Arabian Peninsula and the Levant region around 106,000–94,000, 89,000–73,000, 59,000–47,000, and 45,000–29,000 years ago (ya). Furthermore, their findings document that global climate swings played a crucial role in shaping Late Pleistocene global population distributions.

Given the current estimates of AMH migrations out of Africa, there are several ideas about when and where AMHs and Neanderthals first interacted. Theories on the initial interactions of AMHs and Neanderthals include the ranges 35–45 kya, 50–75 kya, and 100–120 kya. It is clear that interactions occurred among AMHs and Neanderthals and led to introgression, but not clear how these interactions might have affected Neanderthal extinction and which interactions led to the genetic contributions from Neanderthals that we see in modern humans today.

Because anatomically modern humans entered southwestern England and parts of Italy 35,000–45,000 ya (during the last modern human migration out of Africa, referred to as the "recent dispersal"), the time estimates suggest that Neanderthals and humans overlapped in southern Europe for at least 5000 years. This time frame would have allowed AMHs and Neanderthals to interact and occasionally breed. The evidence for introgression 35–45 kya is substantiated by the 40,000-year-old remains of an early AMH found in Romania that contained Neanderthal DNA, indicating that at least some introgression occurred during the 5000-year overlap of humans and Neanderthals in Europe.[10]

Some investigators have suggested the short overlap between Neanderthals and AMHs 35–45 kya may not have been enough time for significant interbreeding to occur, and that interaction occurred much earlier. Evidence for earlier migration of AMHs comes from investigators who have reported evidence for long-term human occupation, around 80 kya, at the site of Dhaba in the Middle Son River Valley of Central India. Their evidence comes from the stone tool industry found at Dhaba that spans the Toba volcanic eruption, which occurred around 74 kya. The lithic industry from Dhaba strongly resembles stone tool assemblages from the African Middle Stone Age and Arabia, and the earliest artifacts from Australia, suggesting that it is likely the product of *Homo sapiens* as they migrated eastward out of Africa.[11]

Other investigators have also found evidence that supports the idea that AMHs reached the Levant around 75 kya before their spread throughout Eurasia.[12] This time period of migration would have resulted in Neanderthals and AMHs overlapping geographically for more than 30,000 years, suggesting that Neanderthals and modern humans likely interacted in the Levant at least intermittently. In addition, evidence for Neanderthals and humans interbreeding during this period is insinuated by Neanderthal portions of the genome carried by non-African individuals today. A key example and possible evidence for early interaction is the proportion of Neanderthal DNA ancestry in East Asians (~12%–20%) compared to European individuals (2%–6%). Studies examining this reduced ancestry in European populations have concluded that a longer-term and more complex interaction between humans and Neanderthals existed in East Asia than that proposed during the modern human migration into southwestern England and parts of Italy 35–45 kya.[13,14]

However, several hypotheses may explain the discrepancy in Neanderthal ancestry between Europeans and East Asians. These include a single episode of gene flow and selective forces that shifted alleles and/or demographics, turning the remaining Neanderthal alleles into the frequencies we see in modern populations. Additional explanations include differential strengths of purifying selection across Eurasia[15] and modern Europeans losing part of their Neanderthal ancestry through interbreeding with a ghost population that was unadmixed.[14,16]

The possible coexistence of modern humans and Neanderthals in the Levant more than 100 kya is seen by a series of remarkable discoveries of early modern human remains in the Mount Carmel Range in what is today northern Israel. The most notable are the modern human fossils in the Skhul (1939) and Qafzeh (1934) sites by teams headed by Arthur Keith and Theodore McCown, and René Neuville, respectively. These AMH bones are from the early Middle Paleolithic, ranging between 100,000 and 120,000 ya. Specifically, these digs are located in the western slopes of Mount Carmel near the lower Sea of Galilee, and they include other caves such as the Tabun, Jamal, and el-Wad Caves. It is thought that these caves contain the earliest modern human remains outside of Africa. Of particular interest is that the Skhul and Qafzeh modern human samples belong to the same strata as Neanderthal bones in the adjacent Tabun Cave. Electron spin resonance and thermoluminescence dating techniques indicated that the Tabun Neanderthal samples were contemporaneous with the modern human remains found in Skhul and Qafzeh, both from about 120,000 ya during the Early Middle

Paleolithic or the Mousterian. This proximity in time and space of these modern humans and Neanderthal fossils suggests a possible coexistence of these two groups. Since the modern humans of Skhul and Qafzeh disappeared about 80,000 ya, and Neanderthal sites in the area, such as Kebara Cave, persisted to 61,000 to 48,000 ya, Neanderthals prevailed longer in the region.[17] It is also possible that Neanderthals inhabited the Near East during two separate occasions, the first from 120,000 to 100,000 ya and the second from 61,000 to 48,000 ya.

The modern human specimens from Skhul and Qafzeh are similar to the Omo and Herto fossils from Ethiopia. The modern human remains in the Israeli caves exhibit a similar mixture of archaic and modern characteristics as the Omo and Herto. Although the Arabian crania possess brow ridges and projecting faces reminiscent of Neanderthals and the Omo and Herto specimens, their features, in general, are less robust. It has been postulated that the Skhul and Qafzeh hominins signal the first migration of modern humans from Ethiopia about 125,000 ya, by way of the Red Sea southern route. The robust features exhibited by the Skhul and Qafzeh hominins represent archaic *Homo sapiens* characteristics that they share with Omo and Herto rather than Neanderthal features resulting from admixture. In fact, it is possible that due to the low sea level and the short distance between lands at the Red Sea crossing, the Ethiopian and Arabian modern humans were part of a continuous population. On the other hand, since close coexistence still occurred, introgression cannot be dismissed.[18,19]

These different time intervals where humans and Neanderthals inhabited similar regions as well as possible interactions between them at different times and locations begs the question: If some form of human-Neanderthal interaction was responsible for the extinction of the Neanderthals, why did it not occur sooner, during their initial interaction in the Levant?

Climate change and geological events

The one theory where AMHs were not involved in Neanderthal extinction is the theory of climate change and geological events. The relevance of climate in the debate over Neanderthal extinction was initially discounted, as Neanderthals disappeared in Oxygen Isotope Stage 3 (OIS 3) when conditions were thought to be relatively stable.[20] Some hypotheses, however, consider climatic instability during the millennia before the last glacial maximum to have been a significant force in Neanderthal extinction. One model postulates that habitat degradation and fragmentation led to the decimation and eventual disappearance of Neanderthal populations before the arrival of AMHs. In this scenario, AMHs would have arrived in areas previously occupied by Neanderthals after Neanderthal extinction, and the two species would never have met in Europe.[21] Of course, DNA evidence suggests that Neanderthals and AMHs did meet in Europe and interbred;

however, the number of Neanderthals may have been greatly reduced by the time of AMH arrival.[22]

A similar model considers the Neanderthal demise as only one of the many Late Pleistocene megafauna extinctions caused by the extensive loss of an environment used by many animals of the time.[23] Support for a significant climatic effect comes from detailed paleoclimatic records, which indicate that OIS 3 was dominated by more unstable climatic conditions than previously thought and may have been precipitated by unusually intense volcanic activity.[24] Modeling the indirect effects of environmental change based on these data found two stress peaks at 65 and 30 kya. The second was more prolonged and severe, and possibly related to Neanderthal extinction.[20,22]

The theory that Neanderthals could not adapt to the dramatic environmental and ecological changes associated with climatic fluctuations suggests that it led to Neanderthal extinction. Climate change around 55 kya would have led to ecological changes in flora and fauna. The changes would have led to fluctuations in grasslands and the development of forests. This, in turn, may have led to significant changes in prey and other food sources. In addition, the cold-adapted physiology of the Neanderthals may have also made it difficult to adapt because of insufficient time to recover from the changes.[25,26]

There is some evidence for new adaptations based on changes in tool types observed during the time period 55 kya. For example, archeological finds have shown changes from crude stone flakes to tools using tree resin to bind stone points and crafting tools from bone and wood. These changes in tools may reflect the attempt to adapt to new prey and hunting strategies.[25,26]

The theory of environmental and ecological changes also assumes that anatomically modern humans had already adapted to various environments after leaving Africa and were better suited to tolerate changing conditions.

Computer simulations of alternating climate change integrating archeological and chronological data show that the geographical range of Neanderthals contracted, whereas the human ranges expanded during the climatic phases of the Upper Paleolithic. This indicates that geographic expansion and competition from the larger populations of humans may have contributed to the Neanderthal demise.[27,28]

The climatic and ecological theory also suggests that climate changes pushed Neanderthals into Southern Europe, but proposes that Neanderthals could not compete for food or space with modern humans once humans occupied Neanderthal territory. The theory points to cultural, mental, and technological differences between humans and Neanderthals, and insinuates that humans may have eliminated Neanderthals because of their inferiority either by violence or by pushing them into regions that were too difficult to survive (similar to the effects of some invading humans on aboriginal groups).

Studies of mineral deposits in caves from the Carpathian Mountains, a range of mountains forming an arc across Central and Eastern Europe, indicated, between 44,000 and 40,000 ya, that a sequence of severely cold and arid conditions caused successive regional

Neanderthal depopulation intervals across Europe. They may have facilitated staggered repopulation by modern humans. The investigators concluded that repetitive depopulation-repopulation cycles during the periods of cold associated with Marine Isotope Stage 3 (Last Glacial Period) may have facilitated multiple genetic turnovers in Europe between 44,000 and 34,000 ya, and may explain the repeated replacement of Europe's population and its genetic composition.[29]

In one study, investigators sequenced five Neanderthal genomes from bones and teeth found in Belgium, France, Croatia, and the Russian Caucasus. The remains were dated from 39,000 to 47,000 years old. The sequences were compared to a prior Neanderthal genome from the Caucasus region and indicated that Neanderthals had split from a common ancestor around 150,000 ya and experienced a significant population turnover approximately 38,000 ya, just before their extinction. The researchers suggest that this could have been due to extreme cold periods, leading to the extinction of local populations and then recolonization from Southern Europe or Western Asia. Overall, the research suggests that Neanderthal populations showed a geographical population structure (different characteristics in different geographic regions). The researchers point out that the genomes they looked at lacked modern human DNA, although four of the genomes were from Neanderthals who lived at a time when modern humans were in Europe.[30]

Another explanation for the slow demise of Neanderthals after humans left Africa is the occurrence of geological events that may have deterred modern humans from moving into Neanderthal territory. Although modern humans had already left Africa, the Campanian Ignimbrite volcanic eruption in southern Italy, about 40,000 ya, may have created a semidesert buffer zone that prevented modern humans from entering southern Iberia. This allowed Neanderthals to have safe refuge in the area until modern humans arrived. This theory proposes that the Neanderthals were not as cognitively advanced as humans and may have had more difficulty competing for similar resources. In this scenario, a failure to adapt and exploit small game would have resulted in a less diverse diet. This inability to shift their prey focus and hunting strategies would have been instrumental in Neanderthal extinction.[26,28]

Others have suggested that environmental stress associated with the eruption of the Campanian Ignimbrite invoked a potential driver for Neanderthals' extinction and broader upheaval in Paleolithic societies. These investigators found that peak cooling and acid deposition after the eruption lasted 1–2 years and that the most intense cooling evaded hominin population centers in Western Europe. They conclude that the environmental effects of the Campanian Ignimbrite eruption alone were insufficient to explain the demise of Neanderthals in Europe. However, significant volcanic cooling during the years following the eruption and between 50 and 250 million tons of sulfur dioxide that was blasted into the stratosphere, resulting in acid rain and cooler temperatures across parts of Europe, Asia, and North America, could have impacted the viability of already small Neanderthal populations and influenced many aspects of daily life for Neanderthals and anatomically modern humans.[31]

The authors point out that the timing of the last known Neanderthal settlements in Europe coincides with the Campanian Ignimbrite eruption, leading to the ongoing debate of the volcano's role in the demise of Neanderthals. However, if the volcanic eruption was to blame, it was only because the Neanderthal population was already so small. In Europe, where hominins were competing for space and resources, the results of the volcanic eruption likely caused temperature declines between three and five degrees centigrade. That would have made life harder, but not unsurvivable, as evidenced by the continued existence of *Homo sapiens*.[31]

AMH competition

It is difficult to accept climate and geologic change as the sole reason for Neanderthal demise, given that they had survived numerous climatic and geological changes during their existence. Furthermore, little association has been found between proposed dates for Neanderthal disappearance and major climatic events, suggesting that Neanderthals did not become extinct by a lone catastrophic climatic event.[2] Many investigators have proposed if climate played a role, it would be a complex series of events involving environmental deterioration and the appearance of AMHs, causing increased competition for limited resources. In this scenario, fluctuating climate, an unstable environment, and competition would lead to eventual Neanderthal extinction.[22]

Hypotheses advocating competition have proposed several possible AMH competitive advantages. These include technological advances, such as: better clothing and shelter; improved hunting techniques and more diverse subsistence strategies, which included the consumption of birds and fish; social differences, such as larger group sizes and more elaborate social networks among modern humans; and demographic factors, possibly including differences in birth and mortality rates or interbirth intervals of the two species. Important differences have been found between Neanderthals and modern humans in their life history and demography, including faster growth and possibly shorter life expectancy in Neanderthals,[32] as well as a much higher population density among Upper Paleolithic modern humans compared to Neanderthals.[22,33]

One argument for competition is the role of climate change and its association with clothing technology. This argument coincides with the demise of Neanderthals and the emergence of AMH behavior. Both can be viewed as reflecting interactions between biological and behavioral cold adaptations through extreme climatic fluctuations during the Upper Pleistocene.[29] Studies have indicated special difficulties climatic changes posed and the need for adequate preadaptations involving complex clothing manufacture. Some have argued that preexisting biological cold adaptations delayed the development of clothing technologies among Neanderthals, ultimately resulting in their extinction, while the greater biological vulnerability of AMHs promoted an early appearance of behavioral adaptations among some AMH groups, resulting in advanced technological clothing.[34]

This argument revolves around the Stage 3 Project and its findings that the sudden, extreme environmental fluctuations late in MIS3 somehow brought Neanderthals to the point of no return.[35] Neanderthals were cold-adapted but only up to a point—they favored mild cold and avoided more severe cold. However, based on physiological considerations and wind chill estimates, researchers are reluctant to blame the cold itself for the extinction of Neanderthals, and instead invoke the lack of clothing technology.[34]

The physiological distinction between complex and simple clothing arises from the thermal effectiveness of clothing—in other words, whether a garment properly fits the body and the number of layers. Draped, single-layer clothing can provide only limited protection, whereas fitted, multilayered clothing assemblages can provide sufficient protection for survival in polar and subpolar environments. The argument is that Neanderthals did not obtain the proper clothing for the changes in environment, which then drove them south but only after significant losses in the population. The loss in numbers, along with other environmental factors (change in food, shelter, and other resources), led to their demise.[34]

In addition to Neanderthals' failure to adapt to quick changes in environment, several investigators have proposed that humans were more socially advanced and had created a culture where the division of labor was the norm. Division of labor in the modern human population created a safer environment for children and the group. It used the skills of hunters and gatherers instead of a more individualized effort used by Neanderthals for survival. It has been shown that in extant hunter-gatherer groups, a complex social structure consisting of kin and metagroups can lead to advanced behavior and support from related and unrelated individuals within a group. This complex behavior in extant hunter-gatherers can be explained when men and women have equal influence in selecting group members, which may have had a transformative effect on social organization and social networking in early modern humans.[36]

Studies of brain fossils suggest that Neanderthals invested more neural tissue to their visual and somatic systems, leaving less neural tissue for areas associated with problem-solving and critical thinking like the prefrontal cortex and parietal lobe. On the other hand, Cro-Magnon brains were organized differently, allotting more neural tissue to regions associated with social adaptations and problem-solving. These data support the notion that anatomically modern humans had increasingly more cognitive abilities and social networks than Neanderthals, potentially contributing to the extinction of Neanderthals after the arrival of Cro-Magnons in Europe. Over time, the slight but continual decrease in human cranial capacity insinuates that our current cranial capacity may have evolved to accommodate our smaller physique.[37]

Some have speculated that when humans did make advances into Neanderthal territory in Western Europe around 45,000 to 35,000 ya, violence and genocide led to the Neanderthal demise. The debate centers on which groups of humans arrived in Western Europe when Neanderthals were present. The Protoaurignacian culture appeared about 42,000 ya in Southwest and South-Central Europe, and is thought to reflect the movement of modern humans from the Near East. Recent studies on dental remains of the Protoaurignacian revealed mtDNA of a modern human type. The teeth were dated at 41,000 years before present and represent the oldest remains of the Aurignacian culture in southern Europe. The DNA studies provide evidence that people of the Aurignacian culture were in the region during the Neanderthal extinction. If the culture and superiority theory is true, this could have triggered the demise of the remaining Neanderthals in the region through either violence or the advanced human cognitive or social structure advantage.[38]

One argument against the competition theory is the behavioral differences of hunting and living arrangements between AMHs and Neanderthals. Several differences have been noted between modern humans and Neanderthals in the Levant. It appears that the two groups used different patterns of occupation of the same caves, and that AMH fossils are associated with single season occupation based on the bones of prey they acquired with the AMH fossils. In other words, AMHs shifted their living quarters seasonally, including camping in the field when the weather was more favorable. On the other hand, the Neanderthal fossils, from about the same time, indicate that Neanderthals practiced multiseasonal occupation. These differences represent fundamental behavioral variation between the two groups and suggest that competition may have been minimal.[39]

The hunting strategies of Neanderthals and modern humans were also different. Due to their more robust anatomy and body proportions, Neanderthals are thought to have practiced a more close-proximity style of hunting. It is theorized that they thrust and forced heavy spears with triangular stone points into the bodies of nearby large prey.[40] This type of game hunting plan would have necessitated specific types of spear points due to the force employed during the actual penetration of the prey. On the other hand, the weapons associated with modern human remains are more fragile wooden spears made to throw as projectiles at prey from a distance and may have involved smaller prey.[40]

In addition to hunting on land, there is evidence that Neanderthals exploited submerged aquatic resources. The collection of pumices common in the Upper Paleolithic was part of Neanderthal behavior well before the arrival of modern humans in Western Europe. Research at Grotta dei Moscerini and L of Grotta del Cavallo in southern Italy, dated between MIS 5 to early MIS 4, are two Italian Neanderthal sites with a large assemblage of retouched shells. Evidence indicates that specimens from these sites were gathered directly from the seafloor as live animals by skin diving Neanderthals. Archeological data from these sites in Italy, as well as France and Spain, confirm that shell fishing and freshwater fishing was a common activity of Neanderthals.[41]

There is a lack of evidence in the archeological record that modern humans outcompeted Neanderthals in terms of subsistence and weaponry. The idea that humans were superior in a wide range of domains is not supported by archeological records of Neanderthals and their modern human contemporaries, as the Neanderthal archeological record is not different enough to explain the demise in terms of inferiority in archeologically visible domains.[42]

Therefore, a third alternative explaining Neanderthal extinction is that the Neanderthals became unfit because of interbreeding with humans. The idea assumes that anatomically modern humans may have benefitted from some introgression, but that Neanderthals did not. Alternatively, it assumes that interbreeding caused the demise of Neanderthals through assimilation, where those individuals who were not hybrids (totally Neanderthal) could no longer compete for a variety of reasons, including disease, discrimination, displacement, or inbreeding due to small population size.[42]

Genetic disparities and Neanderthal extinction

The amount of Neanderthal nuclear DNA in modern humans of European descent suggests evidence for limited admixture: a small portion (up to ~4%) of the genomes of non-Africans derive from Neanderthals, indicating that interbreeding probably occurred in the Near East during the earliest dispersal of modern humans out of Africa, but before they arrived in Europe.[43]

Research by Liu et al.[44] has determined that modern human bones from the Zhiren Cave in China dating to about 100,000 ya suggest that the early incursions of AMHs from Northeast Africa into the Levant may not have died in Arabia but were fertile in the peopling of the world, reaching into the Far East with possible interbreeding with Neanderthals.[44]

Further evidence for Neanderthal-AMH interactions comes from examining the Neanderthal DNA segments in modern human genomes, reflecting a time frame for admixture of 50,000–60,000 ya (Skoglund, P. and Mathieson) before the diversification of East Asian and European lineages. In addition, the genome of Ust'-Ishim, an ancient individual related to modern East Asians and Europeans, has similar levels of Neanderthal ancestry to modern Eurasians, consistent with an admixture episode occurring ~52–58 kya.[10]

Some investigators contend that Neanderthals did not become extinct but instead were assimilated into the modern human gene pool. Unfortunately, the fossil record is ambiguous on this point, although several European Upper Paleolithic AMH specimens have been identified as Neanderthal-AMH hybrids.

However, the demographic modeling of admixture, combined with territorial expansion, indicates that the level of introgression would be produced under meager (<2%) interbreeding rates and substantial barriers to reproduction between Neanderthals and modern humans, arguing against assimilation.[45] Moreover, this low Neanderthal contribution to the modern human gene pool would most likely have occurred because of low Neanderthal population size, evidenced by the Neanderthal fossil record across Europe vanishing abruptly.

Due to potential biological reproductive barriers, fertile offspring may have been compromised. Biological reproductive barriers include physiological or behavioral differences that interfere with interbreeding and cross-fertilization between populations. These can consist of temporal isolation, environment isolation, behavioral isolation, and mechanical isolation. Of these, temporal and environmental are the most likely in this scenario—temporal since the time of ovulation would play a role in reproduction and environmental since populations would need to come in contact with each other in order for AMH-Neanderthal introgression to occur. On the other hand, mechanical differences would not occur between human species, and ancient human sexual behavior would not seem to be a restriction.

Postzygotic barriers (barriers after two species have mated) would include genetic incompatibility, zygotic mortality, inviable hybrids, hybrid sterility, and hybrid breakdown (results in the mating of two F1 fertile offspring resulting in sterile F2 generation offspring). It is difficult to assume which of these barriers might have been effective due to introgression, but it is clear that the offspring from Neanderthal females and AMH males underwent some type of postzygotic barrier. This is because we do not see any Neanderthal mitochondria DNA in modern-day humans,[46] which would indicate no successful interbreeding of female Neanderthal and human males leading to fertile offspring. Therefore, the Neanderthal nuclear DNA found in modern humans must have come from female humans mating with Neanderthal males. After introgression, it appears that selection was more efficient in removing deleterious genes from the human population. Over time, Neanderthal nuclear DNA sequences were slowly removed from the human genome by selection pressure, leading to the pattern of Neanderthal genes that we see today in modern humans. If selection in Neanderthals was slower in removing deleterious genes from the population, this could have contributed to their extinction.[22]

Disease transmission

The transmission of unique diseases from AMHs to Neanderthals has also been suggested to have led to Neanderthals' demise or at least contributed to their extinction. Evidence from animal studies indicates that a genetic diversity difference upon contact between different groups could produce significant epidemics in the less diverse population due to immunological consequences of reduced diversity in the Major Histo-Compatability (MHC) region.[47,48] This hypothesis, pointing to disproportionate susceptibility to pathogens, may also have resulted from differences between AMHs and Neanderthals in genetic diversities. Ancient genomes suggest that AMH genetic diversity exceeded that of Neanderthals, so that the latter may have been exposed to a greater number of epidemics.[30,49]

Migrating out of Africa, AMHs reached the Levant tens of thousands of years prior to their further spread throughout Eurasia, whereas Neanderthals seem never to have spread south of the Levant. Neanderthals and AMHs both occupied the Levant for thousands of years before the spread of modern humans into the rest of Eurasia but remained geographically localized for thousands of years. Studies by Greenbaum et al.[50] have proposed that infectious-disease forces can explain the localization and persistence of this interspecies boundary that kept the species separated. These investigators suggest that after several hundred thousand years of mostly independent evolution, Neanderthals and AMHs likely acquired immunity and tolerance to different pathogens—possibly a Neanderthal temperate pathogen tolerance and an AMH tropical pathogen tolerance. Once contact between the two species was reestablished in the Levant, each species would be exposed to novel pathogens, resulting in the spread to the new susceptible hosts. This event would place a disease burden on both species.

Greenbaum et al.'s studies, using dynamical-systems models, support the idea that introgression resulted in the transmission of alleles related to the immune system that would have gradually diminished barriers, leading to more persistent interspecies interaction and eventually the breakdown of geographic barriers. Differences between the species in their characteristic pathogens and/or their genetic diversity could have allowed AMHs to overcome disease burdens earlier than Neanderthals, giving AMHs an advantage in their subsequent spread into Eurasia.[50]

Several genomic studies have recorded potential signatures of events in which pathogens were transmitted between AMHs and Neanderthals or between AMHs and other archaic humans.[51–53] Furthermore, studies have identified signatures of positive selection on putatively introgressed Neanderthal genes in AMHs, particularly in genomic regions such as the MHC complex, which are associated with the immune system.[52–55] These findings suggest that disease burden due to interspecies pathogen transmission was significant and could have been involved with Neanderthal extinction.[50]

Other investigators have invoked pathogens related to cannibalism (Kuru Model) as leading to Neanderthal extinction through the acquisition of transmissible spongiform encephalopathies.[56] This, however, has been debated since there are only a handful of instances where cannibalism has been documented in Neanderthals.[57]

Small populations and inbreeding depression

Although the replacement of Neanderthals by AMHs has typically been attributed to environmental pressure and/or AMH superiority, others have presented models that suggest that no such factors might be needed to account for the demise of Neanderthals, starting from the observation that Neanderthal populations were small before the arrival of AMHs.[58]

Genetic studies indicate that Neanderthal effective population size—the size of the ideal population that would undergo the same amount of random genetic drift as the actual population—was already small by ~400 kya, amounting to no more than 3000–3900 individuals, a level that was sustained almost up until the species' extinction.[55]

To investigate what might have spelled the end for the Neanderthals, other than competition, the researchers modeled how Neanderthal populations might have fared over 10,000 years, taking three distinct factors into account. The first was inbreeding, which harms the fitness of the population. The second involved so-called Allee effects, where small populations fail to grow because of limited mate choice and have too few people to hunt, protect food from other animals, and raise the group's children. The third factor was natural fluctuations in birth rates, deaths, and sex ratios.

The investigators suggest that the relatively small populations alone would have contributed to their demise: even if Neanderthals had been identical to modern humans in their cognitive, social, and cultural traits, even in the absence of interspecific competition. In addition, they suggest that if AMHs contributed to the demise of Neanderthals, that contribution might have had nothing to do with resource competition, but rather with how the incoming AMH populations geographically restructured the resident Neanderthal populations, in a way that reinforced poor fitness, and the effects of inbreeding and stochasticity.[59]

This means that by the time the last Neanderthals reached Southern Europe, they were already inbred due to the small population size. In addition, investigators have suggested that their fertility was declining, perhaps due to a lack of food, as infertility can result from decreasing body fat. Researchers have proposed that even a slight change in the fertility of young females could have had a dramatic impact on population growth rate and thus on Neanderthals survival.[60,61]

Conclusion

Forty thousand years ago in Europe, we were not the only human species alive, and fossil evidence indicates that there were at least three other human species. One of these species was the Neanderthals, who were remarkably similar to us and lived in many regions of Europe for more than 300 ky. Evidence of their existence vanishes around 28–30 kya, with fossil evidence indicating the last surviving individuals in Southern Europe.

Some paleo-archeologists have hypothesized that Neanderthal extinction was due to Neanderthal's inability to reproduce fast enough to keep up with AMHs. Others suggest that modern humans were cognitively more advanced and outcompeted the Neanderthals, slaughtered any bands of Neanderthal they came across, or infected them with novel diseases. And some suggest that an environmental catastrophe, like a volcanic eruption in Europe or extreme climate change, killed off many plants and animals, leading to the Neanderthal demise through starvation.

Over the last several decades, investigators have concluded that to suggest that the demise of Neanderthals was simply due to cognitive differences between AMHs and Neanderthals is analogous to suggesting that the numerous indigenous communities and states taken over by European colonization was due to some inherent cognitive deficiencies within those populations. Initially, investigations implied that Neanderthals were cognitively inferior; however, new data has concluded that Neanderthals were similar to AMHs. This argument is strengthened by archeological data and genetic insights indicating that we share 99.5% of the same DNA with Neanderthals, some of which modern humans still carry today. Across thousands of genomes, researchers have identified a combined total of 20% Neanderthal DNA in modern humans.

Climatic evidence shows that Neanderthals were exposed to an increasingly hostile environment toward the end of their existence. Extreme climatic changes drastically changing from relatively warm to bitterly cold in other parts of Europe pushed them further south. This difference in climate may have led to differences in food types and other resources. For example, Neanderthals' close-combat hunting tactics, which had served them well for larger game, may have made it much more challenging to catch smaller game to sustain them when larger animals were in short supply—implying that modern humans had more technologies to catch smaller prey items, using nets or traps.

In addition to moving south, there is evidence that Neanderthal populations were decreasing in size, meaning that by the time the last Neanderthals reached their final destination, they were very inbred. This may have led to inbreeding depression affecting their fertility. In addition, fertility may have been declining due to a lack of food and decreasing body fat.

It has also been proposed that humans migrating out of Africa brought diseases (pathogens or parasites) that were deadly to Neanderthals. There are some suggestions that once contact between the two species was established, each species would be exposed to novel pathogens, placing a disease burden on both species. Differences between the species in their characteristic pathogens and/or their genetic diversity could have allowed AMHs to overcome disease burdens earlier than Neanderthals, giving AMHs an advantage and contributing to the demise of Neanderthals.

The interest in Neanderthal extinction is driven by our need for understanding how their end might relate to our own demise. Knowing what factors may have contributed to Neanderthal extinction and what might have been avoided could provide lessons for our future and our ability to prevent our extinction.

The reasons for the Neanderthal extinction are still not obvious, and evidence is beginning to point to a complex series of events other than just a simple single explanation. Because of the limitations imposed by genetic analysis, fossil and archeological evidence, radiocarbon dating, and geological events, it will never be clear what factors were directly involved in Neanderthal extinction. Questions will always remain as to whether it resulted from one or more contemporary theories or some mechanism(s) that have not been explored. Although other human species went extinct, many argue that meeting and interacting with other early humans may have been crucial for us to become the successful species we are today.

References

1. Klein RG. Whither the Neanderthals. *Science*. 2003;299:1525–1527.
2. Tzedakis PC, Hughen KA, Cacho I, Harvati K. Placing late Neanderthals in a climatic context. *Nature*. 2007;449:206–208.
3. Gilpin W, Feldman MW, Aoki K. An ecocultural model predicts Neanderthal extinction through competition with modern humans. *Proc Natl Acad Sci U S A*. 2016;113:2134–2139.
4. Kolodny O, Feldman MW. A parsimonious neutral model suggests Neanderthal replacement was determined by migration and random species drift. *Nat Commun*. 2017;8:1040.
5. Finlayson C, et al. Late survival of Neanderthals at the southernmost extreme of Europe. *Nature*. 2006;443:850–853.
6. Higham TFG, et al. The timing and spatio-temporal patterning of Neanderthal disappearance. *Nature*. 2014;512:306–309.
7. Higham, et al. The earliest evidence for anatomically modern humans in northwestern Europe. *Nature*. 2011;479:521–524.
8. Jennings RP, et al. The greening of Arabia: multiple opportunities for human occupation of the Arabian Peninsula during the Late Pleistocene inferred from an ensemble of climate model simulations. *Quat Int*. 2015;382:181–199.
9. Timmermann A, Friedrich T. Late Pleistocene climate drivers of early human migration. *Nature*. 2016;538:92–95. https://doi.org/10.1038/nature19365.
10. Fu Q, et al. An early modern human from Romania with a recent Neanderthal ancestor. *Nature*. 2015. https://doi.org/10.1038/nature14558.
11. Clarkson C, Harris C, Li B, et al. Human occupation of northern India spans the Toba super-eruption ~74,000 years ago. *Nat Commun*. 2020;11:961. https://doi.org/10.1038/s41467-020-14668-4.
12. Karmin M, et al. A recent bottleneck of Y chromosome diversity coincides with a global change in culture. *Genome Res*. 2015;25:459–466.
13. Hershkovitz I, et al. The earliest modern humans outside Africa. *Science*. 2018;359:456–459.
14. Villanea FA, Schraiber JG. Multiple episodes of interbreeding between Neanderthal and modern humans. *Nat Ecol Evol*. 2019;3:39–44.
15. Sankararaman S, et al. The genomic landscape of Neanderthal ancestry in present-day humans. *Nature*. 2014;507:354–357.
16. Lazaridis I, et al. Genomic insights into the origin of farming in the ancient Near East. *Nature*. 2016;536:419–424.
17. Grun R, et al. U-series and ESR analyses of bones and teeth relating to the human burials from Skhul. *J Hum Evol*. 2005;49:316–334.
18. Coppa A. Newly recognized Pleistocene human teeth from Tabun Cave, Israel. *J Hum Evol*. 2005;49 (3):301–315. https://doi.org/10.1016/j.jhevol.2005.04.005.
19. Coutinho-Nogueira D, et al. Qafzeh 9 early modern human from Southwest Asia: age at death and sex estimation re-assessed. *Homo*. 2021;72(4). https://doi.org/10.1127/homo/2021/1513.
20. Stringer C, et al. Climatic stress and the extinction of Neanderthals. In: van Andel T, Davies W, eds. *Neanderthals and Modern Humans in the European Landscape During the Last Glaciation*. Cambridge, UK: McDonald Institute for Archaeological Research; 2003:233–240.
21. Finlayson C. *Neanderthals and Modern Humans*. Cambridge, UK: Cambridge University Press; 2000.
22. Harvati K. What happened to the Neanderthals? *Nat Educ Knowl*. 2012;3(10):13.

23. Stewart JR. The ecology and adaptation of Neanderthals during the non-analogue environment of Oxygen Isotope Stage 3. *Quat Int*. 2005;137:35–46.
24. Golovanova LV, et al. Significance of ecological factors in the Middle to Upper Paleolithic transition. *Curr Anthropol*. 2010;51:655–691. https://doi.org/10.1086/656185.
25. Wong K. Our inner neanderthal. *Sci Am*. 2012;22:76–81.
26. Fa JE, et al. Rabbits and hominin survival in Iberia. *J Hum Evol*. 2013;64:233–241.
27. Banks W, et al. Neanderthal extinction by competitive exclusion. *PLoS ONE*. 2008;3. https://doi.org/10.1371/journal.pone.0003972.
28. Seddon C. *Humans: From the Beginning*. San Bernardino, CA: Glanville Publications; 2015:112–115.
29. Staubwasser M, et al. Impact of climate change on the transition of Neanderthals to modern humans in Europe. *PNAS*. 2018;115:9116–9121. https://doi.org/10.1073/pnas.1808647115.
30. Hajdinjak M, et al. Reconstructing the genetic history of late Neanderthals. *Nature*. 2018;555:652–656.
31. Black BA, et al. Campanian ignimbrite volcanism, climate, and the final decline of the Neanderthals. *Geology*. 2015;43:411–414. https://doi.org/10.1130/G36514.1.
32. Smith TM, et al. Dental evidence for ontogenetic differences between modern humans and Neanderthals. *Proc Natl Acad Sci U S A*. 2010;107(49):20923–20928. https://doi.org/10.1073/pnas.1010906107.
33. Mellars P, French JC. Tenfold population increase in western Europe at the Neandertal-to-modern human transition. *Science*. 2011;333:623–627.
34. Gilligan I. Neanderthal extinction and modern human behaviour: the role of climate change and clothing. *World Archaeol*. 2007;39(4):499–514. https://doi.org/10.1080/00438240701680492.
35. van Andel TH, Davies W, eds. *Neanderthals and Modern Humans in the European Landscape during the Last Glaciation: Archaeological Results of the Stage 3 Project*. Cambridge: McDonald Institute for Archaeological Research; 2003.
36. Dyble M, et al. Sex equality can explain the unique social structure of hunter-gatherer bands. *Science*. 2015;348:796–798.
37. Pearce E, et al. New insights into differences in brain organization between Neanderthals and anatomically modern humans. *Proc R Soc B Biol Sci*. 2007. https://doi.org/10.1098/rspb.2013.0168.
38. Benazzi S, et al. The makers of the Protoaurignacian and implications for Neanderthal extinction. *Science*. 2015;348:793–796.
39. Lieberman DE. *Neandertal and Early Modern Human Mobility Patterns*. Cham, Switzerland: Springer International Publishing; 2002.
40. Shea JJ. Neanderthals meet modern humans. Modern human origins and Neanderthal extinctions in the Levant. *Athena Rev*. 1997;2:21–32.
41. Villa P, et al. Neandertals on the beach: use of marine resources at Grotta dei Moscerini (Latium, Italy). *PLoS ONE*. 2020;15(1). https://doi.org/10.1371/journal.pone.0226690, e0226690.
42. Villa P, Roebroeks W. Neandertal demise: an archaeological analysis of the modern human superiority complex. *PLoS ONE*. 2014. https://doi.org/10.1371/journal.pone.0096424.
43. Durand EY, et al. Testing for ancient admixture between closely related populations. *Mol Biol Evol*. 2011;28:2239–2252.
44. Liu W, et al. Human remains from Zhirendong, South China, and modern human emergence in East Asia. *Proc Natl Acad Sci*. 2010. https://doi.org/10.1073/pnas.1014386107.
45. Currat M, Excoffier L. Strong reproductive isolation between humans and Neanderthals inferred from observed patterns of introgression. *Proc Natl Acad Sci U S A*. 2011;108(37):15129–15134. https://doi.org/10.1073/pnas.
46. Ghirotto S, et al. No evidence of Neandertal admixture in the mitochondrial genomes of early European modern humans and contemporary modern humans. *Am J Phys Anthropol*. 2011;146:242–252.
47. Black F. Why did they die? *Science*. 1992;258:1739–1940.
48. Curtis VA. Infection-avoidance behaviour in humans and other animals. *Trends Immunol*. 2014;35:457–464.
49. Theodosopoulos AN, Hund AK, Taylor SA. Parasites and host species barriers in animal hybrid zones. *Trends Ecol Evol*. 2019;34:19–30.

50. Greenbaum G, Getz WM, Rosenberg NA, et al. Disease transmission and introgression can explain the long-lasting contact zone of modern humans and Neanderthals. *Nat Commun*. 2019;10:5003. https://doi.org/10.1038/s41467-019-12862-7.
51. Chen Z, et al. Ancient evolution and dispersion of human papillomavirus type 58 variants. *J Virol*. 2017;91:e01285–e012817.
52. Enard D, Petrov DA. Evidence that RNA viruses drove adaptive introgression between Neanderthals and modern humans. *Cell*. 2018;175:360–371.
53. Houldcroft CJ, Ramond JB, Rifkin RF, Underdown SJ. Migrating microbes: what pathogens can tell us about population movements and human evolution. *Ann Hum Biol*. 2017;44:397–407.
54. Prüfer K, et al. The complete genome sequence of a Neanderthal from the Altai Mountains. *Nature*. 2014;505:43–49. 24352235.
55. Prüfer K, et al. A high-coverage Neandertal genome from Vindija Cave in Croatia. *Science*. 2017;358:655–658.
56. Underdown S. A potential role for transmissible spongiform encephalopathies in Neanderthal extinction. *Med Hypotheses*. 2008. https://doi.org/10.1016/j.mehy.2007.12.014.
57. Riel-Salvaore J. Mad Neanderthal disease? Some comments on, a potential role for transmissible spongiform encephalopathies in Neanderthal extinction. *Med Hypotheses*. 2008. https://doi.org/10.1016/j.mehy.2008.04.015.
58. Rogers AR, et al. Early history of Neanderthals and Denisovans. *Proc Natl Acad Sci U S A*. 2017;114:9859–9863. https://doi.org/10.1073/pnas.1706426114.
59. Vaesen K, et al. Inbreeding, Allee effects and stochasticity might be sufficient to account for Neanderthal extinction. *PLoS ONE*. 2019;14(11). https://doi.org/10.1371/journal.pone.0225117, e0225117.
60. Degioanni A, et al. Living on the edge: was demographic weakness the cause of Neanderthal demise? *PLoS ONE*. 2019;14(5). https://doi.org/10.1371/journal.pone.0216742, e0216742.
61. Skoglund P, Mathieson I. Ancient genomics of modern humans: the first decade. *Annu Rev Genomics Hum Genet*. 2018;19:381–404.

CHAPTER 15

The Denisovans

True wisdom comes to each of us when we realize how little we understand about life, ourselves, and the world around us.

***Socrates*[1]**

In the beginning

Numerous species of the genus *Homo* evolved in Africa, and recent evidence points to different hominin species inhabiting Africa, Asia, and Europe at different times. African specimens include *H. habilis*, *H. ergaster*, *H. erectus*, *H. antecessor*, *H. heidelbergensis*, and *H. naledi*.

Seven million-year-old (myo) hominin fossils from a potential cousin (*Sahelanthropus tchadensis*) found in West Africa, two myo fossils from Malapa Nature Reserve in South Africa (*H. naledi*), 315 thousand-year-old (kyo), and 260 kyo human fossils from Morocco (at Jebel Irhoud) and South Africa (at Florisbad), respectively, as well as genetic evidence from extant individuals in South Africa, indicates that we did not originate in East Africa and in fact resulted from a conglomeration of *Homo* species that introgressed around various regions in Africa.[2] Furthermore, the discovery of *H. heidelbergensis*, an intermediate form between *H. erectus* and *H. sapiens*, suggests that they were the species that ultimately evolved into modern-day humans in Africa.

Homo habilis, *H. erectus*, and *H. heidelbergensis* were presumably the first to leave Africa, with the European form of *H. heidelbergensis* later evolving into *H. neanderthalensis* in Europe, and Denisovans in Northern Asia (Fig. 1).

The evidence for *H. heidelbergensis* giving rise to Neanderthals, Denisovans, and humans comes from the many fossils found in Europe, Eurasia, and Africa. Over the years, the focus on human evolution and technological advances has come from European and North American institutions. However, recent interest from East Asia has revealed several unique transitional fossils found in modern-day China that are different from European and African *H. heidelbergensis*. The findings may indicate that other intermediate forms of *Homo* may have existed or that another group of hominins descended from *H. erectus* gave rise to East Asian populations. This scenario would have involved a species like *H. erectus* interbreeding with groups from Africa and Eurasia, ultimately resulting in modern East Asians.[3,4]

Sex and Cohabitation Among Early Humans
https://doi.org/10.1016/B978-0-12-819086-9.00016-6

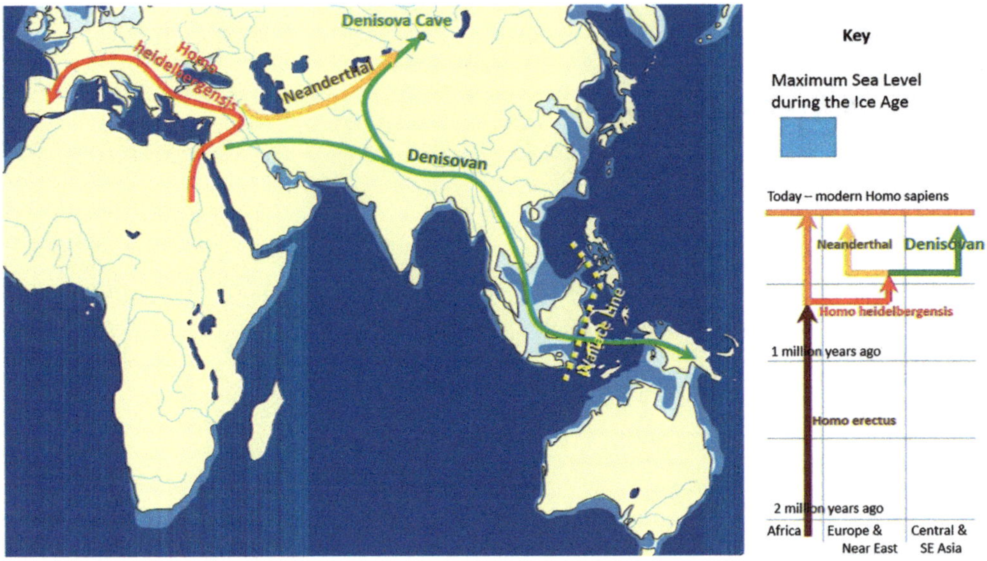

Fig. 1 Comparison of the geographic spread of Denisovans, Neanderthals, and *H. heidelbergensis*. https://en.wikipedia.org/wiki/Denisovan#/media/File:Spread_and_evolution_of_Denisovans.jpg.

The Denisovans

The Denisovans are an extinct species or subspecies of human that ranged across Asia during the Lower and Middle Paleolithic. Denisovans are known from only a few fossil remains, so most of what is known about them comes from DNA evidence.

Denisovans were the first group of humans to be discovered based on their DNA alone, when a German geneticist (Johannes Krause) extracted mtDNA from what he thought was a Neanderthal finger bone (Fig. 2) from Denisova Cave in Southern

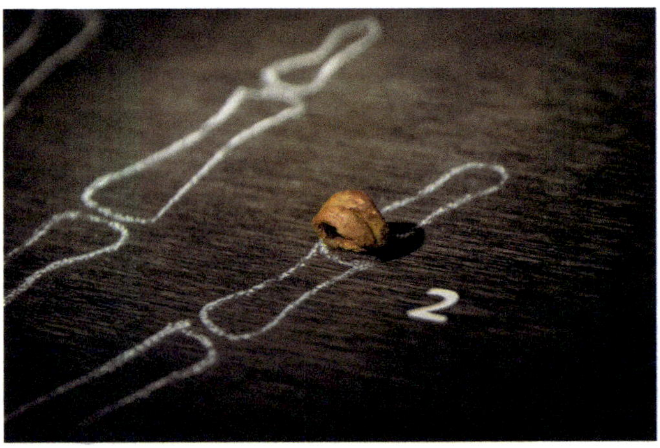

Fig. 2 Remnants of a finger bone found in the Denisova cave. https://en.wikipedia.org/wiki/Denisovan#/media/File:Denisova_Phalanx_distalis.jpg.

Fig. 3 The Denisova Cave where the first fossils of Denisovans were found. https://en.wikipedia.org/wiki/Denisovan#/media/File:Таинственная_пещера.jpg.

Siberia (Fig. 3). However, when Krause sequenced the mtDNA, he found the DNA was not from a Neanderthal; instead, he had found a new human lineage, the Denisovans. Since then, some experts have determined that Denisovans are an entirely new species or subspecies of the genus *Homo*, while others have argued that they are merely eastern Neanderthals.[5,6]

In 2012, a new single-stranded DNA library preparation method allowed researchers to reconstruct a high-coverage Denisovan genome approximately 30 times greater than what could be done for modern humans at the time.[7] The high-coverage sequencing allowed Meyer and colleagues[8] to determine that Denisovans had 23 chromosomes. The genome was from a girl with dark skin, brown hair and eyes, and there were about 100,000 recent changes in the modern human genome that occurred since we split from the Denisovans.

The evidence collected so far suggests that Denisovans were the contemporaries of Neanderthals and *Homo sapiens*. While scientists don't have a complete Denisovan skeleton, they have uncovered several small bone fragments, a few teeth, and one finger bone (Figs. 4 and 5).

Based on fossil and DNA evidence, the Denisovans lived in Siberia and East Asia, but it is challenging to know the exact time period when they existed because very few Denisovan fossils have been discovered. However, dating the fossils found in the Denisova Cave and DNA analysis indicate Denisovans inhabited the cave 50–200 kya. In addition, DNA evidence suggests that Neanderthals and Denisovans both lived in Denisova Cave, but not simultaneously.

Fig. 4 Arm or leg bone fragments found in the Denisova Cave. https://en.wikipedia.org/wiki/Denisovan#/media/File:Denisova-111.jpg.

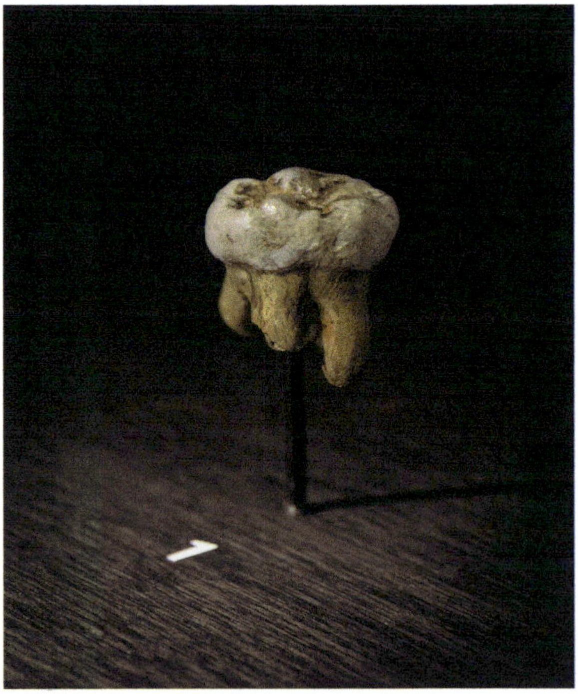

Fig. 5 Replica of a Denisovan molar from the Denisova Cave. https://en.wikipedia.org/wiki/Denisovan#/media/File:Denisova_Molar.jpg.

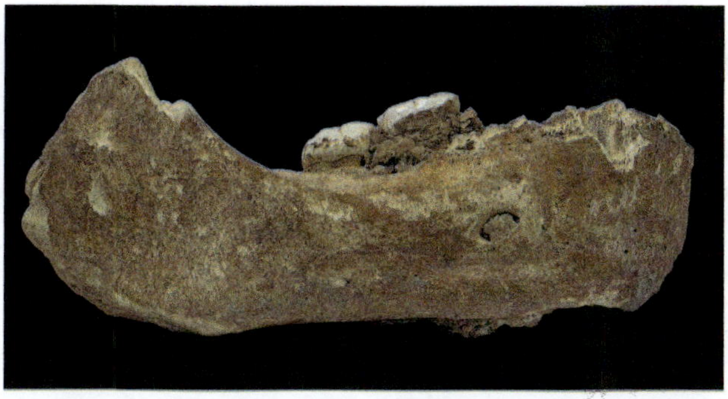

Fig. 6 Mandible with two attached molars. The mandible was discovered in the 1980s and was confirmed to be Denisovan in 2019. https://en.wikipedia.org/wiki/Denisovan#/media/File:Xiahe_mandible.jpg.

Besides the findings at Denisova Cave, a partial mandible (Fig. 6) was discovered in the Tibetan Plateau's Baishiya Karst Cave and was analyzed in 2019. The mandible was determined to be more than 160 kyo, and while no ancient DNA could be extracted from it, protein analyses indicated that it was Denisovan.[9]

Many suspect that several East Asian hominin fossils are also Denisovan that have been misclassified as other species. However, without the ability to isolate and sequence DNA or perform protein analyses and only a few Denisovan fossils for comparison, it will be difficult to determine if these Asian fossils are of Denisovan origin.[5]

Since the initial discovery of the finger bone at the Denisova Cave, excavations at the site have yielded small and highly fragmented remains of 12 hominins. These remains have enabled genome-wide data to be obtained from both Neanderthal and Denisovan remains, while cave sediments allow comparisons to be made between the two groups.

Douka et al.[10] obtained 50 radiocarbon measurements from the late Middle and Upper Paleolithic layers of the Denisova Cave and obtained 1 mtDNA sequence from 1 of 3 hominin fragments. Their analysis estimates the age of the oldest Denisovan fossil at the site to be 195 kyo. All Neanderthal fossils and a Denisovan-Neanderthal hybrid[11] were dated between 80 and 140 kya. The youngest Denisovan was dated 52–76 kya. Douka et al. reported that their radiocarbon dating of tooth pendants and bone points (43 kya and 49 kya, respectively) provided evidence for the earliest production of these artifacts in Northern Eurasia. The authors determined that the artifacts were associated with the Denisovan population based on current archaeological evidence. Because no ancient anatomically modern human (AMH) remains have been identified in the Altai region, it was impossible to determine if ancient AMHs were involved in producing the artifacts[10] (Fig. 7).

Early in the investigation of Denisovans, genetic evidence suggested that Denisovans split from the modern human lineage about 800 kya and from Neanderthals around

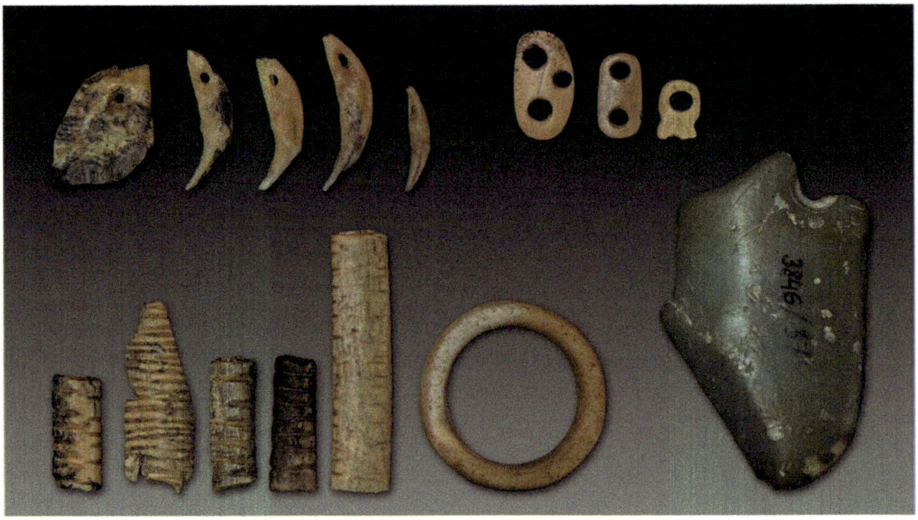

Fig. 7 Tooth pendants and bone points thought to have been produced by the Denisovans. https://en.wikipedia.org/wiki/Denisovan#/media/File:Denisova_Cave_pendants_notched_bones.jpg.

600 kya. Geneticists also concluded that based on genomic DNA sequencing, Denisovans were more closely related to Neanderthals than to modern humans. Further, genomic sequences indicated that modern humans, Neanderthals, and Denisovans shared a common ancestor, most likely *H. heidelbergensis*.[12,13]

However, mtDNA and nuclear DNA collected from AMHs, Neanderthals, and Denisovans are somewhat contradictory in that the DNA from the different sources suggests different divergent timelines and puzzling phylogenetic discrepancies. Autosomal genomes from the three groups show that Neanderthals and Denisovans are sister groups that split from a common ancestor with modern humans between 550 and 765 kya.[14,15] In contrast, mtDNAs of Neanderthals are more similar to the mtDNAs of modern humans than those of Denisovans.

To add to the confusion, DNA from the Sima de los Huesos Cave in Spain provided mixed results. The cave contained a unique assemblage of 28 hominin individuals of what archaeologists thought belonged to 400 kyo Neanderthals, considering the anatomical characteristics of the fossils. While the Sima de los Huesos hominins shared some derived morphological features with Neanderthals, the mitochondrial genomes retrieved from two individuals were more closely related to the mtDNA of Denisovans than to that of Neanderthals. However, since mtDNA does not reveal the full relationships among populations, Meyer et al.[16] examined the nuclear DNA from two specimens. The nuclear DNA sequences showed that the Sima de los Huesos hominins were related to Neanderthals rather than to Denisovans, indicating that the divergence between Neanderthals and Denisovans predates the fossils dated 430 kya, after the separation from the modern human lineage. Based on these results, it has been suggested that Neanderthals originally carried Denisovan-like mtDNA, which was later completely replaced via gene flow from an early lineage related to modern humans.[17,18]

Early investigations determined that Denisovans had a very extensive range encompassing a region from Southeast Asia and Near Oceania to Western Eurasia in Atapuerca in the Sima de los Huesos complex; however, as pointed out previously, nuclear DNA evidence suggested that the remains found in the cave were from Neanderthals who only carried mtDNA similar to Denisovans. No remains from Denisovans have been found in Northeastern Spain.

Ancient genomic DNA and mtDNA

MtDNA is an essential source for ancient DNA analysis. Since mitochondria can be present hundreds to thousands of times in a single cell, mtDNA is much more likely to exist in ancient samples than nuclear DNA. Furthermore, when studying ancient fossil remains in many cases, no nuclear DNA was left, leaving mitochondrial DNA analysis as a powerful alternative. In the past, ancient mtDNA has been used to trace historical civilizations and ancient trade routes, and to uncover geographical domestication origins and lineage relationships.

The mtDNA extracted from a Denisovan fossil in 2008 differs from that of modern humans by 385 nucleotide bases out of 16,659 bases in the mitochondrial genome. In contrast, the difference between mtDNA of modern humans and Neanderthals is around 202 bases. These numbers agree with the theory that modern humans were more closely related to Neanderthals than to Denisovans. Using the mtDNA data and the estimated rate of mutations (molecular clock), Krause et al. suggested that Denisovan mtDNA diverged from that of modern humans and Neanderthals about 1.3 mya–780 kya, whereas modern human and Neanderthal mtDNA diverged 618–321 kya. Based on these estimates, Krause et al.[6] concluded that Denisovans were the descendants of an earlier migration of *H. erectus* out of Africa, completely distinct from modern humans and Neanderthals.

The presence of mitochondrial DNA (mtDNA) sequences in nuclear genomes have been widely reported in various eukaryotic organisms.[19] The transfer of mtDNA into the nuclear DNA of humans appears to be an ongoing evolutionary process[20] resulting in various fixed nuclear mitochondrial DNA segments, referred to as NUMTs. These sequences appear in present-day human genomes, and polymorphic NUMTs have been detected in human populations worldwide. NUMTs vary in size and can consist of almost the entire mitochondrial genome.[21] In general, NUMTs behave as noncoding sequences in the nuclear genome and appear to evolve deletions, duplications, and inversions without any functional constraints.[22,23]

Bücking et al.[23] attempted to identify archaic NUMTs residing in modern human nuclear genomes. They analyzed 221 high-coverage genomes from Oceania and Indonesia, identifying reads that map both to the nuclear and mitochondrial DNA. They classified reads according to the source of the mtDNA and found one NUMT of Denisovan origin present in 15 human genomes. Further analysis of the NUMT flanking region suggested that this insertion was more likely to have happened in a Denisovan individual and introgressed into modern humans with the Denisovan nuclear DNA, rather than in a descendant of a Denisovan female and a modern human male. Bücking et al. concluded that further discovery of archaic NUMTs in modern humans could be used to detect interbreeding between archaic and modern humans and could reveal new insights into the nature of interbreeding events.

Further support for the Denisovan range in Southeast Asia and Near Oceania comes from genetic evidence indicating that 3%–5% of DNA in some populations from Melanesia, New Guinea, and Australia is of Denisovan origin. The genetic data provides evidence for introgression and a Denisovan presence in the region, even though currently there is no fossil record of Denisovans in Island Southeast Asia (ISEA). With this extensive geographical expanse, it was not unexpected to find that Denisovan exhibits considerable mtDNA diversity. This wide distribution of Denisovans, Neanderthals, and modern humans and introgression are indicative of the complex interactions among these hominins.

Genetic evidence now shows that a Philippine Negrito ethnic group has inherited the most Denisovan ancestry of all. Indigenous people known as the Ayta Magbukun have around 5% of their DNA from Denisovans. This finding of differences in Denisovan DNA ancestry in ISEA fits an evolutionary scenario in which two or more Denisovan populations independently reached various Southeast Asian islands, including the Philippines and what are now Papua New Guinea, Australia, and Tasmania.[24]

The hominin fossil record of ISEA indicates that at least two endemic archaic species were present around the time AMHs arrived in the region, more than 50 kya. These two species, *Homo luzonensis* and *H. floresiensis*, were present, but there is no evidence that they introgressed with AMHs. However, contemporary human populations across ISEA carry distinct genomic traces of interbreeding with Denisovans.

Denisovans may have encountered *H. luzonensis* and two other hominids whose fossils are found on different Southeast Asian islands, *H. floresiensis* on Flores and *H. erectus* on Java.[25] The idea that Denisovans could have interacted with these ancient hominins is based on *H. floresiensis* surviving from at least 100 kya to around 60 kya while *H. erectus* arrived on Java about 1.6 million years ago and died out between 117,000 and 108,000 years ago.[26]

One idea is that perhaps either *H. luzonensis* and/or *H. floresiensis* is the source of the Denisovan ancestry in modern human genomes in ISEA; however, the anatomical attributes of both of these extinct ISEA hominin species (*H. luzonensis* and *H. floresiensis*) are not readily reconcilable with the few confirmed specimens of Denisovans from Altai and the Tibetan Plateau. In addition, morphological and archaeological data indicate that both *H. floresiensis* and *H. luzonensis* had been in the region long before the estimated emergence of the Denisovans. Because of this, *H. floresiensis* and *H. luzonensis* are described as two distinct superarchaic hominin species that evolved independently on their respective island locales. The other possible admixture source, Indonesian *H. erectus*, is excluded because of its last appearance in the region around 108 kya. Thus, the source(s) of Denisovan introgression into modern human genomes in ISEA remains elusive.[25,27–29]

To address the apparent disparity between the lack of Denisovan fossil and genetic evidence, in ISEA, Teixeira et al.[25] performed a comprehensive search for superarchaic introgression in more than 400 modern human genomes, including more than 200 from ISEA. Their results substantiate widespread Denisovan ancestry in modern ISEA populations, but failed to detect any archaic admixture signals compatible with the endemic fossil record of ISEA. Teixeira et al. concluded that geographic ancestry patterns on Southeastern Asian islands and in Australia suggest that a genetically distinct Denisovan population from southern parts of mainland East Asia settled this region.

Interbreeding

Although the commonly held view is that modern human ethnicity and background are products of a single common ancestor, from a genetic standpoint, we are a product of many ancestors with alleles coming and going from different populations and a mix of those alleles through recombination.

However, we know that *Homo sapiens* mated with Denisovans and that introgression benefited some modern humans. For example, the EPAS1 gene variant that modern Tibetans and Sherpas inherited from Denisovans makes them better adapted for living at high altitudes, protecting them from hypoxia (a condition where tissues in the body are deprived of oxygen). Similarly, scientists have even discovered that some modern populations in Oceania have an immune system that's partly coded (and bolstered) by DNA acquired from Denisovans.[5]

In Eurasia, interbreeding between Neanderthals and Denisovans with modern humans took place several times. The introgression events into modern humans are estimated to have happened about 47,000–65,000 years ago with Neanderthals and about 44,000–54,000 years ago with Denisovans.

The comparison of the high-coverage sequence to 11 present-day human populations showed that approximately 6% of contemporary Papuan DNA comes from Denisovans. Australian Aborigines and people from Southeast Asian islands also show Denisovan DNA, and about 0.2% of the DNA of mainland Asians and Native Americans is Denisovan. The DNA evidence points to a patchwork of people with and without Denisovan DNA. Estimates from the DNA analysis conclude that interbreeding between AMHs and Denisovans occurred in Asia about 40 kya.[24]

Comparison of the X chromosome from Papuans and Denisovans revealed that there is less Denisovan DNA on the X chromosome of the human New Guinea population compared with the rest of the genome. One possible explanation is that mating was predominately between male Denisovans and female humans, because the male only contributes one X chromosome. Another explanation could have been that predominately human female migrants encountered Denisovans, thereby diluting the Denisovan component of the X chromosome. Both explanations assume intermixing between the two female X chromosomes, one human and one Denisovan, which may have eliminated some of the Denisovan X chromosome DNA while retaining the human components. There could also have been detrimental genes on the Denisovan X chromosome that would have been removed by natural selection.[7]

In 2014, Prufer et al.[14] sequenced the genome of a Neanderthal from the Caucasus and compared archaic genomes with 25 present-day human genomes. They demonstrated that several interbreeding events occurred among Neanderthals, Denisovans, and early modern humans, possibly including gene flow into Denisovans from an unknown archaic group.

The unknown archaic group contributed between 2.7% and 5.8% of the Denisovan genome.[30] The researchers estimated that the archaic DNA contributor split from humans, Neanderthals, and Denisovans between 900,000 and 4 million years ago, long before the groups diverged from each other. They suggest that the ancestor was possibly *H. erectus* because it is believed to be the predecessor of all modern humans. There are no known indications that the archaic group that introgressed with Denisovans interbred with humans or Neanderthals. The data further suggested that at least 0.5% of the Denisovan genome came from Neanderthals; however, no Denisovan DNA was detected in Neanderthals. The group was unable to determine whether any gene flow from modern humans to Neanderthals or Denisovans happened, most likely because the Neanderthal and Denisovan DNA that has been sequenced existed at a time before modern humans were around.[30]

As discussed previously, introgression and adaptation appear to have significantly shaped some modern human immune systems. In one study, Abi-Rached et al.[31] used whole-genome comparisons and identified introgression from Denisovans to modern humans. The analysis of human leukocyte antigen class I (HLA-I), a set of immune system components, revealed how modern humans acquired the HLA-B*73 allele in West Asia through admixture with Denisovans. Abi-Rached et al. identified archaic HLA haplotypes from Denisovans and Neanderthals carrying functionally distinctive alleles that were introgressed into modern Eurasian and Oceanian populations. Several of the alleles, which encode unique or strong ligands for natural killer cell receptors, represent more than half the HLA alleles of modern Eurasians, and they speculated that the alleles were introduced later into Africans.[31]

A 50,000-year-old skull and rib and a 35,000-year-old finger bone found in the Strashnaya Cave in the Altai Mountains of Central Asia in August of 2015 may shed more light on the links between Denisovans, Neanderthals, and modern humans. The finger is thought to be from modern humans, whereas the skull and rib may be from a Neanderthal or Denisovan.

So far, genetic analysis has not been successful. However, the addition of specimens over time, as well as improved genetic analysis methods may continue to enhance our understanding of the interactions between these groups of early hominids.[32]

Because only a few fossilized fragments have been found, we still do not know what Denisovans looked like, but by comparing modern human, Neanderthal, and Denisovan genomes, researchers have identified more than 31,000 genetic changes that distinguish humans from Neanderthals and Denisovans. The changes may be associated with human survival, and a number of the genetic changes have been linked to brain development.[16]

What did the Denisovans look like?

In 2019, Gokhman et al.[33] attempted to determine the facial features of a Denisovan female. Researchers used the genomic DNA sequences from the original finger bone found in the Denisova Cave and applied a molecular technique that was introduced in 2014 based on epigenetics. The technique uses a chemical modification called methylation, which silences gene expression. In methylated DNA, the cytosine nucleotide degrades over thousands of years into a different end product. By tracking degradation in an ancient genome, scientists created a methylation map. They used this method to examine the DNA extracted from the Denisova Cave finger bone. They compared the girl's methylation map with similar maps of modern humans, Neanderthals, and chimpanzees, focusing on areas where the degree of methylation differed by more than 50%.[34]

To determine how Denisovans' unique methylation patterns might influence their facial features, the researchers examined genes recorded in the Human Phenotype Ontology database. These genes in the database are known to cause specific anatomical changes in modern humans when they are missing or defective. Because DNA methylation is known to turn off genes, they may have effects comparable to methylated genes in the database, making it possible for researchers to infer Denisovan anatomy.[34]

The researchers discovered 56 Denisovan anatomical features that may have differed from humans or Neanderthals, 34 of them in the skull, resulting in facial features that looked fairly similar to those of a Neanderthal. Similarly, the researchers reported the data indicating a flat cranium, protruding lower jaw, and sloping forehead. Yet she also had key differences. For example, the reconstructed face was notably wider than that of a modern human or Neanderthal, and the arch of teeth along the jawbone was longer.[33]

A test of the model came when another team concluded that it belonged to a Denisovan based on ancient proteins in the Baishiya jawbone. They discovered their model was a close fit: the jawbone was wider than that of either humans or Neanderthals, and there were hints that it protruded about as much as in Neanderthals but more than in modern humans. The team's predictions also match skull fragments from Xuchang, China, that some argue belong to a Denisovan.[34]

In addition, recent genetic evidence has shown that *Homo sapiens* admixed with Neanderthals in the Middle East, Denisovans in Southeast Asia, and another unidentified group. Through genetics, linguistics, and archaeology, we have been able to follow the migration patterns of humans, and today we define our species and our genetic diversity as a result of ancestry, migration, admixture, and natural selection.

Acquisition of human characteristics

All the recognized archaic *Homo* species that preceded *Homo sapiens* eventually went extinct, leaving *H. sapiens* as the only *Homo* species to survive. Archaeology, modern technology, and computational tools in biology have led to a better understanding of human evolution. Previous chapters described some of the molecular genetic evidence that allows us to infer that the first anatomically modern humans (AMHs) arose due to introgression with other hominins in various parts of Africa 200–400 kya. This eventually led to the African form of *H. heidelbergensis*, which evolved into AMHs. AMHs then migrated out of Africa from regions in Southeast Africa and possibly northwest Africa on several different occasions. Humans, however, were not the first to occupy areas outside of Africa. They followed other species such as *H. habilis*, *H. erectus*, Neanderthals, and Denisovans. Based on DNA evidence, the ancestors of Neanderthals and Denisovans split from the AMH shared ancestor about 600 kya, most likely in Africa. They expanded into Eurasia, where the Neanderthals moved west while the Denisovans moved east. *Homo sapiens* populations that left Africa around 40–50 kya replaced the Neanderthals in Europe and the Denisovans in Northern Asia, but not before inbreeding with Neanderthals, Denisovans, and possibly another *Homo* species.

Various forms of molecular genetic analysis have provided clues to how closely related AMHs, Neanderthals, and Denisovans were to each other and the differences between these *Homo* genera and other primates. The developmental similarity between humans, Neanderthals, Denisovans, and chimpanzees, for example, makes it easy to see why their genetic makeups are so similar. Chimpanzees and humans share many of the same phenotypic characteristics (i.e., similar structural polarity; placement of the head, arms, hands, legs; five digits on each limb with nails not claws; similar ear anatomy; and similar mandible). In 2005, researchers sequenced the chimp genome and showed that humans share about 98.8% of our DNA with chimpanzees, making them our closest living relatives.

The Neanderthals are considered today to be our closest extinct relatives, and we shared a common ancestor about 550–765 kya before diverging about 381–473 kya. Genomic DNA sequencing analysis showed that Neanderthal DNA is 99.7% identical to present-day human DNA (humans share a similar percentage of genomic DNA with Denisovans). The 0.3% difference occurred after the split between humans and Neanderthals, resulting in two archaic human genomes, and although we share many of the same genes, new mutations gave rise to new derived alleles. Next-generation sequencing (NGS) of archaic and extant human genomes indicates that some modern humans can have genetic material from Neanderthals, or Denisovans, and another unidentified group.[12,35]

To examine more carefully the relationships between Neanderthals, Denisovans, and modern humans, Shaefer et al.[36] sequenced and compared genomes from 279 modern humans from people all over the world and then compared those genomes to ancient genomes from one Denisovan and two Neanderthals. They developed a new ancestral recombination graph inference algorithm that scaled to large genome-wide datasets. From the data collected, they generated a genome-wide ancestral recombination graph including human and archaic hominin genomes. This allowed them to create a map within human genomes of archaic ancestry and genomic regions not shared with archaic hominins by either admixture or incomplete lineage sorting. They found that only 1.5%–7% of the modern human genome is unique to modern-day humans. In addition, they found evidence of multiple bursts of adaptive changes specific to modern humans within the past 600 ky involving genes related to brain neural development and function. The study found that most of the uniquely human genes stem from two distinct bursts of evolution: 600 kya and 200 kya. It is unclear what role the unique genes played in evolution, but the authors speculate that they may have been involved with communication and/or the development of social behavior.

The algorithm developed by Shaefer and colleagues[36] also highlighted genes that humans inherited from an even older ancestor (ghost sequences), one that lived around 500 kya, which eventually gave rise to our species as well as Neanderthals and other hominins. The study supported the idea that a mysterious population of human ancestors that scientists haven't discovered yet may have interbred with Neanderthals and Denisovans before these species mixed with modern humans.

Prior studies have identified sequences thought to be unique to humans. These so-called human accelerated regions (HARs) were first described in 2006. HARs are a set of 49 segments of the human genome that are conserved in *Homo sapiens* but absent in all other primate, mammalian, or vertebrate species tested so far. The main characteristic of these variants is that they are found in regions of high evolutionary conservation. These HAR sequence variations include single nucleotide substitutions, deletions, and segmental duplications. The vast majority of these single nucleotide substitutions are found in noncoding, likely regulatory regions. Many genes associated with these human-specific alleles, often through novel enhancer activity, were shown to be implicated in the development of certain brain areas, including the prefrontal cortex.[37]

What other factors could have caused the differences between AMHs, Neanderthals, and Denisovans? To address this question, Weyer and Paabo[38] compared transcription factor binding sites (TFBSs) in humans, Denisovans, and Neanderthals, and revealed 25 sites that were human-specific yet remained in the ancestral state in Denisovans and Neanderthals. Additionally, when investigators examined the ancestral and derived forms of the TFBSs using reporter constructs in 3 neuronal cell lines, the activity of 12 of the derived versions of TFBSs differed from the respective ancestral variants. This suggested that most of this class of evolutionary differences between modern humans and Neanderthals might affect gene expression in at least some tissues or cell types.[38]

Copy number variations (CNVs) refer to a type of DNA structural variation with abnormal copy number changes that involve DNA fragments that are typically longer than 1 Kb and result in gains (duplication or insertional transpositions), losses (deletion), or complex rearrangements of the genome. Increasing evidence indicates that CNVs are likely to contribute significantly to human diversity and play an essential role in disease susceptibility and evolution. CNV sequences that increase or decrease the number of nucleotides between different species are sometimes called copy number differences (CNDs).[39–41]

McLean et al.[42] identified 510 deletions in noncoding regions that were human-specific, while these same regions were highly conserved (not deleted) in chimpanzees and other mammals. Moreover, these deletions were enriched in sequences near genes involved in steroid signaling and neural function. For example, one deletion, found by McLean et al.[43] in a noncoding region, was associated with a gene (*GADD45G*) involved in growth arrest and correlated with the expansion of specific brain regions in humans. Similarly, studies have shown that the hydrocephalus-inducing homolog (HYDIN) found in mice has an additional human homolog on chromosome 1. This gene is thought to be associated with the regulation of brain size. Deletions of the human HYDIN homolog led to microcephaly, while duplications led to macrocephaly.

Another human-specific CNV (deletion) was located in the enhancer of the human androgen receptor (*AR*) gene. The enhancer is responsible for the development of sensory vibrissae (whiskers) and penile spines (small keratinized surface projections on the penis). A follow-up study by Reno et al.[44] found that the same *AR* enhancer deletion removes these anatomical features. The deletion in all humans also existed in the Neanderthal and Denisovan genomes, demonstrating that the enhancer deletion is a characteristic of the human lineage. There has been speculation that this may have also been involved with changes in our reproductive behavior due to less painful intercourse and increased intimacy among humans.[45]

Sexual behavior between modern and archaic humans (Neanderthals and Denisovans) has been examined using DNA sequencing and statistical methods to discover where and how often humans, Neanderthals, and Denisovans mated. Investigators have determined that East Asians have three Neanderthal ancestors in their lineage, Europeans two, and Melanesians one. In addition, due to recent evidence, it is estimated that Denisovans mated with Melanesians at least once, and there may have been two different Denisovan groups that mated with AMHs. Denisovans, Neanderthals, and modern humans all bred together at some point, but it is difficult to tell how often and to trace introgression that did not produce offspring. Neanderthal and Denisovan DNA sequences are concentrated in certain regions of the human genome and missing from others.[46,47]

Extinction

By 10,000 years ago, the Denisovan population was gone. The disappearance of the species resembles a mass extinction, but there's no obvious environmental catastrophe (volcanic eruptions, earthquake, tsunami, climate change, asteroid impact) that appears to be involved with their disappearance. Researchers are not sure exactly when the Denisovans went extinct; however, DNA evidence suggests they may have survived in New Guinea or its surrounding islands until 15,000–30,000 years ago.[48]

It is unknown why they disappeared, and because there is very little fossil evidence for their existence and few artifacts, it may never be known why or how they went extinct. It's certainly possible that *Homo sapiens* outcompeted the Denisovans, but there is no evidence of that occurring.

Conclusion

Denisovans are known from only a few fossil remains (small bone fragments, a few teeth, and one finger bone), so most of what is understood about them comes from DNA evidence. Genetic evidence suggests that Denisovans split from the modern human lineage about 800 kya and from Neanderthals around 600 kya. Based on genomic DNA sequencing, Denisovans were more closely related to Neanderthals than to modern humans. Further, genomic sequences indicated that modern humans, Neanderthals, and Denisovans shared a common ancestor, most likely *H. heidelbergensis*.

The evidence collected so far suggests that Denisovans were the contemporaries of Neanderthals and *Homo sapiens*. The Denisovans lived in Siberia and East Asia, based on fossil and DNA evidence. DNA sequencing of AMHs showed that approximately 6% of contemporary Papuan DNA comes from Denisovans. Australian Aborigines and people from Southeast Asian islands also show Denisovan DNA, and about 0.2% of the DNA of mainland Asians and Native Americans is Denisovan. The DNA evidence points to a patchwork of people with and without Denisovan DNA, and estimates from DNA analysis conclude that interbreeding between AMHs and Denisovans occurred in

Asia about 40 kya. It is clear that interbreeding between AMHs and Denisovans contributed to both detrimental and advantageous genes. Because there is very little fossil evidence for their existence and few artifacts, it may never be known why or how they went extinct.

References

1. Socrates. https://www.brainyquote.com/quotes/socrates_391046.
2. Wong K. New evidence of mysterious Homo naledi raises questions about how humans evolved. *Sci Am*. 2017;(May).
3. Qiu J. The forgotten continent: fossil finds in China are changing ideas about the evolution of modern humans and our closest relatives. *Nature*. 2016;535:218–220.
4. Li Z-Y, et al. Late Pleistocene archaic human crania from Xuchang, China. *Science*. 2017;355:969–972.
5. Al-Shamahi E. *What Did the Denisovans Look like? How Did they Go Extinct? Full Guide to the Ancient Humans*; 2021. https://www.sciencefocus.com/science/denisovans/.
6. Krause J, et al. The complete mitochondrial DNA genome of an unknown hominin from southern Siberia. *Nature*. 2010;464:894–897. https://doi.org/10.1038/nature08976.
7. Reich D, et al. Denisova admixture and the first modern human dispersals into Southeast Asia and Oceania. *Am J Hum Genet*. 2011;89:1–13.
8. Meyer M. A high-coverage genome sequence from an archaic Denisovan individual. *Science*. 2012;338:222–226.
9. Chen F, et al. A late Middle Pleistocene Denisovan mandible from the Tibetan plateau. *Nature*. 2019;569:409–412. https://doi.org/10.1038/s41586-019-1139-x.
10. Douka K, Slon V, Jacobs Z, et al. Age estimates for hominin fossils and the onset of the Upper Palaeolithic at Denisova Cave. *Nature*. 2019;565:640–644. https://doi.org/10.1038/s41586-018-0870-z.
11. Slon V, et al. Neanderthal and Denisovan DNA from Pleistocene sediments. *Science*. 2017;356. https://doi.org/10.1126/science.aam9695.
12. Green, et al. A draft sequence of the Neanderthal genome. *Science*. 2010;328:710–722. https://doi.org/10.1126/science.1188021.
13. Rogers AR, et al. Early history of Neanderthals and Denisovans. *Proc Natl Acad Sci U S A*. 2017;114:9859–9863. https://doi.org/10.1073/pnas.1706426114.
14. Hawks J. Neanderthals and Denisovans as biological invaders. *PNAS*. 2017;114(37):9761–9763. https://doi.org/10.1073/pnas.1713163114.
15. Prüfer K, Racimo F, Patterson N, et al. The complete genome sequence of a Neanderthal from the Altai Mountains. *Nature*. 2014;505:43–49. https://doi.org/10.1038/nature12886.
16. Meyer M, et al. Nuclear DNA sequences from the Middle Pleistocene Sima de los Huesos hominins. *Nature*. 2016;531:504–507.
17. Posth C, et al. Deeply divergent archaic mitochondrial genome provides lower time boundary for African gene flow into Neanderthals. *Nat Commun*. 2017;8:16046. https://doi.org/10.1038/ncomms16046.
18. Petr M, et al. The evolutionary history of Neanderthal and Denisovan Y chromosomes. *bioRxiv*. 2020. https://doi.org/10.1101/2020.03.09.983445.
19. Bensasson D, et al. Mitochondrial pseudogenes: evolution's misplaced witnesses. *Trends Ecol Evol*. 2001;16(6):314–321. https://doi.org/10.1016/S0169-5347(01)02151-6. Accessed 2018-11-05.
20. Hazkani-Covo E, Zeller RM, Martin W. Molecular poltergeists: mitochondrial DNA copies (numts) in sequenced nuclear genomes. *PLoS Genet*. 2010;6(2). https://doi.org/10.1371/journal.pgen.1000834, 1000834.
21. Dayama G, Emery SB, Kidd JM, Mills RE. The genomic landscape of polymorphic human nuclear mitochondrial insertions. *Nucleic Acids Res*. 2014;42(20):12640–12649. https://doi.org/10.1093/nar/gku1038.
22. Perna NT, Kocher TD. Mitochondrial DNA: molecular fossils in the nucleus. *Curr Biol*. 1996;6(2):128–129.

23. Bücking R, Cox MP, Hudjashov G, et al. Archaic mitochondrial DNA inserts in modern day nuclear genomes. *BMC Genomics*. 2019;20:1017. https://doi.org/10.1186/s12864-019-6392-8.
24. Larena M, et al. Philippine Ayta posses the highest level of Denisovan ancestry in the world. *Curr Biol*. 2021. https://doi.org/10.1016/j.cub.2021.07.022.
25. Teixeira JC, et al. Widespread Denisovan ancestry in Island Southeast Asia but no evidence of substantial super-archaic hominin admixture. *Nat Ecol Evol*. 2021;5:616. https://doi.org/10.1038/s41559-021-01408-0.
26. Rizal Y, et al. Last appearance of *Homo erectus* at Ngandong, Java, 117,000–108,000 years ago. *Nature*. 2019. https://doi.org/10.1038/s41586-019-1863-2.
27. Détroit F, et al. A new species of *Homo* from the Late Pleistocene of the Philippines. *Nature*. 2019;568:181–186.
28. Brown P, Maeda T. Liang Bua *Homo floresiensis* mandibles and mandibular teeth: a contribution to the comparative morphology of a new hominin species. *J Hum Evol*. 2009;57:571–596.
29. Argue D, Groves CP, Lee MSY, Jungers WL. The affinities of *Homo floresiensis* based on phylogenetic analyses of cranial, dental, and postcranial characters. *J Hum Evol*. 2017;107:107–133.
30. Callaway E. Mystery humans spiced up ancients' sex lives. *Nature*. 2013. https://doi.org/10.1038/nature.2013.14196.
31. Abi-Rached L, et al. The shaping of modern human immune systems by multiregional admixture with archaic humans. *Science*. 2011;334:89–94.
32. Stewart W. Could a skull found in Siberia give new clues about human evolution? 50,000-year-old remains may be a Neanderthal or Denisovan. *Daily Mail (Lond Engl)*. 2015;(August 14).
33. Gokhman D, et al. Reconstructing Denisovan anatomy using DNA methylation maps. *Cell*. 2019;179(1):180–192.e10. https://doi.org/10.1016/j.cell.2019.08.035.
34. Price M. Ancient DNA puts a face on the mysterious Denisovans, an extinct cousin of Neanderthals. *Sci Adv*. 2019. https://doi.org/10.1126/science.aaz5536.
35. Reich D, et al. Genetic history of an archaic hominin group from Denisova cave in Siberia. *Nature*. 2010;464:1053–1060.
36. Schaefer M, et al. An ancestral recombination graph of human, Neanderthal, and Denisovan genomes. *Sci Adv*. 2021. https://doi.org/10.1126/sciadv.abc0776.
37. Levchenko A, et al. Human accelerated regions and other human-specific sequence variations in the context of evolution and their relevance in brain development. *Genome Biol Evol*. 2018;10:166–188.
38. Weyer S, Paabo S. Functional analysis of transcription factor binding sites that differ between present-day and archaic humans. *Mol Biol Evol*. 2016;33:316–322.
39. Prufer K, et al. The bonobo genome compared with the chimpanzee and human genomes. *Nature*. 2012;486:527–531.
40. Jiang Z, et al. Ancestral reconstruction of segmental duplications reveals punctuated cores of human genome evolution. *Nat Genet*. 2007;39:1361–1368.
41. Gokucumen O, et al. Refinement of primate copy number variation hotspots identifies candidate genomic regions evolving under positive selection. *Genome Biol*. 2011. https://doi.org/10.1186/gb-2011-12-5-r52.
42. McLean CY, et al. GREAT improves functional interpretation of cis-regulatory regions. *Nat Biotechnol*. 2010;28:495–501.
43. McLean CY, et al. Human-specific loss of regulatory DNA and the evolution of human-specific traits. *Nature*. 2011;471:216–219.
44. Reno PL, et al. A penile spine/vibrissa enhancer sequence is missing in modern and extinct humans but is retained in multiple primates with penile spines and sensory vibrissae. *PLoS ONE*. 2013;8, e84258.
45. Reno PL. Missing links. *Sci Am*. 2017;42–47.
46. Gibbons A. Rich sexual past between modern humans and Neanderthals revealed. *Science*. 2016. https://doi.org/10.1126/science.aaf4207.
47. Harris K, Nielsen R. The genetic cost of Neanderthal introgression. *Genetics*. 2016;203:881–891.
48. Choin J, et al. Genomic insights into population history and biological adaptation in Oceania. *Nature*. 2021;592:583–589. https://doi.org/10.1038/s41586-021-03236-5.

CHAPTER 16

Interwoven evolution of *Homo erectus*, *Homo floresiensis*, and *Homo sapiens*

> *We know from the study of evolution that, again and again, various branches of animal stock have become overspecialized, and that overspecialization has led to their extinction. Present-day* Homo sapiens *is in many physical respects still very unspecialized... But in one thing man, as we know him today, is overspecialized. His brainpower is very overspecialized compared to the rest of his physical make-up, and it may well be that this overspecialization will lead, just as surely, to his extinction. ... if we are to control our future, we must first understand the past better.*[1]
>
> **Louis Leakey**

Continental East Asia

Discoveries in archeology, anthropology, and genetics have provided insight into the intertwined evolutionary history of *Homo erectus* and *Homo sapiens* in Eastern Eurasia. The association of these two species during part of their existence is particularly apparent in East Asia and Indonesia/Australia.[2] In East Asia, instances of possible coexistence of *Homo erectus* and *Homo sapiens* are evident in a number of sites including Xujiayao (~260–370 ka), Majuangou (~1.66 Ma), Shangshazui (~1.6–1.7 Ma), Yuanmou (~1.7 Ma), Gongwangling (~1.63 Ma), Yunxian (~0.936 Ma), Zhoukoudian (~0.27–0.78 Ma), Chenjiawo (~0.6 Ma), Nanjing (~0.580 Ma), Hexian (~0.412 Ma), Chaoxian (~310–360 ka), Jinniushan (~260 ka), Hualongdong (~275–331 ka), and Dali (~270 ka).[3] The approximate ages of their existence are provided in parenthesis. Fig. 1 provides the locations of most of the abovementioned fossil sites. As depicted in this figure, the areas are distributed throughout what is today east China.

These fossils as a group are characterized by numerous cases of mosaicism in which combinations of ancestral and derived traits are found throughout various body parts. The East Asia section of Chapter 7 describes the combination of ancestral and derived traits that typically characterize these fossils from east China. A number of the fossils mentioned above exhibit such extreme cases of admixture of ancestral and derived traits that it has been difficult to assign them to specific species with any degree of certainty, prompting scholars to state, for example, as is the case of the Dali cranium, that the skull may belong to a late

Sex and Cohabitation Among Early Humans
https://doi.org/10.1016/B978-0-12-819086-9.00002-6

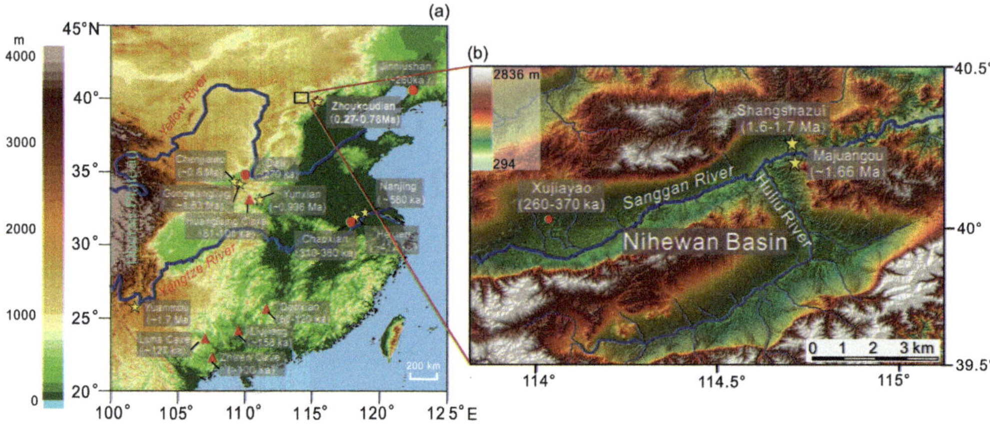

Fig. 1 Locations and ages of sites of potential coexistence of *Homo erectus* and *Homo sapiens* in continental East Asia (https://www.sciencedirect.com/science/article/pii/S0047248417300349#fig1).

Homo erectus, *Homo heidelbergensis*, or archaic *Homo sapiens*.[4] Typically, East Asian *Homo* fossils with mixed *Homo erectus* and modern human features, which cannot be affiliated to classic *Homo erectus* and modern *Homo sapiens*, have been classified as archaic *Homo sapiens* because scholars are not sure what they represent. For example, the Xujiayao *Homo* fossils have mixed characteristics reminiscent of European Neanderthals, Asian *Homo erectus*, and modern *Homo sapiens*.[3] Today the Xujiayao hominins are referred to as archaic *Homo sapiens*. Based on this extreme degree of admixture of traits among hominin groups coexisting in time and space, it is possible that some of the hominin populations were *Homo erectus*, some were archaic *Homo sapiens*, and some were made up of hybrids.

Island Southeast Asia

Kalinga, Luzon

Recently a number of discoveries on the island of Luzon in the Philippines have occupied center stage in the field of human evolution as a series of hominin remains have been unearthed. The oldest fossils were discovered in Kalinga in northern Luzon. It included stone tools dated between 777,000 and 631,000 years ago (ya) by electron-spin resonance applied to tooth enamel and fluvial quartz of clay deposits.[5] The lithics unearthed included 6 cores, 49 flakes, and 2 hammer stones. All tools lacked intentional retouch and were made by various techniques resulting in different final products. In addition, an almost complete (75% complete) skeleton of *Rhinoceros philippinensis*, an extinct species, and other fossils including stegodon (extinct elephant-like species), brown deer, freshwater turtle, and monitor lizard were found as part of the assemblage. Some of the fossil remains exhibit cut marks and signs of hammering in an attempt to reach the bone marrow.

Based on the available evidence, it is not clear how and from where the hominins at Kalinga arrived at northern Luzon in the Philippines. Considering that at the time the lithics were made, around 700,000 ya, the Strait of Taiwan was a dry land bridge, it is possible that the hominins that made the tools and killed megafauna in Kalinga arrived from East Mainland Asia via Taiwan. Today the Strait of Taiwan is a shallow body of water less than 150 m deep and about 130 km wide at its narrowest part. Yet this route would have necessitated crossing the Strait of Luzon, which separated southern Taiwan from the northern Philippines and some sort of intentional (e.g., raft) or unintentional (e.g., tsunami) navigation. Currently, this body of water is a very treacherous divide up to 4000 m deep and 250 km wide. Considering the known tenure of *Homo erectus* and Denisovans in East Asia, it is likely that the toolmakers at Kalinga were one of these two groups. It is also possible these toolmakers were ancestors of the nearby Callao hominin, *Homo luzonensis* that were dated to 67,000 ya.[6]

Callao Cave, Luzon

Unlike the Kalinga site, the Callao Cave has provided hominin fossils including upper right premolars and molars from one individual, a left premolar, a right third molar, a right third metatarsal bone of the foot, two phalanges of the fingers, two phalanges of the foot and a femoral shaft of a juvenile. In total, the assemblage represents the remains of three individuals.[6] The metatarsal was dated to 67,000 ka using U-series analysis. Analyses of the bones (Fig. 2) reveal that the individuals were small in stature, averaging 4 ft 8 in. for females and 4 ft 11 in. for males, likely due to insular dwarfism, and they possessed a mixture of ancestral traits reminiscence of australopithecines and *Homo erectus* as well as derived characteristics typical of modern humans.[7]

The premolars possess characteristics that resemble those of australopithecines and early *Homo erectus* while the enamel-dentine juncture is similar to that of Asian *Homo erectus*. Also as in Asian *erectus*, some premolars (e.g., P3) have robust and divergent roots, a condition rare in *sapiens*. Multivariate statistical analysis of P3–M2 dental diameters resulted in clustering with Asian *Homo erectus*.[6] The finger bones are long, narrow, and curved as in australopithecines. They are compressed from the palm to the back of the hand, as seen in australopithecines and *Homo habilis*. The foot bones are morphologically identical to those of *Australopithecus africanus* and Australopithecus *afarensis*. Overall, the fossils exhibit a combination of ancestral (*Australopithecus*-like) and derived (*sapiens*-like) traits different from the ones seen in other species in the genus *Homo* (including *Homo floresiensis* and *Homo sapiens*). Remarkably, anatomical manual and pedal similarities are observed between the Callao fossils dated to 67,000 ka and the more ancient australopithecines [tenure range 4.5–1.2 million years ago (mya)].

Based on their unique combination of traits, the remains were given the name *Homo luzonensis* after the island of Luzon. It has been postulated that *Homo luzonensis* may have been an ancestor of *Homo erectus* or some other earlier *Homo* species in Southeast Asia and related to the 777,000–631,000-year-old toolmakers from nearby Kalinga.[8] Extraction of good quality DNA for sequencing has failed.[6]

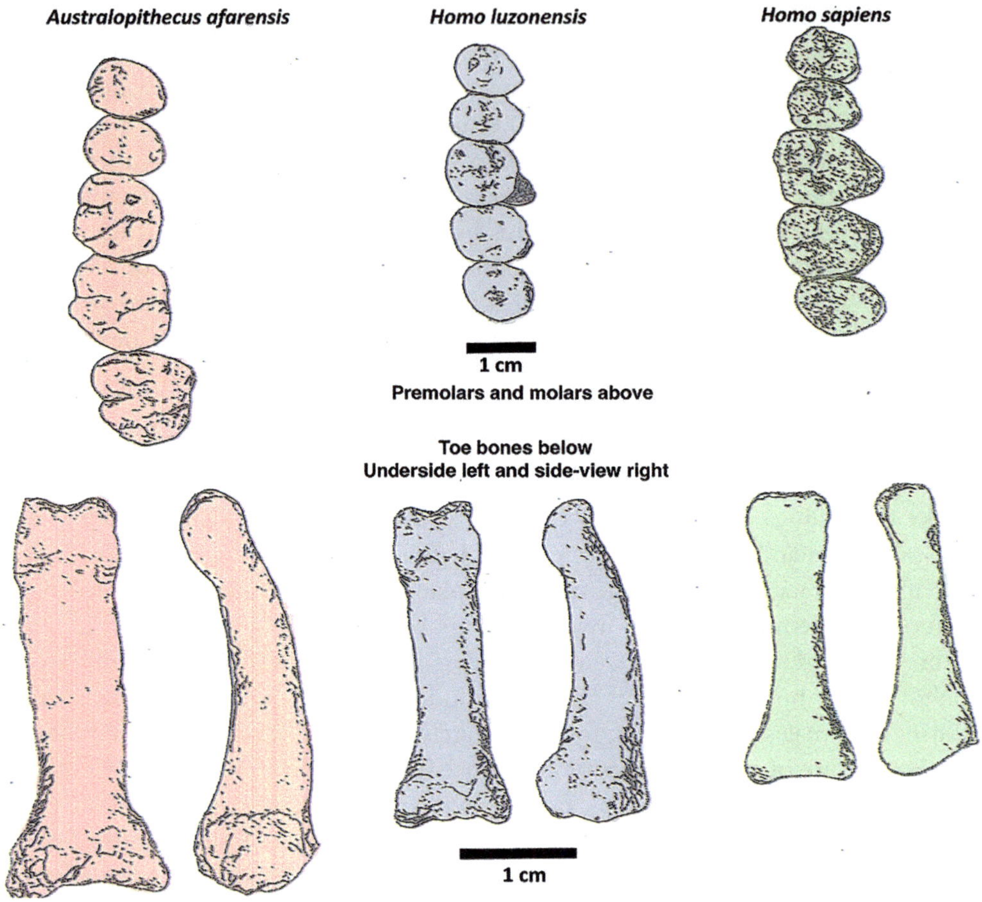

Fig. 2 Comparison of dental and toe bone morphology among *Australopithecus afarensis*, *Homo luzonensis*, and *Homo sapiens* (https://upload.wikimedia.org/wikipedia/commons/a/a6/Homo_luzonensis_teeth_and_foot_bones_diagram.jpg).

Homo luzonensis ate mainly (90%) Philippine deer and used tools for butchering them, as seen in one deer tibia possessing cut marks. Inside the Callao Cave, remains of Philippine warty pig (*Sus philippensis*) and an extinct bovid were present.[8]

Indonesia-Australia

Java

In the Indonesia-Australia region, the continuity of traits linking the *Homo erectus* fossils of Sangiran (1.6–1.0 mya) (Fig. 13, Chapter 7) and Ngandong (109,000–106,000 ya) (Fig. 14, Chapter 7), both from Java, to prehistoric and recent aboriginal Australians have been well documented by a number of scholars since the original research of Australian anthropologist Alan Gordon Thorne (1939–2012).[9] Chapter 7 provides a description of the specific traits in common between the Indonesian *Homo erectus* and the Australian prehistoric *Homo sapiens* from Kow Swamp. The Kow Swamp fossil is 13,000–9000 years old[10] and initially it was thought to be from *Homo erectus*, not modern *Homo sapiens*.[9] Although contested by some over the years, the anatomical parallelisms between *sapiens* from Kow Swamp and *erectus* have been well-documented by many scholars since the original research. This continuity of traits has suggested to a number of investigators that Indonesian *Homo erectus* introgressed into archaic *Homo sapiens*.

In Java, the remains at Ngandong probably represent one of the last populations of *Homo erectus* prior to the extinction of the species. The finding encompasses 12 calvaria (skull caps) and 2 tibiae (lower leg bones) that were found in a layer about 20 m above the Solo River (Central Java). The ages of these late *Homo erectus* fossils since their discovery in 1931 and 1933[11] have been heavily debated and currently indirect (from nonhominin remains), and direct (from *H. erectus* skulls) ages fluctuate between 117 and 108 ka[12] and 60–70 ka with minimum age estimates of around 40 ka.[13] As such, this population of *Homo erectus* may have been contemporaneous with the earliest *Homo sapiens*, which arrived in Southeast Asia and Australasia about 60,000 to 45,000 ya.[14]

Sulawesi

About 1000 km from the Ngandong site in Central Java across the Java Sea is the southern coast of the island of Sulawesi, where recently a number of lithic artifacts and fossils of megafauna have been unerarthed.[15] The megafauna (pygmy proboscideans, giant tortoise, large endemic suid, crocodile, *Bubalus*, *Stegodon*, and *Celebochoerus*) and stone tools were found in four sites in the Cabenge region between 2007 and 2012. No hominin remains have been found so far. One open-air site goes by the name of Talepu (Fig. 3), and the stone tools were securely dated to a period between 194,000 and 118,000 ya. The value of this finding is the continuous period of time that the stone tools and megafauna represent. Before this discovery, scholars were of the opinion that *Homo sapiens* arrived on this island as part of the colonization of Australia about 50,000 ya. Furthermore, this timeline indicates that prior to the arrival of *Homo sapiens* to Sulawesi, the island was colonized by premodern hominins. At the earliest dates of this time period, the tools were not likely made by *Homo sapiens*. Likely candidates that were around the area at the time include *Homo erectus*, *Homo floresiensis*, and Denisovans. At the other end of the timeline, about 118,000 ya, a fourth candidate can be added to the list, *Homo sapiens*. The Zhirendong specimens from south China clearly indicate that *sapiens* were in South East Asia 120,000 ya.[16] In addition, fossil teeth from a cave in Daoxian in Hunan province of southern China date from about 80,000–120,000 are *sapiens*'.[17] If archaic *sapiens* lived in east China about 120,000 ya, contemporary cohabitation of Java and even Sulawesi by

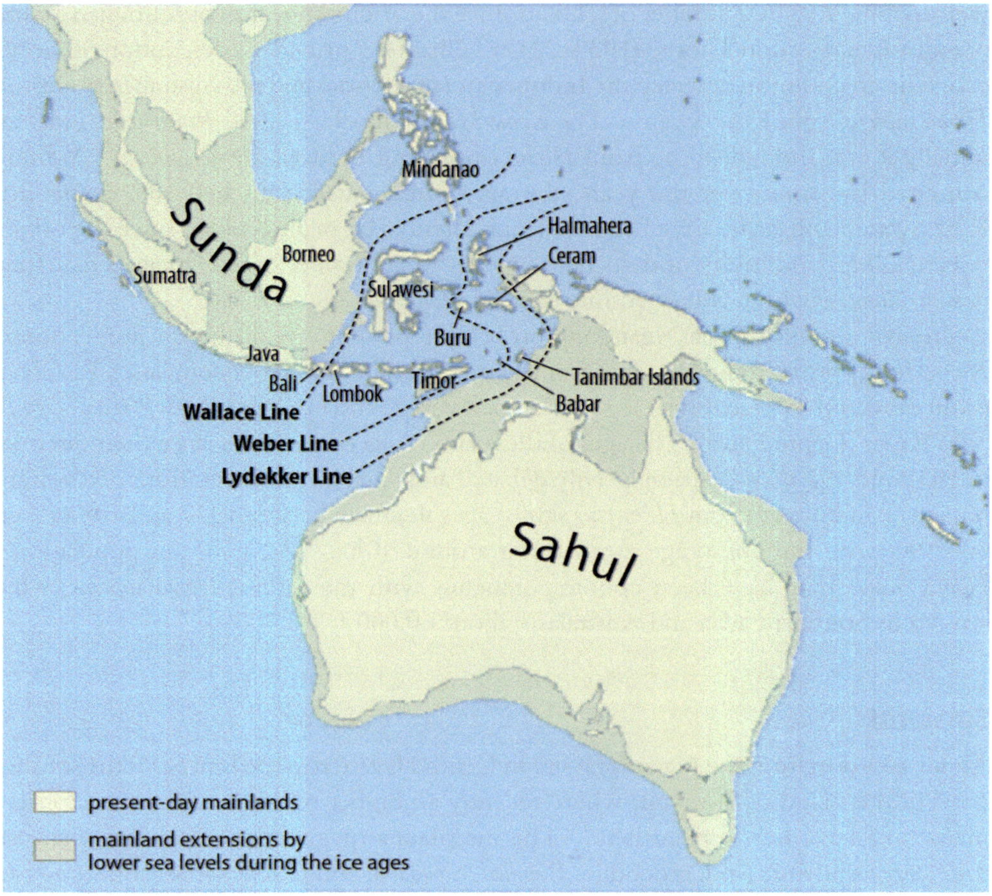

Fig. 3 Indonesia. Two water crossings were present during periods of low sea level (−120 m): Wallace's Line to the west and Lydekker's Line to the east. Exposed land during periods of low sea level is shaded (https://upload.wikimedia.org/wikipedia/commons/5/5f/Map_of_Sunda_and_Sahul.png).

erectus, Denisovans, and *sapiens* is not far-fetched. If East Asia was home to *Homo erectus*-like groups, they might have reached Sulawesi as part of a continuous population. *Homo erectus* inhabited Southeast Asia from about 1.5 mya to perhaps as recently as half a million years ago or even 50,000 ya.[18]

Yet, unlike the hominin residents of Ngandong and Sangiran in Jaba, a part of the contiguous continental landmass of Sunda, whoever made the stone tools at Talepu had to cross the ocean water from Borneo to reach Sulawesi. During glacial periods this crossing (today known as the Makassar Strait) entailed about 19 km of treacherous ocean,[19] and although several species of terrestrial mammals, including macaques and pigs, made the crossing during the Pleistocene (2.5 my–12,000 ya), it is not clear how hominins got to Sulawesi.

About 270 stone artifacts were found in the Cabenge area. The lithics do not follow any known style (industry), and therefore the investigators could not associate them to a specific culture or species. Overall, core reduction was not extensive, and although there is patterning in the flaking techniques, there is little indication that the stoneworkers were creating tools with a specific form or purpose in mind. Rather, they were just stone-flaking to generate sharp-edged flakes for general purpose, as a source of flakes. Dating using the uranium-series method and the chronology afforded by the megafauna fossils, the Talepu site exhibits continuous habitation from >200 kyr and <780 kyr in age. These results also suggest that the Talepu cultural sequence ends about 100 kya.

Flores

*Liang Bua (*Homo floresiensis*)*

Flores is one of Indonesia's Lesser Sunda Islands known for the location of the Liang Bua cave, where the first specimen of *Homo floresiensis* was discovered. In addition to Liang Bua, the island is known for a number of hominin archeological sites (Fig. 4). Interest in archaic hominins in the island of Flores (the name given by the Portuguese to the Lesser Sunda Island in the eastern half of Indonesia) started when the late New Zealand-born Australian rock-art archeologist Michael John Morwood (1950–2013) (Fig. 5) became intrigued in the 1960s by the assertions of Dutch priest (Jesuit missionary) and amateur archeologist Theodorus Lambertus Verhoeven (1907–90), who reported finding 750,000-year-old stone tools in central Flores.[20] After discovering 800,000-year-old fossils of stegodontids (dwarf elephants), Morwood reached the conclusion that the islands of Wallacea were settled by *Homo erectus* before modern humans got there.[21] Morwood was

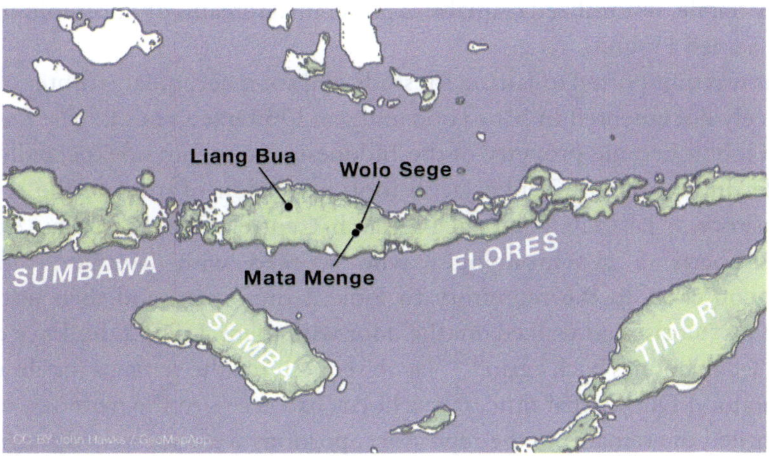

Fig. 4 Excavation sites in the Island of Flores (https://medium.com/@johnhawks/this-is-where-scientists-may-find-the-next-hobbits-c41318c61f64).

Fig. 5 Michael John Morwood (1950–2013) (https://www.nature.com/articles/500401a).

interested in proving that Verhoeven was right and wanted to provide evidence of early incursions by hominins, likely *Homo erectus* that found their way into Australia by way of Indonesia. With this in mind, he enlisted the help of collaborators at the Pusat Survei Geologi, the geological survey institute in Bandung, Indonesia. His team visited several of the sites reported by Verhoeven and, in the process, found additional stone tools that extended hominin presence in Flores to 1 mya, providing the first evidence of an early *Homo* species east of Wallace's line (the faunal transitional zone between Asia and Australia). In 2001, the team started excavations in one site, a limestone cave in western Flores. Two years later, while Morwood was in Java, Thomas Sutikna (a team member), while completing the excavation of a 6-m-deep hole at the site, found unexpectedly the partial very fragile, unfossilized cranial and postcranial remains of a hominin that was subsequently named Hobbit.

Hobbit was transported to Jakarta for study and soon became the center of a scientific scandal nicely documented in Morwood and van Oosterzee's book, *The Discovery of the Hobbit.*[22] Hobbit became property of the Indonesian state, but unexpectedly, the Indonesian paleoanthropologist Teuku Jacob took possession of most of the remains from their repository at Jakarta's National Research Centre of Archaeology. According to Richard Roberts (a geochronologist who worked with Morwood in the dig), "Teuku Jacob came to the repository to look at the bones, and then just placed the remains in his suitcase and walked out the door with them. I didn't think we'd ever actually get to see the bones back again."[23] Jacob sequestered the remains for about 3 months in his laboratory. During that time, few scholars had access to the material, and when he finally returned the remains to the repository, portions were severely damaged, and two leg bones were missing.[24] Jacob, however, denied doing anything wrong. At the time,

Hobbit was at the center of a heated controversy with a group of scientists proposing that the small hominin represented a unique species, *Homo floresiensis*, and an opposing camp insisting that it was just the bones of a diseased individual, possibly one with Down syndrome. Jacob was a strong advocate of the second group and came to international prominence by claiming that the remains were those of a microcephalic modern human. Some scientists still insist that Jacob's irresponsible actions were motivated by his desire to prove his point.

The discovery of Hobbit was unexpected in different ways. First, Thomas Sutikna, the collaborator in charge of the dig at the Liang Bua cave while Morwood was in Java, was not looking specifically for *Homo erectus*-like remains. Second, the anatomy of the specimen discovered was so bizarre that they could not make sense of it. The discovery occurred on September 2, 2003, and Morwood and his team were looking for evidence of the original migration of AMHs from Asia to Australia. Instead, Sutikna found an almost complete, small-statured skeleton that they designated LB1. LB1 possesses a nearly complete skull that belonged to a 30-year-old female. Fig. 6 provides an image of a reconstruction of *Homo floresiensis*. In addition to the name Hobbit, people know her as "Little Lady of Flores" or "Flo." The bones were not fossilized and were so fragile that the team had to wait for them to dry prior to excavation.

Initially, the assemblage was dated to 12,000 ya.[25] This initial date estimation was challenging to reconcile with the age expectation of such an archaic-looking hominin. This issue was partially ameliorated when a different set of dates provided an age of 60,000–100,000 ya for the skeletal remains and 50,000–190,000 ya for the associated lithics.[25] During 2003–04, 7 more partial skeletons were unearthed, and from 2009 to 2015, additional bones and teeth for a total of 15 individuals, all of very small stature, were found.[26] Stone artifacts proportional to the size of the specimens were also recovered from the cave. The style of the lithics associated with the bones is similar to those manufactured by AMHs in East Timor.[27] In addition to the *Homo floresiensis*' bones, two teeth that belonged to modern humans were found in 2010 and 2011.[28] The upper premolar and the lower molar teeth were dated indirectly from associated charcoal to about 46,000 ya. The lower molar was *sapiens*-like whereas the premolar seems slightly more ancestral, possibly belonging to *Homo erectus*.

Today scholars wonder if the disappearance of *Homo floresiensis* and the appearance of *sapiens* in Flores were coincidental or whether it had something to do with the extinction of the former. It is also possible that *floresiensis* survived beyond 50,000 ya and that more recent remains are waiting to be discovered. In connection with this possibility are the stories that are part of the folklore of the inhabitants of Flores, which describe small, hairy, cave dwellers similar in size to *Homo floresiensis* known as Ebu Gogo ("grandparent who eats anything").[29]

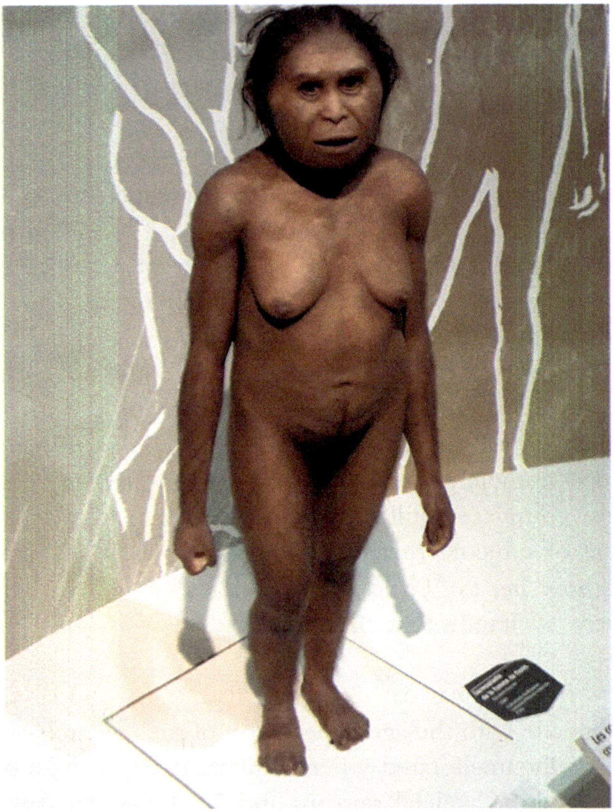

Fig. 6 Reconstruction of a 3ft tall 30-year-old woman who lived 50,000 years ago on the island of Flores, *Homo floresiensis* (https://australian.museum/learn/science/human-evolution/homo-floresiensis).

The anatomy of Flo and the other member of her group are difficult to explain. All of the specimens recovered represent individuals with small bodies (about 1.1 m or 3 ft 7 in.). The shortest average members of our species, the Mbenga and Mbuti of Africa, are 1.5 m (4 ft 11 in.) and the Twa (African Pygmy) and the Semang (Negritos of the Malay Peninsula) are 1.37 m (4 ft 6 in.).[30] Other than their extremely small stature, many experts believe that Hobbit is a version of *Homo erectus* from Flores, both groups overlapped in Indonesia.[31] Currently, most scholars are of the opinion that *Homo floresiensis* evolved to a miniature version of *Homo erectus* by a process known as insular dwarfism, where size decreases as a response to limited resources in an island ecosystem. Reductions in body sizes resulting from island isolation are not unprecedented in the animal kingdom.[32]

LB1 exhibits a small cranial capacity of about 380 cm^3 (23 cu. in.) within the range of contemporary chimpanzees. Fig. 7 provides a comparison of representative hominin

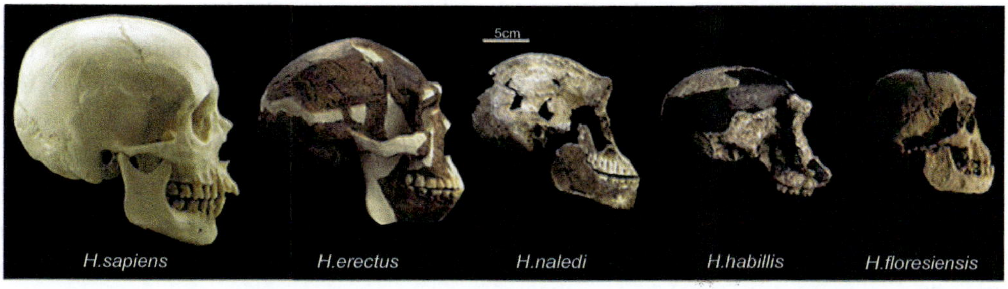

Fig. 7 Comparisons of *Homo* skulls (https://thenaturalhistorian.com/2017/06/07/tricksy-hominin-fossils-hobbitses-are-human-but-homo-naledi-is-not).

skulls illustrating the extremely small size of *floresiensis*' skull. Based on 3D-morphometrics, the skull of LB1 differs significantly from normal, small-bodied individuals and microcephalic skulls of AMHs, resembling more the skull of *Homo erectus*.[33] Flo's brain/body size ratio resembles that of australopithecines, but its endocast shape resembles *Homo erectus*. LB1's skull exhibits a *Homo erectus* morphology, including a flat and sloping forehead, thick cranial bones, as well as a short and flat face and no chin. In addition, Flo's dentition shows a number of similarities to *Homo erectus*, which could mean that *erectus* was the ancestor species.[34] She possessed frontal and temporal lobes and a lunate sulcus (fissure in the occipital lobe) in the position seen in advanced hominins consistent with high cognitive processing.[35] Among the stone artifacts found in the cave were flakes, points, perforators, blades, and microblades, many associated with remains of juvenile Stegodon (extinct elephants). These types of tools in association with small Stegodon suggest the population was targeting this type of easy prey just like the nearby population of *Homo erectus* at Soa Basin 50 km from the Liang Bua cave in Flores.[31]

The postcranial skeleton of *floresiensis* is also unique. Among the main anatomical differences relative to *sapiens* was a humeral head that twisted 120 degrees, while in AMHs is twisted between 145 and 165 degrees to the plane of the elbow joint.[31] Together with their relatively short clavicle, they would have looked somewhat scrunched down. The wrist anatomy resembles those of australopithecines' missing advanced features that evolved about 800,000 ya.[36] They had flat feet that were large relative to the size of their body.[37]

Mata Menge

About 70 km east of the Liang Bu cave in central Flores is the site of Mata Menge. This area has been the source of stone tools and megafauna fossils since the 1950s. Interest in the fossils remains of the area started with the pioneering work of Theodorus Lambertus Verhoeven.[38] From the region of Mata Menge, archeologists have unearthed hundreds of fossils from extinct animals (e.g., dwarf elephants or stegodontids) and simple stone tools.

Some of these lithics were found under volcanic layers deposited at least 1 mya. Although no hominin remains were found at the time, the presence of the tools indicated that they somehow reached Flores about 1 mya. Over the years since the initial discoveries of Verhoeven, archeologists remained hopeful that eventually, the hominin toolmakers would be discovered.

In October 2014, a mandible and six teeth were unearthed.[39] At least three hominin individuals contributed to this assemblage: one adult and two children. The fossils were dated to about 700,000 BP (before present).[40] The age of the remains was accessed by $^{40}Ar/^{39}Ar$ and fission track of volcanic ash and pyroclastic deposits as well as uranium-series and electron spin resonance of the fossil teeth. At the time, the climate in the region was hot and dry, and the immediate area where the fossils were found was savannah open grassland with some wetland.

Associated with the hominin remains, megafauna including stegodon (*Stegodon florensis insularis*), Komodo dragon (*Varanus komodoensis*), giant rat (*Hooijeromys nusatenggara*), and giant tortoise (*Geochelone*) were unearthed. The direct ancestor of *Stegodon florensis insularis* is the larger-bodied *Stegodon florensis florensis* that became extinct during the Early Pleistocene about 900,000 ya during a local volcanic eruption-related faunal replacement.[41] Although the *Stegodon florensis insularis* later found in Liang Bua belongs to the same subspecies as the ones in Mata Menge about 600,000 before, the former were about 30% smaller.[41] At least five elephant-like lineages of Stegodons experienced insular dwarfism in the region, attesting to the impact of island isolation/limited resources.

A total of 507 stone tools were recovered from Mata Menge. The lithic raw material was transported to the site in the form of flake blanks. The more complex tools found exhibit a removal of material centripetally or radially on the blanks.[42] At Mata Menge, hominins made the tools by striking flakes from volcanic stones they collected in riverbeds and lakeshore cobble deposits.[43] The stones at Liang Bua were flaked similarly, suggesting a technological connection between the two groups of hominins.[43] The stone tool's similarities indicate a technological continuity spanning more than 600,000 years of hominin evolution on the island of Flores. Although the lithics found at Mata Menge and Liang Bua show that the toolmakers were skillful at striking flakes, they were not very sophisticated. In both sites the tools were amorphous and undifferentiated, suggesting limited cognitive planning.[41] It seems as if these hominins were primarily interested in generating stone tools with sharp cutting edges and not specific utensils for specific functions.[43]

The mandible discovered at Mata Menge (Fig. 8) belonged to an individual about 20% shorter than Flo, and it possessed features closer to *Homo erectus* than the earlier *Homo habilis*.[39,40] This earlier Mata Menge adult individual may have been from a population of smaller hominins than the later Liang Bua, or it could be an isolated extra-small specimen. Additional adult individuals would need to be examined to discriminate between the two possibilities. The size and shape of the teeth recovered at Mata Menge closely resemble those from Liang Bua, except that the former appears more ancestral. While the six teeth

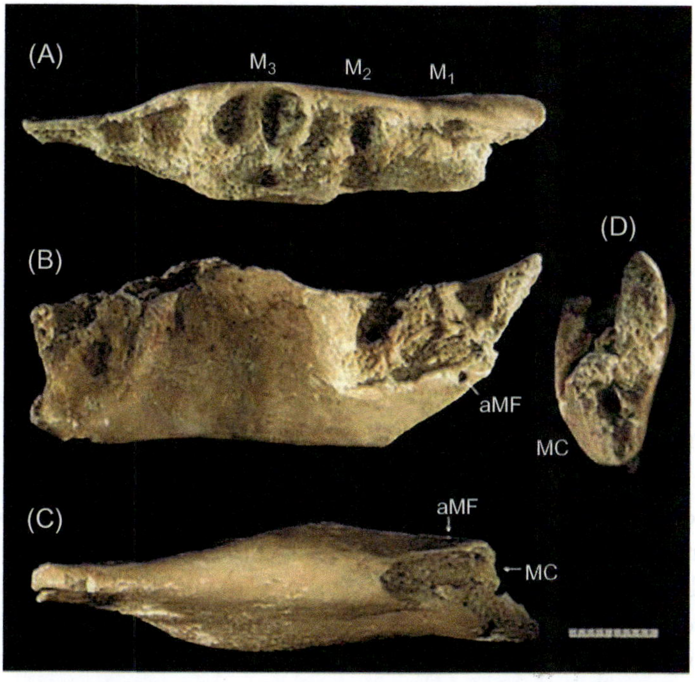

Fig. 8 Mandible from Mata Menge (https://johnhawks.net/weblog/fossils/flores/mata-menge-van-den-Bergh-remains-2016.html).

exhibit a mosaic of ancestral and derived features, they seem more similar to *Homo erectus* than to other earlier hominins. With the technological similarities in the toolmaking tradition, most scholars agree that the 700,000 ya hominin population at Mata Menge may have been the ancestors of *Homo floresiensis* at Liang Bua. It is noteworthy that the individual (and most likely the population) at Mata Menge has already undergone an extreme version of dwarfism about 600,000 years prior to the group at Liang Bua. This means that if the Mata Menge dwarf population descended from a *Homo erectus* population that arrived in Flores around 1 mya, the effects of insular dwarfism generated a *Homo floresiensis*-like species in only 300,000 years.

Wolo Sege

The Wolo Sege site is just 500 m away from Mata Menge in an open cattle-yard field. No faunal remains have been found, and only 48 in-situ stone artifacts, including some Acheulean-like tools, have been discovered (Fig. 9). Over the stone artifacts, there is a volcanic layer of ignimbrite with a $^{40}Ar/^{39}Ar$, registering a volcanic eruption date of 1.02 mya. The dating of 1.02 mya based on the $^{40}Ar/^{39}Ar$ isotopes ratio represents the minimum age for hominins on Flores. Therefore, it is likely that hominins arrived on Flores much earlier than 1.02 mya.[44]

Fig. 9 *Wolo Sege* (http://news.bbc.co.uk/2/hi/science/nature/8574037.stm).

It is not clear if a population of miniature *Homo floresiensis* already existed in Java around 1.02 mya, or if a full-size *Homo erectus* coexisted with a diminutive *Homo floresiensis*-like population in Java around 1 mya or earlier and one or both migrated from Java to Bali, or if, once in Bali, *Homo erectus* evolved into a dwarf population in just 300,000 years. What is known is that the Wolo Sege stone artifacts are predominantly small and morphologically undifferentiated flakes struck from cobbles by direct hard-hammer percussion (Fig. 9), and they include a bifacially and centripetally worked radial core, similar to those characteristic of the Mata Menge assemblage of stone artifacts. The similarities in some of the tools in Wolo Sege and Mata Menge have prompted some investigators to postulate that the later population descended from the former.[44] Yet, by looking at the manufacturing tradition used to make the lithics, it is difficult to assess definitely the species that created the tools at Wolo Sege.

Yet combining all the evidence available, an evolutionary chronology can be created in which *Homo erectus* settled on Flores and then shrank to a small size. Stone stools have been found on Flores dating back as early as 1 mya, around the same time *Homo erectus* was living on nearby Java.[41] It is also possible that a diminutive *Homo floresiensis*-like population survived somewhere on the island of Flores for some time after 50,000 ya, the date Flo lived in the vicinity of Liang Bua cave in western Flores. It is likely that this population became extinct more recently, giving rise to the Ebu Gogo folklore of the little hairy people who live on the mountains of Flores.[29] It is also possible that *Homo floresiensis* or a small hominin encountered Denisovans in Flores as the latter were dispersing through insular Southeast Asia, as some recent data suggest.[45]

One million years of technological continuity in Flores

Examination of the stone tools found in the various sites of the island of Flores illustrates similar lithic traditions (Fig. 10). The lithic artifacts found in Wolo Sege, Mata Menge,

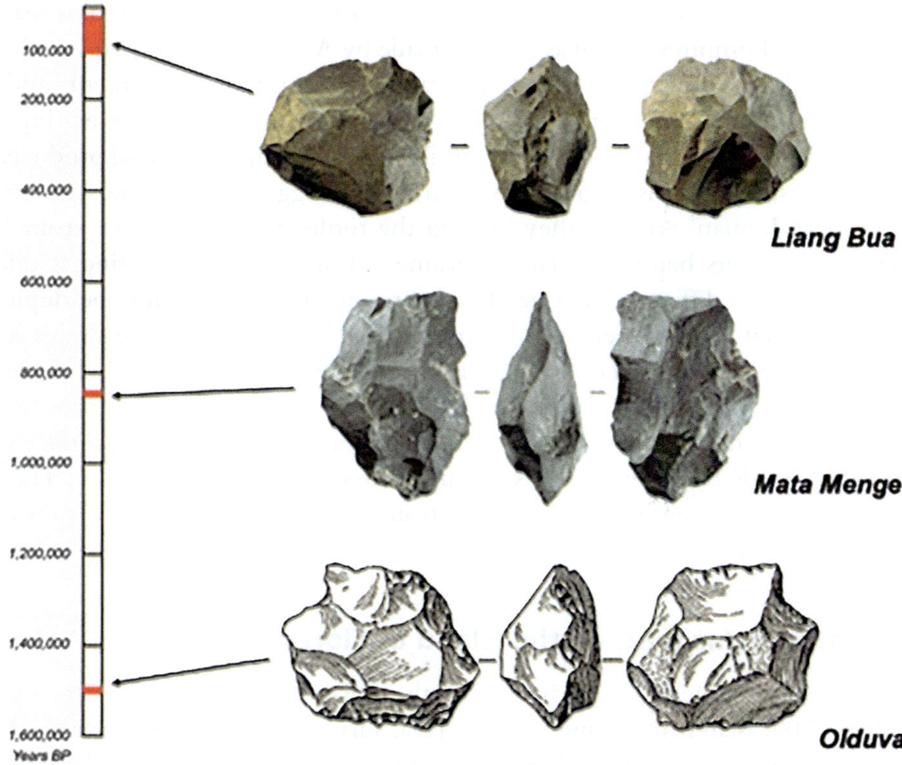

Fig. 10 Similar stone tools from Liang Bua, Mata Menge, and Olduvai Gorge, Africa (after Leakey 1971), arranged chronologically (copyright Mark Moore 2009) (https://www.peterbrown-palaeoanthropology.net/Stone%20Tools%20on%20Flores%201.htm).

and Liang Bua in the island of Flores date to occupations as early as 1.02 mya, 1.00 mya, and 0.19 mya, respectively. Therefore, non*sapiens* exhibit a technological uniformity that lasted for at least 1 my. Remarkably the observed similarities within the island can be traced back to the earliest African hominin lithic traditions.

The stone tools at the three sites in Flores were flaked in identical ways, producing the same type of cores and flakes.[46] At Liang Bua, rounded cobbles and boulders of volcanic tuff were collected in a riverbed some meters away from the cave. The cores were struck outside the cave to generate flakes. Then, the large flakes were taken into the cave to make smaller ones using a hammerstone. Although the hominins in all the sites were skillful in string flakes, the process was not sophisticated. The tools were generally amorphous, and lack differentiation and specialization.[47] It seems that the aim of these hominins throughout their 1-my tenure was just to make tools with sharp cutting edges. They were not generating specific tools for specific functions. When compared to early African lithics, a similar level of simplicity and lack of sophistication is seen.[48]

The cave at Liang Bua is of particular interest since it has provided tools made by both non*sapiens* hominins as well as artifacts made by AMHs.[43] Both AMHs and *Homo floresiensis* produced large flakes outside the cave and then transported them inside the cave and reduced the lithic further, using the same flaking technique generating stone products and by-products of remarkably similar size and shape. The end products created are very similar. This degree of similarity has suggested to some scholars that when modern humans arrived, they learned the toolmaking techniques from *Homo floresiensis* toolmakers before the latter became extinct. Researchers find it difficult to explain how the different steps needed to produce the lithics could be duplicated by *sapiens* without direct observation or interaction.[43] Yet the tools made by AMHs differ from those created by *Homo floresiensis* in many ways. AMHs used chert instead of the volcanic stones used by *floresiensis*. AMHs used the tools created to process plant material as well as meat. This is indicated by the silica residue left on the surface of the stone artifacts made by *sapiens*. AMHs also used the tools to manipulate fire. The tools made by *sapiens* are often burned, probably from scuffing them into hearths or cooking ovens inside the cave.[43]

How *Homo erectus* reached the island of Flores

The Indonesian Archipelago extends from Southeast Asia to Near Oceania.[49] It consists of more than 17,000 islands, including Sumatra, Java, Sulawesi, and parts of Borneo and New Guinea, covering 1,904,569 km^2 (735,358 sq. mi). Indonesia is the world's largest island country. Currently, at an interglacial period, many of the islands are separated by treacherous ocean channels that partition the archipelago biodiversity, creating biogeographical zones. During the Pleistocene Ice Age, which included the time period between 2,580,000 and 11,700 ya, several islands of the archipelago became a continuous landmass with Asia, and together they became a supercontinent, the Sunda supercontinent (Fig. 3). These land bridges between the islands resulted from the sequestration of ocean water in frozen form at the poles as the earth temperature dropped. Similarly, a number of islands (e.g., New Guinea, which consists of Papua New Guinea and Western New Guinea) at the extreme east of the Indonesian Archipelago established land bridges with each other and Australia. Thus, all this formed a continuous landmass that is known today as the Sahul supercontinent. *Homo erectus*' tenure included this time period from 2 mya to 100,000 ya.[12]

During this epoch of cold and dry weather, the sea level dropped about 150 m below the current sea level. For example, Bali was connected to Java and Sumatra and to the mainland of Asia, sharing Asian fauna.[19] Currently, Java and Bali are islands separated by only 2.4 km, the Bali strait. On the island of Java, *Homo erectus* sites are known from Sangiran, Mojokerto, Sembiran, and Trunyan.[50]

Yet the island of Flores has always been separated from Asia and the islands to the west by at least two channels. Even during glacial periods, when most of Indonesia was linked together into two large landmasses (Sunda and Sahul) and the sea level in the region was as low as 150 m below today's levels, a number of dangerous bodies of water needed to be crossed to reach Flores from the west. The first one that *erectus* needed to travel was an open water gap between east Bali and Lombok. To reach the island of Lombok, migrants needed to navigate this 19 km channel, which is notorious for its swift currents that run perpendicular to the shortest distance between the islands. Although, on a clear day, *erectus* was likely able to see the island of Lombok at a distance while standing on Bali (Fig. 3), the details of how *erectus* managed such a crossing are not clear. In this channel, any floating material will rapidly travel north or south with the current, away from the islands into the open ocean. Under these conditions, the likelihood of crossing the Strait of Lombok from the island of Bali to the island of Lombok would have been improbable. The second crossing from west to east in the direction of Flores involved navigating between the islands of Sumbawa and Flores. This gap of ocean during the Pleistocene Ice Age was a 9 km gap of treacherous water capable of moving objects rapidly latitudinally into Open Ocean. The crossings of the Strait of Lombok and the Sumba Channel may represent the most ancient seafaring events undertaken by hominins. Living within the constraints of small-island resources, it is expected that faunal migrations may have occurred across geographical barriers such as water channels. Especially in situations in which resources are depleted within islands due to ecological collapse, and nearby lands are visible, individuals of a species may have attempted to reach an island on the horizon. *Homo erectus* in Indonesia had building materials, including bamboo and plant fibers, to make rafts, yet it is not clear if they had the wits to execute the task of building rafts. Free swimming the Lombok and the Sumba Channels is not impossible, but the likelihood of success is not high.

In all the scenarios mentioned above, there is an element of purpose in traveling between islands. Yet there are many possibilities that do not require a plan; crossing the Indonesian channels may have happened accidentally. For example, in the Indonesian Archipelago, seismic and volcanic activity is frequent. Under these conditions, organisms are subject to a number of geological events that could have facilitated the transportation of plants and animals among islands. Take, for example, the sudden occurrence of a tsunami. A clan of *Homo erectus* searching for shellfish close to the shoreline could be attracted by the receding shoreline only to be pulled suddenly into the open sea by the receding tide in the direction of the nearby island. It is not far-fetched to envision that these types of sudden catastrophic events could have deposited hominins from one island to the next nearby landmass as unintentional swimmers or piggybacked on vegetation.

The facts that *Homo erectus*, Denisovans, and *Homo sapiens* somehow were able to settle isolated islands in Indonesia and that they cohabitated and possibly admixed have motivated scholars to investigate and search for evidence of admixture and introgression. In the next section of this chapter, data suggesting admixture involving hominin groups are presented.

How is ancient admixture/introgression determined?

One of the intriguing questions regarding human evolution is whether present-day humans derived their genetic material from a single lineage or whether *sapiens* is the result of an admixture of populations that at some point contributed to the genetic makeup of the species. In recent decades, although there is no consensus on the specific level of contribution made by specific archaic species of *Homo*, instances of introgression involving archaic species have been reported.[51] Suspected cases of archaic introgression have become routinely reported in the literature as extinct, and extant hominin species have been systematically compared using several molecular biology and phylogenetic techniques. When addressing the question of ancient introgression, investigators are primarily concerned with whether archaic populations died out in the past or whether they interbred with AMHs, passing some of their genes to present-day humans.

Two main methods are employed to investigate introgression among species. In one, involving AMHs investigators perform direct comparisons between homologous archaic and contemporary human DNA sequences. Using this approach, the introgressed nucleotide sequences and the homologous *Homo sapiens* DNA are part of the same genome. A different method utilizes only sequence polymorphism data from extant humans. This approach analyzes patterns of linkage disequilibrium (LD) to detect the presence of ancient admixture.[52] Sequence divergence and lack of recombination have been useful in identifying LD distribution in the human lineage dating back to more than 2 mya. Typically, the sequence of events begins when there is an evolutionary split in the hominin lineage resulting in only one branch leading to humans. Over time, involving 1 my or more, these two lineages diverge independently, becoming increasingly different. At times after a protracted separation of the lineages, interbreeding occurs, leading to the incorporation of DNA from the nonhuman lineage. The introgressed DNA is unique from the DNA in the human lineage and often is not capable of pairing with its homologous hominin sequence during meiosis, leading to a decrease in recombination.

One of the early studies that investigated introgressed sequences by identifying regions in LD was published in 2005 and describes a 17,500 bp (base pair) sequence at the Xp21.1 location (i.e., position 21.1 on the short arm of the X chromosome) that has been in LD for at least 1 my.[53] These patterns of deep haplotype divergence and long-range linkage disequilibrium are best explained by a prolonged period of ancestral population separation followed by relatively recent interbreeding. This inference supports human evolution models that incorporate admixture between divergent branches of the genus *Homo*.

Other locations that show evidence of ancient splitting in the hominin lineage, followed by differentiation of the branches by independent mutations and recent introgression resulting in LD, have been described. These include the African San Bushmen (hunter-gatherers of South Africa), the Xhosa (a Bantu group from the

Eastern Cape region of South Africa), and the Mbuti (forest pygmy population of eastern Congo).[54] These African populations possess DNA fragments in LD that suggest that the populations separated from an unknown population about 1.0–1.5 mya. What is unclear in this evolutionary scenario is how, after such a prolonged separation and deep haplotype divergence, lineages that lived in reproductive isolation for more than 1 my were able to interbreed and establish gene flow. A related issue is how species that evolved independently after separation could mate 1–2 my after the lineages split. The expectations of such a protracted separation would be species with very different physical and behavioral characteristics, which might hinder communication and mating and bring about reproductive isolation.

Additional data supporting archaic gene flow, possibly from *erectus*, has been detected from the natives of the Andamanese Islands.[55] Andamaneses are natives of an archipelago located in the southeastern part of the Bay of Bengal in Southeast Asia, and they are considered a Negrito population based on their dark skin and diminutive stature. They are viewed as an amalgamation of archaic populations that settled the region around the last glacial maximum about 26,000 ya. Genomic analysis of Andamanese individuals has provided direct evidence of introgression of an archaic DNA postulated to be of *Homo erectus* origins. The results provided by Mondal et al.[55] indicate that the Andamaneses, Papuans, and other populations of mainland India possess about 2%–3% fewer African alleles than Europeans or East Asians. Furthermore, aboriginal Australians have about 6%–7% fewer African alleles than Europeans or East Asians. Another way to express these findings is that Andamaneses, Papuans, and other populations of mainland India possess a proportion of DNA from an extinct hominin, which is absent in Europeans and East Asians.

Additional studies examining *erectus*-derived introgression in worldwide populations have identified a slight amount of excess unique genetic variation in Island Southeast Asia (ISEA) and Papua populations that is not shared with other human groups.[45] This residual variability in ISEA and Papua is observed after accounting for all the variability seen in all the other populations worldwide. This "left over" variability is thought to derive from a group of *erectus*-like hominins that introgressed into the Denisovan lineage. In turn, the Denisovan lineage traveled to ISEA and Papua and admixed with early AMHs. It was estimated that the amount of *erectus*-like DNA introgressed via Denisovans into ISEA and Papua populations is about 1%.

Conclusions

Homo erectus was the first of our hominin ancestors to have *sapiens*-like body proportions, including shorter arms and longer legs relative to its torso. It was also the first hominin to migrate out of Africa. It is theorized that *erectus* was a runner.[19] It survived for more than 1.5 my. In terms of longevity as a species, its tenure was by far the longest of all our human relatives.

The earliest fossils of *erectus* date back to approximately 2 mya. *Homo erectus* was the first hominin to reach Southeast Asia, with a continental range extending from Western Europe to Oceania. In addition, it is generally acknowledged that *erectus* were the ancestors of several African species such as *Homo heidelbergensis* and *Homo antecessor*, with the former generally considered to have been the ancestor to Neanderthals and Denisovans, and *Homo sapiens*. Furthermore, Asian populations of *erectus* are likely ancestral to *Homo floresiensis* of the island of Flores in Indonesia and *Homo luzonensis* of the Philippines.[6] Of particular interest in the discussion of this chapter is that *Homo erectus* is considered a chronospecies, and as such, the time of the disappearance of *Homo erectus* is a matter of contention. Thus, especially in regions where *erectus* coexisted with several other hominin groups, it is often difficult to identify if some were *erectus* subspecies or unique species. In those instances of coexistence, the likelihood of interbreeding and introgression were likely responsible for many of the mosaicisms observed. In addition, there are also several proposed subspecies of *erectus* with varying levels of uniqueness. The last known population of *H. erectus* is *Homo e. soloensis* from Java, dated to around 117,000–108,000 ya.[12]

The first sections of this chapter dealt with the intertwined evolutionary relationship that existed during the Pleistocene in continental East Asia and ISEA among various populations of *Homo erectus* and other hominin groups. The distribution of these sites in the region is rather uniform, and the specimens present multiple combinations of ancestral and derived traits. Often, specific fossils exhibit such extreme cases of ancestral and derived types of mosaicism that scholars cannot decide to which species they belong to or whether they represent cases of introgression or hybridization or unique species.

Various archeological sites have been described in ISEA exhibiting fossils with different combinations of ancestral and derived morphological characteristics. As a group, the individuals in these populations are characterized by their small size and capacity to settle individual islands by crossing Open Ocean. It is not clear how these miniature hominins could have reached the various islands of ISEA and when they evolved into their small size. Although some of the fossils exhibit traits in common with *Homo erectus*, some of the island-specific hominins are unique in some of their anatomies, and some even show similarities to australopithecines. It is possible that some forms of *erectus* coexisted with *sapiens* in ISEA.

A number of investigations have reported suspected cases of introgression involving *Homo erectus* and *Homo sapiens*. Although there is no consensus on the level of contribution that archaic humans made to the contemporary gene pool, in recent years, several specific cases of ancient introgression have demonstrated that ancient gene flow involving *Homo erectus* has taken place. Two main methods are employed to investigate introgression among species. In one, investigators performed direct comparisons between homologous archaic and contemporary human DNA sequences. Using this approach, the introgressed nucleotide sequences and the homologous *Homo sapiens* DNA are part of the same genome. A different method utilizes only sequence polymorphism data from extant humans. This approach analyzes patterns of LD to detect the presence of ancient admixture. Sequence divergence and lack of recombination have been useful in identifying LD distribution in the human lineage dating back to 2 mya. The patterns of LD are used to identify ancient cases of introgression and/or interbreeding.

References

1. Leaky L. *Adam's Ancestors: The Evolution of Man and His Culture.* Gloucester, MA, USA: Peter Smith Pub Inc; 1980.
2. Zaim Y, Ciochon RL, Polansk JM, et al. New 1.5 million-year-old *Homo* erectus maxilla from Sangiran (Central Java, Indonesia). *J Hum Evol.* 2011;61:363–376.
3. Ao H, Liu C-R, Roberts AP, et al. An updated age for the Xujiayao hominin from the Nihewan basin North China: implications for Middle Pleistocene human evolution in East Asia. *J Hum Evol.* 2017;106:54–65.
4. Wu XZ. A well-preserved cranium of an archaic type of early Homo sapiens from Dali, China. *Sci Sinica.* 1981;24:530–541.
5. Ingicco T, van den Bergh GD, Jago-on C, et al. Earliest known hominin activity in the Philippines by 709 thousand years ago. *Nature.* 2018;557:233–237.
6. Détroit F, Mijares AS, Corny J, et al. A new species of *Homo* from the late Pleistocene of the Philippines. *Nature.* 2019;568:181–186.
7. Fleming N. Unknown human relative discovered in Philippine cave. *Nature News.* 2019;(April 10).
8. Tocheri MW. Previously unknown human species found in Asia raises questions about early hominin dispersals from Africa. *Nature News.* 2019;568:176–178.
9. Thorne AG, Macumber PG. Discoveries of Late Pleistocene man at Kow Swamp. *Nature.* 1972;238:316–319.
10. Brown P. Still flawed: a reply to Pardoe (1994) and Sim and Thorne (1994). *Aust Archaeol.* 1995;41:26–29.
11. Huffman OF, de Vos J, Berkhout AW, et al. Provenience reassessment of the 1931–1933 Ngandong *Homo erectus* (Java), confirmation of the bonebed origin reported by the discoverers. *PaleoAnthropology.* 2010;2010:1–60.
12. Rizal Y, Westaway KE, Zaim Y, et al. Last appearance of *Homo erectus* at Ngandong, Java, 117,000–108,000 years ago. *Nature.* 2020;577:381–385.
13. Yokoyama Y, Falgueres C, Semah F, et al. Gamma-ray spectrometric dating of late Homo erectus skulls from Ngandong and Sambungmacan, Central Java, Indonesia. *J Hum Evol.* 2008;55:274–277.
14. Westawa MC. The first hominin fleet. *Nat Ecol Evol.* 2019;3:999–1000.
15. van den Bergh GD, Li B, Brumm A, et al. Earliest hominin occupation of Sulawesi, Indonesia. *Nature.* 2016;529:208–211.
16. Liu W, Jin C-Z, Zhang Y-Q, et al. Human remains from Zhirendong, South China, and modern human emergence in East Asia. *PNAS.* 2010;107:19201–19206.
17. Liu W, Martinón-Torres M, Cai YJ, et al. The earliest unequivocally modern humans in southern China. *Nature.* 2015;526:696–699.
18. Indriati E, Swisher CC, Letre C, et al. The age of the 20 meter Solo River terrace, Java, Indonesia and the survival of *Homo erectus* in Asia. *PLoS ONE.* 2011;6, e21562.
19. Herrera RJ, Bertrand GB. *Ancestral DNA, Human Origins, and Migrations.* London, UK: Academic Press; 2018.
20. Roberts R, Sutikna T. Michael John Morwood (1950–2013). *Nature.* 2013;500:401.

21. Verhoeven T. Proto-Negrito in den Grotten auf Flores (Indonesie). *Anthropos*. 1958;53:229–232.
22. Morwood M, Oosterzee P. *The Discovery of the Hobbit: The Scientific Breakthrough That Changed the Face of Human History*. New York: Random House; 2007.
23. Callaway E. The discovery of *Homo floresiensis*: tales of the hobbit. *Nature*. 2014;514:422–426.
24. Powledge TM. Flores hominid bones returned. *Sci Mag*. 2005.
25. Morwood MJ, Soejono RP, Roberts RG, et al. Archaeology and age of a new hominin from Flores in eastern Indonesia. *Nature*. 2004;431:1087–1091.
26. Kaifu Y, Kono RT, Sutikna T, et al. Descriptions of the dental remains of *Homo floresiensis*. *Anthropol Sci*. 2015;123:129–145.
27. Marwick B, Clarkson C, O'Connor S, et al. Early modern human lithic technology from Jerimalai, East Timor. *J Hum Evol*. 2016;101:45–64.
28. Callaway E. Human remains found in hobbit cave. *Nature*. 2016. https://doi.org/10.1038/nature.2016.20656.
29. Wong K. The littlest human. *Sci Am*. 2006;16:48–57.
30. Fix AG. Malayan paleosociology: implications for patterns of genetic variation among the Orang Asli. *Am Anthropol*. 1995;97:313–323.
31. Morwood M, Brown P, Jatmiko, et al. Further evidence for small-bodied hominins from the Late Pleistocene of Flores, Indonesia. *Nature*. 2005;437:1012–1017.
32. Van Den Bergh GD, Due Awe R, Morwood MJ, et al. The youngest *Stegodon* remains in Southeast Asia from the late Pleistocene archaeological site Liang Bua, Flores, Indonesia. *Quat Int*. 2008;182:16–48.
33. Lyras GA, Dermitzakis DM, Van Der Geer AAE, et al. The origin of Homo floresiensis and its relation to evolutionary processes under isolation. *Anthropol Sci*. 2008;117:33–43.
34. Kaifu Y, Reiko RT, Sutikna T, et al. Unique dental morphology of *Homo floresiensis* and its evolutionary implications. *PLoS ONE*. 2015;10, e0141614.
35. Falk D, Hildebolt C, Smith K, et al. The brain of LB1, *Homo floresiensis*. *Science*. 2005;308:242–245.
36. Tocheri MW, Orr CM, Larson SG, et al. The primitive wrist of *Homo floresiensis* and its implications for hominin evolution. *Science*. 2007;317:1743–1745.
37. Jungers WL, Larson SG, Harcourt-Smith W, et al. Descriptions of the lower limb skeleton of *Homo floresiensis*. *J Hum Evol*. 2008;57:538–554.
38. Verhoeven T. Proto-Negrito in den Grotten auf Flores (Indonesie). *Anthropos*. 1958;53:229–232.
39. van den Bergh GD, Kaifu Y, Kurniawan I, et al. Homo floresiensis-like fossils from the early Middle Pleistocene of Flores. *Nature*. 2016;534:245–248.
40. Brumm A, van den Bergh GD, Storey M, et al. Age and context of the oldest known hominin fossils from Flores. *Nature*. 2016;534:249–253.
41. Larick R, Ciochon RL. Early hominin biogeography in island Southeast Asia. *Evol Anthropol*. 2015;24:185–213.
42. Brown P, Sutikna T, Morwood MJ, et al. A new small-bodied hominin from the late Pleistocene of Flores, Indonesia. *Nature*. 2004;431:1055–1061.
43. Moore MW, Sutikna T, Morwood MJ, et al. Continuities in stone flaking technology at Liang Bua, Flores, Indonesia. *J Hum Evol*. 2009;57:503–526.
44. Brumm A, Jensen GM, van den Bergh GD, et al. Hominins on Flores, Indonesia, by one million years ago. *Nature*. 2010;464:748–752.
45. Jacobs GS, Hudjashov G, Saag L, et al. Multiple deeply divergent Denisovan ancestries in Papuans. *Cell*. 2019;177:1010–1021.
46. Culotta E. Tools link Indonesian 'hobbits' to earlier Homo ancestor. *Science*. 2006;312:1293.
47. Moore MW. Lithic design space modelling and cognition in Homo floresiensis. In: Shalley A, Khlentzos D, eds. *Mental States: Nature, Function and Evolution*. Amsterdam, Netherlands: John Benjamins; 2007.
48. Moore MW, Brumm A. Homo floresiensis and the African Oldowan. In: Hovers E, Braun DR, eds. *Interdisciplinary Approaches to the Oldowan*. New York, USA: Springer; 2009.
49. van Imhoff E, Beets GA. Demographic history of the Indo-Dutch population, 1930–2001. *J Popul Res*. 2004;21:47–72.
50. Theunissen B. *Eugène Dubois and the Ape-Man from Java*. Boston, USA: Academic Publishers; 1989.

51. Wall JD, Hammer MF. Archaic admixture in the human genome. *Curr Opin Genet Dev.* 2006;16:606–610.
52. Plagnol V, Wall JD. Possible ancestral structure in human populations. *PLoS Genet.* 2006;2, e105.
53. Garrigan D, Mobasher Z, Kingan SB, et al. Deep haplotype divergence and long-range linkage disequilibrium at Xp21.1 provide evidence that humans descend from a structured ancestral population. *Genetics.* 2005;170:1849–1856.
54. Hammer MF, Woerner AE, Mendez FL, et al. Genetic evidence for archaic admixture in Africa. *Proc Natl Acad Sci U S A.* 2011;108:15123–15128.
55. Mondal M, Casals F, Xu T, et al. Genomic analysis of Andamanese provides insights into ancient human migration into Asia and adaptation. *Nat Genet.* 2016;48:1066–1070.

Index

Note: Page numbers followed by *f* indicate figures, and *t* indicate tables.

B

C

D

G

H

I

J

K

N

S

9780128190869